Applied Sedimentology
Second Edition

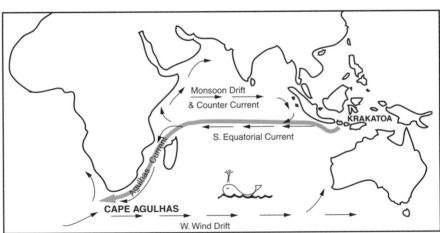

(Upper) Pumice pebbles from Cape Agulhas, the southernmost part of Africa, coin is 2 cm in diameter. These pebbles geochemically match the ejectamenta of the 27 May 1883 eruption of Krakatoa in the Sunda Straits between Java and Sumatra. (Frick, C. and Kent, L. E., 1984. Drift pumice in the Indian and Atlantic Oceans. *Trans. Geol. Soc. South Africa,* **87,** 19–33.) Contemporary accounts describe the Indian Ocean covered with pumice-bergs, some with their own little ecosystems. Glorious sunsets were enjoyed worldwide for years, due to volcanic dust in the upper atmosphere. (Lower) Map showing the voyage of the Cape Agulhas pumice pebbles, some 15,000 km. Now that is sediment transport.

Applied
Sedimentology
Second Edition

Richard C. Selley

Royal School of Mines
Imperial College of Science, Technology, and Medicine
London, United Kingdom

ACADEMIC PRESS

A Harcourt Science and Technology Company

San Diego   San Francisco   New York   Boston
London   Sydney   Tokyo

Cover photo credit: Upward coarsening genetic sequence of basin plain shales, turbidite and grain flow sands, Ecca Group (Permo-Carboniferous), Great Karoo, South Africa. The Karoo sediments, and their equivalents around Gondwanaland, contain major deposits of coal, oil shale, and uranium. Photo courtesy of the author.

This book is printed on acid-free paper. ∞

Academic Press
A Harcourt Science and Technology Company
525 B Street, Suite 1900, San Diego, California 92101-4495, USA
http://www.apnet.com

Academic Press
24-28 Oval Road, London NW1 7DX, UK
http://www.hbuk.co.uk/ap/

Harcourt/Academic Press
A Harcourt Science and Technology Company
200 Wheeler Road, Burlington, Massachusetts 01803
http://www.harcourt-ap.com

Library of Congress Catalog Card Number: 99-68192

International Standard Book Number: 0-12-636375-7

PRINTED IN THE UNITED STATES OF AMERICA
00 01 02 03 04 05 MM 9 8 7 6 5 4 3 2 1

Contents

Part II   Sediment Sedimented

4 Transportation and Sedimentation

5 Sedimentary Structures

6 Depositional Systems

Part III Sediment to Rock

7 The Subsurface Environment

8 Allochthonous Sediments

9 Autochthonous Sediments

10 Sedimentary Basins

Preface

This is the second edition of *Applied Sedimentology,* and effectively the fourth edition of its predecessor *Introduction to Sedimentology,* first published in 1976.

Geology in general, and sedimentology in particular, has changed much during the last 25 years. When the first edition appeared, sediments were studied by geologists, who were bearded, suntanned individuals, with well-developed wilderness-survival skills and an interesting range of tropical diseases. Today, sediments are largely studied by remote sensing, in one form or another, by adept mouse-masters skilled in office survival. Satellite photos, multidimensional seismic surveys, and borehole imagery enable sedimentary rocks to be studied in an air-conditioned office without ever actually being looked at.

To correctly interpret and understand remotely sensed sediments, however, it is essential to understand the fundamentals of sedimentology, preferably by having studied sediments in the wild. Modern earth science graduates are familiar with the latest theological debate pertaining to sequence stratigraphy, and the latest computer program for basin modeling, but are often unable to tell granite from arkose or to explain the formation of cross-bedding. Worthy attempts by teachers to lead students to the margin of the subject before absorbing its foundations exact a high price.

On my last field trip I used the guidebook of the local petroleum exploration society. I visited an outcrop of what the guide described as a carbonatite plug. This proved to be a raised sea stack of white skeletal limestone. At another locality shales were described as containing horizons of volcanic bombs. These were actually rusty, round fossiliferous siderite concretions. The guidebook's account of the complex tectonics of the region was scrambled through failure to use turbidite bottom structures.

Applied Sedimentology is therefore unashamedly about sensual sedimentology. It attempts to provide a firm foundation on which to interpret sediments remotely sensed by various geophysical tools.

Applied Sedimentology is divided into three parts: Rock to Sediment, Sediment Sedimented, and Sediment to Rock, reflecting the holistic nature of the sedimentary cycle. An introductory chapter outlines the field of sedimentology, relates it to the fundamental sciences, and discusses its applications. Part I, Rock to Sediment, consists of two chapters. Chapter 2 outlines weathering, showing not only how weathering gives rise to the terrigenous sediments, but also how it mobilizes and concentrates diverse residual ore deposits. Chapter 3, Particles, Pores, and Permeability, describes the texture of sediments and shows how these are related to porosity and permeability.

Part II, Sediment Sedimented, takes the story a stage further. Chapters 4 and 5 describe sedimentary processes and sedimentary structures, respectively. Chapter 6 outlines the major depositional systems and discusses how their products may serve as petroleum reservoirs and hosts for ore bodies.

Part III, Sediment to Rock, completes the cycle. Chapter 7 describes the subsurface environment within which sediment is turned into rock. Chapter 8 describes clays, sands, and gravels and details their diagenesis and porosity evolution. Chapter 9 does the same for the chemical rocks, describing the mineralogy, composition, and diagenesis of carbonates, evaporites, sedimentary ironstones, coal, phosphates, and chert.

The book concludes with a chapter on sedimentary basins. This describes the mechanics of basin formation, the various types and their sedimentary fill, and their petroleum and mineral potential. It also describes the evolution of basin fluids through time.

Applied Sedimentology is written principally for senior undergraduate and postgraduate students of earth science and engineering. I also hope, however, that it may prove useful to more mature readers who explore and exploit the sedimentary rocks for fossil fuels and mineral deposits.

Richard C. Selley

1 Introduction

1.1 INTRODUCTION AND HISTORICAL REVIEW

The term **sedimentology** was defined by Wadell (1932) as "the study of sediments." Sediments have been defined as "what settles at the bottom of a liquid; dregs; a deposit" (*Chambers Dictionary,* 1972 edition). Neither definition is wholly satisfactory. Sedimentology is generally deemed to embrace chemical precipitates, like salt, as well as true detrital deposits. Sedimentation takes place not only in liquids, but also in gaseous fluids, such as eolian environments. The boundaries of sedimentology are thus pleasantly diffuse.

The purpose of this chapter is twofold. It begins by introducing the field of sedimentology and placing it within its geological context and within the broader fields of physics, chemistry, and biology. The second part of the chapter introduces the applications of sedimentology in the service of mankind, showing in particular how it may be applied in the quest for fossil fuels and strata-bound minerals.

It is hard to trace the historical evolution of sedimentology. Arguably among the first practitioners must have been the Stone Age flint miners of Norfolk who, as seen in Grimes cave, mined the stratified chert bands to make flint artifacts (Shotton, 1968). Subsequently, civilized man must have noticed that other useful economic rocks, such as coal and building stone, occurred in planar surfaces that cropped out across the countryside in a predictable manner. It has been suggested that the legend of the "Golden Fleece" implies that sophisticated flotation methods were used for alluvial gold mining in the fifth century B.C. (Barnes, 1973). From the Renaissance to the Industrial Revolution the foundations of modern sedimentary geology were laid by men such as Leonardo da Vinci, Hutton, and Smith. By the end of the nineteenth century the doctrine of uniformitarianism was firmly established in geological thought. The writings of Sorby (1853, 1908) and Lyell (1865) showed how modern processes could be used to interpret ancient sedimentary textures and structures.

Throughout the first half of the twentieth century, however, the discipline of sedimentology, as we now understand it, lay moribund. The sedimentary rocks were either considered fit only for microscopic study or as sheltered accommodation for old fossils. During this period heavy mineral analysis and point counting were extensively developed by sedimentary petrographers. Simultaneously, stratigraphers gathered fossils, wherever possible erecting more and more refined zones until they were too thin to contain the key fossils.

Curiously enough, modern sedimentology was not born from the union of petrography and stratigraphy. It seems to have evolved from a union between structural geology and oceanography. This strange evolution deserves an explanation. Structural geologists have always searched for criteria for distinguishing whether strata in areas of tectonism were overturned or in normal sequence. This is essential if regional mapping is to delineate recumbent folds and nappes. Many sedimentary structures are ideal for this purpose, particularly desiccation cracks, ripples, and graded bedding. This approach reached its apotheosis in Shrock's volume *Sequence in Layered Rocks,* written in 1948. On a broader scale, structural geologists were concerned with the vast prisms of sediments that occur in what were then called geosynclinal furrows. A valid stratigraphy is a prerequisite for a valid structural analysis. Thus it is interesting to see that it was not a stratigrapher, but Sir Edward Bailey, doyen of structural geologists, who wrote the paper "New Light on Sedimentation and Tectonics" in 1930. This seminal paper defined the fundamental distinction between the sedimentary textures and structures of shelves and those of deep basins. This paper also contained the germ of the turbidity current hypothesis.

The concept of the turbidity flow rejuvenated the study of sediments in the 1950s and early 1960s. While petrographers counted zircon grains and stratigraphers collected more fossils, it was the structural geologists who asked "How are thick sequences of flysch facies deposited in geosynclines?" It was modern oceanography that provided the turbidity current as a possible mechanism (see Section 4.2.2). It is true to say that this concept rejuvenated the study of sedimentary rocks, although in their enthusiasm geologists identified turbidites in every kind of facies, from the Viking sandbars of Canada to the alluvial Nubian sandstones of the Sahara.

Another stimulus to sedimentology came from the oil industry. The search for stratigraphically trapped oil led to a boom in the study of modern sediments. One of the first fruits of this approach was the American Petroleum Institute's "Project 51," a multidisciplinary study of the modern sediments of the northwest Gulf of Mexico (Shepard *et al.,* 1960). This was followed by many other studies of modern sediments by oil companies, universities, and oceanographic institutes. At last, hard data became available so that ancient sedimentary rocks could be interpreted by comparison with their modern analogs. The concept of the sedimentary model was born as it became apparent that there are, and always have been, a finite number of sedimentary environments that deposit characteristic sedimentary facies (see Section 6.3.1). By the end of the 1960s sedimentology was firmly established as a discrete discipline of the earth sciences. Through the 1960s the main focus of research was directed toward an understanding of sedimentary processes. By studying the bedforms and depositional structures of recent sediments, either in laboratory flumes or in the wild, it became possible to interpret accurately the environment of ancient sedimentary rocks (Laporte, 1979; Selley, 1970, 1996; Reading, 1978, 1996). Through the 1970s and 1980s sedimentological research expanded in both microscopic and macroscopic directions. Today a distinction is often made between macrosedimentology and microsedimentology. Macrosedimentology ranges from the study of sedimentary facies down to sedimentary structures. Microsedimentology covers the study of sedimentary rocks on a microscopic scale, what was often termed **petrography.** The improved imaging of sediments by scanning electron microscopy and cathodoluminescence brought about greater understanding of the physical properties of sediments. Improved analytical techniques, including isotope geochemistry and fluid inclusion analysis, gathered new chemical data which improved understanding of sedimentology in general, and diagenesis in particular (Lewis and McConchie, 1994).

The renaissance of petrography enhanced our understanding of the relationship between diagenesis and pore fluids and their effects on the evolution of porosity and permeability in sandstones and carbonates. Similarly, it is now possible to begin to understand the relationships between clay mineral diagenesis and the maturation of organic matter in hydrocarbon source beds.

Macrosedimentology has undergone a revolution in the last 30 years due to geophysics and sequence stratigraphy. Geophysical techniques can image sedimentary structures around a borehole, determine paleocurrent direction, and identify and measure the mineralogy, porosity, and pore fluid composition of sediments. On a larger scale, seismic data can now image the subsurface, revealing reefs, deltas, sandbars, and submarine fans several kilometers beneath the surface. As mentioned in the preface, however, this book is devoted to sensual sedimentology, not remotely sensed sedimentology. A firm grounding in sensual sedimentology is essential before endeavoring to interpret remotely sensed images of borehole walls and polychromatic 3D seismic pictures.

1.2 SEDIMENTOLOGY AND THE EARTH SCIENCES

Figure 1.1 shows the relationship between sedimentology and the basic sciences of biology, physics, and chemistry. The application of one or more of these fundamental sciences to the study of sediments gives rise to various lines of research in the earth sciences. These are now reviewed as a means of setting sedimentology within its context of geology. Biology, the study of animals and plants, can be applied to fossils in ancient

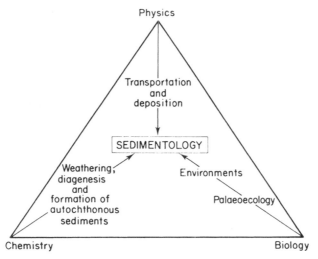

Fig. 1.1. Triangular diagram that shows the relationship between sedimentology and the fundamental sciences.

sediments. Paleontology can be studied as a pure subject that concerns the evolution, morphology, and taxonomy of fossils. In these pursuits the fossils are essentially removed from their sedimentological context.

The study of fossils within their sediments is a fruitful pursuit in two ways. Stratigraphy is based on the definition of biostratigraphic zones and the study of their relationship to lithostratigraphic units (Shaw, 1964; Ager, 1993; Pearson, 1998). Sound biostratigraphy is essential for regional structural and sedimentological analysis. The second main field of fossil study is aimed at deducing their behavior when they were alive, their habitats, and mutual relationships. This study is termed **paleoecology** (Ager, 1963). Where it can be demonstrated that fossils are preserved in place they are an important line of evidence in environmental analysis. Environmental analysis is the determination of the depositional environment of a sediment (Selley, 1996). This review of sedimentology has now moved from the purely biological aspect to facets that involve the biological, physical, and chemical properties of sedimentary rocks. To determine the depositional environment of a rock, it is obviously important to correctly identify and interpret the fossils that it contains. At a very simple level a root bed indicates a terrestrial environment, a coral reef a marine one. Most applied sedimentology, however, is based on the study of rock chips from boreholes. In such subsurface projects it is micropaleontology that holds the key to both stratigraphy and environment. The two aspects of paleontology that are most important to sedimentology, therefore, are the study of fossils as rock builders (as in limestones) and micropaleontology.

Aside from biology, environmental analysis is also based on the interpretation of the physical properties of a rock. These include grain size and texture as well as sedimentary structures. Hydraulics is the study of fluid movement. Loose boundary hydraulics is concerned with the relationship between fluids flowing over granular solids. These physical disciplines can be studied by theoretical mathematics, experimentally in laboratories, or in the field in modern sedimentary environments. Such lines of analysis can be applied to the physical parameters of an ancient sediment to determine the fluid processes that controlled its deposition (Allen, 1970). Environmental analysis also necessitates applying chemistry to the study of sediments. The detrital minerals of terrigenous rocks indicate their source and predepositional history. Authigenic minerals can provide clues to both the depositional environment of a rock as well as its subsequent diagenetic history. Environmental analysis thus involves the application of biology, physics, and chemistry to sedimentary rocks.

Facies analysis is a branch of regional sedimentology that involves three exercises. The sediments of an area must be grouped into various natural types or facies, defined by their lithology, sedimentary structures, and fossils. The environment of each facies is deduced and the facies are placed within a stratigraphic framework using paleontology and sequence stratigraphy. Like environmental analysis, facies analysis utilizes biology, chemistry, and physics. On a regional scale, however, facies analysis involves the study of whole basins of sediment. Here geophysics becomes important, not just to study the sedimentary cover, but to understand the physical properties and processes of the crust in which sedimentary basins form. One particular contribution of physics to sedimentology has been the application of geophysics, specifically the seismic method to sedimentary rocks. Improvements in the quality of seismic data in the 1970s lead to the development of a whole new way of looking at sediments, termed sequence stratigraphy (Payton, 1977). While it had long been recognized that sediments tended to be deposited in organized, and often cyclic, packages, modern high-quality seismic data enabled these to be mapped over large areas. Sequence boundaries could be identified and calibrated with well and outcrop data.

Referring again to Fig. 1.1 we come to the chemical aspects of sediments. It has already been shown how both environmental and facies analyses utilize knowledge of the chemistry of sediments. Petrology and petrography are terms that are now more or less synonymously applied to the microscopic study of rocks (Tucker, 1991; Blatt, 1992; Boggs, 1992; Raymond, 1995). These studies include petrophysics, which is concerned with such physical properties as porosity and permeability. More generally, however, they are taken to mean the study of the mineralogy of rocks. Sedimentary petrology is useful for a number of reasons. As already pointed out, it can be used to discover the provenance of terrigenous rocks and the environment of many carbonates. Petrography also throws light on diagenesis: the postdepositional changes in a sediment. Diagenetic studies elucidate the chemical reactions that took place between a rock and the fluids which flowed through it. Diagenesis is of great interest because of the way in which it can destroy or increase the porosity and permeability of a rock. This is relevant in the study of aquifers and hydrocarbon reservoirs. Chemical studies are also useful in understanding the diagenetic processes that form the epigenetic mineral deposits, such as the lead–zinc sulfide and carnotite ores. Lastly, at the end of the spectrum the pure application of chemistry to sedimentary rocks is termed sedimentary geochemistry. This is a vast field in itself (Krauskopf and Bird, 1995; Faure, 1998). It is of particular use in the study of the chemical sediments, naturally, and of microcrystalline sediments that are hard to study by microscopic techniques. Thus the main contributions of sedimentary geochemistry lie in the study of clay minerals, phosphates, and the evaporite rocks. Organic geochemistry is primarily concerned with the generation and maturation of coal, crude oil, and natural gas. Organic geochemistry, combining biology and chemistry, brings this discussion back to its point of origin. The preceding analysis has attempted to show how sedimentology is integrated with the other geological disciplines. The succeeding chapters will demonstrate continuously how much sedimentology is based on the fundamental sciences of biology, physics, and chemistry.

1.3 APPLIED SEDIMENTOLOGY

Sedimentology may be studied as a subject in its own right, arcane and academic; an end in itself. On the other hand, sedimentology has a contribution to make to the exploitation of natural resources and to the way in which man manipulates the environment. This book has been written primarily for the reader who is, or intends to be, an industrial geologist. It is not designed for the aspiring academic. It is relevant, therefore, to consider the applications of sedimentology. Table 1.1 documents some of the applications of sedimentology. Specific instances and applications will be discussed throughout the book. Most of the intellectual and financial stimulus to sedimentology has come from the oil industry and, to a lesser extent, the mining industry. The applications of sedimentology in these fields will be examined in some detail to indicate the reasons for this fact.

Table 1.1
Some of the Applications of Sedimentology

	Application	Related fields
I. Environmental	Sea-bed structures Pipelines Coastal erosion defenses Quays, jetties, and harbors	Oceanography

	Opencast excavations
	  and tunneling
		Identification of
		  nuclear waste sites
		Engineering geology

	Foundations for
	  motorways
	Airstrips and tower
	  blocks
		Soil mechanics and
		  rock mechanics

II. Extractive	Sand and gravel aggregates	

	A. Whole
	    rock
	    removed { Clays
	Limestones
	Coal
	Phosphate
	Evaporites
	Sedimentary ores
		Quarrying
		
		Mining geology

	B. Pore
	    fluid
	    removed { Water
	Oil
	Gas
		Hydrology
		Petroleum geology

First, however, some of the other uses of sedimentology are briefly reviewed. The emphasis throughout this book is on sedimentology and its relationship to ancient lithified sedimentary rocks. It is important to note, however, that a large part of sedimentology is concerned with modern sediments and depositional processes. This is not just to better interpret ancient sedimentary rocks. These studies are of vital importance in the manipulation of our environment (Bell, 1998). For example, the construction of modern coastal erosion defenses, quays, harbors, and submarine pipelines all require detailed site investigation. These investigations include the study of the regime of wind, waves, and tides and of the physical properties of the bedrock. Such studies also include an analysis of the present path and rate of movement of sediment across the site and the prediction of how these will alter when the construction work is completed. It is well known how the construction of a single pier may act as a trap for longshore drifting sediment, causing coastal erosion on one side and beach accretion on the other. The important interplay between sediments and the fluids that flow through them was mentioned earlier. This topic is of great environmental importance. Porosity and permeability are typically found in sediments, rather than in igneous and metamorphic rocks. Thus hydrogeology is principally concerned with evaluating the water resources of sedimentary aquifers (Ingebritsen and Sanford, 1998). An increasingly important aspect of hydrogeology is the study of the flow of contaminants from man-made surface pollution into the subsurface aquifer (Fetter, 1993).

Turning inland, studies of modern fluvial processes have many important applications. The work of the U.S. Army Corps of Engineers in attempting to prevent the Mississippi from meandering is an example. Studies of fluvial channel stability, flood frequency, and flood control are an integral part of any land utilization plan or town development scheme.

Engineering geology is another field in which sedimentology may be applied. In this case, however, most of the applications are concerned with the physical properties of sediments once they have been deposited and their response to drainage, or to the stresses of foundations for dams, motorways, or large buildings. These topics fall under the disciplines of soil mechanics and rock mechanics.

Thus before proceeding to examine the applications of sedimentology to the study of ancient sedimentary rocks, the previous section demonstrates some of its many applications in environmental problems concerning recent sediments and sedimentary processes (Murck *et al.,* 1996; Evans, 1997). Most applications of sedimentology to ancient sedimentary rocks are concerned with the extraction of raw materials. These fall into two main groups: the extraction of certain strata of sediment, and the extraction of fluids from pores, leaving the strata intact.

Many different kinds of sedimentary rock are of economic value. These include recent unconsolidated sands and gravels that are useful in the construction industry. Their effective and economic exploitation requires accurate definition of their physical properties such as size, shape, and sorting, as well as the volume and geometry of individual bodies of potentially valuable sediment. Thus in the extraction of river gravels it is necessary to map the distribution of the body to be worked, be it a paleochannel or an old terrace, and to locate any ox-bow lake clay plugs that may diminish the calculated reserves of the whole deposit.

Similarly, consolidated sandstones have many uses as aggregate and building stones. Clays have diverse applications and according to their composition can be used for bricks, pottery, drilling mud, and so forth. Limestones are important in the manufacture of cement and fertilizer and as a flux in the smelting of iron. The use of all these sedimentary rocks involves two basic problems. The first is to determine whether or not the rock conforms to the physical and chemical specifications required for a particular purpose. This involves petrography and geochemistry. The second problem is to predict the geometry and hence calculate the bulk reserves of the economic rock body. This involves sedimentology and stratigraphy. Here geology mingles with problems of quarrying, engineering, and transportation. Geology is nevertheless of extreme importance. It is no use building a brand new cement works next to a limestone crag if, when quarrying commences, it is discovered that the limestone is not a continuous formation, but a reef of local extent.

Coal is another sedimentary rock of importance to the energy budget of most industrial countries. Coal technology is itself a major field of study. Like the other economic sedimentary rocks, coal mining hinges on two basic geological problems: quality and quantity. The quality of the coal is determined by specialized petrographic and chemical techniques. The quantitative aspects of coal mining involve both problems of structural geology and mining engineering as well as careful facies analysis. Classic examples of ancient coal-bearing deltaic rocks have been documented in the literature (e.g., the Circulars of the Illinois State Geological Survey). These studies have been made possible by a combination of closely spaced core holes and data from modern deltaic sediments. Using this information, facies analysis can delineate the optimum stratigraphic and geographic extent of coal-bearing facies. Detailed environmental studies can then be used to map the distribution of individual coal seams. Coal can form in various deltaic subenvironments, such as interdistributary bays, within, or on the crests of channel sands, as well as regionally uniform beds. The coals that form in these various subenvironments may be different both in composition and areal geometry.

Evaporite deposits are another sedimentary rock of great economic importance, forming the basis for chemical industries in many parts of the world. The main evaporite minerals, by bulk, are gypsum ($CaSO_4 \cdot 2H_2O$) and halite (NaCl). Many other salts are rarer but of equal or greater economic importance. These include carbonates, chlorides, and sulfates. The genesis of evaporite deposits has been extensively studied both in nature and experimentally in the laboratory. The study of evaporites is important, not just because of the economic value of the minerals, but because of the close relationship between evaporites and petroleum deposits (Melvin, 1991). It is appropriate to turn now from the applications of sedimentology to the extraction of rock, to consider those uses where only the pore fluids are sought and removed.

Porosity and permeability are typically found in sediments, rather than in igneous and metamorphic rocks. Thus hydrogeology is principally concerned with evaluating the water resources of sedimentary aquifers (Ingebritsen and Sanford, 1998). The world shortage of potable water may soon become more important than the quest for energy. It has been argued that hydrogeology is "merely petroleum geology upside down." This is a shallow statement, yet it contains much truth. Many sedimentary rocks are excellent aquifers and sedimentology and stratigraphy may be used to locate and exploit these. Hydrogeology, like petroleum geology, is largely concerned with the quest for porosity and permeability. It is necessary for an aquifer, like an oil reservoir, to have both the pore space to contain fluid and the permeability to give it up. Whereas an oil reservoir requires an impermeable cap rock to prevent the upward dissipation of oil and gas, an aquifer requires an impermeable seal beneath to prevent the downward flow of water. Despite these fundamental differences, there are many points in common between the search for water and for petroleum. Both use regional stratigraphic and structural analyses to determine the geometry and attitude of porous beds. Both can use facies studies and environmental analysis to fulfill these objectives.

1.3.1 Petroleum

As mentioned earlier, the search for petroleum has been the major driving force behind the rapid expansion in sedimentology during the last half century. Almost all commercial petroleum accumulations are found within sedimentary rocks. The majority of petroleum geoscientists believe, therefore, that petroleum originates in sedimentary rocks. Many Russians, however, from Mendele'ev (1877) to Porfir'ev (1974) and Pushkarev (1995), argue that petroleum comes out of the mantle, a view championed in the west by Gold (1979, 1999). Petroleum geology is beyond the scope of this volume, but Chapter 7 provides an introduction before embarking such texts as Selley (1998) and Gluyas and Swarbrick (1999). Hydrocarbons are complex organic compounds that are found in nature within the pores and fractures of rocks. They consist primarily of hydrogen and carbon compounds with variable amounts of nitrogen, oxygen, and sulfur, together with traces of elements such as vanadium and nickel. Hydrocarbons occur in solid, liquid, and gaseous states (see Section 7.3.2). Solid hydrocarbons are variously known as asphalt, tar, pitch, and gilsonite. Liquid hydrocarbons are termed crude oil or simply "crude." Gaseous hydrocarbons are loosely referred to as natural gas, ignoring inorganic natural gases such as those of volcanic origin.

It is a matter of observation that hydrocarbons occur in sedimentary basins, not in areas of vulcanism or of regional metamorphism. Most geologists conclude, therefore, that hydrocarbons are both generated and retained within sedimentary rocks rather than in those of igneous or metamorphic origin. (See Hunt, 1996, for expositions of the Western orthodox view, but see the references cited earlier for an exposition of the abiogenic theory.) The processes of hydrocarbon generation and migration are complex and controversial. Detailed analyses are found in the references previously cited. Almost all sedimentary rocks contain some traces of hydrocarbons. The source of hydrocarbons is generally thought to be due to large accumulations of organic matter, vegetable or animal, in anaerobic subaqueous environments. During burial and compaction this source sediment becomes heated. Hydrocarbons are formed and migrate out of the source rock into permeable carrier beds. The hydrocarbons will then migrate upward, being lighter than the pore water. Ultimately the hydrocarbons will be dissipated at the earth's surface through natural seepages. In some fortunate instances, however, they are trapped by an impervious rock formation. They form a reservoir in the porous beds beneath and await discovery by the oil industry. This brief summary of a complex sequence of events shows that a hydrocarbon accumulation requires a source rock, a reservoir rock, a trap, and a cap rock. Any rock with permeability is a potential reservoir. Most of the world's reservoirs are in sandstones, dolomites, and limestones. Fields also occur in igneous and metamorphic rocks, commonly where porosity has been induced by weathering beneath unconformities, and where fracture porosity has been induced tectonically. The impermeable cap rocks that seal hydrocarbons in reservoirs are generally shales or evaporites. Less commonly "tight" (nonpermeable) limestones and sandstones may be cap rocks. Sedimentology assists in the search for oil in all types of traps, and in the subsequent effective exploitation of a reservoir.

The exploitation of hydrocarbons from an area falls into a regular time sequence. Initially, broad regional stratigraphic studies are carried out to define the limits and architecture of sedimentary basins. With modern offshore exploration this is largely based on geophysical data. Preliminary magnetic and gravity surveys are followed by more detailed seismic shooting. This information is used to map marker horizons. Though the age and lithology of these strata are unknown at this stage, the basin can be broadly defined and prospective structures located within it. The first well locations will test structural traps such as anticlines, but even at this stage seismic data can detect such sedimentary features as delta fronts, reefs, growth faults, and salt diapirs. Today it is possible to directly locate petroleum accumulations because seismic reflections of petroleum: water contacts can sometimes actually be observed. The first wells in a new basin provide a wealth of information. Regardless of whether these tests yield productive hydrocarbons, they give the age and lithology of the formations previously mapped seismically. Geochemical analysis tells whether source rocks are present and whether the basin has matured to the right temperature for oil or gas generation. As more wells are drilled in a productive basin, so are more sedimentological data available for analysis.

The initial main objective of this phase is to predict the lateral extent and thickness of porous formations adjacent to potential source rocks and within the optimum thermal envelope. Regional sedimentological studies may thus define productivity fairways such as a line of reefs, a delta front, or a broad belt of shoal sands. The location of individual traps may be structurally controlled within such fairways. Sedimentology becomes more important still when the structural traps have all been seismically located and drilled. The large body of data now available can be used to locate subtle stratigraphic oil fields. This is the major application of sedimentology in the oil industry and is discussed at length in the next section.

Concluding this review of the role of sedimentology in the history of the exploitation of a basin, it is worth noting the contribution it can make not just in finding fields, but in their development and production. Few reservoir formations are petrophysically isotropic. Most oil fields show some internal variation not only in reservoir thickness, but also in its porosity and permeability (see Fig. 3.14). These differences can be both vertical or horizontal and, in the case of permeability, there is often a preferred azimuth of optimum flow (see Fig. 3.28). These variations are due either to primary depositional features or to secondary diagenetic changes.

Primary factors are common in sandstone reservoirs. Gross variations in porosity within a reservoir formation may relate to the location of discrete clean sand bodies such as channels or bars, within an overall muddy sand. Variations in the direction of maximum flow potential (i.e., permeability) may relate to a gross sand body trend or to sand-grain orientation. In carbonate reservoirs, on the other hand, depositional variations in porosity and permeability tend to be masked by subsequent diagenetic changes. Hence the interest shown by oil companies in carbonate diagenesis. It is important to understand the petrophysical variations within a reservoir. This assists in the development drilling of a field by predicting well locations that will produce the maximum amount of petroleum and the minimum amount of water. Subsequently, secondary recovery techniques can also utilize this knowledge. Selection of wells for water or gas injection should take into account the direction of optimum permeability within the reservoir formation.

Concluding this review of the applications of sedimentology through the evolution of a productive oil basin, it is important to note how close integration is necessary with geophysics. First, to elucidate gross structure and stratigraphy of a basin. Subsequent seismic surveys establish drillable prospects and may image petroleum accumulations. The final phase of development drilling and production necessitates close liaison between geophysicists, geologists, and engineers. The relationship between petrography and petrophysics is most important at this time. The preceding account of the applications of sedimentology in the search for petroleum needs to be put in perspective. A lot of the stuff was found before most oil men could even spell "sedimentology."

1.3.2 Sedimentary Ores: General Aspects

Sedimentology has never been used by the mining industry to the extent that it has been employed in the search for hydrocarbons (Parnell *et al.*, 1990; Evans, 1995). There are two good reasons for this. First, many metallic ores occur within, or juxtaposed to, igneous and metamorphic rocks. In such situations sedimentology can neither determine

Table 1.2
Summary of the Main Sedimentary Ores and Their Common Modes of Origin

Name	Process	Examples
I. Syngenetic	Originated by direct precipitation during sedimentation	Manganese nodules and crusts, some oolitic ironstones
II. Epigenetic	Originated by postdepositional diagenesis	Carnotite, copper–lead–zinc assemblages, some ironstones
III. Placer	Syndepositional detrital sands	Alluvial gold, cassiterite and zircon

the genesis, nor aid the exploitation of the ore body. Secondly, direct sensing methods of prospecting have long been available. Ores can be located by direct geochemical and geophysical methods. Geochemical techniques began with the old-style panning of alluvium gradually working upstream to locate the mother lode. This is now superseded by routine sampling and analysis for trace elements. Direct geophysical methods of locating ore bodies include gravity, magnetometer, and scintillometer surveys. For these reasons sedimentology is not as useful in searching for sedimentary ores as it is for petroleum. Its relevance in mining is twofold. First, it has a large contribution to make to the problems of ore genesis in sedimentary rocks. Secondly, it is useful as a backup tool in the search for and development of ore bodies. Three main processes are generally believed to be responsible for the genesis of sedimentary ores (Table 1.2). Detrital placer deposits are formed where current action winnows less dense quartz grains away to leave a lag concentrate of denser grains. These heavy minerals may sometimes be of economic importance.

Syngenetic sedimentary ores form by direct chemical precipitation within the depositional environment. Epigenetic ores form by the diagenetic replacement of a sedimentary rock, generally limestone, by ore minerals. The origin of sedimentary ores has always excited considerable debate. This generally hinges on two criteria. First, how may epigenetic and syngenetic ores be differentiated? Secondly, to what extent are the mineralizing fluids of epigenetic ores derived from normal sedimentary fluids, and to what extent are they hydrothermal in origin? In recent years it has been increasingly realized that ordinary sedimentary processes can generate both syngenetic, and epigenetic ores in sediments. Conversely, less credence is now given to a hydrothermal origin for many epigenetic ores. The following account of detrital, syngenetic, and epigenetic sedimentary ore attempts to show how sedimentology helps to decipher problems of ore genesis and exploitation.

1.3.2.1 Placer Ores

Detrital sediments are generally composed of particles of varying grain density. For example, most sandstones contain a certain amount of clay and silt. The sand particles are largely grains of quartz, feldspar, and mafic minerals with specific gravities of between

2.0 and 3.0. At the same time most sands contain traces of minerals with specific gravities between 4.0 and 5.0. The heavy minerals, as these are called, commonly consist of opaque iron ores, tourmaline, garnet, zircon, and so on. More rarely they include ore minerals such as gold, cassiterite, or monazite. It is a matter of observation that the heavy mineral fraction of a sediment is much finer grained than the light fraction. There are several reasons for this. First, many heavy minerals occur in much smaller crystals than do quartz and feldspar in the igneous and metamorphic rocks in which they form. Secondly, the sorting and composition of a sediment is controlled by both the size and density of the particles — this is spoken of as their hydraulic ratio. Thus, for example, a large quartz grain requires the same current velocity to move it as a small heavy mineral. It is for this reason that sandstones contain traces of heavy minerals which are finer than the median overall grain size of the less dense grains (see Fig. 8.15). In certain flow conditions, however, the larger, less dense grains of a sediment are winnowed away to leave a residual deposit of finer grained heavy minerals. The flow conditions that cause this separation are of great interest to the mining industry, both as a key to understanding the genesis of placer ores, and in the flotation method of mineral separation in which crushed rock is washed to concentrate the ore. Placer deposits occur in nature in alluvial channels, on beaches and on marine abrasion surfaces (Fig. 1.2A). Placer sands are frequently thinly laminated with occasional slight angular disconformities and shallow troughs. This is true of both fluvial and beach-sand placers. In Quaternary alluvium placers occur in the present channel system, notably on the shallow riffles immediately downstream of meander pools, and also in river terrace deposits and buried channel-fill alluvium below the floor of the present river bed. Gold is a characteristic alluvial placer ore mineral occurring in the Yukon and Australia. The Pre-Cambrian gold- and uranium-bearing fluvial conglomerates of the Witwatersrand appear to be paleoplacers deposited on a braided alluvial fan though this has been disputed (see Section 6.3.2.2.4). Recent marine placers, like their fluvial counterparts, occur at different topographic levels due to Pleistocene sea level changes (Fig. 6.46).

Essentially, placer ores are controlled by the geochemistry of the sediment source, by the climate, and hence depth of weathering, by the geomorphology, which controls the rate of erosion and topographic gradient, and by the hydrodynamics of the transportational and depositional processes. The genesis of placer deposits is seldom in dispute because the textures and field relationships of the ores clearly indicate their detrital origin. Furthermore, the processes of placer formation can be observed at work on the earth today. The origins of syngenetic and epigenetic sedimentary ores are more equivocal.

1.3.2.2 Syngenetic Ores

Syngenetic ores are those that were precipitated during sedimentation. This phenomenon has been observed on the earth's surface at the present time. One of the best known examples of this process is the widespread formation of manganese nodules and crusts on large areas of the beds of oceans, seas, and lakes (see Section 6.3.2.10.1). Goethite ooliths are forming today in Lake Chad in Central Africa (Lemoalle and Dupont, 1971). More spectacular examples of modern sea floor metallogenic processes occur associated with volcanic activity as at Santorini and in the "hot holes" of the Red Sea (Puchelt, 1971; Degens and Ross, 1969).

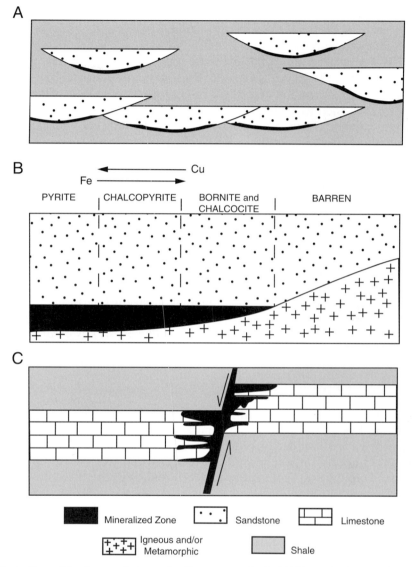

Fig. 1.2. Illustrations of the three main types of sedimentary ore body. (A) Placer detrital ores (e.g., diamond, gold, and other heavy minerals) on channel floors. (B) Syngenetic ores, precipitated during sedimentation, such as those found in the mineralized shale of the Roan Group sediments (Pre-Cambrian) in the Zambian copper belt. Note the correlation between the type of mineralization and paleogeography. (C) Epigenetic replacement of limestone by lead–zinc sulfides in Mississippi Valley-type ore body.

Many stratiform ore bodies have been attributed to a syndepositional or syngenetic origin, though an epigenetic (postdepositional replacement) origin has often been argued too. The criteria for differentiating these origins are examined later. Particularly cogent arguments have been advanced to support a syngenetic origin for ores of copper, manganese, and iron in bedded sedimentary rocks (Bernard, 1974; Wolf, 1976). Classic examples of syngenetic copper ores include the Permian Kupferschiefer of Germany and the Pre-Cambrian Copper belt of Katanga and Zambia. The Kupferschiefer is a thin black organic radioactive shale that is remarkable for its lateral uniformity across the North Sea basin. It overlies the eolian sands of the Rotliegende, and is itself overlain by the Zechstein evaporites. Locally the Kupferschiefer is sufficiently rich in copper minerals such as chalcopyrite and malachite to become an ore. A syngenetic origin for the ore is widely accepted (e.g., Gregory, 1930; Brongersma Sanders, 1967).

In Zambia, relatively unmetamorphosed Pre-Cambrian sediments of the Roan Group overlie an igneous and metamorphic basement. Extensive copper mineralization in the Roan Group sediments includes pyrite, chalcopyrite, bornite, and chalcocite. These occur within shallow marine shales and adjacent fluvial sandstones, but not in eolian sandstones or basement rocks. The mineralization is closely related to a paleocoastline. The ore bodies are mineralogically zoned with optimum mineralization in sheltered embayments (Fig. 1.2B). Early workers favored a hydrothermal epigenetic origin for the ore due to intrusion of the underlying granite. It was demonstrated, however, that the irregular granite/sediment surface is due to erosion, not intrusion. Criteria now quoted in favor of a syngenetic origin for the ore include its close relationship with paleotopography and facies, and the occurrence of reworked ore grains in younger but penecontemporaneous sandstones (Garlick, 1969; Garlick and Fleischer, 1972). Analogous deposits at Kamoto in Katanga have been interpreted as shallow marine in origin, but extensive diagenesis is invoked as the mineralizing agent (Bartholome *et al.*, 1973). Though the source of the Copper belt ores is unknown, a syngenetic origin is now widely accepted, with the proviso that subsequent diagenesis has modified ore fabrics and mineralogy (Guilbert and Park, 1986).

1.3.2.3 Epigenetic Sedimentary Ores

Epigenetic or exogenous ores in sedimentary rocks are those that formed later than the host sediment. Epigenetic ores can be generated by a variety of processes. These include the concentration of disseminated minerals into discrete ore bodies by weathering, diagenetic, thermal, or metamorphic effects. Epigenesis also includes the introduction of metals into the host sediment by meteoric and hydrothermal solutions resulting in the replacement of the country rock by ore minerals.

Criteria for differentiating syngenetic and epigenetic ores have been mentioned in the preceding section. Epigenetic ores are characteristically restricted to modern topography, to unconformities or, if hydrothermal in origin, to centers of igneous activity. The ore may not contain a dwarfed and stunted fauna as in syngenetic deposits. Isotopic dating will show the ore to postdate the host sediment (Bain, 1968).

Once upon a time, epigenetic ores were very largely attributed to hydrothermal emanations from igneous sources, except for obvious examples of shallow supergene enriched ores. Within recent years it has been shown that many epigenetic ores lie far distant from any known igneous center and show no evidence of abnormally high geothermal gradients (Sangster, 1995). It has been argued that these ores formed from concentrated chloride-rich solutions derived from evaporites and from residual solutions derived from the final stages of compaction of clays (Davidson, 1965; Amstutz and Bubinicek, 1967). The two main groups of ore attributed to these processes are the Mississippi Valley type lead–zinc sulfides (Fig. 1.2C), and the uranium "roll-front" ores. These are described in more detail in Section 6.3.2.8 and Section 6.3.2.2.4, respectively.

1.3.2.4 Ores in Sediments: Conclusion

The preceding account shows that many ore bodies occur intimately associated with sedimentary rocks. They originated as detrital placers, syngenetically by direct precipitation, and epigenetically by diagenetic precipitation and replacement from fluids of diverse and uncertain origins. Sedimentology throws considerable light on these problems of ore genesis, notably by utilizing its geochemical and petrophysical aspects. It can be argued that these problems of ore genesis are peripheral to the actual business of locating workable deposits. But if one believes, for example, that lead–zinc sulfide mineralization is an inherent feature of reef limestones, then the search can be extended beyond areas of known hydrothermal mineralization. Theories of metallogenesis thus play a part in deciding which areas may be prospective for a particular mineral.

Regardless of prevailing prejudices of metallogenesis, however, sedimentology can still be used as a searching tool. Facies analysis can map a complex of mineralized reefs, define a minette iron ore shoreline, or locate permeability barriers in carnotite-rich alluvium. Sedimentologic surveys can be carried out at the same time as direct geochemical and geophysical methods and should form an integral part of the total exploration effort.

Finally, Table 1.3 shows the wide range of sedimentary environments in which mineral

Table 1.3
Occurrence of Mineral Deposits in Various Sedimentary Environments[a]

Main lithology	Environment	Mineralization
Sandstones	Fluvial	Placers Uranium roll-front ores Cu–Fe red bed ores
	Deltas	Coal Sideritic ironstones
	Lakes	Oil shales Evaporites Coal "Bog" iron ores
	Beaches	Placers
Carbonates	Continental shelves	Phosphates Sedimentary iron ores Evaporites
	Reefs	Pb–Zn Mississippi Valley ores
Shale and chert	Pelagic	Cu–Zn sulfides Manganese nodules Barytes

[a] Examples are discussed at appropriate places in the ensuing text, principally in Chapter 6.

deposits occur. A knowledge of sedimentology in general, and of environmental interpretation in particular, is useful to locate and exploit such mineral deposits.

SELECTED BIBLIOGRAPHY

Bell, F. G. (1998). "Environmental Geology." Blackwell, Oxford. 594pp.
Evans, A. M. (1997). "Introduction to Economic Geology and its Environmental Impact." Blackwell Science, Oxford. 352pp.
Ingebritsen, S. E., and Sanford, W. E. (1998). "Groundwater in Geological Processes." Cambridge University Press., Cambridge, UK. 341pp.
Krauskopf, K. B., and Bird, D. K. (1995). "Introduction to Geochemistry," 3rd ed. McGraw-Hill, London. 640pp.
Murck, B., Skinner, B. J., and Porter, S. C. (1996). "Environmental Geology." Wiley, Chichester. 558pp.

REFERENCES

Ager, D. V. (1963). "Principles of Paleoecology." McGraw-Hill, New York. 371pp.
Ager, D. V. (1993). "The Nature of the Stratigraphical Record," 3rd ed. Wiley, London. 151pp.
Allen, J. R. L. (1970). "Physical Processes of Sedimentation." Allen & Unwin, London. 248pp.
Amstutz, G. C., and Bubinicek, L. (1967). Diagenesis in sedimentary mineral deposits. In "Diagenesis in Sediments" (S. Larsen and G. V. Chilingar, eds.), pp. 417–475. Elsevier, Amsterdam.
Bailey, E. B. (1930). New light on sedimentation and tectonics. Geol. Mag. 67, 77–92.
Bain, G. W. (1968). Syngenesis and epigenesis of ores in layered rocks. Int. Geol. Congr., Rep. Sess., 23rd, Prague. Sec. 7, pp. 119–136.
Barnes, J. W. (1973). Jason and the Gold Rush. Proc. Geol. Assoc. 84, 482–485.
Bartholome, P., Evrard, P., Katekesha, F., Lopez-Ruiz, J., and Ngongo, M. (1973). Diagenetic ore-forming processes at Komoto, Katanga, Republic of Congo. In "Ores in Sediments" (G. C. Amstutz and A. J. Bernard, eds.), pp. 21–41. Springer-Verlag, Heidelberg.
Bell, F. G. (1998). "Environmental Geology." Blackwell, Oxford. 594pp.
Bernard, A. J. (1974). Essai de revue des concentration metalliféres dans le cycle sedimentaires. Geol. Rundsch. 63, 41–51.
Blatt, H. (1992). "Sedimentary Petrology," 2nd ed. Freeman, New York. 514pp.
Boggs, S. (1992). "Petrology of Sedimentary Rocks." Prentice Hall, Hemel Hempstead, England. 707pp.
Brongersma-Sanders, M. (1967). Permian wind and the occurrence of fish and metals in the Kupferschiefer and Marl Slate. Proc. Int.-Univ. Geol. Congr., 15th, Leicester, pp. 61–71.
Davidson, C. F. (1965). A possible mode of strata-bound copper ores. Econ. Geol. 60, 942–954.
Degens, E. T., and Ross, D. A., eds. (1969). "Hot Brines and Recent Heavy Metal Deposits in the Red Sea." Springer-Verlag, Berlin. 600pp.
Evans, A. M., ed. (1995). "Introduction to Mineral Exploration." Blackwell Science, Oxford. 420pp.
Evans, A. M. (1997). "Introduction to Economic Geology and its Environmental Impact." Blackwell Science, Oxford. 352pp.
Faure, G. (1998). "Principles and Applications of Geochemistry." Prentice Hall, Hemel Hempstead, England. 625pp.
Fetter, C. W. (1993). "Contaminant Hydrogeology." Prentice Hall. Hemel Hempstead, England. 458pp.
Garlick, W. G. (1969). Special features and sedimentary facies of stratiform sulphide deposits in arenites. In "Sedimentary Ores Ancient and Modern" (C. H. James, ed.), Spec. Publ. No. 7, pp. 107–169. Geol. Dept., Leicester University.

Garlick, W. G., and Fleischer, V. D. (1972). Sedimentary environment of Zambian copper deposition. *Geol. Mijnbouw* **51,** 277–298.

Gluyas, J., and Swarbrick, R. E. (1999). "Petroleum Geoscience." Blackwell, Oxford. 288pp.

Gold, T. (1979). Terrestrial sources of carbon and earthquake outgassing. *J. Pet. Geol.* **1,** 3–9.

Gold, T. (1999). "The Deep Hot Biosphere." Springer-Verlag, New York. 235pp.

Gregory, J. W. (1930). The copper-shale (Kupferschiefer) of Mansfeld. *Trans. — Inst. Min. Metall.* **40,** 3–30.

Guilbert, J. M., and Park, C. F. (1986). "Ore Deposits." Freeman, New York. 985pp.

Hunt, J. H. (1996). "Petroleum Geochemistry and Geology," 2nd ed. Freeman, San Francisco. 743pp.

Ingebritsen, S. E., and Sanford, W. E. (1998). "Groundwater in Geological Processes." Cambridge University Press, Cambridge, UK. 341pp.

Krauskopf, K. B., and Bird, D. K. (1995). "Introduction to Geochemistry," 3rd ed. McGraw-Hill, London. 640pp.

Laporte, L. F. (1979). "Ancient Environments," 2nd ed. Prentice-Hall, Englewood Cliffs, NJ. 163pp.

Lemoalle, J., and Dupont, B. (1971). Iron-bearing oolites and the present conditions of iron sedimentation in Lake Chad (Africa). *In* "Ores in Sediments" (G. C. Amstutz and A. J. Bernard, eds.), pp. 167–178. Springer-Verlag, Heidelberg.

Lewis, D. W., and McConchie, D. (1994). "Analytical Sedimentology." Chapman & Hall, London. 197pp.

Lyell, C. (1865). "Elements of Geology." John Murray, London. 794pp.

Melvin, J. L., ed. (1991). "Evaporites, Petroleum and Mineral Resources." Elsevier, Amsterdam. 556pp.

Mendele'ev, D. (1877). Entsehung und Vorkommen des Minerols. *Ber. Dtsch. Chem. Ges.* **10,** 229.

Murck, B., Skinner, B. J., and Porter, S. C. (1996). "Environmental Geology." Wiley, Chichester. 558pp.

Parnell, J., Lianjun, Y., and Changming, E., eds. (1990). "Sediment-hosted Mineral Deposits." Blackwell, Oxford. 240pp.

Payton, C. E., ed. (1977). "Seismic stratigraphy: Applications to Hydrocarbon Exploration, Mem. No. 26. Am. Assoc. Pet. Geol., Tulsa, OK. 516pp.

Pearson, P. N. (1998). Evolutionary concepts in biostratigraphy. *In* "Unlocking the Stratigraphic Record" (P. Doyle and M. R. Bennett, eds.), pp. 123–144. Wiley, London.

Porfir'ev, V. B. (1974). Inorganic origin of petroleum. *AAPG Bull.* **58,** 3–33.

Puchelt, H. (1971). Recent iron sediment formation at the Kameni Islands, Santorini (Greece). *In* "Ores in Sediments" (G. C. Amstutz and A. J. Bernard, eds.), pp. 227–245. Springer-Verlag, Heidelberg.

Pushkarev, Y. D. (1995). Gas-petroliferous potential of the Precambrian basement under European platforms according to radiogenic isotope geochemistry of oils and bitumens. *In* "Precambrian of Europe: Stratigraphy, Structure, Evolution and Mineralization," pp. 90–91. Russian Academy of Sciences, St. Petersburg.

Raymond, L. A. (1995). "Sedimentary Petrology." McGraw Hill, Maidenhead, England. 768pp.

Reading, H. G. (1978). "Sedimentary Environments and Facies." Blackwell Scientific, Oxford. 557pp.

Reading, H. G. (1996). "Sedimentary Environments and Facies," 3rd ed. Blackwell Scientific, Oxford. 688pp.

Sangster, D. F. (1995). What are Irish-type deposits anyway? *Ir. Assoc. Econ. Geol. Annu. Rev.,* pp. 91–97.

Selley, R. C. (1970). "Ancient Sedimentary Environments." Chapman & Hall, London. 237pp.

Selley, R. C. (1996). "Ancient Sedimentary Environments," 4th ed. Chapman & Hall, London. 300pp.

Selley, R. C. (1998). "Elements of Petroleum Geology," 2nd ed. Academic Press, San Diego, CA. 470pp.

Shaw, A. B. (1964). "Time in Stratigraphy." McGraw-Hill, New York. 365pp.

Shepard, F. P., Phleger, F. B., and van Andel, T. H., eds. (1960). "Recent Sediments, Northwest Gulf of Mexico." Am. Assoc. Pet. Geol., Tulsa, OK. 394pp.

Shotton, F. W. (1968). Prehistoric man's use of stone in Britain. *Proc. Geol. Assoc.* **79,** 477–491.

Shrock, R. R. (1948). "Sequence in Layered Rocks." McGraw-Hill, New York. 507pp.

Sorby, H. C. (1853). On the oscillation of the currents drifting the sandstone beds of the southeast of Northumberland, and their general direction in the coalfield in the neighbourhood of Edinburgh. *Rep. Proc. Geol. Polytech. Soc.,* W. Riding, Yorkshire, *1852,* pp. 232–240.

Sorby, H. C. (1908). On the application of quantitative methods to the study of the structure and history of rocks. *Q. J. Geol. Soc. London* **64,** 171–233.

Tucker, M. (1991). "Sedimentary Petrology," 2nd ed. Blackwell Science, London. 268pp.

Wadell, H. A. (1932). Sedimentation and sedimentology. *Science* **77,** 536–537.

Wolf, K. A., ed. (1976). "Handbook of Strata-Bound and Stratiform Ore Deposits," Vols. 1–7. Elsevier, Amsterdam.

Part I

Rock to Sediment

The entire mass of stratified deposits in the earth's crust is at once the monument and measure of the denudation which has taken place.

Charles Lyell (1838)

2 Weathering and the Sedimentary Cycle

2.1 INTRODUCTION

Before proceeding to the analysis of sedimentary rocks, their petrography, transportation, and deposition, it is appropriate to analyze the genesis of sediment particles. A sedimentary rock is the product of provenance and process. This chapter is concerned primarily with the provenance of sediment; that is to say the preexisting rocks from which it forms and the effect of weathering on sediment composition. First, however, it is apposite to consider the place of sediment particles within the context of what may be called, for want of a better term, "the earth machine." Consider a geologist sitting in a shady Saharan wadi idly hammering a piece of Continental Mesozoic "Nubian" sandstone. A piece of that sandstone falls from the cliff to the wadi floor. Within a second, a sample of Mesozoic Nubian sandstone has suddenly become part of a Recent alluvial gravel. The pebble deserves closer inspection. This sandstone clast is composed of a multitude of quartz particles. The precise provenance of these cannot be proved, but locally this formation overlies Upper Paleozoic sandstones and infills ancient wadis cut within them. The Mesozoic sand grains are clearly, in part, derived from the Upper Paleozoic sandstones. Similarly it can be shown that the Upper Paleozoic sediment has been through several previous cycles of erosion and deposition. The sand grains first formed from the weathering of Pre-Cambrian granite.

This simple review introduces the concept of the sedimentary cycle and shows how sediments are frequently polycyclic in their history. Now move from the specific event to the other extreme. The concept of plate tectonics is reviewed in Chapter 10. At this point, however, it is necessary to briefly introduce it, to better place sand grains in their broader perspective. The basic thesis of plate tectonics is that the earth's crust is formed continuously along linear zones of sea floor spreading. These occupy the midoceanic rift valleys, areas of extensive seismic and volcanic activity. New crust, formed by vulcanicity, moves laterally away from the zone of sea floor spreading, forming a rigid lithospheric plate. Simultaneously, at the distal edge of the plate, there is a linear zone of subduction marked by earthquakes and island arcs. Here old crust is drawn down and digested in the mantle. Sediment forms initially, therefore, from volcanic rocks on the midocean ridges. It may then be recycled several times, as seen in the Sahara, but inexorably individual particles are gradually drawn to the edge of the plate to descend and be destroyed in the mantle. The history of a sediment particle may be likened to a ball bouncing lethargically over many millions of years on a conveyor belt. The sand grain's history is one major cyclic event composed of many subsidiary ones. We now examine the sedimentary cycle in more detail.

2.2 THE SEDIMENTARY CYCLE

This section looks more closely at the sedimentary cycle in the smaller scale, leaving aside the major cycle of plate formation and destruction. Classically the sedimentary cycle consists of the phases of weathering, erosion, transportation, deposition, lithifaction, uplift, and weathering again (Fig. 2.1).

Weathering is the name given to the processes that break down rock at the earth's surface to form discrete particles (Ollier, 1969). **Erosion** is the name given to the processes that remove newly formed sediment from bedrock. This is followed generally by transportation and finally, when energy is exhausted, by deposition. The processes and products of weathering are examined more closely in the next section. It is sufficient at this point to state that weathering is generally divided into biological, chemical, and physical processes. Chemical weathering selectively oxidizes and dissolves the constituent minerals of a rock. Physical processes of weathering are those that bring about its actual mechanical disaggregation. Biological weathering is caused by the chemical and physical effects of organic processes on rock.

Erosion, the removal of new sediment, can be caused by four agents: gravity, glacial action, running water, and wind. The force of gravity causes the gradual creep of sediment particles and slabs of rock down hillsides, as well as the more dramatic avalanches. Glacial erosion occurs where glaciers and ice sheets scour and abrade the face of the earth as they flow slowly downhill under the influence of gravity. Moving water is a powerful agent of erosion in a wide spectrum of geomorphological situations ranging from desert flash flood to riverbank scouring and sea cliff undercutting. The erosive action of wind, on its own, is probably infinitesimal. Wind, however, blowing over a dry desert, quickly picks up clouds of sand and sandblasts everything in its path for a height of a meter or so. Eolian sandblasting undercuts rock faces, carving them into weird shapes, and expedites the erosion of cliffs by gravity collapse and rainstorm.

It is important to note that it is an oversimplification to place erosion after weathering in the sedimentary cycle. Weathering processes need time for their effects to be noticeable on a rock surface. In some parts of the earth, notably areas of high relief, erosion may occur so fast that rock is not exposed to the air for a sufficient length of time to undergo any significant degree of weathering (Fig. 2.2). This point is amplified in the next section.

Returning to the role of gravity, ice, water, and wind, it is apparent that these are the agents both of erosion and of subsequent transportation of sediment. The physical processes of these various transporting media are described in Chapter 4. At this point, however, it is appropriate to point out the role played by these agents in the segregation of sediments. The products of weathering are twofold: solutes and residua. The solutes are the soluble fraction of rocks that is carried in water. The residua are the insoluble products of weathering, which range in size from boulders down to colloidal clay particles. It is interesting to note the competency of the various transporting media to

Fig. 2.1. The Rocks Display'd (from Wilson, in Read, 1944), illustrating that the sedimentary cycle is a small part of the whole crustal cycle of the dynamic earth. Individual sedimentary grains of stable minerals, principally quartz, may be recycled several times before being destroyed by metamorphism. Courtesy of the Geologists' Association.

handle and segregate the products of weathering. Gravity and ice, as seen in the work of avalanches and glaciers, are competent to transport all types of weathering product, solutes, and residual particles. They are however, both inefficient agents of sediment segregation. Their depositional products, therefore, are generally poorly sorted boulder beds and gravels. Water, by contrast, is a very efficient agent for carrying material in solution; it is less efficient, however, in transporting residual sediment particles. Current velocities are seldom powerful enough to carry boulders and gravels for great distances. For this same reason running water segregates sands from gravels and the colloidal clays from the detrital sands. The sediments deposited from running water, therefore,

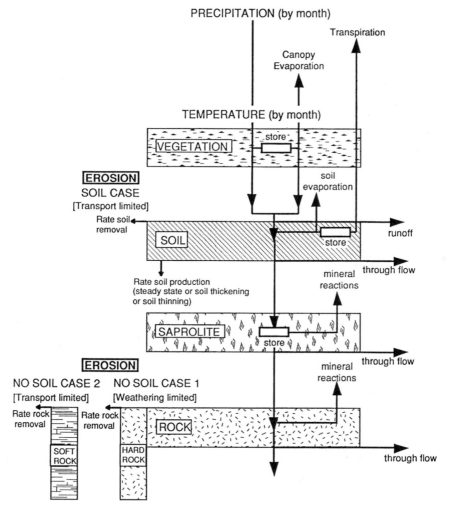

Fig. 2.2. Flow diagram showing the interreactions between water, vegetation, and bedrock during soil formation. Rate of erosion and, hence, soil removal will vary with the intensity of the weathering process, the intensity of the erosive process (ice, water, or wind), and the slope. (From Leeder *et al.,* 1998, *Basin Res.* **10,** 7–18. Courtesy of Blackwell Science Ltd.)

include sands, silts, and clays. Finally wind action is the most selective transporting agent of all. Wind velocities are seldom strong enough to transport sediment particles larger than about 0.35 mm in diameter. Eolian sediments are generally of two types, sands of medium-fine grade, which are transported close to the ground by saltation, and the silty "loess" deposits, which are transported in the atmosphere by suspension.

This review of the agents of sediment transport shows that sediments are segregated into the main classes of conglomerates, sands, shales, and limestones, by natural processes. This is worth bearing in mind during the discussions of sedimentary rock nomenclature and classification that occur in subsequent chapters. Before proceeding to these, however, it is necessary to examine weathering processes in further detail.

2.3 WEATHERING

Weathering, as already defined, includes the processes that break down rock at the earth's surface to produce discrete sediment particles. Weathering may be classified into chemical, physical, and biological processes. Chemical processes lead essentially to the destruction of rock by solution. Physical processes cause mechanical fracture of the rock. Biological weathering is due to organic processes. These include both biochemical solution, brought about largely by the action of bacteria, and humic acids derived from rotting organic matter, as well as physical fracturing of rock such as may be caused by tree roots (Fig. 2.3). Many animals also contribute to weathering by their burrowing activity. Burrows in terrestrial sediments have been described from far back in the stratigraphic record. Modern burrowing animals range from aardvarks to petrivorous porcupines (Gow, 1999).

2.3.1 Biological Weathering and Soil Formation

Soil is the product of biological weathering (Martini and Chesworth, 1992). It is that part of the weathering profile which is the domain of biological processes (Fig. 2.2). Soil consists of rock debris and humus, which is decaying organic matter largely of plant origin. Humus ranges in composition from clearly identifiable organic debris such as leaves and plant roots, to complex organic colloids and humic acids. It is doubtful that soils, as defined and distinguished from the weathering profile, existed much before the colonization of the land by plants in the Devonian.

The study of soils, termed **pedology,** is of interest to geologists insofar as it affects rock weathering and sediment formation. Pedology is, however, of particular importance to

Fig. 2.3. Weathering of rock by the invasion of plant roots, a mixture of biological and physical weathering. Chalk (Upper Cretaceous) invaded by pine tree roots, Dorking, England.

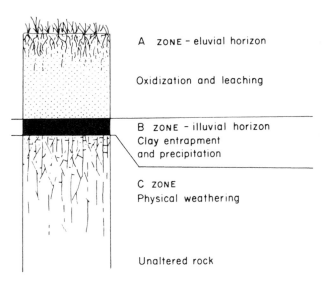

A ZONE – eluvial horizon

Oxidization and leaching

B ZONE – illuvial horizon
Clay entrapment
and precipitation

C ZONE
Physical weathering

Unaltered rock

Fig. 2.4. Terminology and processes through a soil profile.

agriculture, forestry, and to correct land utilization in general. Pedologists divide the vertical profile of a soil into three zones (Fig. 2.4). The upper part is termed the "A zone," or eluvial horizon. In this part of the profile organic content is richest and chemical and biochemical weathering generally most active. Solutes are carried away by groundwater. The fine clay fraction percolates downward through the coarser fabric supporting grains.

Below the A zone is the "B zone," or illuvial horizon. At this level downward percolating solutes are precipitated and entrap clay particles filtering down from the A zone. Below the illuvial horizon is the "C zone." This is essentially the zone where physical weathering dominates over chemical and biological processes. It passes gradually downward into unweathered bedrock. The thickness of a soil profile is extremely variable and all three zones are not always present. Thus soil thickness depends on the rate of erosion, climatic regime, and bedrock composition. As already seen, in areas of high relief, erosion can occur so fast that weathering and soil formation cannot develop. By contrast, in humid tropical climates granite can be weathered for nearly 100 m. This forms what is known as "granite wash," which passes, with the subtlest transition, from arkosic sand down to fresh granite. Ancient granite washes occasionally make good hydrocarbon reservoirs because they may be highly porous in their upper part. The Augila oil field of Libya is a good example (Williams, 1968). Epidiagenesis, the formation of porosity by weathering at unconformities, is discussed in greater detail when sandstone diagenesis and porosity are described (see Section 8.5.3). Returning to biological weathering and soils, it is known that soil type is closely related to climate. If erosion is sufficiently slow for a soil profile to evolve to maturity, there is a characteristic soil type for each major climatic zone, irrespective of rock type (Fig. 2.5). Modern soils and their ancient counterparts are now reviewed.

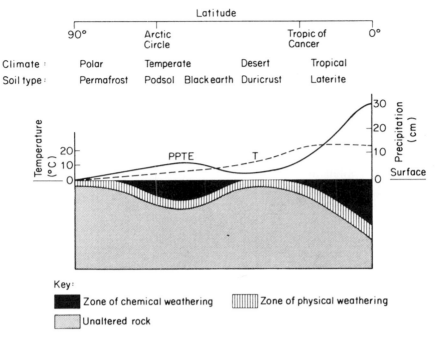

Fig. 2.5. Diagrammatic profile of the modern hemisphere showing the relationship between climate, soil, and weathering. (For sources, see Strakhov, 1962, and Lisitzin, 1972.)

2.3.1.1 Modern Soils

In polar climates true soil profiles do not develop due to the absence of organisms. A weathering mantle may be present, but this is frozen for much, if not all of the year. This is called permafrost. In temperate climates leaching plays a dominant role. The A zone is intensely weathered, though it may support an upper peaty zone of plant material. The high pH of such soil inhibits or delays bacterial decay. The B zone may be deep but is typically well developed as a limonitic or calcareous hard-pan that inhibits drainage. Soils of this general type include the "podsols" of cool temperate climates and the humus rich "tchernozems," or "black earths," of warm temperate zones.

In arid climates, by contrast, the downward percolation of chemicals by leaching is offset by the upward movement of moisture by capillary attraction. Precipitation of solute occurs, therefore, at or close to the land surface. Organic content is very low. By this means are formed the "duricrusts" that are found at or close to the surface in many modern deserts (Woolnough, 1927). These hard crusts commonly show mottled, nodular, pisolitic, and concretionary structures as the minerals are precipitated in colloform habits (see Plate 5A). Sometimes cylindrical and anastomosing tubular concretions occur in ancient soil profiles. These are attributed to the precipitation of minerals around plant root systems that have subsequently dissolved. Such structures are referred to as "rhizoconcretions" or "dikaka" (Glennie and Evamy, 1968). Modern rhizoconcretions occur around the roots of palm trees in modern desert oases. They are composed of the curious range of minerals that result from the evaporation of camel urine. Analogous rhizoconcretions due to dinosaur micturation are avidly sought. Duricrusts are of several chemical types. The most common are the "kankars" and "caliches" composed of calcite and dolomite; these are sometimes termed **calcretes** (Chapman, 1974; Semeniuk, 1986; Wright and Tucker, 1990). Less common are the siliceous duricrusts, termed **silcrete,** generally composed of chalcedony (Milnes and Thiry, 1992). Ferruginous duricrusts are termed **ferricrete.** In low-lying waterlogged areas in arid climates the duricrust is formed of evaporite minerals. These occur in salt marshes (sabkhas, see Section 6.3.2.7.4) both inland and at the seaside. Humid tropical climates form soils that are very rich in iron and kaolinite. The composition and origin of **laterite,** as these ferruginous soils are called, are described in more detail shortly in Section 2.33 on chemical weathering.

This review of modern soils, though brief, should be sufficient to show how significant climate is in weathering and how it therefore helps to determine which type of sediment is produced in a particular area.

2.3.1.2 Fossil Soils

Ancient soil horizons occur throughout the geological record (Thiry and Simon-Coincon, 1999). **Paleosols,** as they are termed, often define sequence boundaries and occur beneath unconformities in all types of rock. Sometimes they are found conformably within continental and coastal sedimentary sequences where they provide invaluable evidence of subaerial exposure of the depositional environment (Fig. 2.6).

The rocks immediately beneath an unconformity are often intensely weathered, but

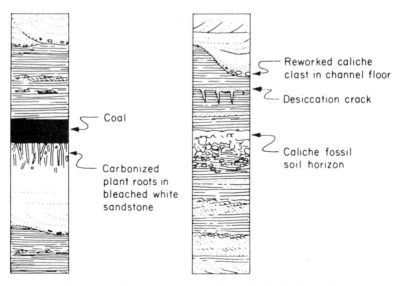

Fig. 2.6. Cross-sections through fossil soil (paleosol) horizons. (Left) Coal bed (black) with seat earth beneath. This is a white, bleached, kaolinitic sandstone pierced by plant roots. Based on examples in the Coal Measures (Carboniferous) of England. (Right) Caliche soil horizon of nodular carbonate in red fluvial siltstone. Reworked caliche pebbles in the overlying channel sand testify to penecontemporaneous formation. Based on examples in the Old Red Sandstone (Devonian) of the Welsh borderlands.

Fig. 2.7. Photograph of carbonate nodular caliche (pale blobs) in overbank floodplain deposits (dark ground mass) of the Old Red Sandstone (Devonian) Freshwater West, Wales. These nodules are an early stage of soil formation. They are known to the aboriginals of these parts as "cornstones," because their lime content fertilizes an otherwise barren modern soil. The occurrence of intraformational cornstone conglomerates on fluvial channel floors testifies to their penecontemporaneous origin.

erosion has generally removed the upper part of the soil profile. An interesting exception to this rule has been described by Williams (1969), who discussed the paleoclimate that weathered the Lewisian gneisses prior to the deposition of Torridonian (Pre-Cambrian) continental sediments in northwest Scotland. Soil horizons within sedimentary sequences are especially characteristic of fluvial and deltaic deposits. Many examples occur beneath coals and lignites. These are pierced by plant rootlets, and often are white in color due to intensive leaching. Such paleosols are often called **seat earths.** Examples are common in the Upper Carboniferous Coal Measures of northern England, in which two types of paleosol have been recognized. **Ganisters** are silica-rich horizons, which are quarried for the manufacture of refractory bricks. **Tonsteins** are kaolinite-rich seat earths (Bohor and Triplehorn, 1993). It has been argued both that these are normal desilicified soil horizons, and also that they may be intensely weathered volcanic ash bands (see Section 8.3.2.1). Examples of soils in fluvial environments include the caliches of the Old Red Sandstone facies of the Devonian in the North Atlantic borderlands. These are sometimes termed "cornstones." They occur as nodules and concretionary bands within red floodplain siltstones (Fig. 2.7). The occurrence of reworked cornstone pebbles in interbedded channel sands testifies to their penecontemporaneous origin (Friend and Moody-Stuart, 1970).

2.3.2 Physical Weathering

Four main types of physical weathering are generally recognized: freeze–thaw, insolation, hydration and dehydration, and stress release. Freeze–thaw weathering occurs where water percolates along fissures and between the grains and crystals of rock. When water freezes, the force of ice crystallization is sufficient to fracture the rock. The two halves of a fracture do not actually separate until the ice thaws and ceases to bind the rock together. Freeze–thaw weathering is most active, therefore, in polar climates and is most effective during the spring thaw. Insolation weathering occurs by contrast in areas with large diurnal temperature ranges. This is typical of hot arid climates. In the Sahara, for example, the diurnal temperature range in winter may be 25°C. Rocks expand and contract in response to temperature. The diverse minerals of rocks change size at different rates according to their variable physical properties. This differential expansion and contraction sets up stresses within rock. When this process occurs very quickly the stresses are sufficient to cause the rock to fracture. This is why insolation weathering is most effective in arid desert climates. In the author's personal experience this process was most dramatically experienced when trying to sleep on the slopes of the volcano Waw en Namus in the Libyan Sahara. Here black basalt sands are cemented by evaporite minerals. Sleep was impossible for several hours after sunset as the rock snapped, crackled, and popped.

In climatic zones that experience alternate wet and dry seasons, a third process of physical weathering occurs. Clays and lightly indurated shales alternatively expand with water and develop shrinkage cracks as they dehydrate. This breaks down the physical strength of the formation; the shrinkage cracks increase permeability, thus aiding chemical weathering from rainwater, while waterlogged clays may lead to landslides.

The fourth main physical process of weathering is caused by stress release. Rocks have elastic properties and are compressed at depth by the overburden above them. As rock is gradually weathered and eroded the overburden pressure decreases. Rock thus expands and sometimes fractures in so doing. Such fracturing is frequently aided by lateral downslope creep. Once stress-release fractures are opened they are susceptible to enlargement by solution from rainwater and other processes.

Stress release, insolation, hydration–dehydration, and freeze–thaw are the four main physical processes of weathering. Stress release is ubiquitous in brittle rocks, insolation is characteristic of hot deserts, hydration–dehydration is typical in savannah and temperate climates, and freeze–thaw of polar climates.

2.3.3 Chemical Weathering

The processes of chemical weathering rely almost entirely on the agency of water. Few common rock-forming minerals react with pure water, evaporites excepted. Groundwater, however, is commonly acidic. This is due to the presence of dissolved carbon dioxide from the atmosphere forming dilute carbonic acid. The pH is also lowered by the presence of humic acids produced by biological processes in soil. The main chemical reactions involved in weathering are oxidation and hydrolysis. Carbonic acid dissolved in groundwater releases hydrogen ions thus:

$$H_2O + CO_2 = H_2CO_3 = HCO_3^- + H^+.$$

The released hydrogen may then liberate alkali and alkali-earth elements from complex minerals, such as potassium feldspar:

$$2KAlSi_3O_8 + 2H^+ + 9H_2O = Al_2Si_2O_5(OH)_4 + 2K^+ + 4H_2SiO_4.$$

This reaction leads to the formation of kaolinite and silica. The weathering reactions are extremely complex and still little understood. (For further details, see Curtis, 1976.) Numerous studies have been made of the rate of chemical weathering of different rock-forming minerals (e.g., Ruxton, 1968; Parker, 1970). This work suggests that the rate of relative mobility of the main rock-forming elements decreases from calcium and sodium to magnesium, potassium, silicon, iron, and aluminum. Rocks undergoing chemical weathering, therefore, tend to be depleted in the first of these elements, with a concomitant relative increase in the proportions of iron oxide, alumina, and silica.

The order in which minerals break down by weathering is essentially the reverse of Bowen's reaction series for the crystallization of igneous minerals from cooling magma (Fig. 2.8). Chemical weathering separates rock into three main constituents: the solutes, the newly formed minerals, and the residuum. The solute includes the elements such as the alkali metals, principally sodium and potassium, and the rare earths, magnesium, calcium, and strontium. These tend to be flushed out of the weathering profile and ultimately find their way into the sea to be precipitated as calcium carbonate, dolomites, and evaporite minerals. The residuum is that part of the rock which, when weathered, is not easily dissolved by groundwater. As Fig. 2.8 shows, the residuum may be expected

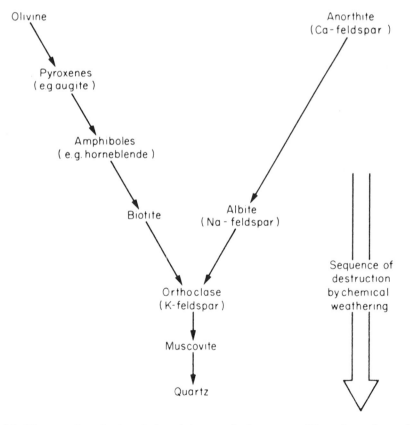

Fig. 2.8. The rate of weathering of minerals is generally the reverse of Bowen's reaction series.

to be largely composed of quartz (silica) and, depending on the degree of weathering, of varying amounts of feldspar and mica.

A further important result of chemical weathering is the formation of new minerals, principally the clays (Wilson, 1999). The mineralogy and deposition of the clays is described in Chapter 8. It is, however, necessary at this point to discuss their formation by weathering. The clay minerals are a complex group of hydrated aluminosilicates. Their chemically fairly stable molecular structures scavenge available cations and attach them to their lattices in various geometric arrangements. The clay minerals are thus classified according to the way in which hydrated aluminosilicate combines with calcium, potassium, magnesium, and iron. During the early stages of weathering the mafic minerals (olivines, pyroxenes, and amphiboles) break down to form the iron- and magnesium-rich chlorite clays. Simultaneously, weathering of the feldspars produces the smectites, illites, and kaolinite clays. As weathering progresses the clays are partially flushed out as colloidal-clay particles, but can also remain to form a residual clay deposit. If weathering continues further still all the magnesium and calcium-bearing minerals are finally leached out. The ultimate residuum of a maturely weathered rock consists, therefore, of quartz (if it was abundant in the parent rock), newly formed kaolinite (the purest clay mineral, composed only of hydrated aluminosilicate), bauxite (hydrated alumina), and limonite (hydrated iron oxide). For such a residuum to form by intense chemical weathering, a warm humid climate is necessary, coupled with slow rates of erosion. These deposits, though thin, can form laterally extensive blankets which are often of considerable economic significance. Three main types of residual deposit are generally recognized and defined according to their mineralogy (Fig. 2.9). These are now briefly described.

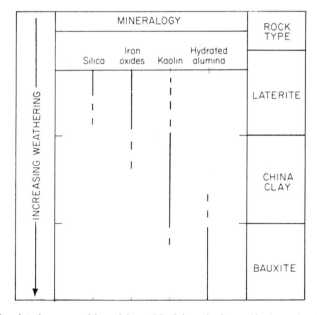

Fig. 2.9. Diagram showing the composition of the residual deposits formed by intensive chemical weathering.

2.3.3.1 Laterite

The term **laterite,** from the Latin *later,* a brick, was originally proposed by Buchanan (1807) to describe the red soils of parts of India, notably those developed on the plateau basalts of the Deccan (Patel, 1987). Subsequently, the precise definition of the term has become somewhat diffuse (Sivarajasingham *et al.,* 1962). As generally understood, laterite is the product of intensely weathered material. It is rich in hydrated iron and aluminium oxides, and low in humus, silica, lime, silicate clays, and most other minerals (Meyer, 1997). Analyses of Ugandan laterites given by McFarlane (1971) show between 40 and 50% iron oxide, and between 20 and 25% of both alumina and silica. In physical appearance laterite is a red-brown earthy material which, though often friable, can harden rapidly on exposure to the atmosphere, a property useful in brick-making. Laterites often show pisolitic and vermiform structures (Plate 5A). The pisolites are rounded, concentrically zoned concretions up to a centimeter or more in diameter. Vermicular laterite consists of numerous subvertical pipes of hard laterite in a friable matrix.

Lateritic soils are widely distributed throughout the humid tropics, occurring throughout India and Africa, as already mentioned, and also in South America. It has been argued already that residual deposits such as laterite require prolonged intensive chemical weathering for them to reach maturity. This will occur fastest in areas of low relief, where rates of erosion are low. High initial iron content in the parent rock is also a further significant factor. Thus laterites are often best developed on plateau basalts and basic intrusive rocks. Thin red laterite bands are commonly found separating fossil basalt lavas.

2.3.3.2 China Clay/Kaolin

China clay, or kaolin, is the name given to the rock aggregate of the mineral kaolinite. This is the hydrated aluminosilicate clay mineral ($Al_2O_3 \cdot 2SiO_2 \cdot 2H_2O$). China clay occurs in three geological situations, only one of which is, strictly speaking, a residual weathering product. First, it is formed by the hydrothermal alteration of feldspars in granite. The granites of southwest England are an example (Bristow, 1968). Secondly, china clay is formed by intensive weathering of diverse rock types, but particularly those which are enriched in aluminosilicates, such as shales, and acid igneous and metamorphic rocks. Thirdly, china clay can be transported short distances from hydrothermal and residual deposits to be resedimented in lacustrine environments. Here the kaolin beds occur associated with sands and coals or lignites. The "ball clays" of Bovey Tracey were deposited in an Oligocene lake, which received kaolin from rivers draining the Dartmoor granites (Edwards, 1976). The residual kaolins are very variable in composition. With increasing iron content they grade into the laterites, and with desilicification they grade into the bauxites. Where kaolin forms as a residue on granitoid rocks, large quantities of quartz may be present that must be removed before the kaolin can be put to use. China clay is economically important in the ceramic and paper-making industry.

2.3.3.3 Bauxite

Bauxite is a residual weathering product composed of varying amounts of the aluminium hydroxides boehmite, chliachite, diaspore, and gibbsite (Valeton, 1972). It takes its name from Les Baux near Arles, in France (Plate 5A). The bauxite minerals are formed by the hydrolysis of clay minerals, principally kaolinite:

$$H_2O + Al_2O_3 \cdot 2SiO_2 \cdot 2H_2O = Al_2O_3nH_2O + 2SiO_2 \cdot 2H_2O$$

water +   kaolinite   aluminium   +   silicic acid
hydroxide

Bauxite thus needs the presence of pure leached clay minerals as its precursor. This occurs in two geological settings. Bauxite is often found overlying limestone formations as, for example, in the type area of southern France, and in Jamaica. In these situations the calcium carbonate of the limestones has been completely leached away to leave an insoluble residue of clays. These have been desilicified to form the bauxite suite minerals (Bardossy, 1987).

In the second situation, as seen in Surinam in South America (Valeton, 1973), bauxite occurs as a weathering product of kaolinitic sediments that forms a thin veneer on Pre-Cambrian metamorphic basement. In both these geological settings it is apparent that the formation of bauxite is the terminal phase of a sequence of weathering that included kaolinization as an intermediate step. Bauxite is of great economic importance because it is the only source of aluminium.

2.3.4 Economic Significance and Conclusion

From the preceding description of weathering it should be apparent that it is necessary to understand the processes of weathering to gain insight into sediment formation. To conclude this chapter it is appropriate to summarize some of the major points and to draw attention to the economic significance of weathering. It is important to remember that weathering separates rock into an insoluble residue and soluble chemicals. The residue, composed largely of quartz grains and clay particles, ultimately contributes to the formation of the terrigenous sands and shales. The soluble products of weathering are ultimately reprecipitated as the chemical rocks, such as limestones, dolomites, and evaporites. Most terrigenous sediments are the product of weathered granitoid rocks and preexisting sediment. Basic igneous and volcanic rocks and limestones are very susceptible to chemical weathering and contribute little material to the insoluble residue. Figure 2.10 shows how, in a terrain of diverse rock types, only a few lithologies contribute to the adjacent sediment apron. Aside from these generalities, weathering is important because of the way it can concentrate minerals into deposits that are worth exploiting economically. The bauxites and china clays have already been mentioned.

There are four main ways in which weathering is economically important in a geological context. Weathered profiles are often buried beneath unconformities. These are commonly zones of enhanced porosity, both in sands (see Section 8.5.3.4) and especially in carbonates (see Section 9.2.7). The product of limestone weathering consists of nothing but calcium and carbonate ions that are carried away in solution. Limestone

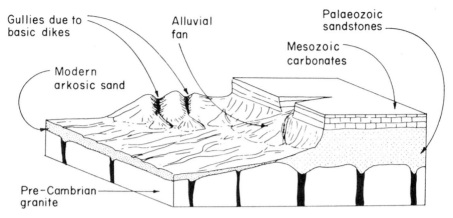

Fig. 2.10. Field sketch of part of Sinai. Mesozoic limestones overlie Nubian sandstones, which in turn overlie Pre-Cambrian granitoid rocks with basic intrusives. Despite this diversity of rock types the sediment derived from this terrain is essentially a feldspathic quartz sand derived from the granites and Nubian sandstones. The limestones and basic intrusives contribute little insoluble residue. Their weathering products are largely carried away in solution.

weathers to a characteristic land form known as "karst," characterized by solution-enlarged joints, vugs, and finally cavernous porosity (Fig. 2.11). Ancient paleokarsts are zones of extensive vuggy, cavernous and fracture porosity (Vanstone, 1998). These weathered zones may, therefore, act as conduits to migrating mineralizing fluids and to petroleum. In certain situations weathered zones can act as reservoirs in truncation

Fig. 2.11. Karstic landscape formed by chemical weathering of Dinantian (Mississippian) limestone. Note the solution-enlarged joint system known to the aborigines of these parts as "grikes," Ribblesdale, England.

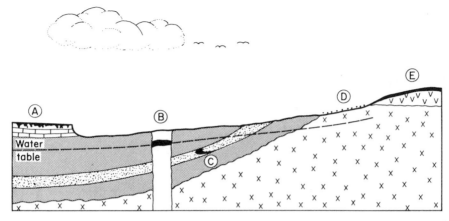

Fig. 2.12. Diagrammatic crustal profile showing the habitats of economic deposits related to weathering processes: (A) Bauxite formed from the weathered residue of limestone. (B) Supergene enrichment of sulfide ore. (C) Uranium "roll-front" ore body. (D) Residual placer formed on weathered basement. (E) Lateritic iron, nickel, and manganese deposits formed on weathered ferromagnesian-rich volcanics.

petroleum traps (see Section 7.3.2). Three types of mineral deposit are caused by weathering processes: residual deposits, supergene-enriched sulfide ores, and uranium roll-front mineralization (Fig. 2.12).

2.3.4.1 Residual Deposits

Residual deposits may include both remobilized and insoluble weathering products. The formation of laterite, kaolinite, and bauxite has already been discussed. Not all laterites are enriched solely by iron. Nickel and manganese concentration may also take place in favorable circumstances. Nickel laterites have been mined in the Pacific Island of New Caledonia since 1875. Intense weathering of ultrabasic igneous rocks (principally peridotite) has generated a soil profile some 20 m deep. The nickel ore occurs between unaltered bedrock and a shallow zone of pisolitic iron laterite. Major high-grade concentrations of nickel ore occur on sloping hillsides (Golightly, 1979, 1981; Meillon, 1978). Morro da Mana in Brazil is an example of a residual manganese deposit. The protore occurs in a complex Pre-Cambrian basement of metasediments that have undergone igneous intrusion and contact metamorphism leading to the formation of tephroite (manganese silicate). As with the New Caledonian example, the manganese ore occurs between a shallow pisolitic laterite and unaltered bedrock (Webber, 1972).

Not all residual deposits are formed by solution and reprecipitation within the weathering profile. Some are residues of minerals that are insoluble in weathering. These are described as eluvial deposits. They grade downslope into alluvial placer ores. Economic concentrations of gold, cassiterite, diamonds, and apatite are known to have formed in this way (Cox and Singer, 1986).

2.3.4.2 Supergene Enrichment

Weathering is also very important in the supergene enrichment of sulfide ores, such as those in Rio Tinto, Spain, and many of the world's porphyry copper deposits (Guilbert

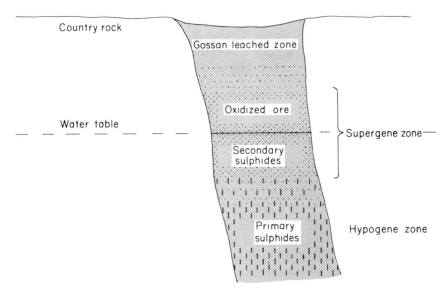

Fig. 2.13. Diagrammatic section through a weathered sulfide-ore body. Secondary ore minerals are concentrated near the water table by oxidation and leaching in the upper gossan zone.

and Park, 1986, pp. 796–831). Four zones can be distinguished in the weathering profile of such ores (Fig. 2.13). At the top is an upper leached zone, or "gossan," composed of residual silica and limonite. This grades down into the "supergene" zone, adjacent to the water table. Immediately above the table oxidized ore minerals are reprecipitated, such as malachite, azurite, and native copper. Secondary sulfide ores, principally chalcocite, are precipitated in the lower part of the supergene zone. This grades down into the unweathered hypogene ore, mainly composed of pyrite and chalcopyrite.

2.3.4.3 Uranium Roll-Front Deposits

Uranium ores occur in several settings. One of the most common is the roll-front ore body. The origin of this type of deposit was once in dispute, but it is now generally agreed that its genesis is intimately connected to weathering. Uranium roll-front ores occur in the Colorado Plateau, Wyoming, New Mexico, and the Texas Gulf Coast of the United States (Turner-Peterson and Fishman, 1986; Granger and Warren, 1969; Galloway and Hobday, 1983; Hobday and Galloway, 1998). In these areas uranium mineralization occurs in rocks ranging in age from Permian to Tertiary. It is generally absent from marine and eolian sands. Typically it occurs in poorly sorted arkosic fluvial sands that are rich in carbonaceous detritus. Plant debris and fossil tree trunks are not uncommon. Pyrite and carbonate cement are typical minor constituents. The host sandstones are generally cross-bedded with conglomeratic erosional bases and typical channel geometries. The usual ore body is in the form of a **roll-front** (Fig. 2.14). This is an irregular cup-shaped mass lying sideways with respect to the channel axis. The sandstone on the concave side of the roll is highly altered and bleached to a white color. Pyrite, calcite, and carbonaceous material are absent; matrix and feldspars are extensively kaolinized.

These observations have suggested to many geologists that mineralization occurred

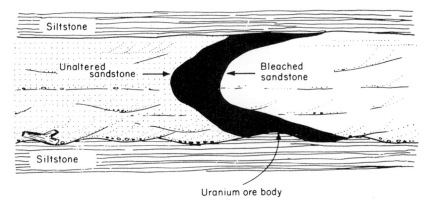

Uranium ore body

Fig. 2.14. Sketch illustrating the occurrence of carnotite roll-front ore bodies in fluvial channel sandstones. The direction of front migration in this case was from right to left.

when a diagenetic front moving along a channel was stabilized by a change in the flow or chemistry of migrating pore fluids (Raffensperger and Garvan, 1995a,b). The original source of the uranium has always been a source of debate but volcanic ashes and tuffs are often taken as the primary source, with subsequent migration and concentration by percolating acid groundwater. This close association of mineral migration and hydrology has led to an understanding of the role of permeability barriers in determining the locus of carnotite precipitation. Thus mineralization is absent in the laterally continuous sands of eolian and marine origin, which possess near uniform permeability. It typically occurs adjacent to permeability barriers, such as unconformities, or in fluvial formations with sand:shale ratios of between 1:1 to 1:4 (see Section 6.3.2.2.4).

This brief review shows how weathering is a process of considerable direct economic importance as well as being one of fundamental geological significance.

SELECTED BIBLIOGRAPHY

Hobday, D. K., and Galloway, W. E. (1998). Groundwater processes and sedimentary uranium deposits. *Hydrogeol. J.* **7,** 127–138.
Leeder, M. R., Harris, T., and Kirkby, M. J. (1998). Sediment supply and climate change: Implications for basin stratigraphy. *Basin Res.* **10,** 7–18.
Meyer, R. (1997). "Paleolaterites and Paleosols: Imprints of Terrestrial Processes in Sedimentary Rocks." A. A. Balkema, Rotterdam. 151pp.
Wilson, M. J. (1999). The origin and formation of clay minerals in soils: Past, present and future perspectives. *Clay Miner.* **34,** 1–27.

REFERENCES

Bardossy, G. (1987). "Karst Bauxites." Elsevier, Amsterdam. 441pp.
Bohor, B. F., and Triplehorn, D. M., eds. (1993). "Tonsteins: Altered Volcanic Ash Layers in Coal-bearing Sequences," Spec. Publ. No. 285. Geol. Soc. Am., Boulder, CO. 48p.
Bristow, C. M. (1968). Kaolin deposits of the United Kingdom. *Int. Geol. Congr. Rep. Sess., 23rd,* Prague, vol. 15, pp. 275–288.

Buchanan, F. (1807). "A Journey from Madras Through the Counties of Mysore, Kanara and Malabar," 3 vols. East India Company, London.

Chapman, R. W. (1974). Calcareous duricrust in Al-Hasa, Saudi Arabia. *Geol. Soc. Am. Bull.* **85**, 119–130.

Cox, D. P., and Singer, D. A. (1986). Mineral deposit models. *Geol. Surv. Bull. (U.S.)* **1693**, 1–379.

Curtis, C. D. (1976). Stability of minerals in surface weathering reactions. *Earth Surf. Processes* **1**, 63–70.

Edwards, R. A. (1976). Tertiary sediments and structure of the Bovey Tracey basin, South Devon. *Proc. Geol. Assoc.* **87**, 1–26.

Friend, P. F., and Moody-Stuart, M. (1970). Carbonate deposition on the river flood plains of the Wood Bay Formation (Devonian) of Spitsbergen. *Geol. Mag.* **107**, 181–195.

Galloway, W. E., and Hobday, D. K. (1983). "Terrigenous Clastic Depositional Systems." Springer-Verlag, Berlin. 423pp.

Glennie, E. K., and Evamy, B. D. (1968). Dikaka: Plants and plant root structures associated with eolian sand. *Palaeogeogr., Palaeoclimatol., Palaeoecol.* **4**, 77–87.

Golightly, J. P. (1979). Nickeliferous laterites: A general description. *In* "International Laterite Symposium" (D. J. I. Evans, R. S. Shoemaker, and H. Veltman, eds.), pp. 3–23. AIME, New York

Golightly, J. P. (1981). Nickeliferous laterite deposits. *Econ. Geol. 75th Anniv. Vol.*, pp. 710–735.

Gow, C. E. (1999). Gnawing of rock outcrop by porcupines. *S. Afr. J. Geol.* **95**, 74–75.

Granger, H. C., and Warren, C. G. (1969). Unstable sulfur compounds and the origin of roll-type uranium deposits. *Econ. Geol.* **64**, 160–171.

Guilbert, J. M., and Park, C. F. (1986). "The Geology of Ore Deposits." Freeman, New York. 985pp.

Hobday, D. K., and Galloway, W. E. (1998). Groundwater processes and sedimentary uranium deposits. *Hydrogeol. J.* **7**, 127–138.

Leeder, M. R., Harris, T., and Kirkby, M. J. (1998). Sediment supply and climate change: Implications for basin stratigraphy. *Basin Res.* **10**, 7–18.

Lisitzin, A. P. (1972). Sedimentation in the World Ocean. *Spec. Publ. — Soc. Econ. Paleontol. Mineral.* **17**, 1–218.

Martini, I. P., and Chesworth, W., eds. (1992). "Weathering, Soils and Paleosols." Elsevier, Amsterdam. 618pp.

McFarlane, M. J. (1971). Lateritization and Landscape development in Kyagwe, Uganda. *Q. J. Geol. Soc. Lond.* **126**, 501–539.

Meillon, J. J. (1978). Economic geology and tropical weathering. *CIM Bull.* **71**, 61–70.

Meyer, R. (1997). "Paleolaterites and Paleosols: Imprints of Terrestrial Processes in Sedimentary Rocks." A.A. Balkema, Rotterdam. 151pp.

Milnes, A. R., and Thiry, M. (1992). Silcretes. *In* "Weathering, Soils and Paleosols" (I. P. Martini and W. Chesworth, eds.), Dev. Earth Surf. Processes, Vol. 2, pp. 349–377. Elsevier, Amsterdam.

Ollier, C. C. (1969). "Weathering." Oliver & Boyd, Edinburgh. 304pp.

Parker, A. (1970). An index of weathering for silicate rocks. *Geol. Mag.* **107**, 501–504.

Patel, E. K. (1987). Lateritization and bentonitization of basalt in Kutch, Gujarat State, India. *Sediment. Geol.* **55**, 327–346.

Raffensperger, J. P., and Garvan, G. (1995a). The formation of unconformity-type uranium ore deposits.1. Coupled groundwater flow and heat transport modelling. *Am. J. Sci.* **295**, 581–636.

Raffensperger, J. P., and Garvan, G. (1995b). The formation of unconformity-type uranium ore deposits.1. Coupled hydrochemical modelling. *Am. J. Sci.* **295**, 639–696.

Read, H. H. (1944). Meditations on Granite 2. *Proc. Geol. Assoc.* **55**, 4–93.

Ruxton, B. P. (1968). Measures of the degree of chemical weathering of rocks. *J. Geol.* **76**, 518–527.

Semeniuk, V. (1986). Calcrete breccia floatstone in Holocene sand developed by storm-uprooted trees. *Sediment. Geol.* **48**, 183–192.

Sivarajasingham, L. T., Alexander, L. T., Cady, J. G., and Cline, M. G. (1962). Laterite. *Adv. Agron.* **14**, 1–60.

Strakhov, N. M. (1962). "Osnovy Teorii Litogeneza," 2nd ed., Vol. 1. Izd. Akad. Nauk SSSR, Moscow. 517pp.

Thiry, M., and Simon-Coincon, R., eds. (1999). "Palaeoweathering, Palaeosurfaces and Related Continental Deposits." Blackwell, Oxford. 408pp.

Turner-Peterson, C. E., and Fishman, N. S. (1986). Geologic synthesis and genetic models for uranium mineralization, Grants Uranium Region, New Mexico. *Stud. Geol. (Tulsa, Okla.)* **22,** 35–52.

Valeton, I. (1972). "Bauxites," Dev. Soil Sci., No. 1. Elsevier, Amsterdam. 226pp.

Valeton, I. (1973). Pre-bauxite red sediments and sedimentary relicts in Surinam bauxites. *Geol. Mijnbouw* **52,** 317–334.

Vanstone, S. D. (1998). Late Dinantian palaeokarst of England and Wales: Implications for exposure surface development. *Sedimentology* **45,** 19–38.

Webber, B. N. (1972). Supergene nickel deposits. *Trans. Soc. Pet. Eng. AIME* **252,** 333–347.

Williams, G. E. (1969). Characteristics and origin of a PreCambrian pediment. *J. Geol.* **77,** 183–207.

Williams, J. J. (1968). The stratigraphy and igneous reservoirs of the Augila field, Libya. *In* "Geology and Archaeology of Northern Cyrenaica, Libya" (F. T. Barr, ed.), pp. 197–207. Pet. Explor. Soc. Libya, Tripoli.

Wilson, M. J. (1999). The origin and formation of clay minerals in soils: Past, present and future perspectives. *Clay Miner.* **34,** 1–27.

Woolnough, W. G. (1927). The duricrust of Australia. *J. Proc. R. Soc. N.S. W.* **61,** 24–53.

Wright, V. P., and Tucker, M. E., eds. (1990). "Calcretes." Blackwell, Oxford. 360pp.

3 Particles, Pores, and Permeability

A sediment is, by definition, a collection of particles, loose or indurated. Any sedimentological study commences with a description of the physical properties of the deposit in question. This may be no more than the terse description "sandstones" if the study concerns regional tectonic problems. On the other hand, it may consist of a multipage report if the study is concerned with the physical properties of a small volume of a sedimentary unit such as a petroleum reservoir. Remember that the analysis of the physical properties of a sedimentary rock should be adapted to suit the objectives of the project as a whole.

The study of the physical properties of sediments is an extensive field of analysis in its own right, and is of wide concern not only to geologists. In its broadest sense of particle size analysis, it is of importance to the managers of sewage farms, manufacturers of leadshot and plastic beads and to egg graders. Enthusiasts for this field of study are directed to Tucker (1988) and Lewis and McConchie (1994).

The following account is a summary of this topic that attempts only to give sufficient background knowledge needed for the study of sedimentary rocks in their broader setting. This is in two parts. The first describes individual particles and sediment aggregates; the second describes the properties of pores — the voids between sediment particles.

3.1 PHYSICAL PROPERTIES OF PARTICLES

3.1.1 Surface Texture of Particles

The surface texture of sediment particles has often been studied, and attempts have been made to relate texture to depositional process. In the case of pebbles, macroscopic striations are generally accepted as evidence of glacial action. Pebbles in arid eolian environments sometimes show a shiny surface, termed "desert varnish." This is conventionally attributed to capillary fluid movement within the pebbles and evaporation of the silica residue on the pebble surface. The folklore of geology records that windblown sand grains have opaque frosted surfaces, while water-laid sands have clear translucent surfaces. Kuenen and Perdok (1962) attributed frosting not to the abrasive action of wind and water but to alternate solution and precipitation under subaerial conditions.

Electron microscopic studies show that there are several types of surface texture on sand grains produced by glacial, eolian, and aqueous processes (Krinsley, 1998). The surfaces of water-deposited sand grains are characterized by V-shaped percussion pits and grooves. Glacial sands show conchoidal fractures and irregular angled microtopography. Eolian sands show a flaky surface pattern.

Unfortunately though, many sand grains bear the imprints of several processes. Consider, for example, a quartz grain in a sand dune on a periglacial outwash plain. Furthermore, at subcrop, and at outcrop in the tropics, these abrasional features may be modified by solution and by secondary quartz cementation. Where extensive diagenesis is absent, however, the original imprint of the depositional process has been seen on quartz grains of all ages from Recent to Pre-Cambrian (Krinsley and Trusty, 1986).

3.1.2 Particle Shape, Sphericity, and Roundness

Numerous attempts have been made to define the shape of sediment particles and study the controlling factors of grain shape. Pebble shapes have conventionally been described according to a scheme devised by Zingg (1935). Measurements of the ratios between length, breadth, and thickness are used to define four classes: spherical (equant), oblate (disk or tabular), blade, and prolate (roller). These four types are shown in Fig. 3.1. More sophisticated schemes have been developed by Sneed and Folk (1958), Harrell (1984), Tough and Miles (1984), and Illenberger (1991).

The shape of pebbles is controlled both by their parent rock type and by their subsequent history. Pebbles from slate and schistose rocks tend to commence life in tabular or bladed shapes, whereas isotropic rocks, such as quartzite, are more likely to generate equant, subspherical pebbles. Traced away from their source, pebbles diminish in size and tend to assume equant or bladed shapes (e.g., Plumley, 1948; Schlee, 1957; Miall, 1970).

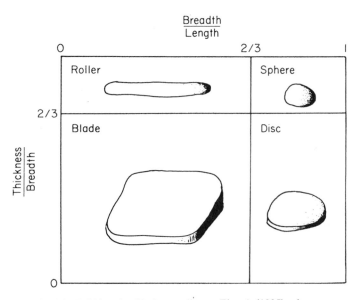

Fig. 3.1. Pebbles classified according to Zingg's (1935) scheme.

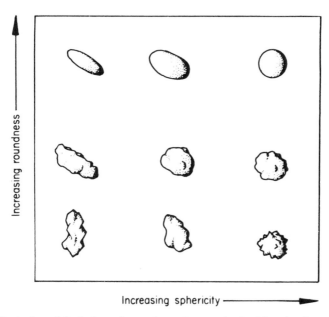

Fig. 3.2. Illustration of the independence of roundness and sphericity of sediment particles.

Attempts have also been made to relate pebble shape to depositional environment (Cailleux and Tricart, 1959; Gale, 1990). Sames (1966) proposed criteria for distinguishing fluvial and littoral pebble samples using a combination of shape and roundness. The samples were restricted to isotropic rocks such as chert and quartzite. It is feasible to measure the axes of sufficient pebbles to analyze a considerable sample statistically. This is seldom practicable with smaller particles. Sand grains can be described and analyzed microscopically, however, with respect to their sphericity and roundness. These are two completely independent parameters as shown in Fig. 3.2.

Sphericity was first defined by Wadell (1932) as an expression of the extent to which the form of a particle approaches the shape of a sphere, that is,

$$\text{Sphericity} = \frac{\text{surface area of the particle}}{\text{surface area of a sphere of equal volume}}.$$

Sneed and Folk (1958) proposed a modified definition of sphericity that is more useful for studying particle settling velocities (see Section 4.1).

Roundness was first defined by Wentworth (1919) as "the ratio of the average radius of the sharpest corner to the radius of the largest inscribed circle." This definition was modified by Wadell (1932) who proposed that it should be the ratio of the average radius of all the corners and edges to the radius of the largest inscribed circle, that is,

$$\text{Roundness} = \frac{\text{average radius of corners and edges}}{\text{radius of maximum inscribed circle}}$$

Wadell's definition is perhaps statistically more sound, but for practical purposes Wentworth's definition is much simpler and faster to use. Additional definitions have been proposed by Russell and Taylor (1937) and Powers (1953).

Considerable attention has been paid to the factors that control the roundness and sphericity of sediment particles. Many studies show the sphericity and roundness of sediments to increase away from their source area (e.g., Wentworth, 1919; Laming, 1966). Kuenen, in a series of papers (1956a,b, 1959, 1960), described the results of the experimental abrasion of both pebbles and sand by various eolian and aqueous processes. These studies indicated that the degree of abrasion and shape change along rivers and beaches was due as much to shape sorting as to abrasion. The experiments themselves showed that eolian action was infinitely more efficient as a rounding mechanism than aqueous transportation over an equivalent distance. There is little evidence for chemical solution to be a significant rounding agent. This is shown by the angularity of very fine sand and silt. Margolis and Krinsley (1971) demonstrated that the high rounding commonly seen in eolian sands is due to a combination of abrasion and of synchronous precipitation of silica on the grain boundary.

3.1.3 Particle Size

Size is perhaps the most striking property of a sediment particle. This fact is recognized by the broad classification of sediments into gravels, sands, and muds. It is easy to understand the concept of particle size. It is less easy to find accurate methods of measuring particles (Whalley, 1972; Lewis and McConchie, 1994).

3.1.3.1 Grade Scales

Sedimentary particles come in all sizes. For communication it is convenient to be able to describe sediments as gravels, sands (of several grades), silt, and clay. Various grade scales have been proposed that arbitrarily divide sediments into a spectrum of size classes. The Wentworth grade scale is the one most commonly used by geologists (Wentworth, 1922). This is shown in Table 3.1, together with the grade names and their lithified equivalents. A common variation of the Wentworth system is the phi (ϕ) scale proposed by Krumbein (1934). This retains the Wentworth grade names, but converts the grade boundaries into phi values by a logarithmic transform:

$$\phi = -\log_2 d,$$

where d is the diameter. The relationship between the Wentworth and phi grade scales is shown in Fig. 3.3.

3.1.3.2 Methods of Particle Analysis

The size of sediment particles can be measured from both unconsolidated and indurated sediment. The most common way is by visual estimation backed up, if necessary, by reference to a set of sieved samples of known sizes. With experience, most geologists can visually measure grain size within the accuracy of the Wentworth grade scale

Table 3.1
Wentworth Grade Scale, Showing Correlation with ϕ Scale and Nomenclature
for Unconsolidated Aggregates and for Lithified Sediments

ϕ values	Particle diameter (mm diam.)	Wentworth grades	Rock name
		Cobbles ⎫	
−6	64	⎬	Conglomerate
		Pebbles ⎭	
−2	4		
		Granules	Granulestone
−1	2		
		Very coarse ⎫	
0	1		
		Coarse	
1	0·5		
		Medium ⎬ sand	Sandstone
2	0·25		
		Fine	
3	0·125		
		Very fine ⎭	
4	0·0625		
		Silt	Siltstone
8	0·0039		
		Clay	Claystone

scheme, at least down to silt grade (Assalay *et al.*, 1998). Silt and clay can be differentiated by whether they are crunchy or plastic between one's teeth, although more sophisticated methods of analysis are available. Well-cemented sediments are measured from thin-section microscope study (Mazzullo and Kennedy, 1985). There is the quick-and-dirty method and the slow enthusiast's method. The first involves no more than measuring the diameter of the field of view of the microscope at a known magnification, counting the number of grains transected by the cross-wires, and dividing by twice the diameter. This process is repeated until tired — or until statistical whims are satisfied. Then calculate the average grain size of the thin section with this formula:

$$\text{Average grain size} = \frac{\Sigma(2d/n)}{N},$$

where n is the number of grains cut by the cross-wires, d is the diameter of the field of view, and N is the total number of fields of view counted. This method is quick and foolproof. It is adequate for most sedimentological studies aimed at broader aspects of facies analysis. It gives only the average grain size and does not describe the sorting of the sample. This method is quite inadequate for detailed granulometric studies. Then it is necessary to measure the length and/or some other size parameter of individual grains. This is not only very time consuming, but care must be taken over the orientation of the thin section and of the selection of sample. It is improbable that the thin section cuts the longest axis of every particle, so there is an inherent bias to record axes that are less

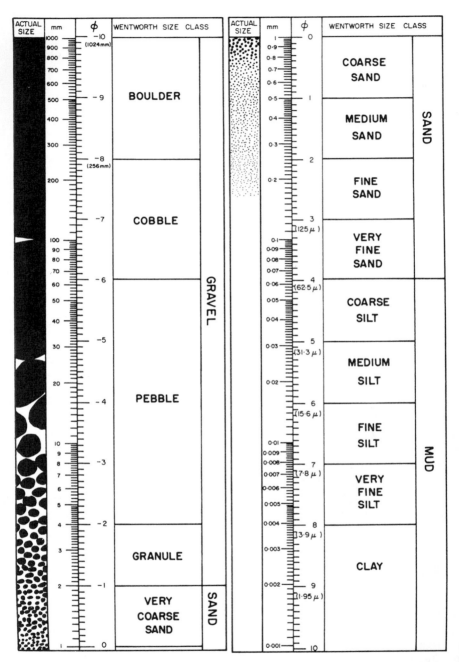

Fig. 3.3. The Udden–Wentworth grade scale for grain sizes, together with the phi (φ) scale conversion chart. (From Lewis and McConchie, 1994, "Analytic Sedimentology." Chapman & Hall, London. With kind permission from Kluwer Academic Publishers.)

than the true long axis (Fig. 3.4). Friedman (1958, 1962) has empirically derived methods for converting grain size data from thin sections to sieve analyses. Ideally, grain size studies of thin sections should be carried out using cathodoluminescence rather than un-

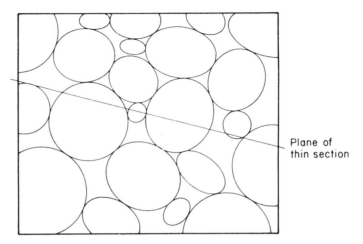

Plane of
thin section

Fig. 3.4. Sketch showing how grain size determined by measuring grain diameters from thin sections will always give too low an average reading, because particles will normally be dissected at less than their true diameters. This problem can be overcome by using the statistical gymnastics described in the text.

der polarized light (see Section 8.5.3.3.2). This will clarify the extent of pressure solution and secondary cementation. These may have considerably modified both grain size and grain shape.

Claystones and siltstones are not amenable to size analysis from an optical microscope. Theoretically their particle size can be measured individually by electron microscope analysis but it is difficult to differentiate detrital from diagenetic clay particles.

Many methods are available for measuring the particle size of unconsolidated sediment. The choice of method depends largely on the particle size. Boulders, cobbles, and gravel are best measured manually with a tape measure or ruler. Sands are most generally measured by sieving. The basic principles of this technique are as follows. A sand sample of known weight is passed through a set of sieves of known mesh sizes. The sieves are arranged in downward decreasing mesh diameters. The sieves are mechanically vibrated for a fixed period of time. The weight of sediment retained on each sieve is measured and converted into a percentage of the total sediment sample (for additional information, see ASTM, 1959). This method is quick and sufficiently accurate for most purposes. Essentially it measures the maximum girth of a sediment grain. Long thin grains are recorded in the same class as subspherical grains of similar girth. This fact is not too important in the size analysis of terrigenous sediments because these generally have a subovoid shape. Skeletal carbonate sands, however, show a diverse range of particle shapes. This factor is overcome by another method of bulk sediment analysis termed **elutriation,** or the settling velocity method. This is based on Stokes' law, which quantifies the settling velocity of a sphere thus:

$$w = \left[\frac{(P_1 - P)g}{18\mu} \right] d^2,$$

where w is the settling velocity, $(P_1 - P)$ is the density difference between the particle and the fluid, g is the acceleration due to gravity, μ is the viscosity, and d is the particle diameter. A particular problem encountered here is the particle diameter. The settling velocity is not only a function of a particle's diameter, but also of its shape. Stokes' law applies when the settling particles are perfect spheres. Most quartz grains actually approach an ovoid shape. Furthermore, a typical sand sample may contain not only ovoid quartz, but also heavy mineral euhedra, mica flakes, and irregular clay floccules. Bioclastic sediments are still more complex. This problem has led to the idea of "hydraulic equivalence" (Briggs *et al.,* 1962). Hydraulically equivalent particles are those that settle out of a column of water at the same speed. Hydraulic equivalence thus takes into account particle density, volume, and shape. This is an important point to consider in any analysis of sediment transportation and deposition (Hand, 1967; White and Williams, 1967). Hydraulic equivalence will be considered later in the context of placer mineral deposits (see Section 6.3.2.2.4). A derivation for Stokes' law can be found in Krumbein and Pettijohn (1938, p. 95). Its significance in sediment transport studies will be returned to later (see Section 4.1).

Grain size analysis by Stokes' law involves disaggregating the sediment, dispersing the clay fraction with an antiflocculent, and tipping it into a glass tube full of liquid. The sediment will accumulate on the bottom in order of decreasing hydraulic equivalence, normally gravel first and clay last. There are several methods of measuring the time of arrival and volume of the different grades. The elutriation method is widely used because it is fast and accurate and can measure particles from granule to clay grade. It necessitates making several assumptions about the effect of particle shape and surface friction on settling velocity.

Table 3.2 summarizes the various methods of particle size analysis.

3.1.3.3 Presentation of Particle Size Analyses

The method of displaying and analyzing granulometric analyses depends on the purpose of the study. Both graphic and statistical methods of data presentation have been developed. Starting from the tabulation of the percentage of the sample in each class (Table 3.3) the data can be shown graphically in bar charts (Fig. 3.5). A more usual

Table 3.2
Methods of Particle Size Analysis

Induration	Sediment grade	Method	
Unconsolidated	Boulders ⎫ Cobbles ⎬ Pebbles ⎭	Manual measurement of individual clasts	
	Granules ⎫ Sand ⎪ Silt ⎬ Clay ⎭	Sieve analysis or elutriation of bulk samples	
Lithified	Boulders ⎫ Cobbles ⎬ Pebbles ⎭	Manual measurement of individual clasts	
	Granules ⎫ Sand ⎪ Silt ⎬ Clay ⎭	Thin section measurement	Quick: grain counts on cross-wires
		X-ray analysis	Slow: individual grain micrometry

Table 3.3
Grain Size Data of Modern Sediments Tabulated by Weight % and Cumulative %

Grain Size	Sample A[a]		Sample B		Sample C		Sample D		Sample E	
	Weight (%)	Cumulative (%)	Weight (%)	Cumulative (%)	Weight (%)	Cumulative (%)	Weight (%)	Cumulative (%)	Weight (%)	Cumulative (%)
Granules	0·00	0·00	0·79	0·79	0·00	0·00	0·00	0·00	3·87	3·87
Very coarse	0·00	0·00	1·10	1·84	0·00	0·00	0·05	0·05	22·75	26·62
Coarse	0·02	0·02	1·66	3·50	0·00	0·00	0·44	0·49	50·68	77·30
Medium	9·00	9·02	3·90	7·40	0·00	0·00	51·89	52·38	20·24	97·54
Fine	90·26	99·28	10·65	18·05	0·00	0·00	47·57	99·95	2·20	99·74
Very fine	0·72	100·00	13·68	31·73	1·09	1·09	0·07	100·02	0·24	99·98
Silt	0·00	100·00	35·47	67·26	45·87	46·96	0·01	100·03	0·02	100·00
Clay	0·00	100·00	32·73	100·00	53·04	100·00	0·00	100·03	0·00	100·00
Total	100·00		99·98		100·00		100·03		100·00	

[a] A, eolian coastal dune, Lincs., England; B, Pleistocene glacial till, Lincs., England; C, abyssal plain mud, Bay of Biscay; D, beach sand, Cornwall, England; E, river sand, Dartmoor, England.

Data supplied by courtesy of G. Evans.

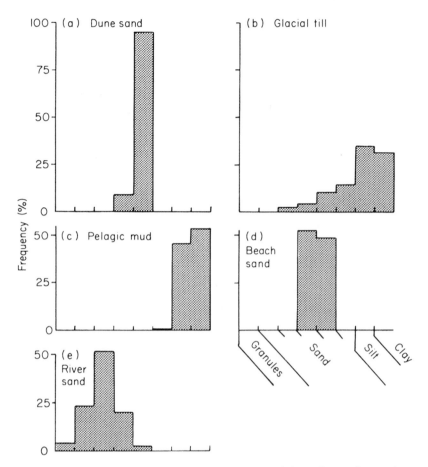

Fig. 3.5. Bar charts or histograms showing the frequency percent of the various sediment classes recorded from the samples in Table 3.3.

method of graphic display is the cumulative curve. This is a graph that plots cumulative percentage against the grain size (Fig. 3.6). Cumulative curves are extremely useful because many sample curves can be plotted on the same graph. Differences in sorting are at once apparent. The closer a curve approaches the vertical, the better sorted it is, because a major percentage of sediment occurs in one class. Significant percentages of coarse and fine end-members show up as horizontal limbs at the ends of the curve. The grain size for any cumulative percent is termed a percentile (i.e., one talks of the 20th percentile and so on). It is a common practice to plot grain size curves on probability paper. This has the great advantage that samples with normal Gaussian grain size distributions plot out on a straight line.

   The sorting of a sediment can also be expressed by various statistical parameters. The simplest of these is the measurement of central tendency, of which there are three commonly used versions: the median, the mode, and the mean. The median grain size is that which separates 50% of the sample from the other, that is, the median is the 50th percentile. The mode is the largest class interval. The mean is variously defined, but a com-

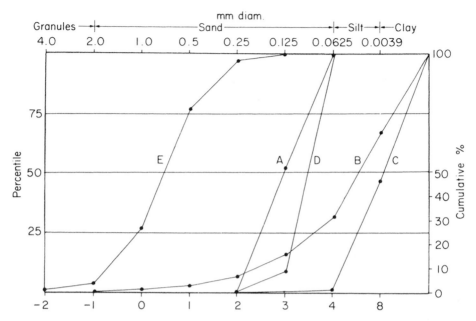

Fig. 3.6. Cumulative percent plotted against grain size for the samples recorded in Table 3.3. This method of data presentation enables quick visual comparisons to be made of both grain size and sorting. The steeper the curve, the better the sample is sorted. Most statistical granulometric analyses require more class intervals to be measured than the Wentworth grade scale classes shown here.

mon formula is the average of the 25th and 75th percentiles. The second important aspect of a granulometric analysis is its sorting or the measure of degree of scatter, that is, the tendency for all the grains to be of one class of grain size. This is measured by a sorting coefficient. Several formulas have been proposed. The classic Trask sorting coefficient is calculated by dividing the 75th percentile by the 25th percentile.

A third property of a grain size frequency curve is termed **kurtosis,** or the degree of peakedness. The original formula proposed for kurtosis is as follows (Trask, 1930):

$$k = \frac{P_{75} - P_{25}}{2(P_{90} - P_{10})},$$

where P refers to the percentiles.

Curves that are more peaked than the normal distribution curve are termed **leptokurtic;** those which are saggier than the normal are said to be **platykurtic** (Fig. 3.7).

The fourth property of a granulometric curve is its skewness, or degree of lopsidedness. The Trask coefficient of skewness is (1930):

$$SK = \frac{P_{25} \times P_{75}}{P_{50}^2}.$$

Samples weighted toward the coarse end-member are said to be positively skewed; samples weighted toward the fine end are said to be negatively skewed (Fig. 3.8).

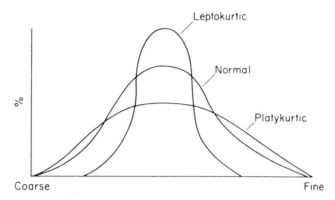

Fig. 3.7. The parameter of "peakedness" (kurtosis) in grain-size distributions.

These are the four statistical coefficients that are commonly calculated for a granulo-metric analysis. In summary they consist of a measure of central tendency, including me-dian, mode, and mean; a measure of the degree of scatter or sorting; kurtosis, the de-gree of peakedness; and skewness, the lopsidedness of the curve.

These concepts and formulas were originally defined by Trask (1930). Additional sophisticated formulas that describe these parameters have been proposed by Inman (1952) and by Folk and Ward (1957). More complex statistical methods of grain size dis-tribution include multivariate techniques, such as factor analysis (Klovan, 1966; Cham-bers and Upchurch, 1979). In most grain size studies particle size frequency is plotted against log particle size. Bagnold and Barndorff-Nielsen (1980) showed that it may be more useful to plot particle size analyses on a log-log scale. Their work shows that many grain size distributions correspond, not to the normal probability function as commonly supposed, but to a hyperbolic probability function. This method has proved capable of differentiating eolian from beach sands (Vincent, 1986).

3.1.3.4 Interpretation of Particle Size Analyses

The methods of granulometry and the techniques of displaying and statistically manip-ulating these data have been described. It is now appropriate to consider the value of

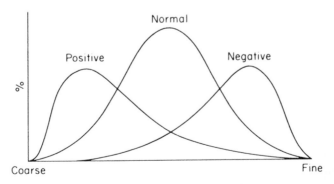

Fig. 3.8. The parameter of skewness (lopsidedness) in grain-size distributions.

this work and its interpretation and application. First of all there are many cases where it is quite inadequate to describe a sediment as "a medium-grained well-sorted sand." Within the sand and gravel industry rigid trade descriptions of marketable sediments are required. This includes hard core, road aggregates, building sands, sewage filter-bed sands, blasting sand, and so on.

Sand and gravel for these and many other uses require specific grain size distributions. These must be described accurately using statistical coefficients such as those described in the preceding section. Within the field of geology, accurate granulometric analyses are required for petrophysical studies that relate sand texture to porosity and permeability (see Section 3.2.3). The selection of gravel pack completions for water wells also requires a detailed knowledge of the granulometry of the aquifer.

Beyond these purely descriptive aspects of granulometry there is one interpretive aspect that has always led geologists on. This is the use of grain size analysis to detect the depositional environment of ancient sediments. Figure 3.9 shows the diversity of modern sands from various recent environments. Glacial outwash is coarse, poorly sorted, and the particles angular (Fig. 3.9A). Fluvial sands are moderately sorted and rounded (Fig. 3.9B). Beach sands are well sorted and rounded, with scattered skeletal debris of low sphericity (Fig. 3.9C). Eolian sands are extremely well sorted and well rounded (Fig. 3.9D).

This much is obvious. The problem is to develop statistical parameters that can differentiate the depositional environment of any given sand sample. This avenue of research has not been notably successful, despite intensive efforts. Modern environments whose sediment granulometry have been extensively studied include rivers, beaches, and dunes. Reviews of this work have been given by Folk and Ward (1957), Friedman (1961, 1967), Folk (1966), Moiola and Weiser (1968), Erlich (1983), and McLaren and Bowles (1985). These reviews show that statistical coefficients can very often differentiate sediments from various modern environments. For example, a number of studies show that beach and dune sands are negatively and positively skewed respectively (e.g. Mason and Folk, 1958; Friedman, 1961; Chappell, 1967). Several complicating factors have emerged, however. Many sediments are actually combinations of two or more different grain size populations of different origins (Doeglas, 1946; Spencer, 1963). These admixtures may reflect mixing of sediments of different environments. It is more likely, however, that the presence of several populations in one sample reflects the action of different physical processes. For example, within Barataria Bay, Louisiana, multivariate analysis defined sediments into populations influenced by wind, wave, current, and gravitational processes (Klovan, 1966). Yet these were all deposited in the same lagoonal or bay environment. Similarly, Visher (1965) demonstrated the variability of sediment type within fluvial channels, and showed how these differences are related to sedimentary structure, that is, to depositional process. The C-M diagrams of Passega (1957, 1964) are another approach to this same problem. Plots of C, the first percentile, which measures the coarsest fraction, against M, the median, are most illuminating. They reveal different fields for pelagic suspensions, turbidites, bed load suspensions, and so on (Fig. 3.10).

A further problem of defining the granulometric characteristics of modern environments is that of inheritance. It has often been pointed out that if a fine-grained sand of uniform grain size is transported into a basin then that is the only granulometric type

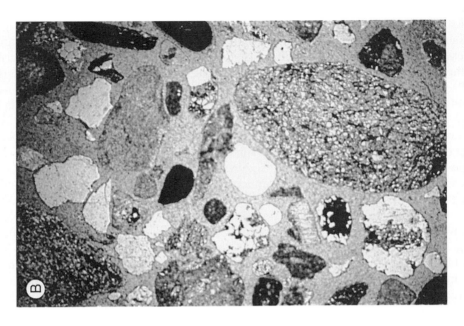

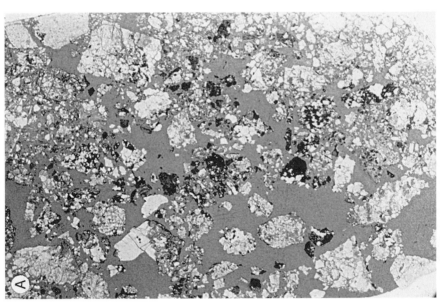

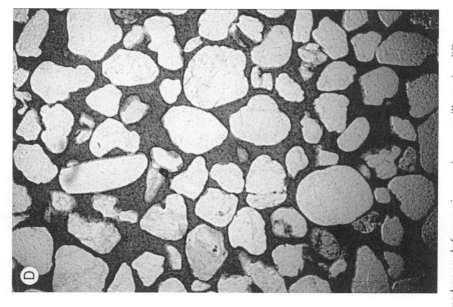

Fig. 3.9. Photomicrographs, in ordinary light, of thin sections of modern sands from various environments illustrating differences in grain size, shape, and sorting. (A) Poorly sorted glacial outwash, Sutherland, Scotland. Note angular lithic grains. (B) Fluvial sand from a Welsh river. Note well-rounded lithic grains. (C) Beach sand, Miami Hilton, Florida. Predominantly quartz, with minor shell debris. (D) Eolian dune sand, Jebel Rummel, Libya. All photographs enlarged 30×.

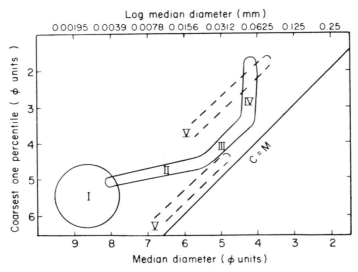

Fig. 3.10. Passega C-M patterns. These plot C, the one percentile, against M, the median 50th percentile. This graph segregates the products of several different sedimentary processes. I, pelagic suspension; II, uniform suspension; III, graded suspension; IV, bed load; V, turbidity currents.

to be deposited, regardless of environment or process. More specific examples of the effect of inheritance on sorting character have been described in studies of modern alluvium from Iran and the Mediterranean (Vita-Finzi, 1971).

Consider now the application of granulometric techniques to the environmental analysis of ancient sedimentary rocks. Several problems are at once apparent. Analyses of modern sediments show that the fine clay fraction is very sensitive to depositional process. In ancient sediments the clay matrix may have infiltrated after sedimentation, or may have formed diagenetically from the breakdown of labile sand grains (see Section 8.5.1). Clay can also be transported in aggregates of diverse size from sand up to clay boulders. Granulometric analyses of a fossil sediment may disaggregate such larger clay clasts to their constituent clay particles. Quartz grains may have been considerably modified by solution and cementation. Couple these factors with the problem of inheritance and the lack of correlation between depositional process and environment. It is at once apparent that the detection of the depositional environment of ancient sediments from their granulometry is extremely difficult. Where other criteria are available, such as vertical grain size profiles, sedimentary structures, and paleontology, it is mercifully unnecessary.

3.2 POROSITY AND PERMEABILITY

3.2.1 Introduction

While it is true that geologists study rocks, much applied geology is concerned with the study of holes within rocks. The study of these holes or "pores" is termed **petrophysics** (Archie, 1950). This is of vital importance in the search for oil, gas, and groundwater,

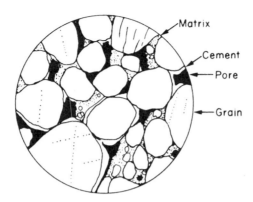

Fig. 3.11. Thin section showing that a sedimentary rock is composed of framework grains and matrix, of syndepositional origin, cement, of postdepositional origin, and pores — sometimes. Pores are the voids unoccupied by any of the three solid components.

and in locating regional permeability barriers that control the entrapment and precipitation of many minerals. It is also necessary for subsurface liquid and radioactive waste disposal and gas storage schemes.

A sedimentary rock is composed of grains, matrix, cement, and pores (Fig. 3.11). The grains are the detrital particles that generally form the framework of a sediment. Matrix is the finer detritus that occurs within the framework. Matrix was deposited at the same time as the framework grains or infiltrated shortly after. There is no arbitrary size distinction between grains and matrix. Conglomerates generally have a matrix of sand, and sandstones may have a matrix of silt and clay. (Note that in the world of the engineer, rocks are only made up of matrix and pores. Failure to be aware of this distinction may cause some confusion.) Cement is postdepositional mineral growth, which occurs within the voids of a sediment. Pores are the hollow spaces not occupied by grains, matrix, or cement. Pores may contain gases, such as nitrogen and carbon dioxide, or hydrocarbons such as methane. Pores may be filled by liquids ranging from potable water to brine and oil. Under suitable conditions of temperature and pressure, pores may be filled by combinations of liquid and gas. The study of pore liquids and gases lies in the scope of hydrology and petroleum engineering. Petrophysics, the study of the physical properties of pores, lies on the boundary between these disciplines and sedimentary geology. The geologist should understand the morphology and genesis of pores and, ideally, be able to predict their distribution within the earth's crust.

3.2.1.1 Definitions

The porosity of a rock is the ratio of its total pore space to its total volume, that is, for a given sample: porosity = total volume − bulk volume. Porosity is conventionally expressed as a percentage. Hence:

$$\text{Porosity} = \frac{\text{volume of total pore space}}{\text{volume of rock sample}} \times 100$$

The porosity of rocks ranges from effectively zero in unfractured cherts to, theoretically, 100% if the "sample" is taken in a cave. Typically porosities in sediments range between 5 and 25%, and porosities of 25–35% are regarded as excellent if found in an aquifer or oil reservoir.

An important distinction must be made between the total porosity of a rock and its effective porosity. Effective porosity is the amount of mutually interconnected pore space present in a rock. It is, of course, the effective porosity that is generally economically important, and it is effective porosity that is determined by many, but not all, methods of porosity measurement. The presence of effective porosity gives a rock the property of permeability. Permeability is the ability of a fluid to flow through a porous solid. Permeability is controlled by many variables. These include the effective porosity of the rock, the geometry of the pores, including their tortuosity, and the size of the throats between pores, the capillary force between the rock and the invading fluid, its viscosity, and pressure gradient.

Permeability is conventionally determined from Darcy's law using the equation:

$$Q = \frac{K \Delta A}{\mu \cdot L},$$

where Q is the rate of flow in cubic centimeters per second, Δ is the pressure gradient, A is the cross-sectional area, μ is the fluid viscosity in centipoises, L is the length, and K is the permeability. Figure 3.12 illustrates the system for measuring the permeability of a rock sample.

This relationship was originally discovered by H. Darcy in 1856 following a study of the springs of Dijon, France. Permeability is usually expressed in darcy units, a term proposed and defined by Wycoff *et al.* in 1934. One darcy, is the permeability which allows a fluid of one centipoise viscosity to flow at one centimeter per second, given a

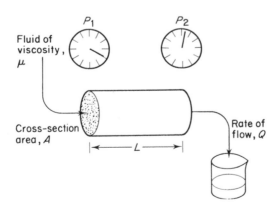

Fig. 3.12. Illustration showing how permeability is measured for a rock specimen. A fluid of viscosity μ is passed through a sample of known cross-sectional area A and length L. The rate of flow is measured, together with the pressure differential recorded on gauges at either end of the sample. Permeability is then calculated according to Darcy's law, as described in the text.

pressure gradient of one atmosphere per centimeter. The permeability of most rocks is considerably less than one darcy. To avoid fractions or decimals, the millidarcy, which is one-thousandth of a darcy, is generally used.

Darcy's law is valid for three assumptions: that there is only one fluid phase present, that the pore system is homogeneous, and that there is no reaction between the fluid and the rock. These conditions are not always satisfied, either in nature or in the laboratory. To illustrate the problem of the first assumption: Many petroleum reservoirs are admixtures of gas, oil, and water, all of which have different viscosities and thus flow rates. To illustrate the problems of the second assumption, that the pore size distribution is uniform: Many rocks have dual pore systems. They may consist, for example, of pores between sand grains and fracture pores that cross-cut the main rock fabric. The former will have a much lower permeability than the latter. To illustrate the problems of the third assumption, that there is no reaction between fluid and rock: Rocks sometimes contain minerals that react with drilling mud and other fluids that are passed through the rock underground or in the laboratory. Halite cement, for example, may be leached out, thus increasing porosity and permeability, or a montmorillonite clay matrix may expand, thus decreasing porosity and permeability (see Section 8.3.2.3). All three of these problems can invalidate the simple assumptions of Darcy's law.

The permeability of rocks is highly variable, both depending on the direction of measurement and vertically up or down sections. Permeabilities ranging from 10 to 100 millidarcies are good, and above that are considered exceptionally high. Figure 3.13 illustrates the concepts of porosity, effective porosity, and permeability in different rock types. Figure 3.14 shows the vertical variations of porosity and permeability that are found in a typical sequence of rock.

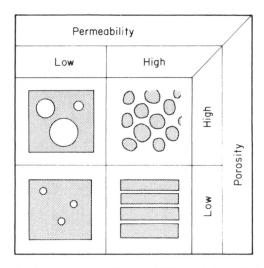

Fig. 3.13. Illustrations showing how porosity and permeability are independent properties of a sediment. Note that permeability is low if porosity is disconnected, whereas permeability is high when porosity is interconnected.

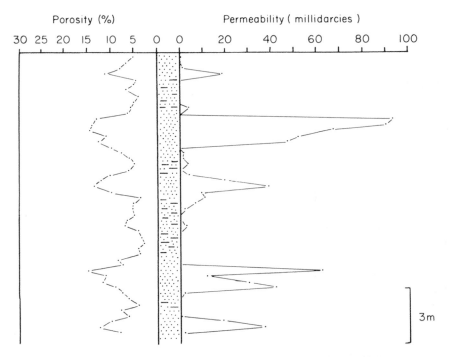

Fig. 3.14. Vertical section of a sandstone sequence showing the variations in porosity and permeability. The observed relationship between increasing porosity and permeability is common in sands with primary inter-granular pore systems.

3.2.1.2 Methods of Measurement of Porosity and Permeability

Several different methods can be used to measure the porosity and permeability of rocks. Many of these require the direct analysis of a sample of the rock in question, either from surface samples or from cores recovered from a borehole. The results from surface outcrop samples are generally deemed unreliable, because they will commonly have been leached to some extent, leading to porosity and permeability values that are higher than their subsurface equivalents.

Seismic data and borehole logs can also be used to measure the porosity and permeability of rocks in the subsurface. It is important to note the way in which the different methods of measurement provide data at different scales. At one extreme, measurement of a core plug provides data for a very small sample. At the other extreme the seismic method provides information for a much larger rock volume (Fig. 3.15). Consideration of the heterogeneity of the sediment then becomes an important consideration.

3.2.1.2.1 Direct methods of porosity measurement

Measurement of porosity by direct methods of porosity measurement requires samples of the rock in question to be available for analysis. These may be hand specimens collected from surface outcrops or they may be borehole cores or small plugs cut from cores.

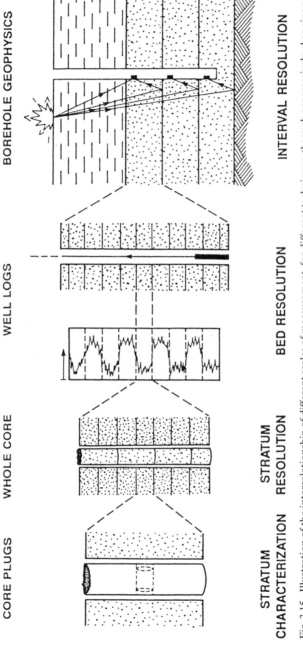

CORE PLUGS WHOLE CORE WELL LOGS BOREHOLE GEOPHYSICS

STRATUM
CHARACTERIZATION

STRATUM
RESOLUTION

BED RESOLUTION

INTERVAL RESOLUTION

Fig. 3.15. Illustration of the interrelationship of different scales of measurement, for different techniques, through progressively changing resolution. (From Worthington, 1991. Copyright © 1991 by Academic Press.)

For all the various methods of directly determining porosity it is necessary to determine both the total volume of the rock sample and either the volume of its porosity or of its bulk volume. Most methods rely on the measurement of the porosity by vacuum extraction of the fluids contained within the pores. Such methods, therefore, measure not total porosity but effective porosity. This is not terribly important because it is the porosity of the interconnected pores that is of significance in an aquifer or hydrocarbon reservoir.

3.2.1.2.2 Indirect methods of porosity measurement

It is often impossible to obtain large enough samples for porosity analysis from underground rocks that hold water, oil, or gas. The porosity of such host rocks must be known in any attempt to assess their economic potential. A number of methods are now available for measuring the porosity of rocks in place when penetrated by a borehole. These are based on measurements of various geophysical properties of the rock by a sonde, a complex piece of electronic equipment that is lowered on a cable down the well bore. Different sondes are designed to measure various properties of the rock. Sondes that can measure the porosity of a rock include the sonic, neutron, and density logs. Of these the sonic method is the least accurate. This is because the acoustic velocity of a rock varies, not only with its porosity, but also with its mineralogy (the acoustic velocity for calcite, for example, is much faster than for quartz). Nonetheless it is described here, because it is so intimately related to the seismic method discussed later. Sonic velocity is recorded continuously by use of an acoustic device in a sonde lowered down a well bore. The sonic velocity of the formation is recorded in microseconds per foot. Given the sonic velocity of the pore fluid and of the pure rock mineral (the sonic velocity of calcite is used for limestones, and of silica for sandstones, etc.) the porosity may be found from the Wyllie equation (Wyllie *et al.*, 1956) thus:

$$\phi = \frac{t_{\log} - t_{ma}}{t_f - t_{ma}},$$

where ϕ is the porosity, $t_{\log}$ is the sonic velocity measured on the log, t_{ma} is the sonic velocity of the matrix (i.e., nonpore rock), and t_f is the sonic velocity of the pore fluid. Additional discussion of indirect methods of measuring porosity will be found in Bateman (1995) and Selley (1998). These geophysical well-logging techniques can give accurate measurements of porosity.

The Wyllie equation can also be used to measure porosity from seismic surveys. The interval velocity of a formation is calculated from the seismic data, and velocities for rock and fluid taken from standard values or, better still, from values measured from real samples of the formations taken from boreholes. The accuracy of the seismic method is as good as the values for fluid and rock velocity. It will work well, for example, in thick formations of uniform limestone, but will be less accurate for heterogeneous clastic formations with rapid lateral and vertical lithological variations.

Detailed discussion of this topic is beyond the scope of this book, but can be found in standard geophysical text books such as Doyle (1995) and Sheriff and Geldert (1995).

3.2.1.2.3 Direct method of permeability measurement

Permeability may be directly measured at outcrop or from a hand specimen or core in the laboratory. Permeability at outcrop may be measured using a probe- or mini-permeameter. This is a portable handheld piece of kit. The technique involves placing a nozzle against the rock face and puffing a fluid, commonly nitrogen, into the rock and measuring the flow rate (Hurst and Goggin, 1995).

Permeability measurement in a laboratory involves pumping gas through a carefully dried and prepared rock sample. This may be either a whole core or a plug cut from a core or hand specimen. The apparatus records the length and cross-sectional area of the sample, the pressure drop across it, and the rate of flow for the test period (refer back to Fig. 3.12). The permeability is then calculated by applying the Darcy equation (Section 3.2.1.1). The viscosity of gas at the temperature of the test can be found in printed tables of physical data. Permeability values from minipermeameter readings are comparable to those from laboratory measurements of whole cores, but are lower than those measured from core plugs (Hurst and Rosvoll, 1991).

3.2.1.2.4 Indirect methods of permeability measurement

Permeability can be calculated by recording the amount of fluid which a known length of borehole can produce over a given period of time. This is applicable in hydrology where, unless the well is artesian, pump tests are carried out. These record the amount of water which can be extracted in a given period, and the length of time required for the water table to return to normal if it was depressed during the test. Both the productivity under testing and the recharge time are measures of the permeability of the aquifer.

Oil and gas reservoirs typically occur under pressure. The permeability of the reservoir can be measured from drill-stem tests and from lengthier production tests. These record the amount of fluid produced in a given period, the pressure drop during this time, and the buildup in pressure over a second time interval when the reservoir is not producing. By turning Darcy's law inside out to apply to the shape of the borehole, it is possible to measure the mean permeability of the reservoir formation over the interval that was tested.

Permeability, unlike porosity, cannot yet be directly measured by using geophysical sondes in boreholes. There are now, however, various geophysical techniques that can measure directional variations in permeability once an actual measurement has been made using the techniques outlined earlier.

3.2.1.3 Capillarity

Fluid movement through sediments is also contingent on capillarity. Consider a trough of liquid within which are arranged vertical tubes of varying diameters (Fig. 3.16). The level of liquid within wide tubes is the same as that without. In narrow tubes, however, liquid is drawn up above the common level. The height is proportional to the tube diameter. This gravity-defying effect, contrary to hydrostatic principles, is due to surface

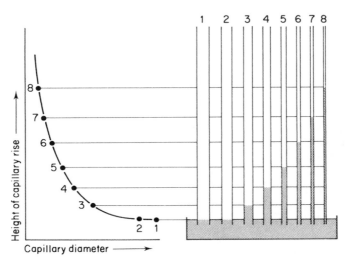

Fig. 3.16. Illustration of the concept of capillary pressure. Note how the height of the liquid column in the the capillary tubes increases with diminishing tube diameter and capillary pressure.

tension at the boundary between the liquid and the atmosphere. This effect is known as **capillarity.**

Capillary pressure may be defined as the pressure difference across an interface between two immiscible fluids. A rigorous mathematical analysis of capillarity will be found in petroleum engineering texts, such as Archer and Wall (1986) and Chierici (1995). The height the liquid is drawn up within a tube is related to the capillary pressure exerted at the boundary between the two fluids, and to the diameter of the tube. The diameter below which capillary flow occurs is obviously important.

Considering sediment particles instead of tubes, there is a critical pore throat radius below which the capillary effect will inhibit fluid flow for a given capillary pressure and a given pressure differential. An important petrophysical parameter of a sediment is its capillary pressure curve. This is measured by plotting increasing pressure against increasing saturation as one fluid is displaced by another. Two critical values for a capillary pressure test are the displacement pressure above which invasion by the new fluid commences, and the irreducible saturation point, above which no more of the new fluid may be injected irrespective of the pressure increase (Fig. 3.17). Because capillarity is a function of the radius of a pore throat, it follows that the capillary pressure curve of a sediment will reflect the size distribution of its constituent pore throat radii. This will be examined when considering the relationship between sediment texture, porosity, and permeability (see Section 3.2.3).

As seen in Chapter 2, capillary flow is important in weathering processes in general, and in the formation of caliche in particular (see Section 2.3.1.1). In the deep subsurface, capillary effects are also very significant, both in terms of connate fluid flow through sediments of varying grain size, and especially where two fluid phases are present, such as in a petroleum reservoir.

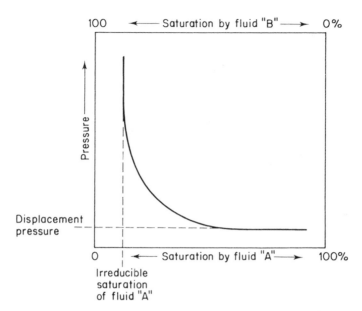

Fig. 3.17. Graph showing the basic parameters of a capillary pressure curve for a rock sample. The sample is initially totally saturated by fluid A. This is gradually displaced by fluid B as pressure is increased.

3.2.2 Pore Morphology

3.2.2.1 Introduction and Classification

Any petrophysical study of a reservoir rock necessitates a detailed description of the amount, type, size distribution, and genesis of its porosity. The classification of the main types of porosity is discussed next, followed by a description of the more common varieties of pores. A large number of adjectives have been used to describe the different types of porosity present in sediments. Choquette and Pray (1970, pp. 244–250) provide a useful glossary of pore terminology.

The pores themselves may be studied by a variety of methods ranging from examination of rough or polished rock surfaces by hand-lens or stereoscopic microscope, through study of thin sections using a petrological microscope, to the use of the scanning electron microscope. Another effective technique of studying pore fabric is to impregnate the rock with a suitable plastic resin and then dissolve the rock itself with an appropriate solvent. Examination of the residue gives some indication, not only of the size and shape of the pores themselves, but also of the throat passages that connect pores (e.g., Wardlaw, 1976). The minimum size of throats and the tortuosity of pore systems are closely related to the permeability of the rock.

These different observational methods show that there are many different types of pore systems. Various attempts have been made to classify porosity types. These range from essentially descriptive schemes, to those which combine descriptive and genetic criteria (e.g., Choquette and Pray, 1970), and those which relate the porosity type to the

Table 3.4
Classification of Porosity Types

	Type	Origin
I. Primary or depositional	(a) Intergranular or interparticle (b) Intragranular or intraparticle	Sedimentation
II. Secondary or postdepositional	(c) Intercrystalline (d) Fenestral	Cementation
	(e) Moldic (f) Vuggy	Solution
	(g) Fracture	Tectonic movement, compaction or dehydration

petrography of the host rock (e.g., Robinson, 1966). The classification shown in Table 3.4 divides porosity into two commonly recognized main varieties (e.g., Murray, 1960). These are primary porosity fabrics, which were present immediately after the rock had been deposited, and secondary or postdepositional fabrics, which formed after sedimentation by a variety of causes. The main porosity types are now described and illustrated (see Plate 1).

3.2.2.2 Primary or Depositional Porosity

Primary or depositional porosity is that which, by definition, forms when a sediment is first laid down. Two main types of primary porosity are recognized: intergranular, or interparticle, and intragranular, or intraparticle.

3.2.2.2.1 Intergranular or interparticle porosity

Intergranular or interparticle porosity occurs in the spaces between the detrital grains that form the framework of a sediment (Fig. 3.18A, Plate 1A). This is a very important

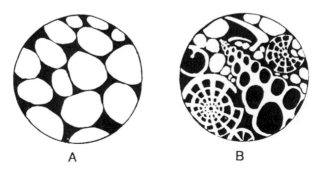

A   B

Fig. 3.18. Sketches of thin sections illustrating primary, depositional porosity. (A) Intergranular (or interparticle) porosity, commonly found in sandstones. (B) Mixed intergranular (or interparticle) and intragranular (or intraparticle) porosity, typical of skeletal sands before burial and diagenesis.

porosity type. It is present initially in almost all sediments. Intergranular porosity is generally progressively reduced by diagenesis in many carbonates, but is the dominant porosity type found in sandstones. The factors that influence the genesis of intergranular porosity and which modify it after deposition are discussed at length in a later section (see Section 3.2.3.1).

3.2.2.2.2 Intragranular or intraparticle porosity

In carbonate sands, particularly those of skeletal origin, primary porosity may be present within the detrital grains. For example, the cavities of mollusks, ammonites, corals, bryozoa, and microfossils can all be classed as intragranular or intraparticle primary porosity (Fig. 3.18B, Plate 1B). This kind of porosity is often diminished shortly after deposition by infiltrating micrite matrix. Furthermore, the chemical instability of the carbonate host grains often leads to their intraparticle pores being modified or obliterated by subsequent diagenesis.

3.2.2.3 Secondary or Postdepositional Porosity

Secondary porosity is that which, by definition, formed after a sediment was deposited. Secondary porosity is more diverse in morphology and more complex in genesis than primary porosity. It is more commonly found in carbonate rocks than in siliciclastic sands. This is because of the greater mobilty of carbonate minerals in the subsurface, compared to quartz. The following main types of secondary porosity are recognizable.

3.2.2.3.1 Intercrystalline porosity

Intercrystalline porosity occurs between the individual crystals of a crystalline rock (Fig. 3.19A, Plate 1C). It is, therefore, the typical porosity type of the igneous and high-grade metamorphic rocks, and of some evaporites. Strictly speaking, such porosity is of primary origin. It is, however, most characteristic of carbonates that have undergone crystallization and is particularly important in recrystallized dolomites. Such rocks are sometimes very important oil reservoirs. The pores of crystalline rocks are essentially planar cavities which intersect obliquely with one another with no constrictions of the boundaries or throats between adjacent pores.

3.2.2.3.2 Fenestral porosity

The term "fenestral porosity" was first proposed by Tebbutt *et al.* (1965) for a "primary or penecontemporaneous gap in rock framework, larger than grain-supported interstices." This porosity type is typical of carbonates. It occurs in fragmental carbonate sands, where it grades into primary porosity, but is more characteristic of pellet muds, algal laminites, and homogeneous muds of lagoonal and intertidal origin. Penecontemporaneous dehydration, lithifaction, and biogenic gas generation can cause laminae to buckle and generate subhorizontal fenestral pores between the laminae (Fig. 3.19B). This type of fabric has been termed **loferite** by Fischer (1964) on the basis of a study

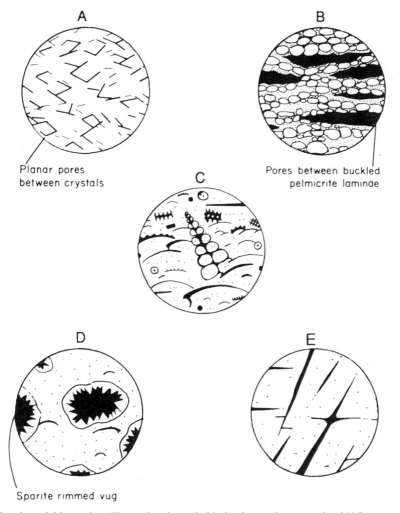

Fig. 3.19. Sketches of thin sections illustrating the main kinds of secondary porosity. (A) Intercrystalline porosity, characteristic of dolomites. (B) Fenestral porosity, characteristic of pelmicrites. (C) Moldic porosity formed by selective leaching, in this case of skeletal fragments (biomoldic). (D) Vuggy porosity produced by irregular solution. (E) Fracture porosity, present in many brittle rocks.

of Alpine Triassic back reef carbonates. Analogous fabrics have been described from dolomite pellet muds in a similar setting from the Libyan Paleocene. The fenestrae in this last case are sometimes partially floored by lime mud, proving their penecontemporaneous origin (Conley, 1971).

A variety of fenestral fabric has long been known as **bird's-eye.** This refers to isolated "eyes" up to 1 cm across that occur in some lime mudstones (Illing, 1954). These apertures have been attributed to organic burrows and also to gas escape conduits. They are frequently infilled by crystalline calcite. As a generalization, rounded "bird's-eye" pores can be attributed to biogenic gas generation. Elongate fenestral pores may be attributed to the rotting of organic material in stromatolites or to buckling of laminated mud during intertidal exposure (Shinn, 1983).

3.2.2.3.3 Stromatactis

Stromatactis is a phenomenon related to fenestral pore systems. This name has been given to irregular patches of crystalline calcite that are common on the flanks of Paleozoic mud mounds around the world (Monty *et al.,* 1995). A stromatactis structure is normally some 10 cm in length and 1–3 cm high. The base is flat, the upper surface domed and irregular. The structures commonly dip radially from the center of the mud mound (Fig. 3.20). Some stromatactis structures have a partial infill of lime mud, the upper surface of which is horizontal, while the overall structure dips down the flank of the mound. This is termed a **geopetal** fabric, and indicates that stromatactis developed as a penecontemporaneous void just below the sea floor. The rest of the void often shows two phases of fill. There is a radiaxial fibrous rim that formed prior to compaction at shallow depth, and a later sparite cement fill (Bathurst, 1982). This structure is, therefore, a variety of secondary porosity.

Stromatactis has been variously attributed to a soft-bodied beast of unknown origin, to bioturbation, to algae, to micrite recrystallization, and to gas bubbles, similar to the birds-eye structure previously mentioned.

The geopetal fabric often seen in stromatactis suggests the downslope slumping of a lithified lime mud crust. The occurrence of stromatactis on the flanks of mud mounds suggests that there may be a causal link. Modern carbonate mud mounds are associated

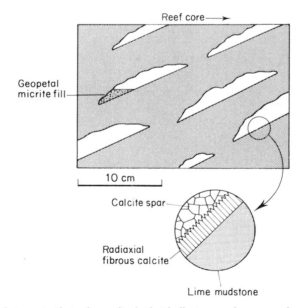

Fig. 3.20. Diagram of stromatactis to show criteria that indicate genesis as pores formed by the downslope slumping of penecontemporaneously lithified lime mud.

with "cold seeps" where gases and liquids, often including methane and petroleum, emerge on the sea bed (see Section 5.3.5.7).

3.2.2.3.4 Moldic porosity

A fourth type of secondary porosity, generally formed later in the history of a rock than fenestrae and stromatactis, is moldic porosity. Molds are pores formed by the solution of primary depositional grains generally subsequent to some cementation. Molds are fabric selective. That is to say, solution is confined to individual particles and does not cross-cut cement, matrix, and framework. Typically in any rock it is all the grains of one particular type that are dissolved. Hence one may talk of oomoldic, pelmoldic, or bio-moldic porosity where there has been selective solution of ooliths, pellets, or skeletal debris (Fig. 3.19C, Plate 1D). The geometry and effective porosity and permeability of moldic porosity can thus be extremely varied. In an oomoldic rock, pores will be sub-spherical and of similar size. In biomoldic rocks, by contrast, pores may be very variable in size and shape, ranging from minute apertures to curved planar pores where shells have dissolved, and cylinders where echinoid spines have gone into solution.

3.2.2.3.5 Vuggy porosity

Vugs are a second type of pore formed by solution and, like molds, they are typically found in carbonates. Vugs differ from molds though because they cross-cut the primary depositional fabric of the rock (Fig. 3.19D, Plate 1E). Vugs thus tend to be larger than molds. They are often lined by a selvage of crystals. With increasing size vugs grade into what is loosely termed "cavernous porosity." Choquette and Pray (1970, p. 244) proposed that the minimum dimension of a cavern is a pore which allows a speleolo-gist to enter or which, when drilled into, allows the drill string to drop by more than half a meter through the rotary table. Large-scale vuggy and cavernous porosity is com-monly developed beneath unconformities where it is referred to as "paleokarst" (see Section 9.2.7). This serves as a petroleum reservoir in a number of fields such as Abqaiq in Saudi Arabia (McConnell, 1951), the Dollarhide field of Texas (Stormont, 1949), and the Casablanca field of offshore Spain (Watson, 1982).

3.2.2.3.6 Fracture porosity

The last main type of pore to be considered is that formed by fractures. Fractures occur in many kinds of rocks other than sediments. Fracturing, in the sense of a breaking of depositional lamination, can occur penecontemporaneously with sedimentation. This often takes the form of microfaulting caused by slumping, sliding, and compaction. Frac-tures in plastic sediments are instantaneously sealed, however, and thus seldom preserve porosity. In brittle rocks fractures may remain open after formation, thus giving rise to fracture porosity (Fig. 3.19E, Plate 1F). This porosity type characterizes rocks that are strongly lithified and is, therefore, generally formed later than the other varieties of po-rosity (Fig. 3.21). It is important to note that fracture porosity is not only found in well-cemented sandstones and carbonates, but may also be present in shales and igneous and metamorphic rocks.

Fracture porosity is much more difficult to observe and analyze than most other pore

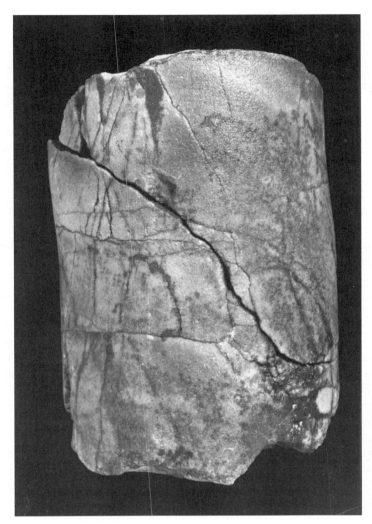

Fig. 3.21. Photograph of core of Gargaf sandstone (Cambro-Ordovician), Sirte basin, Libya, illustrating fracture porosity.

systems. Though fractures range from microscopic to cavernous in size, they are difficult to study in cores.

Fracture porosity can occur in a variety of ways and situations. Tectonic movement can form fracture porosity in two ways. Tension over the crests of compressional anticlines and compactional drapes can generate fracture porosity. Fracture porosity is also intimately associated with faulting and some oil fields show very close structural relations with individual fault systems. The Scipio fields of southwest Michigan are a case in point. These occur on a straight line of about 15 km. Individual fields are about 0.5 km wide. The oil is trapped in a fractured dolomitized belt within the Trenton limestone (Ordovician). This fracture system was presumably caused by movement along a deep fault in the basement (Levorsen, 1967, p. 123).

Fracture porosity can also form from atectonic processes. It is often found immediately beneath unconformities. Here fractures, once formed by weathering, may have been enlarged by solution (especially in limestone paleokarst) and preserved without subsequent loss of their porosity (see Section 9.2.7). Fracture porosity is extremely important in both aquifers and petroleum reservoirs. This is because a very small amount of fracture porosity can give a very large permeability; the fractures connecting up many pores of other types that might otherwise be ineffective. There is, therefore, a very large literature on fractures (e.g., Aguilera, 1980; Reiss, 1981; Van Golfracht, 1982; Nelson, 1986; Atkinson, 1987).

3.2.2.3.7 Summary

The preceding account shows something of the diverse types and origins of pores. Essentially there are two main genetic groups of pore types. Primary porosity is formed when a sediment is deposited. It includes intergranular or interparticle porosity, which is characteristic of sands, and intraparticle porosity found in skeletal carbonate sands.

Secondary porosity forms after sedimentation by diagenetic processes. Recrystallization, notably dolomitization, can generate intercrystalline porosity. Solution can generate moldic, vuggy, and cavernous pores. Because such pores are often isolated from one another permeability may be low. Fractures form in both unconsolidated and brittle sediments. In the first instance the fractures remain closed, but in brittle rocks fracture porosity may be preserved, enlarged by solution, or diminished by cementation. Fracture porosity occurs not just in indurated sediments, but also in igneous and metamorphic rocks.

It is important to note that many sedimentary rocks contain more than one type of pore. The combination of open fractures with another pore type is of particular significance. The problems generated by such dual pore systems negating the assumptions of Darcy's law have been noted earlier. Fine-grained rocks, such as shales, microcrystalline carbonates, and fine sands, have considerable porosity. They often have very low permeabilities because of their narrow pore throat diameter and concomitant high capillary pressure. The presence of fractures, however, can enable such rocks to yield their contained fluids. The success of many oil and water wells in such formations often depends on whether they happen to penetrate an open fracture. Recognition of the significance of fractures in producing fluids from high-porosity, low-permeability formations has led to the development of artificial fracturing by explosive charges that simultaneously wedge the fractures open with sand, glass beads, etc. Similarly, the productivity of fractured carbonate reservoirs can be increased by the injection of acid to dissolve and enlarge the fractures. Figure 3.22 summarizes the relationship between pore type, porosity, and permeability.

3.2.3 The Origin of Primary Porosity

The porosity and permeability of a sediment are controlled by its textural characteristics at the time of deposition and by subsequent diagenetic changes, including compaction, cementation, and solution.The effects of diagenesis on the porosity and permeability of sandstones and carbonates are discussed in Chapters 8 and 9, respectively.

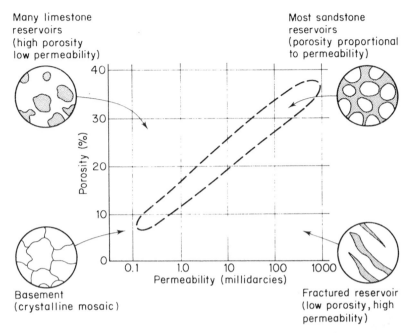

Fig. 3.22. Graph showing the relationship between pore type, porosity, and permeability. For intergranular porosity seen in most sandstones, porosity generally plots as a straight line against permeability on a logarithmic scale. Shales, chalks, and vuggy rocks tend to be porous but impermeable. Fractures increase permeability with generally little increase in porosity.

The rest of this chapter is concerned with the factors that control porosity and permeability at the time of deposition.

3.2.3.1 Relationships between Porosity, Permeability, and Texture

Beard and Weyl (1973) showed that the porosity of a newly deposited sediment is a result of five variables: grain size, sorting, grain shape (sphericity), grain roundness (angularity), and packing. A considerable amount of work has been done on the way that these five factors affect primary porosity. This work includes theoretical mathematical studies (Engelhardt and Pitter, 1951) and experimental analyses of artificially made spheres, unconsolidated modern sediment, and even ancient rocks. The results of this work are summarized here for the five parameters listed.

3.2.3.1.1 Effect of grain size on porosity and permeability

Theoretically, porosity is independent of grain size. A mass of spheres of uniform sorting and packing will have the same porosity, regardless of the size of the spheres. The volume of pore space varies in direct proportion to the volume of the spheres (Fraser, 1935). Rogers and Head (1961), working with synthetic sands, showed that porosity is independent of grain size for well-sorted sands. This "ideal" situation is seldom found in nature. Pryor (1973) analyzed nearly 1000 modern sands and showed that porosity decreased with increasing grain size. River sands were the reverse, however, possibly due to packing differences. This trend is probably due to a number of factors that are only indirectly linked with grain size. Finer sands tend to be more angular and to be able to support looser packing fabrics, hence they may have a higher porosity than coarser sands.

Whatever the cause of the relationship, it has been shown empirically that porosity generally increases with decreasing grain size for unconsolidated sands of uniform grain size. This relationship is not always true for lithified sandstones, whose porosity often increases with grain size. This may be because the finer sands have suffered more from compaction and cementation. Permeability, by contrast, increases with increasing grain size (Fraser, 1935; Krumbein and Monk, 1942; Pryor, 1973). This is because in finer sediments the throat passages between pores are smaller and the higher capillary attraction of the walls inhibits fluid flow. This relationship is found in both unconsolidated and lithified sand.

3.2.3.1.2 Effect of sorting on porosity and permeability

A number of studies of sediments have shown that porosity increases with increased sorting (Fraser, 1935; Rogers and Head, 1961; Pryor, 1973; Beard and Weyl, 1973). Krumbein and Monk (1942) and Beard and Weyl (1973) demonstrated that increased sorting correlates with increased permeabilities. A reason for these relationships is not hard to find. A well-sorted sand has a high proportion of detrital grains to matrix. A poorly sorted sand, on the other hand, has a low proportion of detrital grains to matrix. The finer grains of the matrix block both the pores and throat passages within the framework, thus inhibiting porosity and permeability, respectively (Fig. 3.23).

Pryor's study (1973) of modern sands from different environments confirmed this relationship for river sands, but showed that beach and dune sands were anomalous in that their permeability increased with decreased sorting. Figure 3.24 summarizes the relationships between porosity, permeability, grain size, and sorting for unconsolidated sands.

3.2.3.1.3 Relationship between grain shape and roundness on porosity and permeability

Undoubtedly the shape and roundness of grains affect intergranular porosity. Little work has been done on this problem, probably because of the time needed to measure these parameters in sufflciently large enough samples to be meaningful.

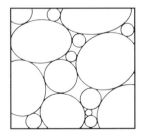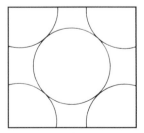

Fig. 3.23. A well-sorted sediment (right) will have better porosity and permeability than a poorly sorted one (left). In the latter the space between the framework grains is infilled, thus diminishing porosity. The heterogeneous fabric diminishes permeability by increasing the tortuosity of the pore system.

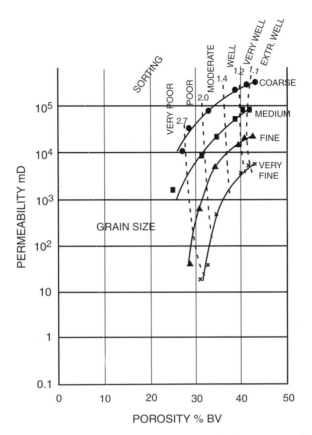

Fig. 3.24. Graph showing the relationship between petrophysics and sediment texture in unconsolidated clay free sand. (From Nagtegaal, 1978. Courtesy of the Geological Society of London.)

Fraser (1935) concluded that sediments composed of spherical grains have lower porosities than those with grains of lower sphericity. He attributed this to the fact that the former tend to fall into a tighter packing than sands of lower sphericity (see also Beard and Weyl, 1973).

3.2.3.1.4 Relationship between fabric, porosity, and permeability

The way in which the particles of a sediment are arranged is termed the **fabric.** There are two elements to fabric: grain packing and grain orientation. Because these are closely related to primary porosity, we discuss them next.

Grain packing. Graton and Fraser (1935) demonstrated how the porosity of a sediment varies according to the way in which its constituent grains are packed. They showed that theoretically there are six possible packing geometries for spheres of uniform size (Fig. 3.25). These range from the loosest "cubic" style, with a theoretical porosity of 48%, to the closest rhombohedral packing with a theoretical porosity of 26% (Fig. 3.26). These ideal situations never occur in nature. More realistic packing models are obtained

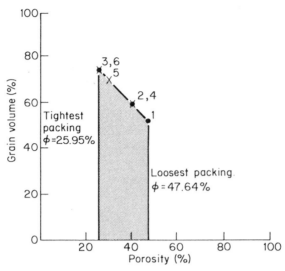

Fig. 3.25. Graph of porosity plotted against grain volume for the six theoretical packing geometries for perfect spheres. Note the range in porosity between the loosest and tightest packing. ●, Cubic layers where 1, cube on cube; 2, cubes laterally offset; 3, cubes obliquely offset. X, Rhombohedral layers where 4, rhomb on rhomb; 5, rhombs laterally offset; 6, rhombs obliquely offset. (Based on data in Graton and Fraser, 1935.)

when the analysis is based not on spheres, but on prolate spheroids which approximate more closely to real sand grains (Allen, 1970).

   Though packing no doubt plays a major part in controlling the primary porosity of a sediment, this parameter has been one of the hardest to analyze in consolidated rocks. The reasons for this are threefold: difficulty of measurement, lack of knowledge of the control of environment and depositional process on packing, and the effect of postdepositional compaction. A number of workers have suggested methods of measuring and quantifying packing. Emery and Griffiths (1954) proposed a packing index that is the product of the number of grain contacts observed in a thin-section traverse and the average grain diameter divided by the length of the traverse. Kahn (1956) renamed this packing index as "packing proximity" and revamped the original formula. Mellon (1964) proposed a horizontal packing intercept, which is the average horizontal distance between framework grains.

   The problem with all of these different packing indices is that it is extremely labori-

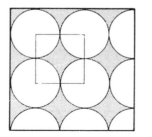
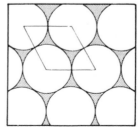

Fig. 3.26. Diagram illustrating the arrangement of (left) the loosest (cubic) packing style, with a porosity of 47.64%, and (right) the tightest (rhombohedral) style, with a porosity of 26.95%.

ous to measure large enough samples, and sufficient specimens, to produce enough data to manipulate. Furthermore, these measures of packing are determined from lithified thin sections. It is possible to impregnate and thin section unconsolidated sand, but it is unlikely that this can be done without disturbing the original packing. In addition, Morrow (1971) has pointed out how the packing of a sediment fabric will change from type to type within and between adjacent laminae. Because of these problems little is known of the relationship between packing and primary depositional porosity.

Intuitively one might expect pelagic muds, turbidites, and grain flows to be deposited with a looser packing than traction deposits. Perhaps cross-bedded sands tend to be more loosely packed than flat-bedded sand. There are few data to support these speculations. Pryor (1973) has shown that modern river sands are more loosely packed than beach and eolian dunes. The fact is that postdepositional compaction so modifies sand grains that packing can have little influence on the porosity of lithified sediment (Lundegard, 1992).

Grain orientation. The orientation of sediment particles is generally discussed with reference to the axis of sediment transport (the flow direction) and to the horizontal plane. The orientation of pebble fabrics is tolerably well known because their large size makes them relatively easy to measure. Sand-grain orientation has, until recently, been less well understood, possibly because it is more complex, but probably because it is harder to measure and quantify.

One of the most common features of gravel fabrics is "imbrication" in which the pebbles lie with their long axis parallel to the flow direction and dipping gently up current. Isolated pebbles on channel floors are often imbricated too (Fig. 3.27). This is a useful paleocurrent indicator (see Section 5.4.1). Dispersed pebbles in diamictites (pebbly mudstones) also commonly show an alignment parallel to the direction of movement. This is true both of mud flow deposits and of glacial tills (see, for example, Lindsay, 1966, and Andrews and Smith, 1970, respectively). Studies of the orientation of pebbles dispersed in fluvial sediments have yielded conflicting results. Many studies (reviewed

Fig. 3.27. Channel showing imbrication fabric of shale conglomerate on the channel floor. The clasts dip upcurrent in opposition to the downcurrent dip direction of the foresets. Cambro-Ordovician, Jordan.

in Schlee, 1957) show that long axes tend to parallel current flow. Some data from cross-bedded sands, however, show this orientation on the foreset, but on the topset and bottomset, the pebbles are elongated perpendicular to the current direction (Johansson, 1965; Sengupta, 1966; Bandyopadhyay, 1971; Gnaccolini and Orombelli, 1971). Studies of sand-grain orientation were held back when it was only possible to measure each grain individually. Indirect methods of orientation measurement have now evolved (Shelton and Mack, 1970; Sippel, 1971; Schafer and Teyssen, 1987) that enable aggregate samples to be measured with speed.

Potter and Pettijohn (1977, pp. 47–55) gave a detailed review of the relationship between grain orientation and current direction. Most studies show that in flat-bedded sands grains are elongated parallel to current direction, and that this axis coincides with maximum permeability. Few sediments are made of spherical particles however, perhaps with oolitic sands being the only example. Most quartz grains are slightly elongated, while platy grains of mica, clay, bioclasts, and plant debris also occur. The shapes of skeletal sand grains are still more complex. Thus when a sediment is deposited the flaky grains settle parallel to the horizon and contribute to the laminated appearance. These flakes will inhibit vertical permeability. In the horizontal plane permeability will be higher, the highest permeability paralleling the long axes of the grains. Thus permeability correlates with paleocurrent direction (Fig. 3.28). This relationship has been seen in flat-bedded sands in fluvial and turbidite deposits (Shelton and Mack, 1970; Martini, 1971; von Rad, 1971; Hiscott and Middleton, 1980). On beaches grains are aligned perpendicular to the shoreface (Curray, 1956; Pryor, 1973).

3.2.3.1.5 Effect of sedimentary structures on porosity and permeability

Finally it is appropriate to move up from consideration of the control of the fabric on the porosity and permeability of a sediment, to consideration of sedimentary structures. Of these the most studied has been cross-bedding. For cross-bedded sands the relationships between permeability and fabric are more complex (Fig. 3.29). Grain-size variations appear to exert a greater influence than grain orientation. Thus in avalanche cross-beds, such as occur on braid bars, grain size and thus permeability will increase downward. In eolian sands, however, the reverse is often seen, with a sandy foreset grading down into a silty toeset. In such units permeability will decrease downward (Fig. 3.30).

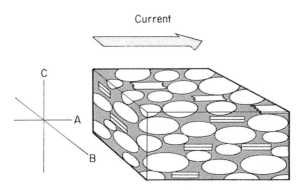

Fig. 3.28. Block of flat-bedded sediment showing the relationship between fabric and current orientation. Permeability is usually lowest in the vertical (C) axis, and highest in the horizontal (A) axis parallel to the current.

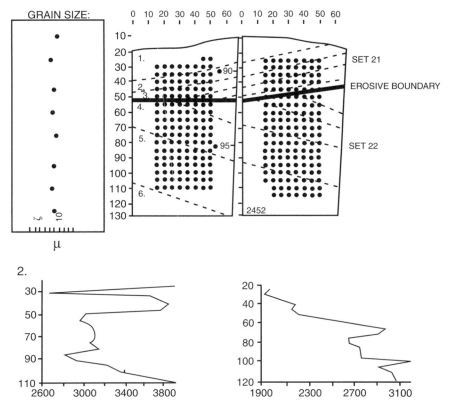

Fig. 3.29. Sketch of cross-bedded sandstone core showing vertical variations in permeability. Vertical scale is centimeters. Horizontal scales on graphs show permeability in millidarcies. (From Hurst and Rosvoll, 1991. Copyright © 1991 Academic Press.)

Vertical permeability variations are not too significant in thin cross-beds, but they become important in cross-beds several meters thick. This is seen, for example, in the Rotliegendes gas reservoirs of the southern North Sea, where eolian cross-beds up to 5 m thick show large vertical permeability variations (Van Veen, 1975). Pryor (1973) and Weber (1982) have addressed the problem of three-dimensional variation in cross-bedded sands. They both conclude that permeability is highest in the deepest and central part of trough cross-beds. This is partly due to the grain-size variations already discussed, but also because the set boundaries are commonly defined by shale laminae that act as permeability barriers.

On a still larger scale there tend to be significant vertical variations in different types of sand bodies. Channels, for reasons described in Section 6.3.2.2.3, tend to show a vertical upward decline in grain size and, therefore, may possess a vertical decline in permeability. Barrier bar and mouth bar sands, however, commonly possess upward-coarsening grain-size profiles (for reasons discussed in Sections 6.3.2.6 and 6.3.2.5.1, respectively). Thus they commonly show an upward increase in permeability.

In a simple world channels trend down the paleoslope, whereas barrier bars are aligned parallel to it (Fig. 3.31). Therefore, most sedimentary deposits have complex regional permeability variations. These variations are economically important for a

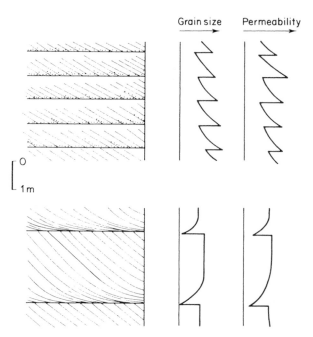

Fig. 3.30. Sketches of cross-bedding to show grain-size and porosity variations. In avalanche tabular-planar cross-bedding (upper), grain size and permeability increase down each set. In some eolian cross-bedding (lower), foreset sands curve down into silty toesets. This correlates with a downward decrease in permeability through each set.

number of reasons. Obviously, water flow through an aquifer will be affected by internal permeability heterogeneities, so too will be the flow of oil and gas in a petroleum reservoir. Similarly, the permeability variations of sedimentary formations affect the flow of mineralizing fluids. These fluids often finally precipitate ores at permeability barriers.

Considerable attention has thus been paid to the continuity of sands, and conversely of shales, in different environments (Harris and Hewitt, 1977; Le Blanc, 1977). Mathematical methods have been devised to quantify the relationship between sand continuity and permeability. Pryor and Fulton (1976) have devised the lateral continuity index (LCI) and the vertical continuity index (VCI) as significant parameters of sedimentary deposits. To calculate the LCI a series of cross-sections is drawn. Maximum and minimum sand continuity are measured and divided by the section lengths. The VCI is calculated by measuring the aggregate sand thickness at various locations and dividing it by the thickness of the thickest sands. In a study of Holocene sands of the Rio Grande, Pryor and Fulton (1976) found wide ranges of continuity when comparing fluvial, fluviomarine, and pro-delta deposits. Weber (1982) has approached the problem from the opposite direction, compiling data on shale continuity in various environments (Fig. 3.32).

Earlier in this chapter, we mentioned that one should note the different volumes of sediment whose porosity or permeability is measured by different techniques, ranging from thin section to seismic. In conclusion one should now also note the hierarchy of different heterogeneities encountered in sedimentary rocks (Fig. 3.33). This chapter has mainly described the controls on microscopic heterogeneities, and at its conclu-

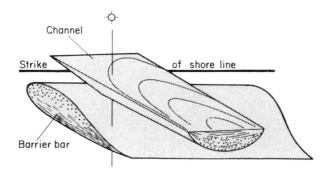

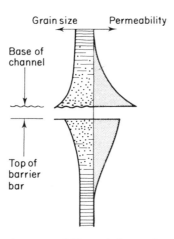

Fig. 3.31. Diagram showing permeability variations and trends in sand bodies. Channels often have upward fining grain-size profiles, so permeability may decline upward within a channel. Barrier sands often have upward-coarsening grain-size profiles, so they often show an upward increase in permeability. On a regional scale channels will tend to trend down the paleoslope, whereas barrier bars will parallel it. These permeability variations affect the flow of petroleum and other fluids, and may control the locus of precipitation of certain types of mineral deposit.

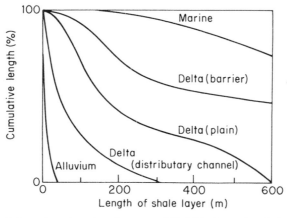

Fig. 3.32. Graph of shale layer length against frequency (%). This shows that shale layers, and thus permeability barriers, increase in continuity from nonmarine to marine deposits. Note, however, the wide range of values in deltaic deposits. (Modified from Weber, 1982.)

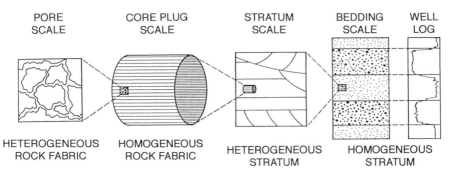

Fig. 3.33. Illustrations of the different scales at which heterogeneity may be encountered in sediments. This chapter has dealt with the depositional controls on a microscopic scale, from fabric up to that of sedimentary structures. Chapter 6 deals with heterogeneity on the macroscopic scale of sedimentary facies. (From Worthington, 1991. Copyright © 1991 Academic Press.)

sion has only briefly considered macroscopic heterogeneities. The latter relate to the depositional environment of the sediment, and are thus considered more fully in this context in Chapter 6.

SELECTED BIBLIOGRAPHY

Hurst, A., and Rosvoll, K. J. (1991). Permeability in sandstones and their relationship to sedimentary structures. *In* "Reservoir Characterization" (L. W. Lake, H. B. Carroll, and T. C. Weston, eds.), Vol. 2, pp. 166–196. Academic Press, San Diego, CA.

Krinsley, D. H. (1998). "Backscattered Scanning Electron Microscopy and Image Analysis of Sediments and Sedimentary Rocks." Cambridge University Press, Cambridge, UK. 324pp.

Lewis, D. W., and McConchie, D. (1994). "Analytical Sedimentology." Chapman & Hall, London. 197pp.

Weber, K. J. (1982). Influence of common sedimentary structures on fluid flow in reservoir models. *JPT, J. Pet. Technol.* **34,** 665–672.

Worthington, P. F. (1991). Reservoir characterization at the Mesoscopic scale. *In* "Reservoir Characterization" (L. W. Lake, H. B. Carroll, and T. C. Weston, eds.), Vol. 2, pp. 123–165. Academic Press, San Diego, CA.

REFERENCES

Aguilera, R. (1980). "Naturally Fractured Reservoirs." Pennwell Books, Tulsa, OK. 703pp.

Allen, J. R. L. (1970). The systematic packing of prolate spheroids with reference to concentration and dilatency. *Geol. Mijnbouw* **49,** 211–220.

Andrews, J. T., and Smith, D. I. (1970). Statistical analysis of till fabric: Methodology, local and regional variability. *Q. J. Geol. Soc. London* **125,** 503–542.

Archer, J. S., and Wall, C. G. (1986). "Petroleum Engineering Principles and Practice." Graham & Trotman, London. 362pp.

Archie, G. E. (1950). Introduction to petrophysics of reservoir rocks. *Bull. Am. Assoc. Pet. Geol.* **34,** 943–961.

Assalay, A. M., Rogers, C. D. F., and Smalley, I. J. (1998). Silt. *Earth Sci. Rev.* **45,** 61–89.

ASTM (1959). "Symposium on Particle Size Measurement," Spec. Tech. Publ. No. 243. Am. Soc. Test. Mater., Philadelphia. 303pp.

Atkinson, B. K. (1987). "Fracture Mechanics of Rocks." Academic Press, London. 600pp.

Bagnold, R. A., and Barndorff-Nielsen, O. (1980). The pattern of natural size distribution. *Sedimentology* **27,** 199–207.

Bandyopadhyay, S. (1971). Pebble orientation in relation to cross-stratification: A statistical study. *J. Sediment. Petrol.* **41**, 585–587.

Bateman, R. M. (1995). "Open-Hole Log Analysis and Formation Evaluation." IHRDC, Boston.

Bathurst, R. G. C. (1982). Genesis of stromatactis cavities between submarine crusts in Palaeozoic carbonate mud buildups. *Q. J. Geol. Soc. London* **139**, 165–181.

Beard, D. C., and Weyl, P. K. (1973). The influence of texture on porosity and permeability on unconsolidated sand. *Am. Assoc. Pet. Geol. Bull.* **57**, 349–369.

Briggs, L. I., McCulloch, D. S., and Moser, F. (1962). The hydraulic shape of sand particles. *J. Sediment. Petrol.* **32**, 645–656.

Cailleux, A., and Tricart, J. (1959). "Initiation à l'Etude des Sables et des Galets." Centre de Documentation, University of Paris.

Chambers, R. L., and Upchurch, S. B. (1979). Multivariate analysis of sedimentary environment using grain size frequency distributions. *J. Math. Geol.* **11**, 27–43.

Chappell, J. (1967). Recognizing fossil strand lines from grain-size analysis. *J. Sediment. Petrol.* **3**, 157–165.

Chierici, G. L. (1995). "Principles of Petroleum Engineering," 2 vols. Springer-Verlag, Berlin. 419 and 398 pp.

Choquette, P. W., and Pray, L. C. (1970). Geologic nomenclature and classification of porosity in sedimentary carbonates. *Am. Assoc. Pet. Geol. Bull.* **54**, 207–250.

Conley, C. D. (1971). Stratigraphy and lithofacies of Lower Paleocene rocks, Sirte Basin, Libya. *In* "Symposium on the Geology of Libya" (C. Gray, ed.), pp. 127–140. University of Libya, Tripoli.

Curray, J. R. (1956). Dimensional grain orientation studies of recent coastal dunes. *Bull. Am. Assoc. Pet. Geol.* **40**, 2440–2456.

Darcy, H. (1856). "Les Fontaines Publiques de la Ville de Dijon." Dalmont, Paris. 674pp.

Doeglas, D. J. (1946). Interpretation of the results of mechanical analyses. *J. Sediment. Petrol.* **16**, 19–40.

Doyle, H. (1995). "Seismology." Wiley, Chichester. 243pp.

Emery, J. R., and Griffiths, J. C. (1954). Reconnaissance investigation into relationships between behaviour and petrographic properties of some Mississippian sediments. *Bull. Earth Miner. Sci. Exp. Stn., Pa. State Univ.* **62**, 67–80.

Engelhardt, W. V., and Pitter, H. (1951). Uber die Zusammenhange zwischen Porosilat, Permeabilitat und Korngrosse bei Sanden und Sandsteinen. *Heidelb. Beitr. Mineral. Petrogr.* **2**, 477–491.

Erlich, R. (1983). Size analysis wears no clothes, or have moments come and gone. *J. Sediment. Petrol.* **53**, 1–31.

Fischer, A. G. (1964). The Lofer cyclothems of the Alpine Triassic. Symposium on Cyclic Sedimentation. *Bull. — Kans., State Geol. Surv.* **169**, 107–150.

Folk, R. L. (1966). A review of grainsize parameters. *Sedimentology* **6**, 73–94.

Folk, R. L., and Ward, W. C. (1957). A study in the significance of grain size parameters. *J. Sediment. Petrol.* **27**, 3–26.

Fraser, H. J. (1935). Experimental study of the porosity and permeability of clastic sediments. *J. Geol.* **43**, 910–1010.

Friedman, G. M. (1958). Determination of sieve-size distribution from thin section data for sedimentary petrological studies. *J. Geol.* **66**, 394–416.

Friedman, G. M. (1961). Distinction between dune, beach and river sands from their textural characteristics. *J. Sediment. Petrol.* **31**, 514–529.

Friedman, G. M. (1962). Comparison of moment measures for sieving and thin section data in sedimentary petrologic studies. *J. Sediment. Petrol.* **32**, 15–25.

Friedman, G. M. (1967). Dynamic processes and statistical parameters compared for size frequency distribution of beach and river sands. *J. Sediment. Petrol.* **37**, 327–354.

Gale, S. J. (1990). The shape of beach gravels. *J. Sediment. Petrol.* **60**, 787–789.

Gnaccolini, M., and Orombelli, G. (1971). Orientazione dei ciottoli in un delta lacustre Pleistocenico della Brianza. *Riv. Ital. Paleontol. Stratigr.* **77**, 411–424.

Graton, L. C., and Fraser, H. J. (1935). Systematic packing of spheres, with particular reference to porosity and permeability. *J. Geol.* **43**, 785–909.

Hand, B. M. (1967). Differentiation of beach and dune sands, using settling velocities of light and heavy minerals. *J. Sediment. Petrol.* **37,** 514–520.

Harrell, J. (1984). Roller micrometer analysis of grain shape. *J. Sediment. Petrol.* **54,** 643–645.

Harris, D. G., and Hewitt, C. H. (1977). Synergism in reservoir management — the geological perspective. *JPT, J. Pet. Technol.* **29,** 761–775.

Hiscott, R. N., and Middleton, G. V. (1980). Fabric of coarse deep-water sandstones: Tourelle Formation, Quebec, Canada. *J. Sediment. Petrol.* **50,** 703–722.

Hurst, A., and Goggin, D. (1995). Probe permeametry: An overview and bibliography. *AAPG Bull.* **79,** 463–473.

Hurst, A., and Rosvoll, K. J. (1991). Permeability in sandstones and their relationship to sedimentary structures. *In* "Reservoir Characterization" (L. W. Lake, H. B. Carroll, and T. C. Weston, eds.), vol. 2, pp. 166–196. Academic Press, San Diego, CA.

Illenberger, W. K. (1991). Pebble shape (and size). *J. Sediment. Petrol.* **61,** 756–767.

Illing, L. V. (1954). Bahaman calcareous sands. *Bull. Am. Assoc. Pet. Geol.* **38,** 1–45.

Inman, D. L. (1952). Measures for describing the size distribution of sediments. *J. Sediment. Petrol.* **22,** 125–145.

Johansson, C. E. (1965). Structural studies of sedimentary deposits. *Geol. Foeren. Stockholm Foerh.* **87,** 3–61.

Kahn, J. S. (1956). The analysis and distribution of the properties of packing in sand size sediments. *J. Geol.* **64,** 385–395.

Klovan, J. E. (1966). The use of factor analysis in determining depositional environments from grain size distributions. *J. Sediment. Petrol.* **36,** 115–125.

Krinsley, D. H. (1998). "Backscattered Scanning Electron Microscopy and Image Analysis of Sediments and Sedimentary Rocks." Cambridge University Press, Cambridge, UK. 324pp.

Krinsley, D. H., and Trusty, P. (1986). Sand grain surface textures. *In* "The Scientific Study of Flint and Chert" (G. de G. Sieveking and M. B. Hart, eds.), pp. 201–207. Cambridge University Press, Cambridge, UK.

Krumbein, W. C. (1934). Size frequency distributions of sediments. *J. Sediment. Petrol.* **4,** 65–77.

Krumbein, W. C., and Monk, G. D. (1942). Permeability as a function of the size parameters of unconsolidated sands. *Am. Inst. Min. Metall. Eng.* **1492,** 1–11.

Krumbein, W. C., and Pettijohn, F. J. (1938). "Manual of Sedimentary Petrography." Appleton-Century-Crofts, New York. 549pp.

Kuenen, P. H. (1956a). Experimental abrasion of pebbles 1: Wet sand blasting. *Leidse Geol. Meded.* **20,** 131–137.

Kuenen, P. H. (1956b). Experimental abrasion of pebbles 2: Rolling by current. *J. Geol.* **64,** 336–368.

Kuenen, P. H. (1959). Experimental abrasion 3: Fluviatile action. *Am. J. Sci.* **257,** 172–190.

Kuenen, P. H. (1960). Experimental abrasion 4: Eolian action. *J. Geol.* **68,** 427–449.

Kuenen, P. H., and Perdok, W. G. (1962). Experimental abrasion 5: Frosting and defrosting of quartz grains. *J. Geol.* **70,** 648–659.

Laming, D. J. C. (1966). Imbrication, paleocurrents and other sedimentary features in the Lower New Red Sandstone, Devonshire, England. *J. Sediment. Petrol.* **36,** 940–959.

Le Blanc, R. (1977). Distribution and continuity of sandstone reservoirs. Parts I and 11. *JPT, J. Pet. Technol.* **29,** 776–792 and 793–804.

Levorsen, A. I. (1967). "The Geology of Petroleum." Freeman, New York. 724pp.

Lewis, D. W., and McConchie, D. (1994). "Analytical Sedimentology." Chapman & Hall. London. 197pp.

Lindsay, J. F. (1966). Carboniferous subaqueous mass-movement in the Manning Macleay basin, Kempsey, New South Wales. *J. Sediment. Petrol.* **36,** 719–732.

Lundegard, P. D. (1992). Sandstone porosity loss — a "big picture" view of the importance of compaction. *J. Sediment. Petrol.* **39,** 12–17.

Margolis, S. V., and Krinsley, D. H. (1971). Submicroscopic frosting on eolian and subaqueous quartz sand grains. *Geol. Soc. Am. Bull.* **82,** 3395–3406.

Martini, I. P. (1971). Grainsize orientation and paleocurrent systems in the Thorold and Grimsby sandstones (Silurian), Ontario and New York. *J. Sediment. Petrol.* **41,** 225–234.

Mason, C. C., and Folk, R. L. (1958). Differentiation of beach dune and eolian flat environments by size analysis; Mustang Island Texas. *J. Sediment. Petrol.* **18,** 211–226.

McConnell, P. C. (1951). Drilling and production techniques that yield nearly 850,000 barrels per day in Saudi Arabia's fabulous Abqaiq field. *Oil Gas. J.,* December 20, p. 197.

McLaren, P., and Bowles, D. (1985). The effects of sediment transport on grain-size distributions. *J. Sediment. Petrol.* **55,** 457–470.

Mellon, G. B. (1964). Discriminatory analysis of calcite and silicate-cemented phases of the Mountain Park sandstone. *J. Geol.* **72,** 786–809.

Miall, A. D. (1970). Devonian alluvial fans, Prince of Wales Island, Arctic Canada. *J. Sediment. Petrol.* **40,** 556.

Moiola, R. J., and Weiser, D. (1968). Textural parameters: An evaluation. *J. Sediment. Petrol.* **38,** 45–53.

Monty, C. L. V., Bosence, D. W. J., Bridges, P. H., and Pratt, B. R., eds. (1995). "Carbonate Mud Mounds." Blackwell, Oxford. 544pp.

Morrow, N. R. (1971). Small scale packing heterogeneities in porous sedimentary rocks. *Am. Assoc. Pet. Geol. Bull.* **55,** 514–522.

Murray, R. C. (1960). Origin of porosity in carbonate rocks. *J. Sediment. Petrol.* **30,** 59–84.

Nagtegaal, P. J. C. (1978). Sandstone-framework instability as a function of burial diagenesis. *J. Geol. Soc. London* **135,** 101–106.

Nelson, R. A. (1986). "Geologic Analysis of Naturally Fractured Reservoirs." Gulf Publishing, Houston, TX. 320pp.

Passega, R. (1957). Texture as characteristic of clastic deposition. *Bull. Am. Assoc. Pet. Geol.* **41,** 1952–1984.

Passega, R. (1964). Grainsize representation by C. M. Patterns as a geological tool. *J. Sediment. Petrol.* **34,** 830–847.

Plumley, W. J. (1948). Blackhill terrace gravels: A study in sediment transport. *J. Geol.* **56,** 526–577.

Potter, P. E., and Pettijohn, F. J. (1977). "Paleocurrents and Basin Analysis," 2nd ed. Springer-Verlag, Berlin. 425pp.

Powers, M. C. (1953). A new roundness scale for sedimentary particles. *J. Sediment. Petrol.* **23,** 117–119.

Pryor, W. A. (1973). Permeability-porosity patterns and variations in some Holocene sand bodies. *Am. Assoc. Pet. Geol. Bull.* **57,** 162–189.

Pryor, W. A., and Fulton, K. (1976). Geometry of reservoir type sand bodies in the Holocene Rio Grande delta and comparison with ancient reservoir analogs. *Soc. Pet. Eng. Prepr.* **7045,** 81–92.

Reiss, L. H. (1981). "Reservoir Engineering Aspects of Fractured Formations." Gulf Publishing, Houston, TX. 112pp.

Robinson, R. B. (1966). Classification of reservoir rocks by surface texture. *Bull. Am. Assoc. Pet. Geol.* **50,** 547–559.

Rogers, J. J., and Head, W. B. (1961). Relationship between porosity median size, and sorting coefficients of synthetic sands. *J. Sediment. Petrol.* **31,** 467–470.

Russell, R. D., and Taylor, R. E. (1937). Roundness and shape of Mississippi River sands. *J. Geol.* **45,** 225–267.

Sames, C. W. (1966). Morphometric data of some recent pebble associations and their application to ancient deposits. *J. Sediment. Petrol.* **36,** 126–142.

Schafer, A., and Teyssen, T. (1987). Size, shape and orientation in sands and sandstones-image analysis applied to rocks in thin sections. *Sediment. Geol.* **52,** 251–272.

Schlee, J. (1957). Fluvial gravel fabric. *J. Sediment. Petrol.* **27,** 162–176.

Selley, R. C. (1998). "Elements of Petroleum Geology," 2nd ed. Academic Press, San Diego, CA. 470pp.

Sengupta, S. (1966). Studies on orientation and imbrication of pebbles with respect to cross stratification. *J. Sediment. Petrol.* **36,** 362–369.

Shelton, J. W., and Mack, D. E. (1970). Grain orientation in determination of paleocurrents and sandstone trends. *Am. Assoc. Pet. Geol. Bull.* **54,** 1108–1119.

Sheriff, R. E., and Geldert, L. P. (1995). "Exploration Seismology," 2nd ed. Cambridge University Press, Cambridge, UK. 450pp.

Shinn, E. A. (1983). Birdseyes, fenestrae, shrinkage pores and loferites: A reevaluation. *J. Sediment. Petrol.* **53,** 619–628.

Sippel, R. F. (1971). Quartz grain orientations—(the photometric method). *J. Sediment. Petrol.* **41,** 38–59.

Sneed, E. D., and Folk, R. L. (1958). Pebbles in the Lower Colorado River, Texas, a study in particle morphogenesis. *J. Geol.* **66,** 114–150.

Spencer, D. W. (1963). The interpretation of grain size distribution curves of clastic sediments. *J. Sediment. Petrol.* **33,** 180–190.

Stormont, D. H. (1949). Huge caverns encountered in Dollarhide Field. *Oil Gas J.,* April 7, pp. 66–68.

Tebbutt, G. E., Conley, C. D., and Boyd, D. W. (1965). Lithogenesis of a distinctive carbonate fabric. *Contrib. Geol.* **4,** No. 1, 1–13.

Tough, J. G., and Miles, R. G. (1984). A method for characterizing polygons. *Compu. Geosci.* **10,** 347–350.

Trask, P. D. (1930). Mechanical analysis of sediment by centrifuge. *Econ. Geol.* **25,** 581–599.

Tucker, M., ed. (1988). "Techniques in Sedimentology." Blackwell, Oxford. 408pp.

Van Golfracht, T. D. (1982). "Fundamentals of Fractured Reservoir Engineering." Elsevier, Amsterdam. 710pp.

Van Veen, F. R. (1975). Geology of the Leman Gas Field. *In* "Petroleum and the Continental Shelf of North West Europe" (A. E. Woodland, ed.), Vol. I, pp. 223–232. Applied Science Publishers, Barking, England.

Vincent, P. (1986). Differentiation of modern beach and coastal dune sands—a logistic regression approach using the parameters of the hyperbolic function. *Sediment. Geol.* **49,** 167–176.

Visher, G. S. (1965). Fluvial processes as interpreted from Ancient and Recent Fluvial deposits. *Spec. Publ.—Soc. Econ. Paleont. Mineral.* **12,** 116–132.

Vita-Finzi, C. (1971). Heredity and environment in clastic sediments: Silt/clay depletion. *Geol. Soc. Am. Bull.* **82,** 187–190.

von Rad, U. (1971). Comparison between "magnetic" and sedimentary fabric in graded and cross-laminated sand layers, southern California. *Geol. Rundsch.* **60,** 331–354.

Wadell, H. (1932). Volume, shape and roundness of rock particles. *J. Geol.* **40,** 443–451.

Wardlaw, N. C. (1976). Pore geometry of carbonate rocks as revealed by pore casts and capillary pressure. *AAPG Bull.* **60,** 254–257.

Watson, H. V. (1982). Casablanca Field offshore Spain, a paleogeomorphic trap. *Mem.—Am. Assoc. Pet. Geol.* **32,** 237–250.

Weber, K. J. (1982). Influence of common sedimentary structures on fluid flow in reservoir models. *J. Pet. Technol.* **34,** 665–672.

Wentworth, C. K. (1919). A laboratory and field study of cobble abrasion. *J. Geol.* **27,** 507–521.

Wentworth, C. K. (1922). A scale of grade class terms for clastic sediments. *J. Geol.* **30,** 377–392.

Whalley, W. B. (1972). The description and measurement of sedimentary particles and the concept of form. *J. Sediment. Petrol.* **42,** 961–965.

White, J. R., and Williams, G. E. (1967). The nature of the fluvial process as defined by settling velocities of heavy and light minerals. *J. Sediment. Petrol.* **37,** 530–539.

Worthington, P. F. (1991). Reservoir characterization at the Mesoscopic scale. *In* "Reservoir Characterization" (L. W. Lake, H. B. Carroll, and T. C. Weston, eds.), vol. 2, pp. 123–165. Academic Press, San Diego, CA.

Wycoff, R. D., Botset, H. G., Muskat, M., and Reed, D. W. (1934). Measurement of permeability of porous media. *Bull. Am. Assoc. Pet. Geol.* **18,** 161–190.

Wyllie, M. R. J., Gregory, A. R., and Gardner, L. W. (1956). Elastic wave velocities in heterogeneous and porous media. *Geophysics* **21,** 41–70.

Zingg, T. (1935). Beitrage zur Schotteranalyse. *Schweiz. Mineral. Petrogr. Mitt.* **15,** 39–140.

Part II

Sediment
Sedimented

Whenever a running stream charged with mud or sand has its velocity checked, as when it enters a lake or sea, or overflows a plain, the sediment, previously held in suspension by the motion of the water, sinks, by its own gravity, to the bottom.

Charles Lyell (1838)

4 Transportation and Sedimentation

4.1 INTRODUCTION

The transportation and deposition of sediments are governed by the laws of physics. The behavior of granular solids in fluids has been extensively studied by physicists and by engineers of various types. Much of this work has been documented and will be found in texts on hydraulics and fluid dynamics (Dailey and Harleman, 1966; Henderson, 1966; Leliavsky, 1959). Accounts of the physical processes of sedimentation seen from a geological standpoint have been given by Bagnold (1966, 1979), J. R. L. Allen (1970, 1985a), P. A. Allen (1997), and Pye (1994).

This chapter introduces some of the fundamental concepts of sedimentation as a means to understanding the fabric and structures of the deposits which they generate. **Sedimentation** is, literally, the settling out of solid matter in a liquid. To the geologist, however, sedimentary processes are generally understood as those which both transport and deposit sediment. They include the work of water, wind, ice, and gravity. The physics of granular solids in fluids is described here. This is followed by accounts of sediment transport and deposition by these four processes.

Matter occurs in three phases: solid, liquid, and gaseous. The physicist considers gases and liquids together as fluids, on the grounds that, unlike solids, they both lack shear strength. The behavior of granular solids in liquids and gases is comparable. The similarity of the bed forms and structures of windblown and water-laid deposits are the root problem of differentiating them in sedimentary rocks. The starting point of an analysis of sediment transport and deposition is Stokes' law. This was introduced when discussing the use of settling velocity as a way of measuring grain size (Chapter 3). Recall that Stokes' law states:

$$W = \left[\frac{(P_1 - P)g}{18\mu} \right] d^2,$$

where W is the settling velocity, $(P_1 - P)$ is the density difference between the particle and the fluid, g is the acceleration due to gravity, μ is the fluid viscosity, and d is the particle diameter. A derivation of Stokes' law is found in Allen (1985a). Stokes' law states that the settling velocity of a particle is related to its diameter, and to the difference between the particle density and the density of the ambient fluid. Considered at its simplest a small particle will settle faster than a larger one of equal density. Conversely, of two particles of equal diameter but different densities, the denser one will settle first.

Stokes' law is only valid, however, for a single sphere. In the real world, settling velocity also varies according to grain shape and to grain concentration, since sedimentation rate will be affected by adjacent particles colliding. Few sediment grains are perfect spheres. Quartz and feldspar particles are normally ovoid, micas are plate-like, and skeletal fragments highly irregular. The idea of sedimentation-equivalent particles has thus been developed to take into account grain shape. Hydraulically equivalent particles settle at the same velocity in water. Aerodynamically equivalent particles settle at the same velocity in air (Friedman, 1961; Hand, 1967). Detrital minerals have a wide range of densities. Terrigenous sands are largely made up of quartz with a density of 2.65 g/cm^3. But they may also contain feldspars, ranging between 2.55 and 2.76 g/cm^3, and micas, ranging from 2.83 (muscovite) to 3.12 g/cm^3 (biotite). Most sands also contain varying amounts of heavy minerals, arbitrarily defined as those with a density greater than 3.0 g/cm^3. These include many economically important minerals such as gold, with a density of 19 g/cm^3. When segregated these valuable heavy minerals are the placer ores. Thus much work has been done to try to understand the processes that segregate sand particles of different densities (see MacDonald, 1983, for a detailed account). Figure 4.1 shows how small heavy mineral particles have the same settling velocities as larger quartz grains. This is why many placers are deposited in high energy environments, with the ore mineral grains disseminated with quartz gravels — the Witwatersrand reef is a case in point (see Section 6.3.2.2.4).

From considering the purely static situation of sediment settling in motionless fluid it is appropriate to consider how particles will behave when the fluid is moving. The

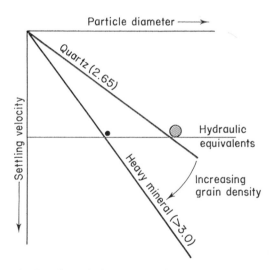

Fig. 4.1. Diagrammatic graph of settling velocity against grain diameter. This shows how a small, heavy mineral grain may have the same settling velocity as a much larger quartz particle. This assumes similar shapes. Mica, though denser than quartz, has a slower settling velocity because of its flaky shape.

Laminar flow, low Reynolds number

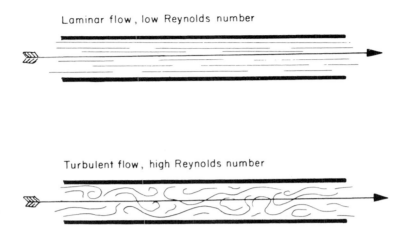

Turbulent flow, high Reynolds number

Fig. 4.2. Illustrative of the difference between laminar and turbulent flow in tubes.

physics of this situation is expressed by the Reynolds equation, from which is derived from a dimensionless coefficient. the **Reynolds number,** thus:

$$R = \frac{Udp}{\mu},$$

where R is the Reynolds number, U is the velocity of the particle, d is the diameter of the particle, p is the density of the particle, and μ is the viscosity of the fluid. For a given situation the Reynolds number can be used to differentiate two different types of fluid behavior at the solid boundary, be it a sphere or a confining surface such as a tube or channel wall. For the low Reynolds numbers the fluid flow is laminar, flow lines running parallel to the boundary surface; for the high Reynolds numbers the flow is turbulent, generating eddies and vortices (Fig. 4.2). For flow in tubes the critical Reynolds number separating laminar and turbulent flow is about 2000. For a particle in a fluid the critical number is about 1. Stokes' law of settling, discussed in Chapter 3, is derived from the Reynolds equation. Note that turbulence is proportional to velocity, but inversely proportional to viscosity.

A second important coefficient of fluid dynamics is the **Froude number.** This is essentially the ratio between the force required to stop a moving particle and the force of gravity; that is, the ratio of the force of inertia and the acceleration due to gravity. Hence:

$$F = \frac{U}{\sqrt{gL}},$$

where U is the velocity of the particle, L is the force of inertia (i.e., the length traveled by the particle before it comes to rest), and g is the acceleration due to gravity.

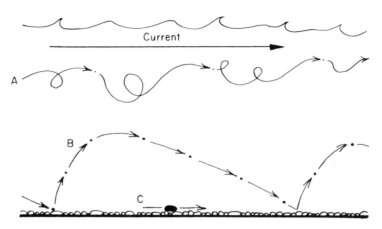

Fig. 4.3. The mechanics of particle movement. (A) Suspension. (B) Bouncing (saltation). (C) Rolling.

For flow in open channels the Froude number is expressed thus:

$$F = \frac{U}{\sqrt{gD}},$$

where D is the depth of the channel and U is the average current velocity.

A Froude number of 1 separates two distinct types of fluid flow in open channels. Each flow regime generates specific bed forms and sediment structures. These are described in detail in the section on aqueous traction sedimentation.

Consider now the mechanics of particle movement. Essentially a grain can move through a fluid (liquid or gaseous) in three different ways: by rolling, by bouncing, or in suspension (Fig. 4.3). In a given situation, the heaviest particles are never lifted from the ground. They remain in contact with their colleagues, but are rolled along by the current. At the same velocity, lighter particles move downcurrent with steep upward trajectories and gentler downward glide paths. This process is known as **saltation.** At the same velocity the lightest particles are borne along by the current in suspension. They are carried within the fluid in erratic but essentially downflow paths never touching the bottom or ground.

In a situation such as a river channel, therefore, gravel will be rolling along the bottom, sand will sedately saltate, and silt and clay will be carried in suspension. Sand and gravel are generally referred to as the traction carpet or the channel bed load. The silt and clay, loosely termed "fines," are referred to as the suspended load. Considerable importance is attached to the critical flow velocity needed to start a particle into motion. The critical flow velocity for a particle is a function of the variables contained in the Froude and Reynolds equations. A number of empirical, experimental, and theoretical studies have been made to determine the critical flow velocity for varying sediment grades, notably by Shields (1936), Vanoni (1964), and Hjulstrom (in Sundborg, 1956). Of these the latter is the better known (Fig. 4.4). As one might expect, the critical fluid flow increases with grain size. An exception to this rule is noted for cohesive clay bottoms. Because of their resistance to friction, they need rather higher velocities to erode

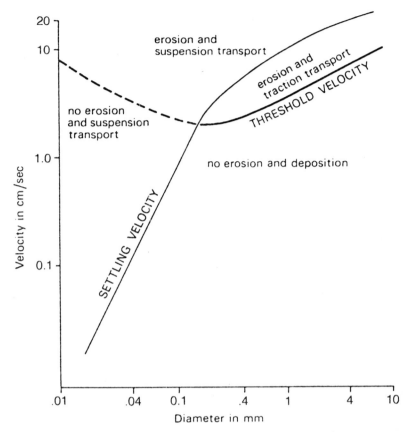

Fig. 4.4. Graph showing the critical velocities required to erode, transport, and deposit sediments of varying grades. (After Orme, 1977. Copyright © 1977 Academic Press.)

them and commence movement than for silt and very fine sand. This anomaly, termed the "Hjulstrom effect," is responsible for the preservation of delicate clay laminae in tidal deposits.

Consider now the sediment types that result from different sorts of fluid flow in both liquid and gaseous states (i.e., in water and air) for the purpose of the geologist. Three types can be recognized: traction deposits, density current deposits, and suspension deposits. Transportation in a traction current is mainly by rolling and saltating bed load. The fabric and structure of sediment deposited from a traction carpet reflect this manner of transport. They are generally cross-bedded sands. Traction currents may be generated by gravity (as for example in a river), or by wind or tidal forces in the sea. Desert sand dunes are also traction deposits.

The deposits of density currents, by contrast, originate from a combination of traction and suspension. Their fabric and structures are correspondingly different from those of traction deposits. They are characterized by mixtures of sand, silt, and clay, which lack cross-bedding and typically show graded bedding. Density currents are caused by differences in density in fluids both liquid or gaseous. These differences may arise from

Table 4.1
Tabulation of Sedimentary Processes and Sediment Types Which They Generate

Subaerial	Traction deposits	Predominantly cross-bedded sands
	Density deposits	*Nuées ardentes,* etc.
	Suspension deposits	Loess
Subaqueous	Traction deposits	Predominantly cross-bedded sands
	Density (turbidity) deposits	Graded sands, silts and clays
	Suspension deposits	Nepheloid clays
Mass gravity transport	Subaerial	Generally unstratified poorly sorted deposits of boulder to clay grade (diamictites)
	Subaqueous	
Glacial transport		

thermal layering, turbidity, or from differences in salinity in liquids. The result is for the denser fluid to flow by gravity beneath the less dense fluid and to traverse the sediment substrate. Geologically the most important density flow is the turbidity current, a predominantly subaqueous phenomenon. Eolian turbid flows include *nuées ardentes* and certain types of high-velocity avalanches and mud flows. They are rare and not volumetrically significant depositional mechanisms. The fundamental differentiation of many sediments into cross-bedded traction-current deposits and graded turbidites was recognized by Bailey in 1930.

The third group of sedimentary deposits includes those that settle out from suspension. These are fine-grained silts and clays and include windblown silts, termed loess, and the pelagic detrital muds or nepheloids of ocean basins.

A fourth major group of sediment types is the **diamictites** (Flint *et al.,* 1960). These are extremely poorly sorted rocks that show a complete range of grain size from boulders down to clay. Diamictites are formed both from glacial processes and also by mud flows, both subaerial and subaqueous. Table 4.1 summarizes the main sedimentary processes and shows how these generate different depositional textures. The various processes are now described in more detail.

4.2 AQUEOUS PROCESSES

4.2.1 Sedimentation from Traction Currents

Consider now one of the most important processes for transporting and depositing sediment. Traction currents are those which, as already defined, move sediment along by rolling and saltation as bed load in a traction carpet. Continuous reworking winnows out the silt and clay particles which are carried further downcurrent in suspension. Finer, lighter sand grains are transported faster than larger, heavier ones. The sediments of a unidirectional traction current thus tend to show a downstream decrease in grain size, termed "size grading." Traction currents may be unidirectional, as in river channels. In estuaries, and in the open sea, however, sediment may be subjected to the to-and-fro action of tidal traction currents or to even more complex systems.

4.2.1.1 Unidirectional Traction Currents

The basic approach to understanding traction-current sedimentation has been through experimental studies of unidirectional flow in confined channels. These can be made in artificial channels, termed flumes, which are now standard equipment in many geological laboratories. The phenomenon now to be described has been documented by Simons *et al.* (1965) and is demonstrated as an integral part of many university courses on sedimentology. The experiment commences with the flume at rest. Sand on the flume floor is flat, current velocity is zero. Water is allowed to move down the flume at a gradually increasing velocity. The sand grains begin to roll and saltate as the critical threshold velocity is crossed. The sand is sculpted into a rippled bed form. Steep slopes face downcurrent, gentle back slopes face upcurrent. Sand grains are eroded from the back slopes, transported over the ripple crests to be deposited on the downstream slope or slip face. Thus the ripples slowly move downstream depositing cross-laminated sand.

With increasing current velocity the bed form changes from ripples to dunes. These are similar to ripples in their shape, mode of migration, and internal structure. They differ in scale, however, being measurable in decimeters rather than in centimeters. After flash floods ephemeral rivers may drain to expose megaripples (dunes) with ripples on their back slopes (Fig. 4.5). Through these phases of ripple and dune formation, the Froude number, introduced in the previous section, is less than 1. With increasing velocity the Froude number approaches 1. This value separates two **flow regimes.** The lower flow regime, with a Froude number of less than 1, generates cross-laminated and cross-bedded sand from ripples and dunes.

As the velocity increases to a Froude number of 1 the dunes are smoothed out and

Fig. 4.5. An ephemeral fluvial channel after a flash flood. Ashquelon, Israel. Looking upstream one can see several megaripples (dunes), with slip faces on their fronts, and ripples on their back sides.

Fig. 4.6. Flat-bedded sands indicative of deposition from a plane bed during shooting flow conditions. Cobbles to right of hammer indicate that the current velocity was far in excess of the velocity required to transport sand. Gargaf Group (Cambro-Ordovician), Jebel Mourizidie, Libyan/Chad border.

the bed form assumes a planar surface. This stage is termed **shooting flow.** Sand is still being transported. Now, however, sand may be deposited in horizontal beds with the grains aligned parallel to the current. Rarely, flat bedded sands contain much larger clasts. This feature proves that the current velocity was far greater than that required to transport the sand particles. It also exceeded the threshold velocity for cobble transportation (Fig. 4.6).

As the current velocity increases further, the Froude number exceeds unity. The plane bed form changes into rounded mounds, termed "antidunes." In contrast to dunes, these tend to be symmetric in cross-section. They may be stable or they may migrate upcurrent to deposit upcurrent dipping cross-beds. Gravel antidunes formed during a flash flood have been described from the modern Burdekin River of Queensland, Australia. These had a wavelength of 19 m and an amplitude of 1 m. Calculations show that the flood discharge was 25,600 m^3 s^{-1} (Alexander and Fielding, 1997).

With decreasing current velocity, the sequence of bed forms is reversed. Antidunes wash out to a plane bed, then dunes form, then ripples, and so back to plane bed and quiescence at zero current velocity (Fig. 4.7).

This experiment elegantly demonstrates the relationship between stream power, bed form, and sedimentary structures. It is, however, difficult to quantify this relationship and to enumerate flow parameters from the study of grain size and sedimentary structures in ancient deposits (Allen, 1985a,b). Experiments have shown that if the fluid parameters are held constant (i.e., velocity and viscosity) then the thresholds at which one bed form changes to another varies with grain size. Most important of all, it is an experimental observation that the ripple phase is absent in sediments with a fall diameter of more than about 0.65 mm. (The fall diameter is a function of the particle diameter

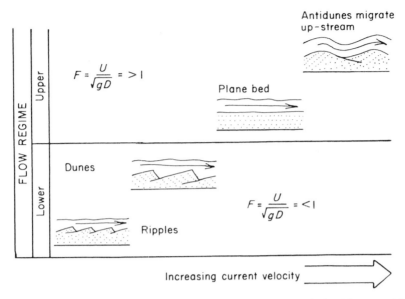

Fig. 4.7. Bed forms and sedimentary structures for different flow regimes. (After Harms and Fahnestock, 1965, and Simons *et al.*, 1965. Courtesy of the Society for Sedimentary Geology.)

and the viscosity of the fluid. The fall diameter of a particle decreases with increasing viscosity. For example, one sand grain may have the same fall diameter as a larger particle in a more viscous fluid.) It is a matter of field observation that cross-lamination is absent in sediments with a particle diameter of over about 0.5 mm.

A second important point to note is the way in which temperature affects sediment structures. Harms and Fahnestock (1965), in their study of reaches of the Rio Grande, showed how, for similar discharges, either plane bed or dunes could be present. The main controlling variable seemed to be temperature. This controlled the fluid viscocity and hence the fall diameter of the sediment.

Figure 4.8 shows the relationship between stream power (current velocity, more or less), fall diameter (grain size, more or less), bed form, and sedimentary structures. Figure 4.9 shows the relationship between grain size and sedimentary structures in a typical ancient fluvial deposit laid down by unidirectional traction currents.

4.2.1.2 Bidirectional Tractional Currents

Unidirectional currents characterize deposition in fluvial channels. In marine environments, however, traction currents are commonly bidirectional (Fleming and Bartholoma, 1995; Black *et al.*, 1998). The periodicity of the current is very variable, ranging from tidal cycles of many hours duration, down to the split second passing of a wave. These are now considered in turn.

A time-velocity graph for a single tidal cycle may be plotted as shown in Fig. 4.10. At high and low tide the current velocity will be zero. Current velocity gradually increases and then decreases as the tide ebbs, and gradually increases and decreases as the tide floods, and so on. As current velocity approaches zero at high and low tide

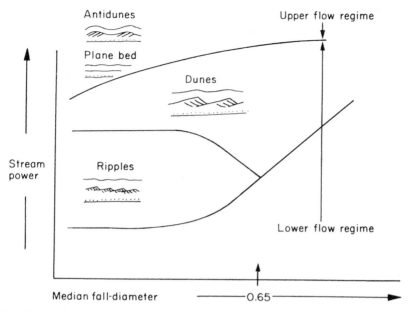

Fig. 4.8. The relationship between stream power, fall diameter, bed form, and sedimentary structure in a uni-directional traction current system. (After Simons *et al.*, 1965. Courtesy of the Society for Sedimentary Geology.)

clay may settle out of suspension. The Hjulstrom effect (see Section 4.1) may allow these clay laminae to be preserved from erosion during the subsequent tidal cycle. Thus clay drapes on ripples and foresets may be indices of past tidal activity. Attempts have even been made to identify tidal cycles in ancient sediments (Visser, 1980; Allen, 1981; Fleming and Bartholoma, 1995). Sand will be transported when the ebb and flood currents are running sufficiently strongly. During these high-velocity intervals, bed forms

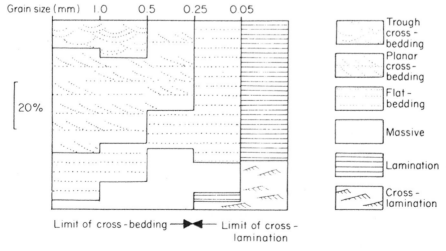

Fig. 4.9. Relationship between grain size and sedimentary structure in alluvial sandstones of the Torridon Group (PreCambrian). Scotland. (Data from Selley, 1966. Courtesy of the Geologists' Association.)

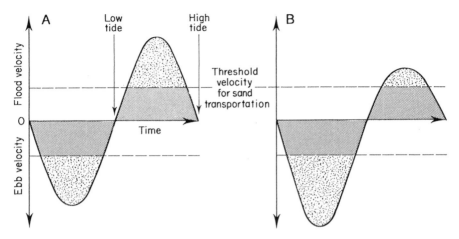

Fig. 4.10. Velocity time graphs for tidal currents. (A) For a symmetric tidal cycle, sand is reworked to and fro without a net transport direction. (B) In the asymmetric cycle, sand is transported in the ebb direction.

and sedimentary structures develop that conform to the flow regime scheme previously outlined.

If the ebb and flood currents are equal in velocity and duration the tidal cycle is symmetrical. Sand will be transported to and fro to deposit herringbone cross-bedding (see Section 5.3.3.4), but there may be no net sediment transport. This is unusual. Normally the tidal cycle is asymmetric, for one of a number of reasons. In a river estuary the river discharge is likely to make the ebb current stronger than the flood (Fig. 4.11). Out at

Fig. 4.11. Tidal estuary, showing megaripples with ebb current (right to left) slip faces. Wells-next-the-Sea, England.

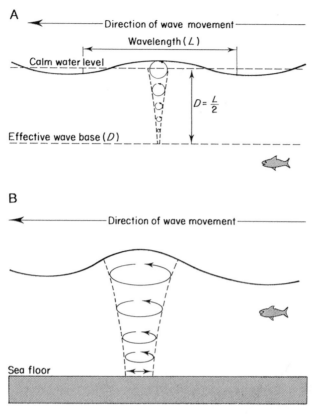

Fig. 4.12. Diagrams showing (A) the orbital motion of open waves and (B) the ellipsoidal motion of shoaling waves.

sea wind-generated currents may also make a tidal cycle asymmetric. Thus in most cases, though sand is constantly reworked, there is a net direction of sediment transport.

Consider now the much shorter cycle of a moving wave. A particle floating on the surface describes an orbit as the wave passes, but there is no net transport. The orbital motions of particles in the water diminish with increasing depth to a surface below which there is no motion. This surface is termed "effective wave base," and is normally half the wavelength (Fig. 4.12A). Below effective wave base there are no wave-generated currents. Sand cannot be mobilized, unless there are currents of other origins; they may only settle out of suspension. As water depth shoals, the shoreward effective wave base impinges on the sea bed. Orbital motion in the water becomes modified into ellipses as the water shallows (Fig. 4.12B). The ellipsoidal motion transports sand to and fro to form current ripple (sometimes also called vortex ripples). These are generally symmetrical and have a chevron-like internal lamination. Unlike current ripples they occasionally bifurcate when viewed from above. Moving up the beach the ripple profiles become asymmetric with the steep face directed landward, and with landward-dipping cross-lamination (Fig. 4.13). These are termed wave current ripples (Reineck and Singh, 1980, pp. 32–36). It has proved possible to define the relationships between current velocity, grain size, and bed form for shallow waves (Fig. 4.14).

Fig. 4.13. Beach showing wave ripples exposed by the ebbing tide. St. Bees, England.

4.2.2 Sedimentation from High-Density Turbidity Currents

The concept of density flow has already been introduced. Where two fluid bodies of different density are mixed, the less dense fluid will tend to move above the denser one. Conversely, the denser fluid will tend to flow downward. Aqueous density flows may be caused by differences of temperature, salinity, and suspended sediment. Glacial melt streams and certain polar currents tend to flow under gravity beneath warmer, less dense water bodies. Water discharged by rivers in temperate latitudes often flows out for considerable distances from the shore above the denser, more saline seawater. Turbid bodies of water with large loads of suspended sediment frequently move as density flows beneath clear water. This particular variety of density current, termed a turbidity current,

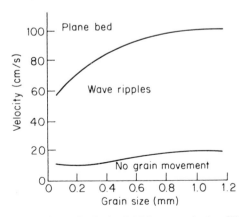

Fig. 4.14. Graph showing the grain size and velocity field for wave ripples. (Simplified from Allen, 1985a.)

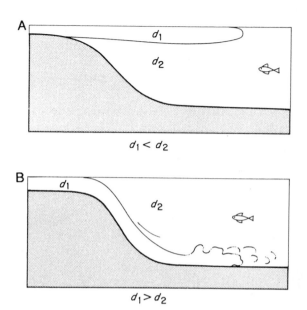

Fig. 4.15. Cross-section through a standing body of water with a channel discharging fluid from the left. (A) When the entering fluid is less dense than that of the standing body of water it will flow out adjacent to the surface. (B) When the situation is reversed, for whatever reason (temperature, salinity or turbidity), it will flow down beneath the less dense fluid. This is a density flow.

is of great interest to geologists (Fig. 4.15). Turbidity currents are widely believed to be a major process for the transportation and deposition of a significant percentage of the world's sedimentary cover. The concept of the turbidity flow was introduced to geology by Bell (1942). The process was originally invoked as an erosional agent capable of scouring the submarine canyons on delta and continental slope margins. Later it was claimed as the generator of what used to be termed flysch deposits (Kuenen and Migliorini, 1950). This facies, typical of geosynclinal troughs — now termed zones of subduction — is characterized by thick sequences of interbedded sand and shale. The sands have abrupt bases and transitional tops, and they tend to fine upward. Sands of this type are often genetically termed "turbidites" (see Sections 5.3.3.3 and 6.3.2.9.1).

This section begins by briefly discussing the hydrodynamics of turbidity flows. The evidence for turbidity currents in the real world is then described, and this is followed by a discussion of the parameters of ancient turbidites. The hydrodynamics of density flows have been studied for a number of years. Attempts to equate the various physical parameters that govern density flows have been based on both theoretical and experimental grounds. The conditions under which a turbidity current will flow down a slope may be expressed thus:

$$S_1 + S_2 = (d_2 - d_1)g.h.a,$$

where S_1 and S_2 are the shear stress between the turbid flow and the floor beneath and the fluid above; d_2 and d_1 are the density of the turbidity flows and of the ambient fluid respectively $(d_2 > d_1)$, g., is the acceleration due to gravity, h., is the height of the flow, and a is the bottom slope. The relationship between the speed of a density flow and other parameters may be expressed as:

$$V = \frac{2(d_2 - d_1)}{d_1} g.h.,$$

where V is the velocity. Derivations and discussions of these equations can be found in Middleton (1966a,b), Allen (1985a), and Baum and Vail (1998). These formulas show that the behavior of a turbid flow is governed by the difference in density between it and the ambient fluid, by the shear stresses of its upper and lower boundary, by its height, and the angle of slope down which it flows. Additional important factors are whether it is a steady flow, such as a turbid river entering a lake, or whether it is a unique event of limited duration as in the case of a liquefied slump.

Turbidity flows have been modeled mathematically by Zeng and Lowe (1997a,b). They have also been studied experimentally in the laboratory and in present-day lakes, seas, and oceans. The early experimental studies of Bell (1942) and Kuenen (1937, 1948) showed that when muddy suspensions of sand were suddenly introduced to a flume they rushed downslope in a turbulent cloud to cover the bottom. Sand settled out first, followed by silt and then clay. Thus beds were deposited with sharp basal contacts that showed an upward-fining grain-size profile from sand to clay in the space of a few centimeters. Additional experiments by Kuenen (1965) produced laminated and rippled graded sands from turbid flows in a circular flume. Experiments by Dzulinski and Walton (1963) showed how small turbidity flows could, in the laboratory, generate many of the erosional features scoured beneath ancient turbidite sands. Experiments such as these have been criticized because particle fall diameter and fluid viscosity were not proportional to each other and to the flow size. Subsequently, however, Middleton (1966a,b, 1967) carried out carefully scaled experiments with plastic beads with a density of 1.52 g/cm^3 and diameters of about 0.18 mm. Suspensions of beads released into a standing body of water generated graded beds similar to those of less scientific experiments.

Turning from the laboratory to the outside world, numerous cases of modern turbidity flows are seen. They have been described from lakes, such as Lake Mead, by Gould (1951) to Norwegian fjords by Holtedahl (1965). In these instances it was possible to demonstrate direct relationships between inflows of muddy river water and extensive layered deposits on the floors of the lakes and fjords. The evidence for the existence of modern marine turbidity flows is equally impressive. A common feature of continental shelves and delta fronts is that they are incised by submarine valleys. Where these terminate at the base of the slope it is common to find a radiating fan-shaped body of sediment. Submarine telegraph cables that cross these regions tend to be broken rather frequently. Daly (1936) postulated that these submarine canyons were eroded by turbidity currents, which snapped the cables. One famous and oft-quoted instance of this was the celebrated Grand Banks earthquake of 1929 (Piper et al., 1999). On 18 November there was an earthquake with an epicenter at the edge of the Grand Banks off Nova Scotia. Within the next few hours, 13 submarine telegraph cables were broken on the slopes and ocean floor at the foot of the Banks. No cables were broken on the Banks themselves.

Subsequently Heezen and Ewing (1952) attributed these breakages to a turbidity current. They postulated that the earthquake triggered slumps on the continental slope of the Grand Banks. These liquefied as they fell and mixed with the seawater until they acquired the physical properties of a turbidity flow. According to the sequence and timing of cable breaks, this flow moved out onto the ocean floor at speeds of up to 100 km/h, ultimately covering an area of some 280,000 km^2. Subsequent coring has revealed an extensive, sharp based, clean graded silt bed over this region.

Many other studies have been published attributing deep-sea sands to turbidity current deposition. These have been described from the Californian Coast (e.g., Hand and Emery, 1964) and the Gulf of Mexico (Conolly and Ewing, 1967) to the Mediterranean (Van Straaten, 1964) and the Antarctic (Payne *et al.*, 1972). Modern bioclastic deep-sea sands have been attributed to turbidite transportation from adjacent carbonate shelves by Rusnak and Nesteroff (1964), Bornhold and Pilkey (1971), and Mullins *et al.* (1984).

These modern turbidite sands show a variety of features. They have abrupt, often erosional bases, but none of the characteristic bottom structures found under ancient turbidites. This may be due to the problems of collecting small cores of unconsolidated sediment. Upward size grading is sometimes but not always present. The sands are frequently clean; an interstitial clay matrix is generally absent. Internally the sands are massive, laminated or cross-laminated. A shallow-water fauna is sometimes present especially in the bioclastic sands; this contrasts markedly with the pelagic fauna of the intervening muds. By analogy with modern experimental and lacustrine turbidites, it can be convincingly argued that deep-sea sands such as these were transported from the continental shelves by turbidity currents. These moved down submarine canyons cut into continental margins and out onto the ocean floors. The decrease in gradient would cause the flow to lose velocity and deposit its load in a graded bed, the coarsest particles settling out first.

This attractive mechanism has been criticized for a number of reasons. First, it has been pointed out that many deep-sea sands do not conform to the ideal turbidite model. Some are clean and well sorted, and are internally cross-laminated. These features could indicate that the sand was deposited by a traction current. Studies of modern continental rises show the existence of currents flowing perpendicular to the slope. These are termed **geostrophic** or **contour currents.** Photographs and cores show that these currents deposit cross-laminated clean sand. Seismic data show the existence of stacked megaripples, tens of meters high, whose axes parallel the slope (Heezen and Hollister, 1971; Hollister and Heezen, 1972; Bouma, 1972). An additional argument against the turbidity current mechanism for deep-sea sand transport is to be found in the submarine canyons down which they are believed to flow. Attempts to trigger turbidites by explosions in canyon heads have been unsuccessful (Dill, 1964). The sediments of the canyons themselves often suggest transportation by normal traction currents aided by some slumping and grain flow (Shepard and Dill, 1966; Shepard *et al.*, 1969). Other critiques of the turbidity current mechanism have been made by Hubert (1964), Van der Lingen (1969) and Simpson (1982).

Turn now from the problems of deciding the role played by turbidity currents in modern deep-sea sands to their ancient analogs. There is a particular sedimentary facies that used to be termed **flysch,** which is described in Chapter 6. Many geologists use this term interchangeably with **turbidite,** implying that these rocks were deposited from tur-

Fig. 4.16. Graded bedding in Khreim Group (Silurian) sandstones. Southern Desert of Jordan.

bidity currents. The sediments termed turbidites show the following features: They are generally thick sequences of regularly interbedded sandstones and shales. These typically occur in orogenic belts or in fault bounded marine basins. The sands have abrupt basal contacts and show a variety of erosional and deformational structures that are more fully described in Chapter 5. These include pear-shaped hollows, termed "flutes," which taper downcurrent, and various erosional grooves and tool marks caused by debris marking the soft mud beneath the turbidite sand. The mud often shows deformation caused by differential movement of the overlying sand. This generates load structures, pseudonodules, slides, and slumps. Internally, the sands tend to show an upward-fining of grain size, termed **graded bedding** (Fig. 4.16). There are several different types of graded bedding. Distribution grading shows a gradual vertical decrease of grain size, while maintaining the same distribution, that is, sorting of the sediment. Coarse tail grading shows a gradual vertical decrease in the maximum grain size. Hence there is a vertical improvement in sorting. These textural differences may relate to different types of density flow (Allen, 1970, p. 194). Compound grading may be present within a single sandstone bed. Reverse grading has been observed. An absence of grading in a turbidite may indicate a source of uniform sediment grade.

The internal structures of turbidite beds are few in number, and tend to be arranged in a regular motif, termed a **Bouma sequence** (Bouma, 1962) (Fig. 4.17). In the ideal model five zones can be recognized from A to E. These have been interpreted in terms of the flow regime by Walker (1965), Harms and Fahnestock (1965), Hubert (1967), and Shanmugam (1997). The scoured surface at the base of a turbidite bed is frequently succeeded by a conglomerate of extraneous pebbles and locally derived mud clasts. This indicates the initial high-power erosive phase of the current. Typically this is overlain by a massive sand, termed the A unit, attributed to sedimentation from antidunes in the upper flow regime. Upcurrent dipping foresets in this unit have been discovered by Walker (1967b) and Skipper (1971). The massive A unit is overlain by the laminated B unit, attributable to a shooting flow regime with deposition from a planar bed form.

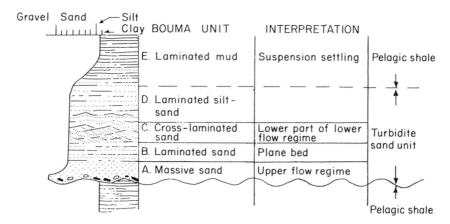

Fig. 4.17. Turbidite unit showing the complete Bouma sequence and its interpretation in terms of flow regimes.

This is succeeded by a cross-laminated C unit, which often shows convolute deformational structures due to penecontemporaneous dewatering (see Section 5.3.4.1). The cross-laminated zone reflects sedimentation from a lower flow regime. This C unit is overlain by a second laminated zone, the D unit, which grades up into the pelagic muds of the E unit, which settle from suspension.

Detailed statistical studies of whole turbidite formations show that sometimes the directional changes are predictable (e.g., Walker, 1970; Pett and Walker, 1971; Parkash, 1970). Moving downcurrent or down-section, the following trends are commonly found: Grain size and sand bed thickness diminish; bottom structures tend to change from channelling to flute marks, groove marks, and finally tool marks. The internal structures of the turbidite appear to undergo progressive truncation from the base upward. First, the massive A unit is overstepped by the laminated B unit. This is in turn overstepped by the cross-laminated C unit, and so on (Fig. 4.18).

Walker (1967a) proposed a statistical coefficient, the P index, as a measure of the proximal (i.e., near source) or distal position of beds within a turbidite formation. Recognition that turbidite formations change their sedimentologic parameters upward and sourceward throws considerable light on their genesis. Field geologists (e.g., Bailey, 1930) are impressed by the great difference between the deposits of traction currents and those of turbidites. The former are cross-bedded, clean, and laterally restricted. Turbidites, on the other hand, are flat-bedded, graded, argillaceous, and often laterally extensive. One tends to contrast the sideways sedimentation of the traction current with the essentially vertical sedimentation of the turbidite. This is an oversimplification and, as further sections in this chapter show, turbidites form part of a continuous spectrum of sedimentary processes that range from a landslide to a cloud of suspended clay.

4.2.3 Sedimentation from Low-Density Turbidity Currents

Fine-grained clay and silt are seldom if ever deposited from traction currents because they tend to be transported in suspension rather than as bed load. A certain amount of

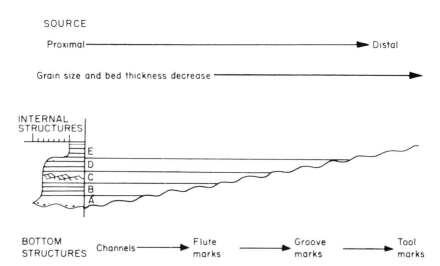

Fig. 4.18. Downcurrent variation in the sedimentary structures of turbidites. (Based on data due to Walker, 1967a,b.)

sand and silt is deposited at the distal end and waning phases of turbidity flows. The bulk of subaqueous silt and clay is transported by a third mechanism, suspension. Suspension-deposited mudrocks can occur interbedded or interlaminated with turbidites or with traction deposits. Three types of suspension can be defined, though the divisions between them are arbitrary.

First, the fine sediments of distal turbidites are essentially suspension deposits. These are thinly interlaminated, laterally extensive laminae of silt and clay. Examples of this sediment type occur in deep marine basins, but are more characteristic of lacustrine environments. Such "varved" deposits, as they are called, are a feature of many Pleistocene glacial lakes (Smith, 1959). Each **varve** consists of a silt-clay couplet and is, by definition, considered as the product of one year's sedimentation. The silt lamina represents suspended load settled out from the summer melt water. The clay lamina, often rich in lime and organic matter, settled from suspension in winter when the lake and its environs were frozen and there was no terrigenous transportation into the lake. Varve-like interlaminated silt and clay also occur in older lake deposits where fossil evidence shows them to have originated in diverse and certainly nonglacial climates. The Tertiary Green River Shale of Wyoming and Utah is a famous example (Bradley, 1931).

A second type of suspension deposit originates from what are termed **nepheloid** layers. These are bodies of turbid water whose density differential with the ambient fluid is not sufficiently large enough for them to sink to the bottom as a conventional turbidity flow, yet they are sufficiently dense to form a cohesive turbid layer suspended within the ambient fluid. Nepheloid layers have been discovered off the Atlantic coast of North America (Ewing and Thorndike, 1965). Bodies of this type may transport clay and organic matter far out into the oceans where the fine sediment settles out of suspension on to the sea bed in the pelagic environment (see Section 6.3.2.10.1). Some nepheloid layers are apparently associated with geostrophic contour currents. These flow in response to differences in both turbidity and temperature. Most geostrophic currents are triggered by the descent of cold polar waters into ocean basins in lower latitudes. Here they rotate according to Coriolis force, clockwise in the Northern Hemisphere, anti-clockwise in the Southern Hemisphere. These geostrophic currents not only transport mud in nepheloid layers, giving rise to muddy **contourites,** but are also responsible for the traction carpet from which are deposited the sandy contourites (Stow and Lovell, 1979; Stow, 1985).

The third main type of suspension deposit occurs where turbid flows enter bodies of water with no significant density difference. This situation, termed "hypopycnal flow" (Bates, 1953), allows a complete mixing of the two water masses. Fine material then settles out of suspension from the admixture of water bodies. The term **hemipelagite** is applied to massive deep marine muds whose absence of evidence of current action suggest that they have settled out of suspension. (They are, however, sometimes bio-turbated.) Hemipelagites grade into the red clays and biogenic muds of the true oceanic oozes, sometimes termed "pelagites." In nearshore environments sedimentation is ac-celerated where muddy freshwater mixes with seawater. The salts cause the clay par-ticles to flocculate and settle more rapidly than if they were still dispersed.

It is easy to establish that a fine-grained deposit settled out of a subaqueous suspen-sion. It is virtually impossible to determine whether the transporting mechanism was the last gasp of turbidity current, a nepheloid layer, or a hypopycnal flow. The end product of all three processes is a claystone with varying amounts of silt. These sediments may be massive or laminated due to vertical variations in grain size or chemical composition. Silt and very fine sand ripples may sometimes be present testifying to occasional trac-tion current activity. Other sedimentary structures include slumps, slides, and synaere-sis cracks. The slides and slumps occur because suspension deposits can form on slopes that are inherently unstable. Sedimentation continues until a critical point is reached at which the mud slides and slumps downslope to be resedimented from suspension in a more stable environment. This may completely disturb the lamination of superficial sediments. Deeper, more cohesive muds may retain their lamination though this may have been disturbed into slump and slide structures due to mass movement downslope. On level bottoms spontaneous dewatering of clays leads to the formation synaeresis cracks (see Section 5.3.5.4), which are themselves infilled by more mud.

4.3 EOLIAN PROCESSES

At the beginning of this chapter it was pointed out that eolian and aqueous transpor-tation and sedimentation shared many features. This is because both processes are es-sentially concerned with the transportation of a granular solid in a fluid medium. Gases and liquids both lack shear strength and share many other physical properties. Eolian processes involve both traction carpets and suspensions (dust clouds). Turbidity flows are essentially unknown except in volcanic gas clouds termed *nuées ardentes.* These are masses of hot volcanic gases with suspended ash and glass shards. These masses move down the sides of volcanoes at great speed. The resultant deposit, termed an "ignim-brite," may have been formed at a sufficiently high temperature for the ash particles to be welded together (Suthren, 1985).

Modern eolian sediment transport and deposition occurs in three situations. They are found in the arid desert areas of the world, such as the Sahara. They are found erratically developed around ice caps, where the climate may have considerable precipitation, but this is often seasonal and ice bound. Dunes also occur on the crests of barrier islands and beaches in diverse climates. The bulk of eolian sediments consists of either traction-deposited sands or suspension-deposited silt. These two types are described next.

4.3.1 Eolian Sedimentation from Traction Carpets

The foundation for the study of sand dunes was laid down in a classic book by Bagnold (1954), updated in 1979. Additional significant work on the physics of eolian sand transport has been described by Owen (1964), Williams (1964), Glennie (1970, 1987), Wilson (1972), McKee (1979), Greeley and Iverson (1985), Pye and Lancaster (1993), and Lancaster (1995). These studies describe how sediment, blown by the wind, moves by sliding and saltation just like particles in water. Silt and clay are winnowed from the traction carpet and carried off in dust clouds. Studies of the threshold velocity needed to commence air movement show that, as with aqueous transport, the threshold velocity increases with increasing grain size. Quartz particles of about 0.10 mm (very fine sand) are the first to move in a rising wind. Silt and clay need velocities as strong as those for fine sand to initiate movement (Horikowa and Shen, in Allen, 1970). This is analogous to the Hjulstrom effect for the threshold of particle movement in aqueous flows.

A relationship between bed form and wind velocity has not been worked out in the same way that the flow regime concept unifies these variables for aqueous flow. Ripples, dunes, and plane beds are all common eolian sand bed forms. It is a matter of observation that ripples are blown out on both dunes and plane beds during sandstorms, to be rebuilt as the wind wanes. The factors that control the areal distribution of sand plains and sand dune field or sand seas are little understood. Attempts have been made, however, to define a model that integrates wind velocity and direction with net sand flow paths (Wilson, 1971, 1972). Particular attention has been paid to the geometry and genesis of sand dunes. Four main morphological types can be defined (Fig. 4.19).

The most beautiful and dramatic sand dune is the **barchan** or lunate dune. This is arcuate in plan, convex to the prevailing wind direction, with the two horns pointing downwind. Barchans have a steep slip face in their concave downwind side (Fig. 4.19A). Barchans are lonely dunes; typically they occur in isolation or as outriders around the edges of sand seas. They generally overlie playa mud or granule deflation surfaces. This suggests that lunate dunes form where sand is in short supply. They are bed forms of transportation, not of net deposition. It is unlikely that they are often preserved in the geological record.

The second type to consider is the stellate, pyramidal, or **Matterhorn** dune (Fig. 4.19B). These consist of a series of sinuous, sharp, rising sand ridges, which merge together in a high peak from which wind often blows a plume of sand, making the dune look as if it were smoking, a dramatic sight in the middle of a desert. These stellate dunes are sometimes hundreds of meters high. They often form at the boundary of sand

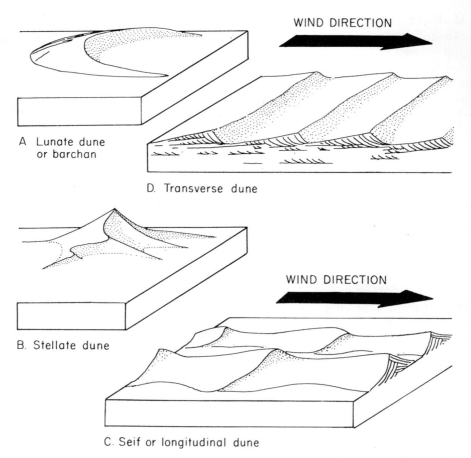

Fig. 4.19. The four main morphological types of sand dune. Probably only the transverse dunes are responsible for net sand deposition. The other three types occur in environments of equilibrium where sand is constantly transported and reworked, but where there is little net sedimentation.

seas and jebels, suggesting an origin due to interference of the wind-stream by resistant topography. This is not the whole story, however, because they also occur, apparently randomly, mixed in with dunes other types.

The third type of dune is the longitudinal or **seif** dune (Figs. 4.19C and 4.20A). These are long, thin, sand dunes with sharp median ridges. Individual dunes may be traced for up to 200 km, with occasional convergence of adjacent seifs in a downwind direction. Individual dunes are up to 100 m high, and adjacent dunes are spread up to 1 or 2 km apart, with flat interdune areas of sand or gravel. The simple gross morphology of seif dunes is often modified by smaller parasitic bed forms. These are sometimes regularly spaced stellate peaks, separated by long gentle saddles, or they may be inclined transverse dunes with steep downwind slip faces. The seif dunes of modern deserts are believed to have formed during the last glacial maximum some 10,000–25,000 B.P., when wind velocities were far higher than they are today. The origin of seif dunes has been discussed by Bagnold (1953), Hanna (1969), Folk (1971), and Glennie (1970, 1987). The consen-

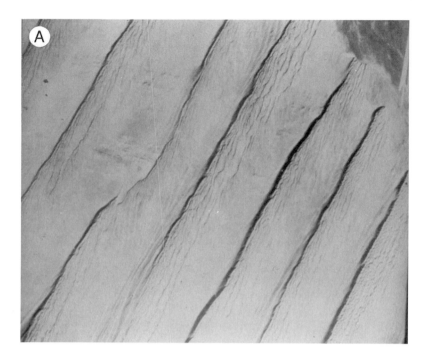

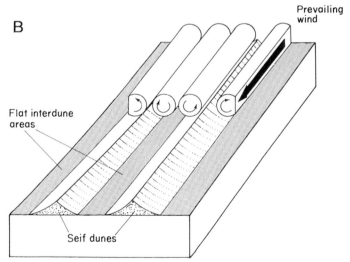

Fig. 4.20. (A) Air photograph of a sand sea composed of seif (longitudinal) dunes aligned parallel to average wind direction. Note the flat interdune areas. (Courtesy of K. Glennie.) (B) Illustration showing the formation of seif (longitudinal) dunes from a helical flow system developed by unidirectional wind flow.

sus of opinion is that they are formed from helicoidal flow cells set up by unidirectional wind systems (Fig. 4.20B).

The fourth type of eolian dune is the **transverse** variety. These are straight, or slightly sinuous, crested dunes that strike perpendicular to the mean wind direction (Figs. 4.19D

Fig. 4.21. Air photograph of a sand sea composed of transverse dunes, with lunate crests. Note the absence of an interdune area. These dunes migrate up the back side of the next dune downwind. Transverse dunes are the type that are preserved in the sedimentary record (compare with the seif dunes in Fig. 4.20). (Courtesy of K. Glennie.)

and 4.21). Their steep slip faces are directed downwind. Transverse dunes seldom occur on deflation surfaces. They are typically gregarious, climbing up the back side of the next dune downwind. This strongly suggests that transverse dunes are the eolian bed form that actually deposits sand. The other three dune types seem largely transportational bed forms. These conclusions are based on several years spent traveling around, through, and sometimes embedded in Arabian and Saharan sand seas.

Considerable attention has been directed toward the study of the internal structure of dunes (McKee and Tibbitts, 1964; McKee, 1966, 1979; Glennie, 1970, 1987; Bigarella, 1972; Ahlbrandt and Fryberger, 1982). Eolian ripples and small dunes contain single cross-laminae and cross-bedded sets that correspond in thickness to the amplitude of the bed form. This has led to the assumption that high-amplitude dunes would generate correspondingly thick cross-beds. Excavated cuttings do not always confirm this assumption. Thick cross-beds in which set thickness conforms to dune height have been observed in dome-shaped "whaleback" dunes both in the White Sands desert of New Mexico (McKee, 1966), and in the Jufara sand sea of Saudi Arabia (Fig. 4.22). Excavations through transverse and lunate dunes, however, have revealed many sets of varying height, separated by erosion surfaces and flat-bedded units (Ahlbrandt and Fryberger, 1982). This heterogeneous bedding reflects erratic variations in gross dune morphology, with alternating phases of erosion and deposition. The cross-beds were deposited on downwind migrating slip faces. The flat and subhorizontal bedded units were deposited on the gentle back slope of the dune.

Seif dunes are internally composed of mutally opposed foresets, which dip perpendicular to the prevailing wind direction. Penecontemporaneous deformation of bedding is quite common in dunes. It takes the form of both fracturing and slipping, as well as various fold structures (McKee, 1971, 1979).

In conclusion, one can see that eolian traction deposits are broadly similar to aqueous eolian ones. They have similar modes of grain transport and thus similar bed forms. Much remains to be learned of the factors that control the morphology of eolian dunes, and of their elusive internal structure.

4.3.2 Eolian Sedimentation from Suspension

Most eolian sediment is deposited by wind blowing over desiccated alluvium. The gravel remains behind. The sand saltates into dunes. The silt and clay are blown away in suspension. This much is known, and there is no question of the competence of wind to transport large amounts of silt and clay in suspension. It has been calculated that between 25 and 37 million tons of dust are transported from the Sahara throughout the longitude of Barbados each year. This quantity of dust is sufficient to maintain the present rate of pelagic sedimentation in the entire North Atlantic (Prospero and Carlson, 1972). Dust is transported in the winds of deserts, but little is actually deposited out of suspension in the way that mud settles on the sea floor. Most silt and clay is finally deposited in playas, following rains and flash floods. Desiccation and cohesion tend to prevent this material from being recycled. The dust suspensions of periglacial deserts are different from those of tropic deserts. Clay is largely absent. They are high in silica particles of medium silt grade produced by glacial action. This material is termed **loess** (Berg, 1964). Extensive deposits of Pleistocene loess occur in a belt right across the Northern Hemisphere to the south of the maximum extent of the ice sheets. Loess occurs in laterally extensive layers, often of great thickness. It is slightly calcareous, massive, and weathers into characteristic polygonal shrinkage cracks. While most authorities agree that loess was transported by eolian suspensions (i.e., as dust clouds), there is some debate as to whether it settled out of free air, or whether it was actually deposited as a result of fluvial action (e.g., Smalley, 1972; Tsoar and Pye, 1987). Some authorities argue for a polygenetic origin of loess, suggesting that it may form by wind action or by in-place cryogenesis of diverse rock types (Popov, 1972).

4.4 GLACIAL PROCESSES

Several types of sedimentary deposit are associated with glaciation (Denoux, 1994; Miller, 1996; Bennett and Glaisner, 1996; Hooke, 1998). These include loess (just described), varved clays, and the sands and gravels of fluvioglacial outwash plains. These deposits, though associated with glaciation, are actually eolian, aqueous suspension and traction current deposits, respectively.

Ice itself transports and deposits one rock type only, termed diamictite (Flint *et al.,* 1960). This is a poorly sorted sediment from boulders down to clay grade. Much of the clay material is composed of diverse minerals, but largely silica, formed by glacial pulverization. Clay minerals are a minor constituent. The boulders show a wide size range, are often angular, and sometimes grooved where ice has caused the sharp corner of one boulder to scratch across a neighbor. Statistical analysis of the orientation of the larger

Fig. 4.22A.

Fig. 4.22. (A) General view and (B) close-up of an excavation through a dune in the Jufara sand sea, Saudi Arabia. This particular dune has deposited a single cross-bed. Internal dune structures are often more complex. (Courtesy of I. Al-Jallal.)

particles shows that their long axes parallel the direction of ice movement (e.g., Andrews and Smith, 1970). Glacial diamictites tend to occur as laterally extensive sheets, seldom more than a few meters thick. They overlie glacially striated surfaces and have hummocky upper surfaces (Fig. 4.23). They also occur interbedded with the periglacial deposits just listed.

Fig. 4.23. Striated glaciated pavement, Findelengletscher, near Zermatt, Switzerland. (Courtesy of J. M. Cohen.)

Fig. 4.24. Gornegletscher, near Zermatt, Switzerland. Note glacially transported debris in median and lateral moraines. (Courtesy of J. M. Cohen.)

Consider now the mechanics of glacial transport and decomposition. Ice, formed from compacted snow, moves both in response to gravity in valley glaciers and in response to horizontal pressures in continental ice sheets. Ice movement is very slow compared with aqueous or eolian currents. On the other hand, it is highly erosive, breaking off boulders from the rocks over which it moves. The detritus that is caught up in the base of a glacier is transported in the direction of ice flow. Material falls off the valley sides to be transported on the surface, and within the glacier (Fig. 4.24). Glacial transport does not sort detritus in the same way that eolian and aqueous currents do. When the climate ameliorates, ice movement ceases and the ice begins to melt in place. Its load of sediment is dumped where it is as a heterogeneous structureless **diamictite.** Modern and Pleistocene diamictites are widely known by a variety of terms such as till, boulder clay, drift, and moraine (see Harland *et al.,* 1966, for a review of terminology). Ancient diamictites occur worldwide at various stratigraphic horizons. Many of these have been interpreted as ancient glacial deposits (**tillites**). As the next section of the book shows, however, not all diamictites are of glacial origin. The genetic name *tillite* should not be applied unless a glacial origin may clearly be demonstrated (Crowell, 1957).

Ancient diamictites for which a glacial origin can be convincingly demonstrated occur at geographically widespread localities at certain geologic times. Late Pre-Cambrian tillites associated with periglacial sediments occur in Canada, Greenland, Norway, Northern Ireland, and Scotland (Reading and Walker, 1966; Spencer, 1971) (Fig. 4.25).

There is also extensive evidence for a glaciation of the Southern Hemisphere in the Permo-Carboniferous (Fig. 4.26). This deposited tillites in South America, Australia, South Africa (the Dwyka tillite), and India (the Talchir boulder beds). Notable descrip-

Fig. 4.25. Geologist overwhelmed at the sight of Late Pre-Cambrian glacial diamictite, Ella Island, East Greenland.

tions of diamictites from the Falkland Islands and New South Wales have been given by Frakes and Crowell (1967) and Whetten (1965), respectively.

Ancient glacial deposits are rare, but they provide a fruitful source for academic controversy out of all proportion to their volume. The economic importance of glacial deposits is similarly equivocal. They lack organic content, so petroleum must enter them from older or younger source rocks. The diamictites may provide regionally extensive seals, and the fluvioglacial sands make excellent petroleum reservoirs, as seen in several Permo-Carboniferous basins of Gondwanaland (Potter *et al.*, 1995). Similarly the Cambro-Ordovician fluvial sands of the southern shores of Tethys may have been laid

Fig. 4.26. A 5-cm-diameter core of Dwyka diamictite, Karoo basin, South Africa. Note the angularity of the clasts and the mud-supported fabric. This is a tillite formed during the Permo-Carboniferous glaciation of Gondwanaland.

down on periglacial outwash plains. These now serve as important petroleum reservoirs in the Sahara and Arabia (see Section 6.3.2.2.4).

4.5 GRAVITATIONAL PROCESSES

The force of gravity is an integral part of all sedimentary processes — aqueous, eolian, or glacial. Gravity can, on its own, however, transport sediment, but for this to acquire a horizontal component it requires some additional mechanism. There is a continuous spectrum of depositional processes graded from pure gravity deposition to the turbidity flows, as already described. This spectrum may be arbitrarily classified into four main groups: rock fall, slides and slumps, mass flows, and turbidites (Dott, 1963).

Rock falls, or avalanches, are examples of essentially vertical gravitational sedimentation with virtually no horizontal transport component (Varnes, 1958). The resultant sediment is a scree composed of poorly sorted angular boulders with a high primary porosity. Subsequent weathering may round the boulders in place, with wind or water

Fig. 4.27. Wastwater screes, England, an example of terrestrial gravity-lead avalanche and debris flow deposits.

transporting finer sediment to infill the voids. Rock falls may occur both on land and under the sea. They may be triggered by earthquakes and, on land, by heavy rain and freeze–thaw action in cold climates. The actual site of a rock fall necessitates a steep cliff from which the sediment may collapse. This may be an erosional feature or a fault scarp (Fig. 4.27).

The second type of gravitational process to consider is the slide or slump. These occur on gentler slopes than rock falls, down to even less than a degree. Slides and slumps can be both subaerial or subaqueous. The sedimentary processes of sliding involve the lateral transportation of sediment along subhorizontal shear planes (Fig. 4.28). Water is generally needed as a lubricant to reduce friction and to permit movement along the slide surfaces. The fabric of the sediment is essentially undisturbed. Slumping involves the sideways downslope movement of sediment in such a way that the original bedding is disturbed, contorted, and sometimes completely destroyed. The detailed morphology of the sedimentary structures that sliding and slumping produce is described in the next chapter.

Sliding and slumping become progressively more effective transporting agents with increasing water content. Many sediments that are slump susceptible are deposited on a slope with loose packing. Once movement is initiated, the sediment packing is disturbed and the packing tends to tighten. Porosity decreases, therefore, and the pore pressure increases. This has the effect of decreasing intergranular friction, allowing the sediment to flow more freely. Thus with increasing water content and, hence, decreasing shear strength, slumping grades into the third mechanism of the spectrum: mass flow or grain flow (Fisher, 1971). This process embraces a wide range of phenomena known by such names as sand flows, grain flows, fluidized flows, mud flows, debris flows, and their resultant deposits, fluxoturbidites, diamictites, and pebbly mudstones. These are defined and described later.

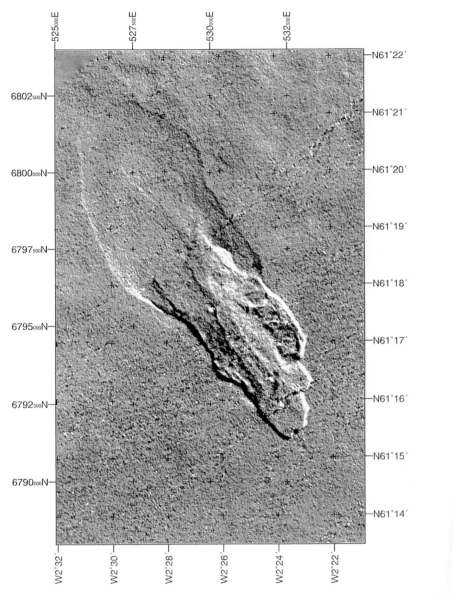

Fig. 4.28. Sea bed image of small (3- × 13-km) debris flow, west of the Shetland Islands. Movement was from the northwest to the southeast. (BD/IPR/20-44 British Geological Survey. © NERC. All rights reserved.)

One of the most celebrated debris flows was shed off the flank of El Hiero in the Canary Islands about 15,000 B.P. The flow originated by rotational collapse in about 400 m of water on negligible slope. It covers an area of 40,000 km², has a volume of 400 km³, and an average thickness of 10 m (Fig. 4.29). The slope failure may have been triggered by one or more of several processes including earthquakes, loading, oversteepening, underconsolidation, rapid sedimentation, or even gas hydrate decomposition (Weaver

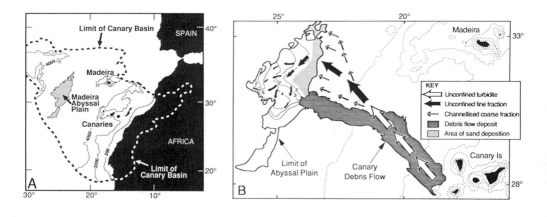

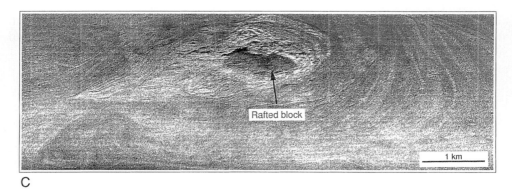

Fig. 4.29. (A) Maps, (B) seismic profile, and (C) TOBI sidescan sonar image illustrating a gravitational flow that moved from the flanks of the Canary islands out into the middle of the Madeira abyssal plain. Note how the debris flow on the slope, with its kilometric clasts, passes out into a turbidity flow on the basin floor. (From Weaver *et al.,* 1994, courtesy of the Geological Society of London.)

et al., 1994; Masson, 1996; Masson *et al.,* 1998). The resultant sediments include deposits of several types, described next.

4.5.1 Debris Flow

A debris flow is defined as a "highly-concentrated non-Newtonian sediment dispersion of low yield strength" (Stow *et al.,* 1996). This includes mud flows, though not all debris flows are muddy. Debris flows occur in a wide range of environments, ranging from deserts to continental slopes (Coussot and Meunier, 1996). In the former situation torrential rainfall is generally required to initiate movement; in the latter, earthquakes, tide, or storm surges are required. In all situations a slope is a necessary prerequisite. In desert environments the debris flow was first described by Blackwelder (1928). This is a mass of rock, sand, and mud which, liquefied by heavy rain, moves down mountain sides. Movement may initially be slow, but with increasing water content, can accelerate to a fast-flowing flood of debris that carries all before it. Large debris flows are a truly catastrophic process destroying animals, trees, and houses (e.g., Scott, 1971). Blackwelder (1928) lists the four prerequisites of a mud flow as an abundance of unconsolidated detritus, steep gradients, a lack of vegetation, and heavy rainfall.

The deposits of debris flows range from boulders to gravel, sand, silt, and clay. Where sediment of only one grade is available at source, the resultant deposit will be of that grade and well sorted. Characteristically, however, debris flow deposits are poorly sorted and massive (Fig. 4.30A). They are variously named debrites, pebbly mudstone (Crowell, 1957), diamictites (Flint *et al.,* 1960), or **fluxoturbidites** (Kuenen, 1958). The last of these, though widely used, is a naughty genetic word for a little-known process, as Walker (1967a) has enthusiastically pointed out. Debris flows may occur both on land and under water. On land, debris flows grade with increasing water content into sheet floods, which are midway between grain flows and traction currents. Sheet floods, or flash floods, deposit subhorizontally bedded coarse sands and gravels with intermittent channelling (Ives, 1936; Hooke, 1967). These characteristically occur on alluvial fans and desert pediments. Subaqueous debris flows are typically found in submarine canyons on delta fronts and continental margins (Stanley and Unrug, 1972). Modern and ancient diamictites from these settings are described in Chapter 6 (see Sections 6.3.2.1 and 6.3.2.9.1, respectively).

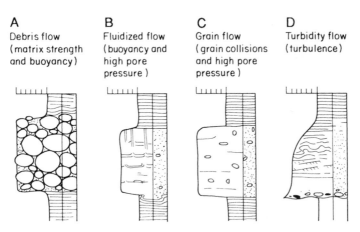

A
Debris flow
(matrix strength
and buoyancy)

B
Fluidized flow
(buoyancy and
high pore
pressure)

C
Grain flow
(grain collisions
and high pore
pressure)

D
Turbidity flow
(turbulence)

Fig. 4.30. Diagrams of the sedimentary sequences produced by gravitational and gravity-related processes. Vertical scales are varied. Debris flows may be tens of meters thick, whereas turbidites are normally less than 1 m thick.

4.5.2 Grain Flows

The concept of the grain flow was expounded by Bagnold (1954, 1966). Grain flows are liquefied cohesionless particle flows in which the intergranular friction between sand grains is reduced by their continuous agitation. This is believed by some to be nonturbulent and to involve considerable horizontal shearing (Lowe, 1976; Middleton and Hampton, 1976). Grain flows have been observed in modern submarine channels. They appear to require a high gradient to initiate them, and a confined space (i.e., a channel) to retain the high pore pressure required for their maintenance. Unlike most debris flows, grain flows are typically well sorted and clay free, though they may contain scattered clasts and astonished marine invertebrates. The *Turbo*-charged grain flows of offshore southern Africa are an example). Internally, grain flows are generally massive with no vertical size grading, though grain orientation may be parallel to flow. Bases are abrupt, often loaded, but seldom erosional. Tops are also sharp. Individual units may be 1 m or so in thickness, but multistory sequences of grain flows may attain tens of meters (see Fig. 6.71).

Hendry (1972) described graded massive marine breccias from the Alpine Jurassic, known locally as **Wild-flysch,** and interpreted them as grain flow deposits. Stauffer (1967) defined the typical features of grain flow deposits and pointed out the differences between them and true turbidity current deposits; namely, grain flow deposits are more typical of the submarine channel, whereas turbidites occur more generally on the fan or basin floor. Individual grain flow beds have erosional bases but lack the suite of sole marks characteristic of turbidites. Grain flow beds are massive or faintly bedded with clasts up to cobble size, scattered throughout them. They are not graded. Turbidites, by contrast, are graded, laminated, and/or cross-laminated and their coarsest clasts are restricted to the base of the bed (Figs. 4.30C and D). Grain flows are commonly derived from continental shelf sands, and are thus normally clean and well sorted and may make excellent petroleum reservoirs (see Section 6.3.2.9.3).

4.5.3 Fluidized Flows

Fluidization of a sand bed occurs when the upward drag exerted by moving pore fluid exceeds the effective weight of the grains. When this upward movement exceeds the minimum fluidization velocity, the bed expands rapidly, porosity increases, and the bed becomes liquefied and fluid supported, rather than grain supported.

The sediments produced by fluidization are similar in many ways to grain flows. They occur in thick nongraded clean sands, with abrupt tops and bottoms. Because of their high porosities, however, fluidized beds frequently contain sand pipes and **dish structures** due to postdepositional dewatering (Fig. 4.30B). Like grain flows, fluidized flows appear to require a slope and trigger to initiate them, and a channel to retain pore pressure. Observations of ancient deep-sea sands show that, once initiated, both grain flows and fluidized flows may move down channels far across basin floors with minimal gradient. For example, sands with typical grain flow characteristics occur in the Paleocene reservoir of the Cod gas field of the North Sea. These sands are now some 200 km from the delta slope from whence they came (Kessler *et al.,* 1980).

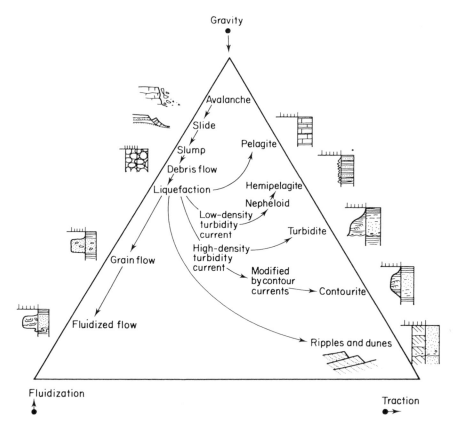

Fig. 4.31. Triangular diagram that attempts to show the relationship between the various sedimentary processes and their deposits. The arrows at the apices of the triangle indicate downward, upward, and lateral grain movement, respectively.

In conclusion it must be stated that gravity-related sedimentary processes are not well understood. They are often observed today, though few observers survive to write about their experiences, and their deposits may be studied. They range from avalanches, via debris flows, grain flows, and/or fluidized flows, to turbidites. This spectrum is marked by decreasing gradient and sand concentration and increasing velocity and water content.

Further detailed discussions of gravitational processes and resultant deposits can be found in Hampton (1972), Middleton and Hampton (1976), Howell and Normark (1982), and Stow (1985, 1986). Figure 4.31 attempts to illustrate the relationships between the various sedimentary processes and their products.

SELECTED BIBLIOGRAPHY

Allen, P. A. (1997). "Earth Surface Processes." Blackwell Science, Oxford. 416pp.

Baum, G. R., and Vail, P. R. (1998). "Gravity Currents: In the Environment and the Laboratory." Cambridge University Press, Cambridge, UK. 244pp.

Bennett, M. R., and Glaisner, N. F. (1996). "Glacial Geology, Ice Sheets and Landforms." Wiley, Chichester. 36pp.

Black, K. S., Paterson, D. M., and Cramp, A., eds. (1998). "Sedimentary Processes in the Inter-tidal Zone." Spec. Publ. No. 139. Geol. Soc. London, London.

Coussot, P., and Meunier, R. (1996). Recognition, classification and mechanical description of debris flows. *Earth Sci. Rev.* **40,** 209–229.

Fleming, B. W., and Bartholoma, A. (1995). "Tidal Signatures in Modern and Ancient Sedi-ments." Blackwell, Oxford. 368pp.

Glennie, E. K. (1987). Desert sedimentary environments, present and past — a summary. *Sedi-ment. Geol.* **50,** 135–166.

Miller, J. M. G. (1996). Glacial sediments. *In* "Sedimentary Environments: Processes, Facies and Stratigraphy" (H. G. Reading, ed.), 3rd ed., pp. 454–484. Blackwell, Oxford.

REFERENCES

Ahlbrandt, T. S., and Fryberger, S. G. (1982). Eolian deposits. Mem.—*Am. Assoc. Pet. Geol.* **31,** 11–48.

Alexander, J., and Fielding, C. (1997). Gravel antidunes in the tropical Burdekin River, Queens-land, Australia. *Sedimentology* **44,** 327–338.

Allen, J. R. L. (1970). "Physical Processes of Sedimentation." Allen & Unwin, London. 248pp.

Allen, J. R. L. (1981). Lower Cretaceous tides revealed by cross-bedding with drapes. *Nature (London)* **289,** 579–581.

Allen, J. R. L. (1985a). "Principles of Physical Sedimentology." Allen & Unwin, London. 272pp.

Allen, J. R. L. (1985b). Loose boundary hydraulics and fluid mechanics: Selected advances since 1961. *In* "Sedimentology Recent Advances and Applied Aspects" (P. J. Brenchley and B. P. J. Williams, eds.), pp. 7–30. Blackwell, Oxford.

Allen, P. A. (1997). "Earth Surface Processes." Blackwell Science, Oxford. 416pp.

Andrews, J. T., and Smith, D. I. (1970). Statistical analysis of till fabric methodology, local and regional variability. *Q. J. Geol. Soc. London* **125,** 503–542.

Bagnold, R. A. (1953). The surface movement of blown sand in relation to meteorology. *In* "Des-ert Research," pp. 89–96. UNESCO, Jerusalem.

Bagnold, R. A. (1954). "The Physics of Blown Sand and Desert Dunes." Methuen, London. 265pp.

Bagnold, R. A. (1966). An approach to the sediment transport problem from general physics. *Geol. Surv. Prof. (U.S.)* **422-I,** 37.

Bagnold, R. A. (1979). Sediment transport by wind and water. *Nord. Hydrol.* **10,** 309–322.

Bailey, Sir E. B. (1930). New light on sedimentation and tectonics. *Geol. Mag.* **67,** 77–92.

Bates, C. C. (1953). Rational theory of delta formation. *Bull. Am. Assoc. Pet. Geol.* **37,** 2119–2162.

Baum, G. R., and Vail, P. R. (1998). "Gravity Currents in the Environment and the Laboratory." Cambridge University Press, Cambridge, UK. 244pp.

Bell, H. S. (1942). Density currents as agents for transporting sediments. *J. Geol.* **50,** 512–547.

Berg, L. S. (1964). "Loess as a Product of Weathering and Soil Formation." Israel Program for Scientific Translations, Jerusalem. 205pp.

Bennett, M. R., and Glaisner, N. F. (1996). "Glacial Geology, Ice Sheets and Landforms." Wiley, Chichester. 36pp.

Bigarella, J. J. (1972). Eolian environments, their characteristics, recognition and importance. *Spec. Publ.— Soc. Econ. Paleontol. Mineral.* **16,** 12–62.

Black, K. S., Paterson, D. M., and Cramp, A., eds. (1998). "Sedimentary Processes in the Inter-tidal Zone," Spec. Publ. No. 139. Geol. Soc. London, London.

Blackwelder, E. (1928). Mudflow as a geologic agent in semi-arid mountains. *Geol. Soc. Am. Bull.* **39,** 465–483.

Bornhold, B. D., and Pilkey, O. H. (1971). Bioclastic turbidite sedimentation in Columbus Basin, Bahama. *Geol. Soc. Am. Bull.* **82,** 1254–1341.

Bouma, A. H. (1962). "Sedimentology of some Flysch Deposits." Elsevier, Amsterdam. 168pp.

Bouma, A. H. (1972). Recent and Ancient turbidites and contourites. *Trans.— Gulf-Coast Assoc. Geol. Soc.* **22,** 205–221.

Bradley, W. H. (1931). Non-glacial marine varves. *Am. J. Sci.* **22,** 318–330.

Conolly, J. R., and Ewing, M. (1967). Sedimentation in the Puerto Rico Trench. *J. Sediment. Petrol.* **37,** 44–59.

Coussot, P., and Meunier, R. (1996). Recognition, classification and mechanical description of debris flows. *Earth Sci. Rev.* **40,** 209–229.

Crowell, J. C. (1957). Origin of pebbly mudstones. *Geol. Soc. Am. Bull.* **68,** 993–1010.

Dailey, J. W., and Harleman, D. R. F. (1966). "Fluid Dynamics." Addison-Wesley, Reading, MA. 454pp.

Daly, R. A. (1936). Origin of submarine canyons. *Am. J. Sci.* **31,** 401–420.

Denoux, M., ed. (1994). "Earth's Glacial Record." Cambridge University Press, Cambridge, UK. 284pp.

Dill, R. F. (1964). Sedimentation and erosion in Scripps submarine canyon head. *In* "Marine Geology" (R. L. Miller, ed.), pp. 23–41. Macmillan, London.

Dott, R. H. (1963). Dynamics of subaqueous gravity depositional processes. *Bull. Am. Assoc. Pet. Geol.* **47,** 104–128.

Dzulinski, S., and Walton, E. K. (1963). Experimental production of sole markings. *Trans. Edinb. Geol. Soc.* **19,** 279–305.

Ewing, M., and Thorndike, E. M. (1965). Suspended matter in deep-ocean water. *Science,* **147,** 1291–1294.

Fisher, R. V. (1971). Features of coarse-grained, high-concentration fluids and their deposits. *J. Sediment. Petrol.* **41,** 916–927.

Fleming, B. W., and Bartholoma, A. (1995). "Tidal Signatures in Modern and Ancient Sediments." Blackwell, Oxford. 368pp.

Flint, R. F., Sanders, J. E., and Rodgers, J. (1960). Diamictite: A substitute term for symmictite. *Geol. Soc. Am. Bull.* **71,** 1809–1810.

Folk, R. L. (1971). Longitudinal dunes of the north-western edge of the Simpson desert, Northern Territory, Australia. 1. Geomorphology and grainsize relationships. *Sedimentology* **16,** 5–54.

Frakes, L. A., and Crowell, J. C. (1967). Facies and palaeogeography of Late Paleozoic diamictite, Falkland islands. *Geol. Soc. Am. Bull.* **78** (1), 37–58.

Friedman, G. M. (1961). Distinction between dune, beach and river sands from their sorting characteristics. *J. Sediment. Petrol.* **31,** 514–529.

Glennie, K. W. (1970). "Desert Sedimentary Environments." Elsevier, Amsterdam. 222pp.

Glennie, K. W. (1987). Desert Sedimentary Environments, present and past — a summary. *Sediment. Geol.* **50,** 135–166.

Gould, H. R. (1951). Some quantitative aspects of Lake Mead turbidity currents. *Spec. Publ.— Soc. Econ. Palaeontol. Mineral.* **2,** 34–52.

Greeley, R., and Iverson, J. D. (1985). "Wind as a Geological Process." Cambridge University Press, Cambridge, UK. 333pp.

Hampton, M. A. (1972). The role of subaqueous debris flow in generating turbidity currents. *J. Sediment. Petrol.* **42,** 775–793.

Hand, B. M. (1967). Differentiation of beach and dune sands using settling velocities of light and heavy materials. *J. Sediment. Petrol.* **37,** 514–520.

Hand, B. M., and Emery, K. O. (1964). Turbidites and topography of north end of San Diego Trough, California. *J. Geol.* **72,** 526–552.

Hanna, S. R. (1969). The formation of longitudinal sand dunes by large helical eddies in the atmosphere. *J. Appl. Meteorol.* **8,** 874–883.

Harland, W. B., Herod, K. N., and Krinsley, D. H. (1966). The definition and identification of tills and tillites. *Earth Sci. Rev.* **2,** 225–256.

Harms, J. C., and Fahnestock, R. K. (1965). Stratification, bed forms and flow phenomena (with an example from the Rio Grande). *Spec. Publ.— Soc. Econ. Paleontol. Mineral.* **12,** 84–155.

Heezen, B. C., and Ewing, M. (1952). Turbidity currents and submarine slumps and the 1929 Grand Banks earthquake. *Am. J. Sci.* **250,** 849–873.

Heezen, B. C., and Hollister, C. D. (1971). "The Face of the Deep." Oxford University Press, Oxford. 659pp.

Henderson, F. M. (1966). "Open Channel Flow." Macmillan, London. 522pp.

Hendry, H. E. (1972). Breccias deposited by mass flow in the Breccia Nappe of the French Pre-Alps. *Sedimentology* **18,** 277–292.

Hollister, C. D., and Heezen, B. C. (1972). Geologic effects of ocean bottom currents: Western North Atlantic. *In* "Studies in Physical Oceanography" (A. L. Gordon, ed.), pp. 37–66. Gordon & Breach, New York.

Holtedahl, H. (1965). Recent turbidites in the Hardangerford, Norway. *In* "Submarine Geology and Geophysics" (W. F. Whittard and R. Bradshaw, eds.), pp. 107–142. Butterworth, London.

Hooke, R. L. (1967). Processes on arid-region alluvial fans. *J. Geol.* **75,** 438–460.

Hooke, R. L. (1998). "Glacier Mechanics." Prentice Hall, Hemel Hempstead, England. 248pp.

Howell, D. G., and Normark, W. R. (1982). Submarine fans. *Mem.—Am. Assoc. Pet. Geol.* **31,** 365–404.

Hubert, J. F. (1964). Textural evidence for deposition of many western N. Atlantic deep sea sands by ocean bottom currents rather than turbidity currents. *J. Geol.* **72,** 757–785.

Hubert, J. F. (1967). Sedimentology of pre-Alpine Flysch sequences, Switzerland. *J. Sediment. Petrol.* **37,** 885–907.

Ives, J. C. (1936). Desert floods in the Sonora Valley, Northern Mexico. *Am. J. Sci.* **32,** 102–135.

Kessler, Z. L. G., Zang, R. D., Englehorn, N. J., and Eger, J. D. (1980). Stratigraphy and sedimentology of a Palaeocene submarine fan complex, Cod Field, Norwegian North Sea. 1. *In* "The Sedimentation of North Sea Reservoir Rocks" (R. H. Hardman, ed.), Paper VII, pp. 1–19. Norw. Petrol. Soc., Oslo.

Kuenen, P. H. (1937). Experiments in convection with Daly's hypothesis on the formation of submarine canyons. *Leidse Geol. Meded.* **8,** 327–335.

Kuenen, P. H. (1948). Turbidity currents of high density. *Int. Geol. Congr. Rep. Sess., 18th,* U.K., Pt. 8, pp. 44–52.

Kuenen, P. H. (1958). Problems concerning course and transportation of flysch sediments. *Geol. Mijnbouw* **20,** 329–339.

Kuenen, P. H. (1965). Experiments in connection with turbidity currents and clay suspensions. *In* "Submarine Geology and Geophysics" (W. F. Whittard and R. Bradshaw, eds.), pp. 47–74. Butterworth, London.

Kuenen, P. H., and Migliorini, C. I. (1950). Turbidity currents as a cause of graded bedding. *J. Geol.* **58,** 91–128.

Lancaster, N. (1995). "Geomorphology of Desert Dunes." Routledge, London. 290pp.

Leliavsky, S. (1959). "An Introduction to Fluvial Hydraulics." Dover, New York. 257pp.

Lowe, D. R. (1976). Grain flow and grain flow deposits. *J. Sediment. Petrol.* **46,** 188–199.

MacDonald, E. H. (1983). "Alluvial Mining." Chapman & Hall, London. 508pp.

Masson, D. G. (1996). Catastrophic collapse of the flank of El Hiero about 15,000 years ago, and the history of large flank collapses in the Canary Islands. *Geology* **24,** 231–234.

Masson, D. G., Canals, M., Alonso, B., Urgeles, R., and Huhnerbach, V. (1998). The Canary Debris Flow: Source area, morphology and failure mechanism. *Sedimentology* **45,** 411–442.

McKee, E. D. (1966). Dune structures. *Sedimentology* **7,** 3–69.

McKee, E. D. (1971). Primary structures in dune sand and their significance. *In* "The Geology of Libya" (C. Grey, ed.), 401–408. University of Libya, Tripoli.

McKee, E. D., ed. (1979). "Global Sand Seas," Prof. Pap. No. 1052. U.S. Geol. Surv., Washington, DC. 421pp.

McKee, E. D., and Tibbits, G. C. (1964). Primary structures of a seif dune and associated deposits in Libya. *J. Sediment. Petrol.* **34** (1), 5–17.

Middleton, G. V. (1966a). Experiments on density and turbidity currents. I. *Can. J. Earth Sci.* **3,** 523–546.

Middleton, G. V. (1966b). Experiments on density and turbidity currents. II. *Can. J. Earth Sci.* **3,** 627–637.

Middleton, G. V. (1967). Experiments on density and turbidity currents. III. *Can. J. Earth Sci.* **4,** 475–505.

Middleton, G. V., and Hampton, M. A. (1976). Subaqueous sediment transport and deposition by sediment gravity flows. *In* "Marine Sediment Transport and Environmental Management" (D. J. Stanley and D. P. J. Swift, eds.), pp. 197–218. Wiley, New York.

Miller, J. M. G. (1996). Glacial sediments. *In* "Sedimentary Environments: Processes, Facies and Stratigraphy" (H. G. Reading, ed.), 3rd ed., pp. 454–484. Blackwell, Oxford.

Mullins, H. T., Heath, K. C., Van Buren, H. N., and Newton, C. R. (1984). Anatomy of modern open-ocean carbonate slope: Northern Little Bahama Bank. *Sedimentology* **31,** 141–168.

Orme, G. R. (1977). Aspects of sedimentation in the coral reef environment. *In* "Biology and Geology of Coral Reefs" (O. A. Jones and R. Endean, eds.), vol. 4, pp. 129–182. Academic Press, San Diego, CA.

Owen, P. R. (1964). Saltation of uniform grains in air. *J. Fluid Mech.* **20,** 225–242.

Parkash, B. (1970). Downcurrent changes in sedimentary structures in Ordovician turbidite greywackes. *J. Sediment. Petrol.* **40,** 572–590.

Payne, R. R., Conolly, J. R., and Abbott, W. H. (1972). Turbidite muds within diatom ooze of Antarctica: Pleistocene sediment variation defined by closely spaced piston cores. *Geol. Soc. Am. Bull.* **83,** 481–486.

Pett, J. W., and Walker, R. G. (1971). Relationship of flute cast morphology to internal sedimentary structures in turbidites. *J. Sediment. Petrol.* **41,** 114–128.

Piper, D. J. W., Cochonatt, P., and Morrison, M. L. (1999). The sequence of events around the epicentre of the 1929 Grand Banks earthquake: Initiation of debris flows and turbidity current inferred from sidescan sonar. *Sedimentology* **46,** 79–97.

Popov, A. I. (1972). Les loess et depots loessoïdes, produit des processes cryolithogènes. *Bull. Peryglaciol.* **21,** 193–200.

Potter, P. E., Franca, A. B., Spencer, C. W., and Caputo, M. V. (1995). Petroleum in glacially-related sandstones of Gondwana: A review. *J. Pet. Geol.* **18,** 397–420.

Prospero, J. M., and Carlson, T. N. (1972). Vertical and areal distribution of Saharan dust over the Western Equatorial North Atlantic Ocean. *J. Geophys. Res.* **77,** 5255–5265.

Pye, K., ed. (1994). "Sediment Transport and Depositional Processes." Blackwell Science, Oxford. 408pp.

Pye, K., and Lancaster, N., eds. (1993). "Aeolian Sediments." Blackwell, Oxford. 176pp.

Reading, H. G., and Walker, R. G. (1966). Sedimentation of Eocambrian tillites and associated sediments in Finmark, Northern Norway. *Paleogeogr., Paleoclimatol., Paleoecol.* **2,** 177–212.

Reineck, H. E., and Singh, I. B. (1980). "Depositional Sedimentary Environments," 2nd ed. Springer-Verlag, Berlin. 549pp.

Rusnak, G. A., and Nesteroff, W. D. (1964). Modern turbidites: Terrigenous abyssal plain versus bioclastic basin. *In* "Marine Geology" (L. R. Miller, ed.), pp. 488–503. Macmillan, New York.

Scott, K. M. (1971). Origin and sedimentology of 1969 debris flows near Glendora, California. *Geol. Surv. Prof. Pap. (U.S.)* **750-C,** C242–C247.

Selley, R. C. (1966). Petrography of the Torridonian Rocks of Raasay and Scalpay, Invernesshire. Proc. Geol. Assoc. Lond. **77,** 293–314.

Shanmugam, G. (1997). The Bouma sequence and the turbidite mind set. *Earth Sci. Rev.* **42,** 199–200.

Shepard, F. P., and Dill, R. F. (1966). "Submarine Canyons and Other Sea Valleys." Rand McNally, Chicago. 381pp.

Shepard, F. P., Dill, R. F., and von Rad, U. (1969). Physiography and sedimentary processes of La Jolla submarine fan and Fan Valley, California. *Am. Assoc. Pet. Geol. Bull.* **53,** 390–420.

Shields, A. (1936). Anwendung der Ahnlichkeitsmechanik und der Turbulenzforschung auf die Geschiebebewegnung: Berlin, Preuss. Versuchsanstalt für Wasser, Erd, und Schiffbau, no. 26, 26p.

Simons, D. B., Richardson, E. V., and Nordin, C. F. (1965). Sedimentary structures generated by flow in alluvial channels. *Spec. Publ.— Soc. Econ. Palaeontol. Mineral.* **12,** 34–52.

Simpson, J. E. (1982). Gravity currents in the laboratory, atmosphere and ocean. *Annu. Rev. Fluid Mech.* **14,** 213–234.

Skipper, K. (1971). Antidune cross-stratification in a turbidite sequence, Cloridorme Formation, Gaspe, Quebec. *Sedimentology* **17,** 51–68.

Smalley, I. J. (1972). The interaction of great rivers and large deposits of primary loess. *Trans. N. Y. Acad. Sci.* [2] **34,** 534–542.

Smith, A. J. (1959). Structures in the stratified late-glacial clays of Windermere, England. *J. Sediment. Petrol.* **29,** 447–453.

Spencer, A. M. (1971). "Late PreCambrian Glaciation in Scotland," Mem. No. 6. Geol. Soc. London, London. 100pp.

Stanley, D. J., and Unrug, R. (1972). Submarine channel deposits, Fluxoturbidites and other indicators of slope and base of slope environments in modern and ancient marine basins. *Spec. Publ.— Soc. Econ. Paleontol. Mineral.,* **16,** 287–340.

Stauffer, P. H. (1967). Grainflow deposits and their implications, Santa Ynez Mountains, California. *J. Sediment. Petrol.* **37,** 487–508.

Stow, D. A. V. (1985). Deep-sea clastics: Where are we and where are we going? *In* "Sedimentology: Recent Developments and Applied Aspects" (P. J. Brenchley and B. P. J. Williams, eds.), pp. 67–94. Blackwell, Oxford.

Stow, D. A. V. (1986). Deep clastic seas. *In* "Sedimentary Environments and Facies" (H. G. Reading, ed.), 2nd ed., pp. 399–444. Blackwell, Oxford.

Stow, D. A. V., and Lovell, J. P. B. (1979). Contourites: Their recognition in modern and ancient sediments. *Earth Sci. Rev.* **14,** 251–291.

Stow, D. A. V., Reading, H. G., and Collinson, J. D. (1996). Deep seas. *In* "Sedimentary Environments: Processes, Facies and Stratigraphy" (H. G. Reading, ed.), 3rd ed., pp. 395–453. Blackwell, Oxford.

Sundborg, Å. (1956). The River Klarälven, a study in fluvial processes. *Geogr. Ann.* v. **38,** 127–316.

Suthren, R. J. (1985). Facies analysis of volcaniclastic sediments: A review. *In* "Sedimentology: Recent Developments and Applied Aspects" (P. J. Brenchley and B. P. J. Williams, eds.), pp. 123–146. Blackwell, Oxford.

Tsoar, H., and Pye, K. (1987). Dust transport and the question of desert loess formation. *Sedimentology* **34,** 139–153.

Van der Lingen, G. J. (1969). The turbidite problem. *N.Z. J. Geol. Geophys.* **12,** 7–50.

Vanoni, V. A. (1964). "Measurements of Critical Shear Stress for Entraining Fine Sediments in a Boundary Layer," W. M. Tech Lab. Rep. K. H.— R — 7. California Institute of Technology, Pasadena. 47pp.

Van Straaten, L. M. J. U. (1964). Turbidite sediments in the southeastern Adriatic Sea. *In* "Turbidites" (A. H. Bouma and A. Brouwer, eds.), pp. 142–147. Elsevier, Amsterdam.

Varnes, D. J. (1958). Landslide types and processes. *Spec. Rep.— Natl Res. Counc., Highw. Res. Board* **29** (544), 20–47.

Visser, M. J. (1980). Neap-spring cycles reflected in Holocene subtidal large-scale bedform deposits: A preliminary note. *Geology* **8,** 543–546.

Walker, R. G. (1965). The origin and significance of the internal sedimentary structures of turbidites. *Proc. Yorks. Geol. Soc.* **35,** 1–32.

Walker, R. G. (1967a). Turbidite sedimentary structures and their relationship to proximal and distal depositional environments. *J. Sediment. Petrol.* **37,** 25–43.

Walker, R. G. (1967b). Upper flow regime bedforms in turbidites of the Hatch Formation, Devonian of New York State. *J. Sediment. Petrol.* **37,** 1052–1058.

Walker, R. G. (1970). Review of the geometry and facies organization of turbidites and turbidite-bearing basins. *Spec. Pap.— Geol. Assoc. Can.* **7,** 219–251.

Weaver, P. P. E., Masson, D. G., and Kidd, R. B. (1994). Slumps, slides and turbidity currents — sea-level change and sedimentation in the Canary basin. *Geoscientist* **4** (1), 14–16.

Whetten, J. T. (1965). Carboniferous glacial rocks from the Werrie Basin, New South Wales, Australia. *Geol. Soc. Am. Bull.* **76,** 43–56.

Williams, G. (1964). Some aspects of the eolian saltation load. *Sedimentology* **3,** 257–287.

Wilson, I. G. (1971). Desert sandflow basins and a model for the development of Ergs. *Geogr. J.* **137,** 180–199.

Wilson, I. G. (1972). Aeolian bedforms-their development and origins. *Sedimentology* **19,** 173–210.

Zeng, K. W., and Lowe, D. R. (1997a). Numerical simulation of turbidity current flow and sedimentation: 1. Theory. *Sedimentology* **44,** 67–84.

Zeng, K. W., and Lowe, D. R. (1997b). Numerical simulation of turbidity current flow and sedimentation: 2. Results and geological application. *Sedimentology* **44,** 85–104.

5 Sedimentary Structures

5.1 INTRODUCTION

This chapter describes the internal megascopic features of a sediment. These are termed **sedimentary structures,** and are distinguished from the microscopic structural features of a sediment, termed the **fabric,** described in Chapter 3. Sedimentary structures are arbitrarily divided into primary and secondary classes. Primary structures are those generated in a sediment during or shortly after deposition. They result mainly from the physical processes described in Chapter 4. Examples of primary structures include ripples, cross-bedding, and slumps. Secondary sedimentary structures are those that formed sometime after sedimentation. They result from essentially chemical processes, such as those which lead to the diagenetic formation of concretions. Primary sedimentary structures are divisible into inorganic structures, including those already mentioned, and organic structures, such as burrows, trails, and borings. Table 5.1 shows the relationships between the various structures just defined. In common with most classifications of geological data, this one will not stand careful scrutiny. The divisions between the various groups are ill defined and debatable. Nevertheless, it provides a useful framework on which to build the analysis of sedimentary structures contained in this chapter.

First, consider the definition of a sedimentary structure more carefully. Colloquially, a sedimentary structure is deemed to be a primary depositional feature of a sediment that is large enough to be seen by the naked eye, yet small enough to be carried by a group of healthy students, or at least to be contained in one quarry. A channel is thus generally considered to be a sedimentary structure — just. A sand dune is generally considered a sedimentary structure; but what about an offshore bar or a barrier island that is just a large and very complex dune? Surely a coral reef is a sedimentary structure? But is it organic or inorganic? These questions highlight the problem of defining exactly what is a sedimentary structure.

Conventionally a sedimentary structure is considered as a smaller scale feature that is best illustrated by the examples of ripples, cross-beds, and slumps already mentioned. This chapter is concerned with such phenomena, both inorganic sedimentary structures and biogenic ones.

The observation, interpretation, and classification of inorganic sedimentary structures are considered first. Sedimentary structures can be studied at outcrop and in cliffs,

Table 5.1
Classification of Sedimentary Structures

I. Primary (physical)	⎰ Inorganic	⎱ Fabric Cross-bedding, ripples, etc.	Microscopic
	⎱ Organic	Burrows and trails	Megascopic
II. Secondary (chemical)	Diagenetic	Concretions, etc.	

quarries, and stream sections. Large-scale channeling and cross-bedding can also be studied using ground-penetrating radar (e.g., Bristow, 1995). They can also be studied in cores taken from wells. Sedimentary structures in cores are the easiest to describe because of the small size of the sample to be observed. Where large expanses of rock are available for analysis, the problem of what is a sedimentary structure becomes more apparent. Cross-beds are seen to be grouped in large units; ripples form an integral part of a bed; a slump structure is composed of contorted beds with diverse types of sedimentary structure. It is at once clear that sedimentary structures do not occur in isolation. Hence the problems of observing, defining, and classifying them.

There are two basic approaches to observing sedimentary structures. The first approach is to pretend the outcrop is a borehole and to measure a detailed sedimentological log. This records a vertical section of limited lateral extent. Every oil company and university geology department has its own preferred scheme and set of symbols. There is a dilemma in developing a method of core logging. The scheme must record sufficient data to allow detailed environmental interpretation. But the scheme must not record so much data as to require an exhaustive training course for novices to apply it and for managers to interpret it. Figure 5.1 illustrates one example, but see Lewis and McConchie (1994) and Goldring (1999) for further details.

The second method is to create a two-dimensional survey of all, or a major part, of the outcrop. This may be recorded on graph paper, using a tape measure and an Abney level for accurately locating inaccessible reference points on cliff faces (Lewis and McConchie, 1994). This method is aided by photography, especially by on-the-spot polaroid photos, on which significant features and sample points can be located. With the advent of the digital camera photos can be easily scanned onto a computer for greater ease and comfort.

Consider now the interpretation of sedimentary structures. They are the most useful of sedimentary features to use in environmental interpretation because, unlike sediment grains, texture, and fossils, they cannot be recycled. They unequivocally reflect the depositional process that laid down the sediment. The interpretation of the origin of sedimentary structures is based on studies of their modern counterparts, on laboratory experiments and on theoretical physics, as dealt with in the previous chapter on sedimentary processes. Examples of these different methods are described in greater detail at appropriate points in this chapter.

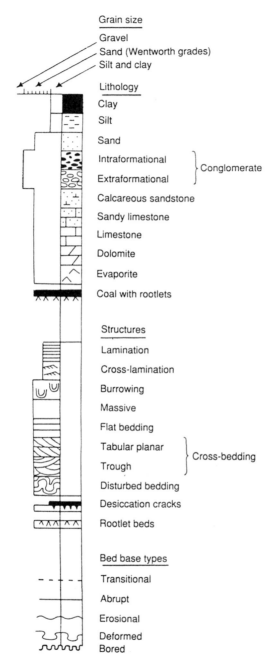

Fig. 5.1. A scheme for logging sedimentary sections. This scheme does not record so many data as to require an exhaustive training course for novices to apply it, nor for managers to interpret it. It does, however, record sufficient data to permit detailed environmental interpretation. (From Selley, R. C. 1996. "Ancient Sedimentary Environments," 4th Ed. Chapman & Hall, London. © 1996 Kluwer Academic Publishers, with kind permission from Kluwer Academic Publishers.)

5.2 BIOGENIC SEDIMENTARY STRUCTURES

A great variety of structures in sedimentary rocks can be attributed to the work of organisms. These structures are referred to as biogenic, in contrast to the inorganic sedimentary structures. Biogenic structures include plant rootlets, vertebrate footprints (tracks), trails (due to invertebrates), soft sediment burrows, and hard rock borings. These phenomena are collectively known as trace fossils and their study is referred to as **ichnology.** Important works on ichnology include those by Crimes and Harper (1970, 1977), Frey (1975), Hantzschel (1975), Gall (1983), Brenchley (1990), and Bromley (1990).

An individual morphological type of trace fossil is termed an **ichnogenus.** One of the basic principles of trace fossil analysis is that similar ichnogenera can be produced by a wide variety of organisms. The shape of a trace fossil reflects environment rather than creator. This means that trace fossils can be very important indicators of the origin of the sediment in which they are found because of their close environmental control. Furthermore, trace fossils always occur in place and cannot be reworked like most other fossils. Sedimentologists therefore need to know something about trace fossils. Some basic principles of occurrence and nomenclature will be described before analyzing the relationship between trace fossils and environments. The various types of ichnogenera cannot be grouped phyllogenetically because, as already pointed out, different organisms produce similar traces. Ichnofossils have been grouped according to the activity which made them (Seilacher, 1964) and according to their topology (Martinsson, 1965). The topological scheme essentially describes the relationship of the trace to the adjacent beds (Fig. 5.2). Table 5.2 equates the two systems side by side. Martinsson's descriptive scheme is easy to apply, whereas Seilacher's necessitates some interpretation. The difference between a feeding burrow and a dwelling burrow, for example, is often subtle (e.g., Bromley, 1975).

The most useful aspect of trace fossils is the broad correlation between depositional environment and characteristic trace fossil assemblages, termed **ichnofacies.** Schemes relating ichnofacies to environments have been drawn up by Seilacher (1964, 1967), Rodriguez and Gutschick (1970), Heckel (1972), Brenchley (1990), and Bromley (1990).

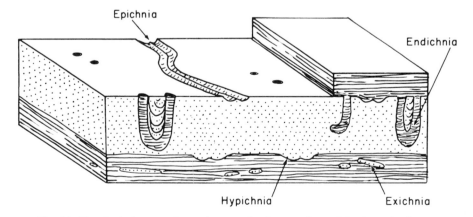

Fig. 5.2. Topological nomenclature for trace fossils according to Martinsson's scheme.

Table 5.2
Nomenclature for Trace Fossil Types

Activity nomenclature (Seilacher, 1964)		Topological nomenclature (Martinsson, 1965)
Repichnia	: crawling burrows ⎤	*Endichnia* and *Exichnia*
Domichnia	: dwelling burrows ⎦	
Fodichnia	: feeding burrows ⎤	
Pascichnia	: feeding trails ⎬	*Epichnia* and *Hypichnia*
Cubichnia	: resting trails ⎦	

Figure 5.3 is a blend of these various schemes. The most landward ichnofacies to be defined consist largely of vertebrate tracks. These include the footprints of birds and terrestrial animals. Dinosaur tracks are particularly well-studied examples (Lockley and Hunt, 1996). The preservation potential of such tracks is low. They are most commonly found on dried-up lake beds, river bottoms, and tidal flats. **Orgasmoglyphs** are produced by rutting dinosaurs. They are commonly found on the upper parts of alluvial fans, where the dinosaurs migrated to breed where the weather was cooler.

Moving toward the sea a well-defined ichnofacies occurs in the tidal zone. This is often named the "Scolithos assemblage" because it is dominated by deep vertical burrows of the ichnogenus *Scolithos* (syn. *Monocraterion, Tigillites,* and *Sabellarifex*). In this environment the sediment substrate is commonly subjected to scouring current action, which often erodes and reworks sediment. Because of this the various invertebrates of the tidal zone — be they worms, bivalves, crabs, etc. — tend to live in crawling, dwelling, and feeding burrows. These burrows exit at the sediment:water interface, but go down deep to provide shelter for the little beasts during erosive phases. The burrows may be simple vertical tubes, like *Scolithos,* vertical U-tubes like *Diplocraterion yoyo* (so named because of its tendency to move up and down), or complex networks of passageways such as *Ophiomorpha.*

In subtidal and shallow marine environments, *Cruziana* and *Zoophycos* ichnofacies have been defined, respectively. In these deeper zones, where marine current action is less destructive, invertebrates crawl over the sea bed to feed in shallow grooves. They also make burrows, but these tend to be shallower and oriented obliquely or subhorizontally. The *Cruziana* ichnofacies is characterized by the bilobate trail of that name (Fig. 5.4). This is generally referred to the action of trilobites. *Cruziana* has, however, been found in post-Paleozoic strata and has been recorded from fluvial formations (e.g., Selley, 1970; Bromley and Asgaard, 1972). The environmental significance of this particular ichnogenus must be interpreted carefully. Camel flies make excellent *Cruziana* trails on modern sand dunes. *Zoophycos* is a trace fossil with a characteristic helical spiral form in plan view. It is generally present at sand–shale interfaces. The detailed morphology of *Zoophycos* and the identity of its creator are a matter for debate (see Crimes and Harper, 1970). Nevertheless, there is general agreement that it occurs in subtidal, shallow marine deposits. Other trace fossils that characterize the *Zoophycos* and *Cruziana* ichnofacies include the subhorizontal burrows *Rhizocorallium* and *Harlania* (syn. *Arthrophycos;* syn. *Phycodes*).

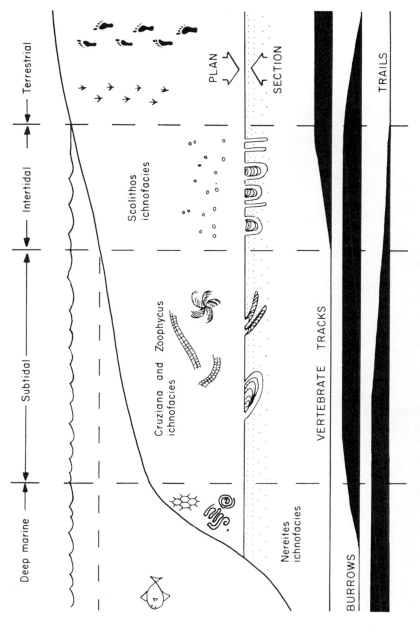

Fig. 5.3. Relationship between ichnofacies and environments, based on schemes proposed by Seilacher (1964, 1967), Rodriguez and Gutschick (1970), and Heckel (1972).

Fig. 5.4. Horizontal burrows of the ichnogenus *Harlania* in shallow marine shales of the Um Sahm Formation (Ordovician), in the Southern Desert of Jordan.

Moving into deep quiet water, a further characteristic ichnofacies is named after *Nereites.* In this environment invertebrates live on, rather than in, the sediment substrate. Burrows are largely absent and surface trails predominate. Characteristic meandriform traces include *Nereites, Helminthoida,* and *Cosmorhaphe.* Polygonal reticulate trails such as *Paleodictyon* are also characteristic of this ichnofacies, though the author has found an excellent specimen of this trace in the fluvial Messak Sandstone (Wealden?) of southern Libya. The Nereites ichnofacies is commonly found in interbedded turbidite sand–shale sequences of "flysch" facies.

This brief review of ichnofacies shows that the concept of environment-restricted assemblages is of great use to sedimentologists. The forms are found in place and are small enough to be studied from subsurface cores as well as at outcrop. When interpreted in their sedimentological context, they are a useful tool in facies analysis. One final point to note about biogenic sedimentary structures is the way in which they disrupt primary inorganic sedimentary structures. Intense burrowing, termed **bioturbation,** leads to the progressive disruption of bedding until a uniformly mottled sand is left. This is particularly characteristic of intertidal and subtidal sand bodies. Vertical burrows in interlaminated sands and shales may increase the vertical permeability of such beds, a point of some significance if they are aquifers or petroleum reservoirs.

5.3 PRIMARY INORGANIC SEDIMENTARY STRUCTURES

5.3.1 Introduction

Before proceeding to the actual descriptions, the classification of primary inorganic sedimentary structures needs to be considered in more detail. The problems of classi-

Table 5.3
Classification of Inorganic Megascopic Primary Sedimentary Structures

Group	Examples	Origin
I. Predepositional (interbed)	Channels Scour-and-fill Flute marks Groove marks Tool marks	Predominantly erosional
II. Syndepositional (intrabed)	Massive Flat-bedding (including parting lineation) Graded bedding Cross-bedding Lamination Cross-lamination	Predominantly depositional
III. Postdepositional (deform interbed and intrabed structures)	Slump Slide Convolute lamination Convolute bedding Recumbent foresets Load structures	Predominantly deformation
IV. Miscellaneous	Rain prints Shrinkage cracks	

fication are already apparent from the preceding discussions. Atlases of sedimentary structures, and attempts at their classification, have been made by Pettijohn and Potter (1964), Gubler (1966), Conybeare and Crook (1968), Harms *et al.* (1982), Collinson and Thompson (1988), and Ricci Lucchi (1995).

Three main groups can be defined by their morphology and time of formation (Table 5.3). The first group of structures is predepositional with respect to the beds that immediately overlie them. These structures occur on surfaces between beds. Geopedants may prefer to term them **interbed structures,** though they were formed before the deposition of the overlying bed. This group of structures largely consists of erosional features such as scour-and-fill, flutes, and grooves. These are sometimes collectively called sole marks or bottom structures.

The second group of structures is syndepositional in time of origin. These are depositional bed forms like cross-lamination, cross-bedding, and flat-bedding. To avoid a genetic connotation, this group may be collectively termed **intrabed structures,** to distinguish them from predepositional interbed phenomena.

The third group of structures is postdepositional in time of origin. These are deformational structures that disturb and disrupt pre- and syndepositional inter- and intrabed structures. This third group of structures includes slumps and slides.

To these three moderately well-defined groups of sedimentary structures must be added a fourth. This last category, named "miscellaneous," is for those diverse structures that cannot be fitted logically into the scheme just defined. The morphology and genesis of the various types of sedimentary structure in each of these four groupings are described next.

5.3.2 Predepositional (Interbed) Structures

Predepositional sedimentary structures occur on surfaces between beds. They were formed before the deposition of the overlying bed. The majority of this group of structures are erosional in origin. Before describing these structures, note two points. The first is one of terminology. When the interface between two beds is split open, the convex structures that depend from the upper bed are termed "casts." The concave hollows in the underlying bed into which these fit are termed "molds" (Fig. 5.5). The second point to note is that the ease and frequency with which these bottom structures are seen is related to the degree of lithification.

Unconsolidated sediments seldom split along interbed boundaries because of their friable nature. This may explain the apparent scarcity of bottom structures recorded from modern deep-sea turbidite sands. On the other extreme, well-lithified metasediments in fold belts tend to split more readily along subvertical tectonic fractures rather than along bedding planes. Thus bottom structures are most commonly seen in moderately well-lithified sediments that split along bedding surfaces.

5.3.2.1 Channels

The largest predepositional interbed structures are channels. These may be kilometers wide and hundreds of meters deep (Fig. 5.6). They occur in diverse environments ranging from subaerial alluvial plains to submarine continental margins. Channeling is initiated by localized linear erosion by fluid flow aided by corrosive bed load. Once a channel is established, however, a horizontal component of erosion develops due to undercutting of the channel bank followed by collapse of the overhanging sector.

The best studied channels are those of fluvial systems (see Section 6.3.2.2.3) and particular attention has been paid to the genesis of channel meandering and the mathematical relationships between sinuosity, channel width, depth, gradient, and discharge (e.g., Schumm, 1969; Rust, 1978). In ancient channels, depth, width, sediment grade, and

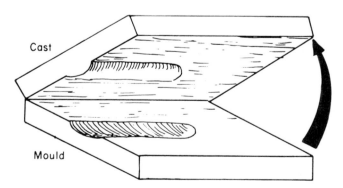

Fig. 5.5. Nomenclature for the occurrence of bed interface sedimentary structures (sole markings).

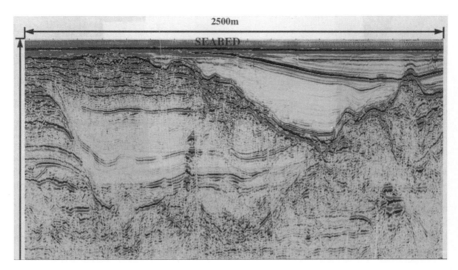

Fig. 5.6. Ultra-high-resolution seismic line showing multistory channel structures. The vertical axis is two-way travel time. The floor of the lower channel is some 100 m below present sea floor.

flow direction can easily be established. Sinuosity can also be measured where there is adequate exposure or well control. Using these data, attempts have been made to calculate the gradient and stream power of ancient channel systems (e.g., Friend and Moody-Stuart, 1972).

Channels are of great economic importance for several reasons. They can be petroleum reservoirs and aquifers, they can contain placer and replacement mineral ore bodies, and they can cut out coal seams. Instances are cited throughout this text. Smaller and less dramatic are the interbed structures termed **scour-and-fill.** These are small-scale channels whose dimensions are measured in decimeters rather than meters. They too occur in diverse environments.

There is a vast nomenclature for the numerous small interbed erosional structures. Reference should be made to the atlases of structures previously cited for exhaustive details of these. The following account describes the three most common varieties: flutes, grooves, and tool marks. Flutes and grooves are scoured by the current alone, aided by granular bed load, whereas tool marks are made by single particles generally of pebble grade.

5.3.2.2 Flute Marks

Flutes are heel-shaped hollows, scoured into mud bottoms. Each hollow is generally infilled by sand, contiguous with the overlying bed (Fig. 5.7). The rounded part of the flute is at the upcurrent end. The flared end points are downcurrent. Flutes are about 1–5 cm wide and 5–20 cm long. They are typically gregarious. Fluting has long been attributed to the localized scouring action of a current moving over an unconsolidated mud bottom. As the current velocity declines, flute erosion ceases and the hollows are buried beneath a bed of sand.

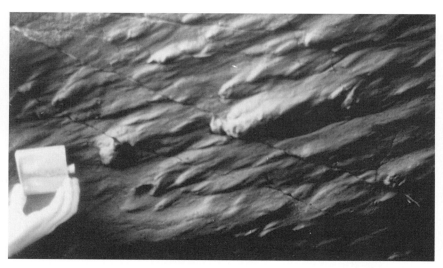

Fig. 5.7. Photograph of flute marks on the base of a turbidite sand. Current flowed from bottom left to top right. Aberystwyth Grits (Lower Silurian), Aberystwyth, Wales.

Allen (1968a, 1969, 1970, p. 82, 1971) has described experiments that explain the hydraulic conditions that generate flutes. The technique used in this and other experiments described in this chapter is as follows. A bed form is carved onto a plaster of Paris (gypsum) surface. Small, regularly spaced pits are marked on this. The slab is then placed in a flume and water passed over it. The pits become elongated in a downcurrent direction. The pit trends show how the current flow, immediately at the bed form: water interface, diverges from the mean flow direction. Allen's experiments show that the flow pattern for flute erosion consists of two horizontal corkscrew vortices that lie beneath a zone of fluid separation at the top of the flute (Fig. 5.8).

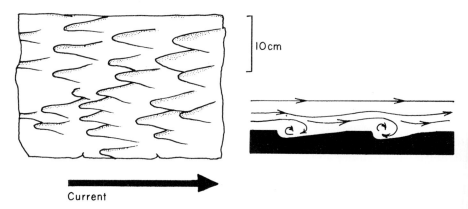

Fig. 5.8. (Left) Sketch of flute marks and (Right) cross-section showing scouring of fluted hollows in soft mud by current vortices.

Fig. 5.9. Photograph of groove marks on the base of a turbidite sand, cut into shale. Hat at top right provides scale (size 8.5). Laingsberg Formation, Ecca Group (Permian), Great Karroo, South Africa.

5.3.2.3 Groove Marks

The second important type of erosional interbed structure is groove marks. These, like flutes, tend to be cut into mud and overlain by sand. They are long, thin, straight erosional marks. They are seldom more than a few millimeters deep or wide, but they may continue uninterrupted for a meter of more (Fig. 5.9). In cross-section, the grooves are angular or rounded. Grooves occur where sands overlie muds in diverse environmental settings. Like flutes, they are especially characteristic of turbidite sands and trend parallel to the current direction determined from flutes and sedimentary structures within the sands. Grooves are seldom associated with flutes, however, and as discussed in Section 4.2.2, they are best developed in a more distal (downcurrent) situation than flutes.

It is clear that grooves are erosional features cut parallel to the current. Their straightness suggests laminar rather than turbulent flow conditions. It has been argued that the grooves are carved by objects borne along in the current. This is not easy to prove, however, if the tools are not found. Furthermore, the linearity of the groove proves that the tools were not saltating or rotating, but that they were transported downcurrent at a constant orientation and at an almost constant elevation with respect to the sediment substrate.

5.3.2.4 Tool Marks

Tool marks are erosional bottom structures that can be attributed to moving clasts. These are erosional features cut in soft mud bottoms like flutes and grooves. They are, however, extremely irregular in shape, both in plan and cross-section, though they are roughly oriented parallel with the paleocurrent. In ideal circumstances it has been possible to find the tool which cut these markings at their downcurrent end. Tools that have been found include pebbles (especially mud pellet stripped up from the bottom), wood and plant fragments, shells, astonished ammonites, and fish vertebrae (e.g., Dzulinksi and Slaczka, 1959).

Flutes, grooves, and tool marks are three of the commonest sole markings found as interbed sedimentary structures. All are erosional and all are best seen in, but not exclusive to, turbidite facies. A variety of other sole markings have been described and picturesquely named. Reviews of these can be found in Potter and Pettijohn (1977), Dzulinski and Sanders (1962), and Dzulinski and Walton (1965). It was pointed out in Section 4.2.2 that several detailed studies of turbidite formations show a correlation between current direction and sedimentary structures. Proximal turbidites tend to occur within channels. Moving downcurrent channels give way to tabular basin floor turbidites under which the sequence of erosional bottom structures grades from scour-and-fill to flute marks to groove marks and finally tool marks at their distal end. These changes are related to the downcurrent decrease in flow velocity of any individual turbidite.

5.3.3 Syndepositional (Intrabed) Structures

Syndepositional structures are those actually formed during sedimentation. They are therefore, essentially constructional structures that are present within sedimentary beds. At this point it is necessary to define and discuss just what is meant by a bed or bedding. **Bedding,** stratification, or layering is probably the most fundamental and diagnostic feature of sedimentary rocks. Layering is not exclusive to sediments. It occurs in lavas, plutonic, and metamorphic rocks. Conversely, bedding is sometimes absent in thick diamictites, reefs, and some very well-sorted sand formations. Nevertheless, some kind of parallelism is present in most sediments. Bedding is due to vertical differences in lithology, grain size, or, more rarely, grain shape, packing, or orientation. Though bedding is so obvious to see it is hard to define what is meant by the terms bed and bedding and few geologists have analyzed this fundamental property (Payne, 1942; McKee and Weir, 1953; Campbell, 1967).

One of the most useful approaches to this problem is the concept of the **sedimentation unit.** This was defined by Otto (1935) as "that thickness of sediment which appears to have been deposited under essentially constant physical conditions." Examples of sedimentation units are a single cross-bedded stratum, a varve, or a mud flow diamictite. A useful rule of thumb definition is that beds are distinguished from one another by lithological changes. Shale beds thus typically occur as thick uninterrupted sequences. Sandstones and carbonates, though they may occur in thick sections, are generally divisible into beds by shale laminae. Here are two more arbitrary but useful definitions:

1. Bedding is layering within beds on a scale of about 1 or 2 cm.
2. Lamination is layering within beds on a scale of 1 or 2 mm.

Using these dogmatic definitions, the synsedimentary intrabed structures are of five categories: massive, flat-bedded, cross-bedded, laminated, and cross-laminated. The morphology and origin of these are now described.

5.3.3.1 Massive Bedding

An apparent absence of any form of sedimentary structure is found in various types of sedimentation unit. It is due to a variety of causes. First, a bed may be massive due to diagenesis. This is particularly characteristic of certain limestones and dolomites that have been extensively recrystallized. Secondly, primary sedimentary structures may be completely destroyed in a bed by intensive organic burrowing.

Genuine depositional massive bedding is often seen in fine-grained, low-energy environment deposits, such as some claystones, marls, chalks, and calcilutites. Reef rock (biolithite) also commonly lacks bedding. In sandstones massive bedding is rare. It is most frequently seen in very well-sorted sands, where sedimentary structures cannot be delineated by textural variations. It has been demonstrated, however, that some sands which appear structureless to the naked eye are in fact bedded or cross-bedded when X-rayed (Hamblin, 1962; Lewis and McConchie, 1994).

Genuine structureless sand beds may be restricted to the deposits of mud flows, grain flows, and the lower (A unit) part of turbidites, though these may be size graded.

5.3.3.2 Flat-Bedding

One of the simplest intrabed structures is flat- or horizontal bedding. This, as its name implies, is bedding that parallels the major bedding surface. It is generally deposited horizontally. Flat-bedding grades, however, via subhorizontal bedding, into cross-bedding. The critical angles of dip that separate these categories are undefined. Flat-bedding occurs in diverse sedimentary environments ranging from fluvial channels to beaches and delta fronts. It occurs in sand-grade sediment, both terrigenous and carbonate.

Flat-bedding is attributed to sedimentation from a planar bed form. This occurs under shooting flow or a transitional flow regime with a Froude number of approximately 1. Sand deposited under these conditions is arranged with the long axes of the grains parallel to the flow direction. Moderately well-indurated sandstones easily split along flat-bedding surfaces to reveal a preferred lineation or graining of the exposed layer (Fig. 5.10). This feature is termed **parting lineation,** or primary current lineation (Allen, 1964). This sedimentary structure provides a paleocurrent indicator, indicating the sense, but not the direction of current flow (Fig. 5.11). It is important to remember that, like many of the bed sole markings previously described, parting lineation will not be seen in friable unconsolidated sands, nor in low-grade metamorphic sediments.

5.3.3.3 Graded Bedding

A graded bed is one in which there is a vertical change in grain size. Normal grading is marked by an upward decrease in grain size (Fig. 5.12). Reverse grading is where the bed coarsens upward. There are various other types (Fig. 5.13). Graded bedding is produced as a sediment settles out of suspension, normally during the waning phase of a turbidity flow (see Section 4.2.2). Though the lower part of a graded bed is normally massive, the upper part may exhibit the Bouma sequence of sedimentary structures (see Fig. 4.17, and Bouma, 1962).

The term "graded bed" is normally applied to beds measurable in centimeters or

Fig. 5.10. Photograph looking down on a bedding surface that exhibits parting lineation. Old Red Sandstone (Devonian), Mitcheldean, England.

decimeters. "Varves," typical of lacustrine deposits, are measurable in millimeters. The term "upward-fining sequence" is normally applied to intervals of several beds whose grain size fines up over several meters.

5.3.3.4 Cross-Bedding

Cross-bedding is one of the most common and most important of all sedimentary structures. It is ubiquitous in traction current deposits in diverse environments. Cross-bedding, as its name implies, consists of inclined dipping bedding, bounded by sub-horizontal surfaces. Each of these units is termed a **set.** Vertically contiguous sets are

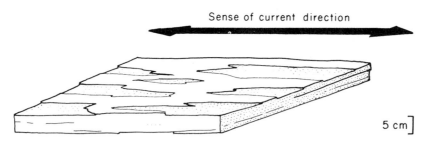

Fig. 5.11. Sketch of current or parting lineation. This appears as a graining on bedding planes of moderately cemented fissile sandstones. It indicates the sense, but not the direction, of the depositing current.

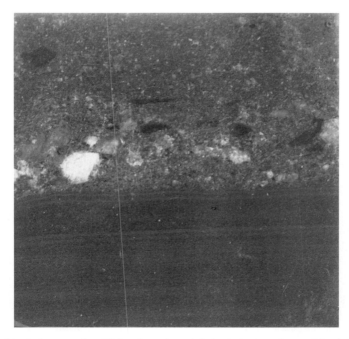

Fig. 5.12. Graded bed of greywacke with basal quartz and shale clasts abruptly overlying interlaminated silt-stone and claystone. Torridon Group (Pre-Cambrian), Raasay, Scotland. Specimen 6 cm high.

termed **cosets** (Fig. 5.14). The inclined bedding is referred to as a **foreset.** Foresets may grade down with decreasing dip angle into a bottomset or toeset. At its top a foreset may grade with decreasing dip angle into a topset. In nature toesets are rare and topsets are virtually nonexistent.

Foresets may be termed heterogeneous if the layering is due to variations in grain size, or homogeneous if it is not. Two other descriptive terms applied to foresets are avalanche and accretion (Bagnold, 1954, pp. 127, 238). Avalanche foresets are planar in vertical section and are graded toward the base of the set. Accretion foresets are un-graded, homogeneous, and have asymptotically curved toesets.

Many workers have recorded the angle of dip of foresets (see Potter and Pettijohn, 1977, p. 101). A wide range of values has been obtained with a mode of between 20 and 25° for ancient sediments. The foreset dip reflects the critical angle of rest of the sand when it was deposited. This will be a function of the grade, sorting, and shape of the sediment as well as the viscosity of the ambient fluid. Legend has it that eolian sands have higher angles of rest than subaqueous sands. There are some data to support this. Dips of 30–35° have been recorded from modern eolian dunes (e.g., McBride and Hayes, 1962; Bigarella, 1972). Dips in modern subaqueous cross-bedded sands seldom appear to exceed 30° (e.g., Harms and Fahnestock, 1965; Imbrie and Buchanan, 1965).

The rather lower angles recorded from ancient sediments may be due to a variety of factors. These include the fact that a set-bounding surface is seldom a valid paleohori-zon datum. They frequently dip upcurrent and thus the maximum measured angle of

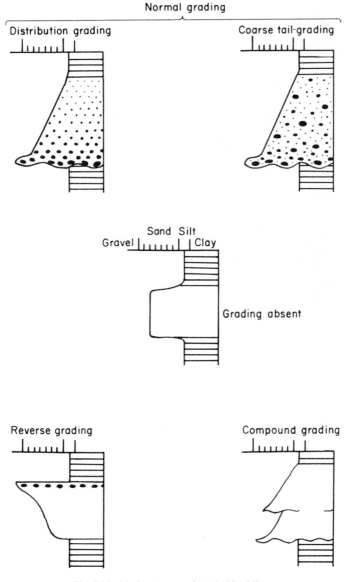

Fig. 5.13. Various types of graded bedding.

foreset dip will be less than the true dip. A second factor is that unless the amount of dip is recorded from a face that is exactly perpendicular to the dip direction, then an apparent dip will be recorded that is less than the true dip. A third factor that may decrease the depositional dip of a foreset is compaction (Rittenhouse, 1972).

The diverse geometric relationships that might exist between foresets and their bounding surfaces has led to a rich blossoming of geological nomenclature and classi-

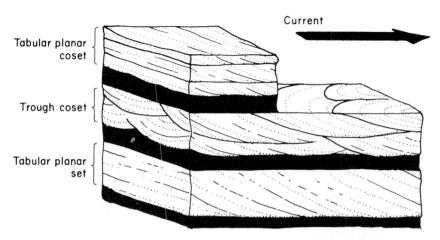

Fig. 5.14. Basic nomenclature of cross-bedding. Tabular planar cross-beds have subplanar foresets. Trough cross-beds have spoon-shaped foresets. Isolated cross-beds are referred to as sets. Vertically grouped foresets constitute a coset.

ficatory schemes (e.g., Allen, 1963). Basically, two main types of cross-bedding can be defined by the geometry of the foresets and their bounding surfaces: tabular planar cross-bedding and trough cross-bedding (McKee and Weir, 1953). In tabular planar cross-bedding, planar foresets are bounded above and below by subparallel subhorizontal set boundaries (Fig. 5.15). In trough cross-bedding, upward concave foresets lie

Fig. 5.15. Tabular planar cross-bedding in fluvial Cambro-Ordovician sandstones, Jebel Gehennah, southern Libya.

Fig. 5.16. Trough cross-bedding in fluvial Cambro-Ordovician sandstones, Jebel Dohone, southern Libya.

within erosional scours which are elongated parallel to current flow, closed upcurrent and truncated downcurrent by further troughs (Fig. 5.16).

Additional details on cross-bedding morphology and nomenclature are given in Potter and Pettijohn (1977, pp. 91–102). Full descriptions of fluvial cross-bedding have been given by Frazier and Osanik (1961), Harms et al. (1963), and Harms and Fahnestock (1965). Cross-bedding in tidal sand bodies has been described by Hulsemann (1955), Reineck (1961), and Imbrie and Buchanan (1965). The last of these studies shows that the internal sedimentary structures of carbonate sands are no different from those of terrigenous deposits.

The genesis of cross-bedding has been studied empirically from ancient and modern deposits and experimentally in laboratories. It appears that much cross-bedding is formed from the migration of sand dunes or megaripples. Flume experiments (described in Section 4.2.1.1) showed how these bed forms migrate downcurrent depositing foresets of sand in their downcurrent hollows. If sedimentation is sufficiently great, then the erosional scour surface in front of a dune will be higher than that of its predecessor and a cross-bedded set of sand will be preserved (Fig. 5.17). Tabular planar cross-bedding will thus form from straight crested dunes. Trough cross-bedding will form in the rounded hollows of more complex dune systems. There are, however, several other ways in which cross-bedding may form and three of these should be noted in particular.

In river channels, especially those of braided type (see Section 6.3.2.2.2), the course consists of an alternation of shoals and pools through which the axial part or parts of the channel (termed the "thalweg") make their path. Where the thalweg suddenly enters a pool there is a drop in stream power and a subaqueous sand delta, termed a braid bar, is built out. Given time, sufficient sediment, and the right flow conditions, this delta may

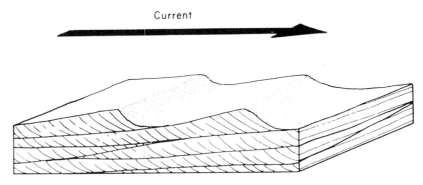

Fig. 5.17. Formation of tabular planar cross-bedding occurs where dunes migrate downcurrent, and where stoss side erosion is less than sedimentation of the foreset. Note that the preserved thickness of each set is less than the height of the dune from which it was deposited.

completely infill the pool with a single set of cross-strata (Jopling, 1965) (e.g., Fig. 5.18; see also Fig. 6.6).

A second important way in which cross-bedding forms is seen in channels. A channel may be infilled by cross-bedding paralleling the channel margin. Alternatively cross-bed deposition occurs on the inner curves of meandering channels synchronous with erosion on the outer curve (see Section 6.3.2.2.3). By this means, a tabular set of cross-strata may be deposited in which the foresets strike parallel to the flow direction (Lyell, 1865, p. 17). This type of lateral cross-bedding is rather larger than most types (sets may be several meters high) and the base is marked by a conglomerate. Close examination of the foresets often shows that they are composed of second-order cross-beds or cross-laminae which do in fact reflect the true current direction (Fig. 5.19).

A third important variety of cross-bedding is that formed by antidunes in upper flow regime conditions. It has been pointed out that at very high current velocities sand dunes develop that migrate upcurrent (Section 4.2.1.1). These deposit upcurrent dipping foresets. The foresets of these antidunes, as they are called, are seldom preserved. As the current wanes prior to net sedimentation, antidunes tend to be obliterated as the bed form changes to a plane bed or dunes. A few instances have been noted. Some have been described from the A unit of turbidites (Skipper, 1971; Hand *et al.,* 1972) as mentioned in Section 4.2.1.1. Alexander and Fielding (1997) have described gravel antidunes in the

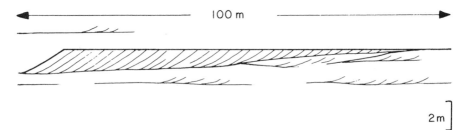

Fig. 5.18. Single large foreset deposited in braided channel chute pool. Cambro-Ordovician, Wadi Rum, Jordan.

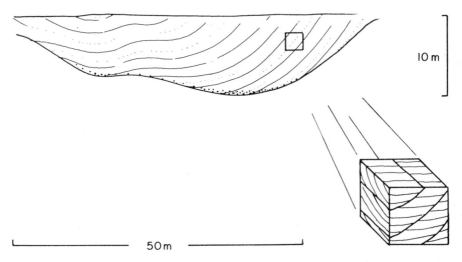

Fig. 5.19. Channel showing complex cross-bedded fill. Major trough cross-sets are themselves composed of smaller tabular planar sets. Marada Formation (Miocene), Jebel Zelten, Libya.

modern Burdekin River of Queensland, Australia. Figure 5.20 illustrates an example from an ancient alluvial environment.

In shallow marine environments it is not uncommon to find **herring-bone** cross-bedding in which bimodally dipping foresets reflect the to-and-fro movement of ebb and flood currents (Fig. 5.21). Allen (1980) has illustrated the spectrum of cross-bedding and associated structures which reflect the degree of symmetry of tidal currents.

The term **hummocky cross-stratification** was first applied by Harms (1975) to a particularly distinctive type of cross-bedding. Each unit contains several sets of irregular convex-up cross-beds, some 10–15 cm thick (Fig. 5.22). Hummocky cross-bedding tends to occur in regular sequences about 0.5 m thick (Dott and Bourgeois, 1982, 1983). The base of each unit is generally a planar erosional surface with a lag gravel, often with bioclasts. The upper contact is sharp or gradational. A cross-laminated and/or bioturbated

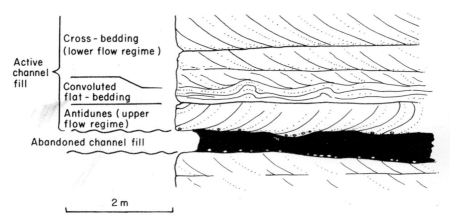

Fig. 5.20. Antidune cross-bedding at base of braided channel sequence. Cambro-Ordovician, Jordan.

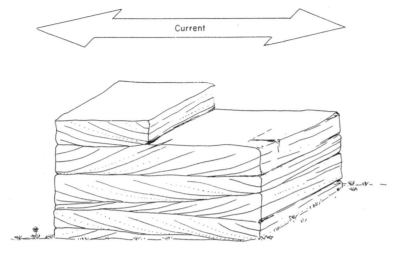

Fig. 5.21. Sketch of herring-bone cross-bedding due to tidal currents.

zone sometimes separates the hummocky cross-bedding from overlying shale. Super-ficially these units look rather like turbidites.

Hummocky cross-stratification normally occurs in vertical sequences overlain by typi-cal cross-bedded shallow marine shelf sands and overlying turbidites and pelagic muds. Walther's law (Section 6.4.1) implies, therefore, that hummocky cross-bedding was de-posited in intermediate water depths. Recent hummocky cross-stratification has been observed on the continental shelf of the northwest Atlantic Ocean in water depths of 1–40 m. It is interpreted as due to a combination of storm-generated and geostrophic currents (Swift *et al.*, 1983). Both observation and deduction suggest, therefore, that hummocky cross-stratified sequences are storm deposits (sometimes referred to as **tempestites** (Ager, 1973). Thus the erosion surfaces mark the height of the storm; cross-stratification reflects the waning storm; cross-lamination and bioturbation indicate interstorm idylls.

It is also possible to integrate hummocky cross-stratification within a sequence of pro-cesses and preservation potentials. Turbidity currents can occur in any water depth, from a shallow lagoon to the ocean floor. In shallow water, however, turbidites are reworked

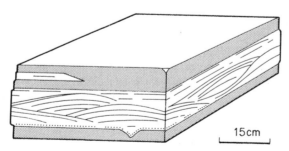

Fig. 5.22. Diagrammatic sketch of hummocky cross-stratification from the Brachina Subgroup (Late Pre-Cambrian), Hallett Cove, South Australia. (Displayed to the author by I. A. Dyson.)

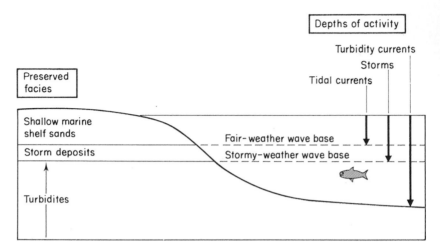

Fig. 5.23. Illustration of the operation, depths, and preserved intervals of turbidite and storm deposits. This shows that hummocky cross-stratification indicates deposition between the fair weather and stormy weather wave bases.

and redeposited by normal traction currents down to a fair weather wave base. Storm deposits can occur from sea level down to a storm weather wave base (Aigner, 1985). They too will be reworked by traction currents above the fair weather wave base, but may in turn rework turbidites (Fig. 5.23).

These descriptions of the various types of cross-bedding show that it is a very complex sedimentary structure. More properly it is a group of structures of diverse morphology and genesis. Particular attention has been paid by geologists to determine depositional environment from the type of cross-bedding. This has not been notably successful because, though the structural morphology is closely related to hydrodynamic conditions, the same set of hydrodynamic parameters can occur in various environments. Hummocky cross-bedding is perhaps the only exception.

Another line of approach has been to try to determine water depth from set height. This has not been very successful either, for several reasons, not the least of which is that the preserved set height is controlled by the degree of erosion that occurred after a set was laid down. Nevertheless, in underwater cross-bedding, water depth cannot have been less than the preserved set height. Set height has also been used to try to distinguish eolian from subaqueous dunes. The folklore holds that eolian dunes deposit higher set heights than subaqueous ones. This is not universally true as the studies cited in Section 4.3.1 show. No satisfactory height limit for subaqueous cross-bedding can be fixed because the internal morphology of submarine dunes is so little known (see Section 6.3.2.7.2). One of the most important things that can be learned from cross-bedding is the flow direction of the currents which deposited them. This can give important clues to the environment, paleogeography, and structural setting of the beds in which they occur. This important topic of paleocurrent analysis is discussed later in the chapter.

5.3.3.5 Ripples and Cross-Lamination

Ripples are a wave-like bed form that occurs in fine sands subjected to gentle traction currents (Fig. 5.24). Migrating ripples deposit cross-laminated sediment. Individual

Fig. 5.24. Ripples in lacustrine Torridon Group (Pre-Cambrian) sandstones. Raasay, northwest Scotland.

cross-laminated sets seldom exceed 2–3 cm in thickness, in contrast to cross-bedding, which is normally >50 cm thick. It is hard to define arbitrarily the set height that separates cross-lamination from cross-bedding. In practice, the problem seldom arises, because sets 5–50 cm thick are rare in nature. Ripple marking in modern and ancient sediments has attracted the interest of many geologists. Studies of historical significance include those of Sorby (1859), Darwin (1883), Kindle (1917), Bucher (1919), and Allen (1968b). The last of these is a definitive work of fundamental importance.

The following account describes the association of ripple bed form and internal cross-lamination and then discusses their origin. Figure 5.25 illustrates the nomenclature of ripples. This particular case shows asymmetric ripples formed by a traction current. In cross-section a ripple consists of a gentle upcurrent stoss side and a steep downcurrent-facing lee side. The highest points of the ripples are the crests. The lowest points are the troughs. The height of the ripple is the vertical distance from trough to crest. The wavelength of a train of ripples (their collective term) is the horizontal distance between two

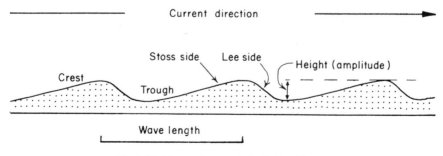

Fig. 5.25. Nomenclature of rippled bed forms.

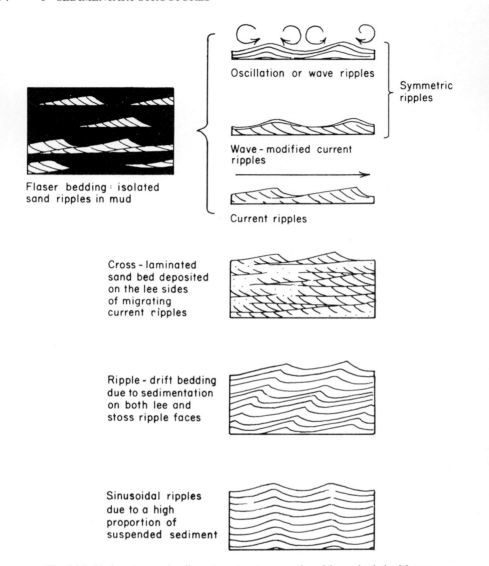

Fig. 5.26. Various types of sedimentary structures produced from ripple bed forms.

crests or troughs. A statistical parameter termed the ripple index (abbreviated to RI) is calculated by dividing the wavelength by the ripple height. Other numerical indices derived from ripple form are described by Tanner (1967) and Allen (1968b). Various types of ripple bedding are defined by a combination of their external shape (bed form) and internal structure (Fig. 5.26).

In cross-section ripples are divisible into those with symmetric and those with asymmetric profiles. Symmetrical ripples, also called oscillation or vortex ripples, are commonly produced in shallow water by the orbital motion of waves (Section 4.2.1.2). In plan view they are markedly subparallel, but occasionally bifurcate. Internally they show laminae that are either concordant with the ripple profile, or chevron-like, or bimodal bipolar. Sometimes symmetrical ripples contain cross-laminae that dip only in a shore-ward direction. These are termed wave-formed current ripples (Fig. 5.26).

Asymmetric ripples, by contrast to symmetric ones, show a clearly differentiated low-angle stoss side and steep-angle lee side. Internally they are cross-laminated, with the cross-laminae concordant with the lee face. Asymmetric ripples are produced by uni-directional traction currents as, for example, in a river channel. It may be hard to in-terpret the origin of some individual ripples. Wave and current action can alternately modify bed forms during a tidal cycle or a fluvial flood phase. Normally it is wave ac-tion that molds a previously formed asymmetric current ripple (Allen, 1979).

Both asymmetric and symmetric ripples can occur with isolated lenses of mudstone. This is termed **flaser bedding** (Reineck and Wunderlich, 1968; Terwindt and Breusers, 1972). With gradually increasing sand content, flaser bedding can grade into beds com-posed entirely of cross-laminated sand in which ripple profiles are absent, though they are sometimes preserved on the top of the bed. Various terms have been proposed for these sedimentary structures, including cross-lamination, climbing ripples, and ripple drift bedding. Jopling and Walker (1968) have defined a spectrum of ripple types, that is related to the ratio of suspended to traction load material which is deposited (Fig. 5.26). Normal traction currents deposit sand on the lee side of the ripple only. With increas-ing suspended load, sedimentation also occurs on the stoss side. This generates a series of ripple profiles whose crests migrate obliquely upward downcurrent. With excessive suspended load, sinusoidal ripple lamination develops from the vertical accretion of symmetric ripple profiles. Jopling and Walker point out that these symmetric ripples that deposit continuous laminae of sediment are distinct from the isolated symmetric ripples formed by wave oscillation.

Particular attention has been paid to trying to differentiate cross-lamination of non-marine and marine origins (e.g., Flemming and Bartholoma, 1995). It has been suggested that draping clay laminae on ripple foresets indicate subtidal deposition; the Hjulstrom effect (see Section 4.1) permits the preservation of the draping laminae formed from clay that settles out at slack water (Visser, 1980). Clay drapes have, however, also been observed on modern intertidal flats (Fenies *et al.,* 1999) and in interdune sabkhas (Glen-nie, 1970, 1987).

Having described ripple morphology in cross-section, now consider them in plan. Ripples seen in modern sediments or exposed on ancient bedding surfaces show a di-versity of shapes. Certain dominant types tend to occur and these have been named (Fig. 5.27). Simplest of all are the straight-crested ripples; these include ripples with both symmetric and asymmetric profiles. Straight-crested or rectilinear ripples can be traced laterally for many times further than their wavelength. They are oriented perpendicu-lar to the direction of wave or current movement that generates them. Sinuous ripples show continuous but slightly undulating crest lines.

The second main group of ripples, as seen in plan, are those whose crest lengths are generally shorter than their wavelength. These are exclusively asymmetric current ripples. Two important varieties can be recognized. Lunate ripples have an arcuate crest, which is convex upcurrent. Linguoid ripples have an arcuate crest, which is convex down-current. In plan view, successive linguoid or lunate ripples may be arranged *en echelon,* out of phase with one another, or in phase, if the crests all lie on the same flow axis. In the same way that trough cross-bedding originates in the migrating hollows of complex

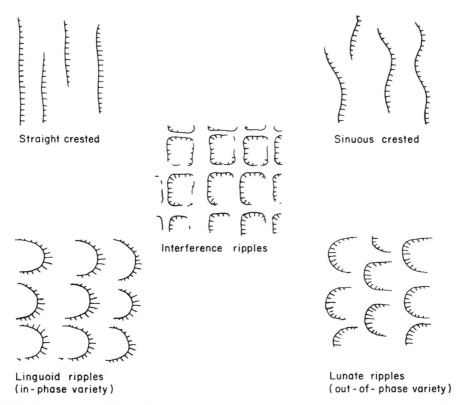

Straight crested

Interference ripples

Sinuous crested

Linguoid ripples
(in-phase variety)

Lunate ripples
(out-of-phase variety)

Fig. 5.27. Nomenclature of rippled bed forms as seen in plan view. Current moves from left to right. (For definitions, see Allen, 1968a.)

dune fields, so does a variety of trough cross-lamination form from migrating complex ripple trains. This structure is picturesquely described as **rib and furrow.**

A third main group of ripples can be recognized from their appearance in plan. These are interference ripples, which, as their name suggests, consist of two obliquely intersecting sets of ripple crests. Interference ripples result from the modification of one ripple train due to one set of conditions by a later train, generated by waves or currents with a different orientation. "Tadpole nests" is a quaint synonym for interference ripples. Having examined their morphology, we now consider the origin of ripples in rather more detail.

It is a matter of observation that ripples do not form in clay or in coarse sand or gravel. They are restricted to coarse silt and sand with a grain size of less than about 0.6 mm diameter. Analysis of traction currents shows that ripple bed forms occur in the lower part of the lower flow regime with a low Froude number (see Chapter 4). Particular attention has been paid to the way in which ripples are actually formed from a plane bed of sand. It has been suggested that ripple trains develop downstream from preexisting irregularities of the sediment substrate (e.g., Southard and Dingler, 1971). An alternative school of thought argues that ripples can form spontaneously on a plane sand bed. Initiation is by random turbulent vortices, which scour the first irregularities (e.g., Williams and Kemp, 1971).

Of more interest to geologists is not so much what initiates ripple formation, but what can be learned from the resultant structure. One would like to gain information about flow conditions, direction, water depth, and environment. Attempts to relate ripple type to flow conditions have not been noticeably successful. Allen (1968a, 1986a,b) has, however, recognized a broad correlation between current ripples and decreasing depth and concomitant increasing velocity. The sequence changes from straight-crested ripples to sinuous crests, and so to lunate and linguoid trains. The water depth in which ripples form has no effect on their height or wavelength except at extremely shallow depths. Flow direction, on the other hand, can be very easily determined from ripples, both from their strike and lee face in plan view, and from the dip direction of their internal cross-lamination. This aspect of paleocurrent analysis is discussed later in the chapter. Ripples occur today in many different environments, ranging from the backs of eolian sand dunes, through rivers and deltas, to the ocean bed. It has already been pointed out that ripples are closely related to a given set of flow conditions and that these may be encountered in diverse environments. One would not, therefore, expect to find a relationship between ripple morphology and environment, as opposed to process. Nevertheless, Tanner (1967, 1971) empirically developed several statistical indices that appear to be capable of differentiating ripples from different sedimentary environments.

5.3.4 Postdepositional Sedimentary Structures

The third main group of sedimentary structures is a result of deformation. These may be termed postdepositional because, obviously, they can only form after a sediment has been laid down. A great variety of deformational structures exist, many of which are ill defined and strangely named. They can be arranged, however, into three main groups arbitrarily defined according to whether the sense of movement was dominantly vertical or dominantly lateral, and according to whether the sediment deformed plastically in an unconsolidated state, or whether it was sufficiently consolidated to shear along slide planes (Table 5.4). These three groups of deformation structures are described next.

5.3.4.1 Vertical Plastic Deformational Structures

Deformational structures that involve vertical plastic movement of sediment are of two main types. One group occurs within sand beds and may be loosely referred to as

Table 5.4
Classification of Postdepositional Deformational Sedimentary Structures

Sense of movement	Structure	Nature of deformation
Dominantly vertical	Load casts and pseudo-nodules, convolute bedding, recumbent foresets, convolute lamination	Plastic (sediments lack shear strength)
Dominantly horizontal	Slumps Slides	Brittle (sediments possess shear strength)

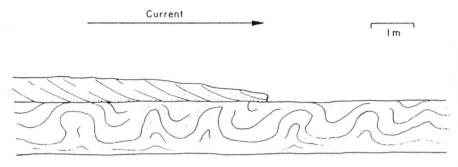

Fig. 5.28. Convolute bedding due to the expulsion of pore water from loosely packed sand. Torridonian (Pre-Cambrian) fluvial sandstones. Raasay, Scotland.

quicksand structures. Structures of the second group develop at the interfaces of sand overlying mud. The simplest type of quicksand structure is seen in vertical section as a series of plastic folds. Typically broad flat synclines separate sharp peaked anticlines. The anticlines are sometimes overturned downcurrent (as shown by cross-bedding in the overlying sand). In plan view the folds are often elongated perpendicular to current direction. This type of quicksand structure involves the deformation of whole beds of sand up to a meter or more thick (Fig. 5.28). It is loosely referred to as **convolute bedding.** This structure is found in many types of sandstone, but is particularly characteristic of fluvial sands (see Selley *et al.,* 1963, and McKee *et al.,* 1967, for ancient and modern examples, respectively).

Convolute bedding is often found associated with deformed cross-bedding. The foresets are overturned downcurrent in the shape of recumbent folds. The axial plane of the fold is commonly tilted downcurrent in any one set (Fig. 5.29). Recumbent foresets, like convolute bedding, are found in diverse traction-deposited sands, and are especially common in the coarse sands of braided alluvium.

Considerable attention has been paid to the origin of convolute bedding and recumbent foresets (e.g., Allen and Banks, 1972; Mills, 1983; van Loon and Brodzikowski, 1987). There is widespread agreement that these structures are caused by the vertical passage of water through loosely packed sand. This water may be due to a hydrostatic head of water, for example, as is seen on an alluvial fan (e.g., Williams, 1970). Alternatively, the water may be derived from the sediment itself. A sand will not compact significantly at the surface, but its grains may be caused to fall into a tighter packing. This results in a decrease in porosity. Excess pore water will be vertically expelled. Labora-

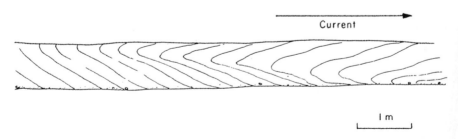

Fig. 5.29. Recumbent foreset deformation in fluvial sandstone. Cambro-Ordovician, Jordan.

tory experiments have shown that this process can indeed generate convolute bedding (Selley, 1969). These experiments showed that the sands could fall into a tighter packing both by vibration and by turbulent eddies in the overlying water.

Convolute bedding has been recorded in modern sediments both as a result of earthquakes and without them (e.g., Barratt, 1966, and McKee *et al.,* 1967, respectively). Allen (1986b) has even established an empirical relationship between the frequency of earthquake-induced deformation, distance from the epicenter, and quake magnitude on the Richter scale. The downcurrent overturning of convolute folds and their association with downcurrent deformed foresets strongly suggest that powerful currents play a significant part in their genesis. This structure is not restricted to aqueously deposited sediment, however, but also occurs in eolian ones (Doe and Dott, 1980).

On a smaller scale, laminated fine sands and silts also show penecontemporaneous vertical deformation structures termed **convolute lamination.** This is similar in geometry to convolute bedding, but occurs in finer grained sediment on a much smaller scale; generally in beds only a decimeter or so high. Convolute lamination is especially characteristic of turbidites, involving deformation of both the laminated and cross-laminated Bouma units. Correlation of fold axes with ripple crests, and the presence of deformed intrabed scour surfaces, suggests that movement was virtually synchronous with deposition. Convolute lamination probably originates, therefore, by the dewatering of the sediment aided by the shear stresses set up by the turbidity flow itself (see also Davies, 1965; Anketell *et al.,* 1970; Visher and Cunningham, 1981).

Convolute bedding, recumbent foresets, and convolute lamination are the three main types of intrabed vertical deformational structures. **Dish structure** is a particular variant of intrasand deformation. This is seen where laminae or bedding planes are intermittently disrupted and upturned like the rim of a dish. Dish structure is a type of dewatering phenomenon that is particularly characteristic of fluidized sand beds (see Section 4.5.3). It testifies to the loose and unstable packing of the sand when first deposited (Lowe and Lopiccolo, 1974; Lowe, 1975). Dish structure is often associated with vertical pipes or pillars that look like organic burrows. The association with dish deformation suggests, however, that the pipes are water escape conduits.

A variety of structures develop where sands overlie muds. The mud:sand interface is often deformed in various ways. Most typically irregular-rounded balls of sand depend from the parent sand bed into the mud beneath. These structures are variously termed **loadcasts,** ball and pillow structures, etc. They are a variety of the broad group of structures termed sole markings or bottom structures. It is important, however, to distinguish deformational bottom structures, like loadcasts, from erosional markings such as grooves and flutes. Sometimes erosional bottom structures become deformed. In extreme cases the sand lobes may become completely detached from their parent bed above. Similarly, thin sand beds may split along their length to form isolated cakes of sand in mud (Fig. 5.30). These discrete bodies of sand in mud are termed **pseudonodules** to distinguish them from normal diagenetic nodules (Macar and Antun, 1949).

Loadcasts and pseudonodules occur at sand:mud interfaces in various environments, both modern and ancient. They are a common feature of turbidite deposits, yet they also occur in deltaic and fluvial sediments. There is general agreement that these structures are generated by the differential loading of a waterlogged sand on an unconsolidated mud. They are easy to make in the laboratory (e.g., Kuenen, 1958, Owen, 1996).

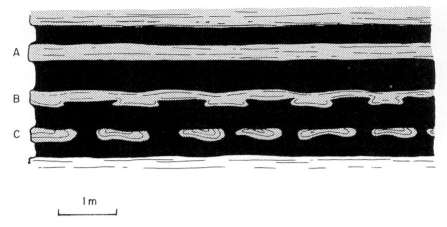

Fig. 5.30. Vertical load structures in interbedded sands and siltstones. Torridonian (Pre-Cambrian) Isle of Fladday, Scotland. Bed A is undeformed, bed B shows well-developed load casts on its lower surface, and bed C has split into discrete pseudonodules.

5.3.4.2 Slumps and Slides

Slump structures, like the structures previously described, involve the penecontemporaneous plastic deformation of sand and mud. Slump folds, however, commonly show clear evidence of extensive lateral movement in a consistent direction (Fig. 5.31). Slump folds are commonly associated with penecontemporaneous faulting and with major low-angle zones of decollement termed "slide planes." Large masses of sediments are lat-

Fig. 5.31. Geologist perplexed by sight of slump bedding in Lisan Marls (Quaternary). Sodom, Israel.

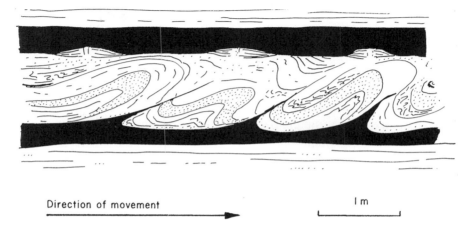

Direction of movement

I m

Fig. 5.32. Slumped beds showing recumbent folds, slide surfaces, and sand volcanoes. The presence of the latter demonstrates not only that movement was penecontemporaneous, but also that it occurred at the sediment–water interface before deposition of the overlying shale. Based on examples in the Carboniferous of County Clare, Ireland.

erally displaced along slide surfaces. In rare, but fascinating cases, the top of a slump bed may be covered by volcanoes of sand complete with axial vents and bedded cones (Fig. 5.32). These are formed from sand carried up during dewatering of the slump after it came to rest (Gill and Kuenen, 1958; Gill, 1979).

When slides and slumps were first studied there was considerable controversy over whether these phenomena were tectonic or whether they were penecontemporaneous soft sediment features. Distinctive criteria for penecontemporaneous movement include the fact that folds and faults are overlain by undisturbed sediment, and their orientations may be unrelated to regional tectonic style and orientation. Disturbed beds may be penetrated by undeformed plant roots or animal burrows, and penecontemporaneous faults lack gangue minerals.

Slumps and slides require for their generation the rapid deposition of muddy sediment on an unstable slope. Lateral movement may be initiated by earthquakes, storms, or perhaps purely spontaneously. These conditions are best met with on delta fronts in actively subsiding basins. Many case histories of sliding and slumping have been recorded from these situations (e.g., Kuenen, 1948; Blanc, 1972; Klein *et al.,* 1972). Nevertheless, slumping occurs on all scales, from the caving of a river bank to the collapse of a continental margin. For example, large slump masses of Pleistocene sediment have been delineated seismically off the Rockall Bank (North Atlantic) and on the continental slope off the east coast of the North Island of New Zealand (Roberts, 1972, and Lewis, 1971, respectively). In the last of these examples, 10- to 50-m-thick beds of Pleistocene sand and silt slumped down slide planes of 1–4°. The Canary Islands slide was described in the previous chapter.

With increasing size a slump block grades into a fault block. On the extreme end of the scale, the coast of southern Africa has experienced extensive gravitational collapse as the Atlantic Ocean opened up (Dingle, 1980), and Barbados slumped into an ocean basin and was then tectonically inverted (Davies, 1971). Thus it is hard to define the boundary between what is a large sedimentary slump and what is a tectonic fault block. Nonetheless extensive lateral sediment displacement is often structurally controlled as it grades into the realms of gravity tectonics.

5.3.5 Miscellaneous Structures

Among the vast number of sedimentary structures that have been observed, many do not fit conveniently into the simple tripartite scheme outlined earlier. These miscellaneous structures include rain prints, salt pseudomorphs, and various vertical dike-like structures of diverse morphology and origin. These include desiccation cracks, synaeresis cracks, sedimentary boudinage, and sand dikes.

5.3.5.1 Rain Prints

Rain prints occur within siltstones and claystones, and where such beds are overlain by very fine sandstones. In plan view, rain prints are circular or ovate if due to windblown rain. They are typically gregarious and closely spaced. Raised ridges are present around each print. Individual craters range from 2 to 10 mm in diameter. Rain prints are good indicators of subaerial exposure but are not exclusive to arid climates, though they may have a higher preservation potential in such conditions.

Care should be taken to distinguish rain prints from pits formed where sand grains impress soft mud. This is sometimes found where fissile shales contain thin coarse sand laminae. Such sand grain imprints lack the raised rim of rain prints (Fig. 5.33).

5.3.5.2 Salt Pseudomorphs

Salt pseudomorphs occur in similar lithological situations to rain prints. They are typically found where claystones or siltstones are overlain by siltstones or very fine sandstones. Salt pseudomorphs are molds formed in soft mud by cubic halite crystals They often show the concave "hopper" habit. The salt crystals grow in mud deposited on the substrate of hypersaline lakes and lagoons. An influx of turbid nonsaline water dissolves the salt crystals and buries the mold beneath a new layer of sediment (Fig. 5.34).

5.3.5.3 Desiccation Cracks

A variety of vertical planar structures have been recognized in sediments, these include shrinkage cracks, sedimentary dikes, and Neptunean dikes. Shrinkage cracks are often

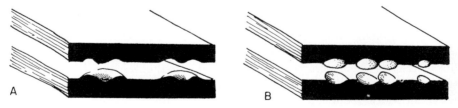

Fig 5.33. (A) Rain prints showing characteristic raised marginal rims. (B) Pits produced by isolated coarse sand grains on shaley partings misidentified as rain prints can cause geo-boobs.

Fig. 5.34. Salt pseudomorphs in hypersaline lagoonal pellet mudstone. Purbeck Beds (Upper Jurassic).

recorded in muddy sediments. They are of two types. Desiccation cracks form subaerially; synaeresis cracks form subaqueously.

Desiccation cracks, also known as sun cracks, are downward tapering cracks in mud, which are infilled by sand. In plan view they are polygonal. Individual cracks are a centimeter or so wide. Polygons are generally about 0.5 m across (Fig. 5.35). The cracks

Fig. 5.35. Desiccation cracks in lacustrine Torridon Group (Pre-Cambrian) shales. Raasay, northwest Scotland.

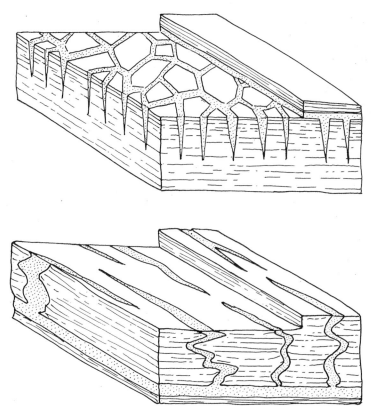

Fig. 5.36. (Upper) Desiccation cracks caused by the contraction of mud to form downward-tapering fissures arranged in polygons. (Lower) Sandstone dikes showing ptygmatic contortions due to compaction and attachment to underlying parent sand bed.

may extend down for an equivalent distance (Fig. 5.36). Picard (1966) described discontinuous linear desiccation cracks that do not join into polygons, but are oriented parallel to the local paleoslope. Even the radius of curvature of desiccation cracks has now been subjected to mathematical analysis (Allen, 1986a). Desiccation cracks may be differentiated from synaeresis cracks if they are associated with rain prints, vertebrate tracks, or other indicators of subaerial exposure.

5.3.5.4 Synaeresis Cracks

Synaeresis cracks are formed in mud by the spontaneous dewatering of clay beneath a body of water (White, 1961; Plummer and Gostin, 1986). They are distinguishable from desiccation cracks because they are infilled by mud similar or only slightly coarser in grade than that in which they grow. Furthermore, synaeresis cracks are generally much smaller than desiccation cracks; typically only 1–2 mm across. Examples have been described from the Torridon Group (Pre-Cambrian) of Scotland (Selley, 1965, p. 373), and from the Devonian Caithness Flags also in Scotland (Donovan and Foster, 1972). The distinction between subaerial desiccation cracks and subaqueous synaeresis cracks is not always easy to make. In particular, the huge polygons of modern playas, into whose cracks a camel may fall, may be due to a combination of subaerial and subaqueous dehydration with complex histories related to Quaternary climatic changes.

5.3.5.5 Sand Dikes

Sand dikes are vertical sheets of sand that have been intruded into muds from sand beds beneath. Though they are sometimes polygonally arranged, they can be distinguished from desiccation cracks by their tendency to die out upward and by the fact that they are rooted to the parent sand bed below. Sand dikes are intruded as liquefied quicksand into water-saturated mud. Like desiccation cracks, they often show ptygmatic compaction effects (Fig. 5.36). A notable example of sand dike intrusion occurs in the Miocene of the Panoche Hills, California, where more than 350 sand dikes and sills intrude the Moreno shale (Smyers and Peterson, 1971). They range from 1 dm to 7 m in width. Individual dikes are up to 1 km in length.

A particular variety of sand dike is the sedimentary boudinage structure. This is morphologically similar to tectonic boudinage. Instead of being due to the necking of a competent limestone or sandstone in incompetent shale, the converse is true. Sedimentary boudinage typically occurs where interbedded, unconsolidated, water-saturated sands and muds are subjected to tension, for example, adjacent to a slump. Clay beds develop necks and are sometimes divided up into blocks by sand intrusion from above and below (Fig. 5.37).

5.3.5.6 Sedimentary Volcanoes

Volcanoes of ejected sediment are sometimes found whose gross morphology of vent and cone are analogous to igneous volcanoes. These sedimentary volcanoes range in scale from diameters of only a few centimeters or so to several kilometers.

Monroes are small (5–10 cm in diameter) mud volcanoes with a mammillated appearance that have been recorded on recent tidal flats. They appear to be generated by the escape of water charged with biogenic gas (Dionne, 1973).

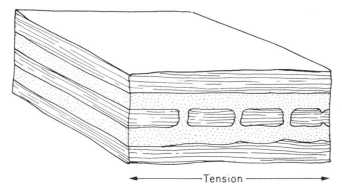

Fig. 5.37. Sedimentary boudinage. Formed by tensional splitting of clay bed accompanied by quicksand injection along the incipient fissures.

Somewhat larger than Monroes are **sand volcanoes.** These are more than 1 m in diameter. They may overlie slumped beds, such as those illustrated (Fig. 5.32) earlier from the Carboniferous slumps of County Clare, Ireland (Gill and Kuenen, 1958; Gill, 1979). A feeder pipe extends down from the center of the volcanic cone into a deeper sand bed. Sometimes the feeder pipe exits at ground level in the form of a depressed pock mark. The feeder, and its concomitant surface expression, is termed a **sand blow.** They have been described from recent sediments from around the world since the eighteenth century (De la Beche, 1851). Sand blows are generally reported in sediments that have recently been subjected to seismic activity (e.g., McCalpin, 1996; Obermeir, 1996). Sand blows on the Mississippi River floodplain, however, are due to flood activity, devoid of recent earthquakes (Li *et al.*, 1996).

At a larger scale still are modern sedimentary volcanoes discovered in deep marine settings, for example, **mud volcanoes** up to 1.5 km in diameter occur in 1800 m of water in the Mediterranean Sea between Greece and Libya. Here many volcanoes have been located above overpressured mud diapirs along a compressive fold belt (Cita *et al.*, 1995). The Gulf of Mexico is a petroleum province where volcanoes ejecting sediments, and assorted gases and liquids, often petroliferous, are ubiquitous. One particularly noteworthy vent occurs 115 km southeast of the Mississippi River delta in over 2100 m of water. This vent is similar in size to the Mediterranean ones (Fig. 5.38). It also occurs above an overpressured mud diapir (Prior *et al.*, 1989).

5.3.5.7 Pock Marks

Pock marks are circular depressions found on the sea bed (Fig. 5.39). They are up to several hundreds of meters in diameter and tens of meters deep. Pock marks have been

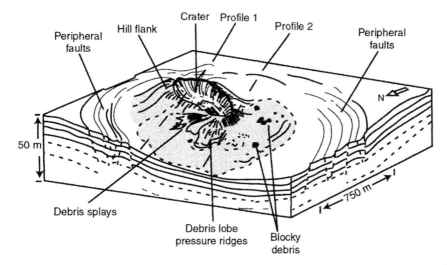

Fig. 5.38. Sketch of a 1-km-diameter sediment volcano imaged by sidescan sonar on the floor of the Gulf of Mexico. [Reprinted with permission from Prior, D. B., Doyle, E. H., and Kaluza, M. J. (1989). Evidence for Sediment Eruption on Deep Sea Floor, Gulf of Mexico. *Science* **243,** 517–518. Copyright © 1989 American Association for the Advancement of Science.]

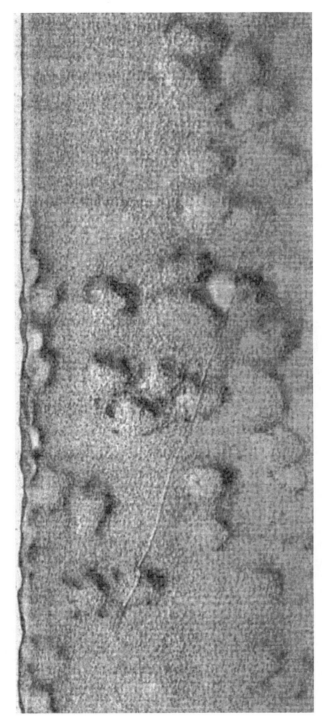

Fig. 5.39. High-resolution sidescan sonar image of closely spaced pock marks from the large gulf between Iran and Arabia. These pock marks are some 10 m wide and 2 m deep. Cementation around the pock marks is suspected from the high backscattering (darker tones) circular rims, and from the presence of epifaunal animals seen on bottom photographs. (Courtesy of N. Kenyon, Southampton Oceanography Centre, England.)

recorded from continental shelves around the world, normally overlying a thick sedimentary succession (Hovland and Judd, 1988). Commonly pock marks overlie petroleum "kitchens" of thick organic-rich overpressured shale. Pock marks often, but not invariably, connect at depth with fault systems that extend down to the "kitchen." Cold seepages of brine and bubbles of carbon dioxide, hydrogen sulfide, methane, and other petroleum gases emerge from modern pock marks. Thus it is often held that pock marks result from the episodic explosive release of petroleum and petroleum-related fluids. Lenses of gas-charged sand have been imaged on subsea seismic data beneath pock marks, suggesting an analogy with the magma chambers beneath igneous eruptive centers (Brooke *et al.*, 1995).

Investigations of a large field of pock marks in the Gulf of Patras, offshore Greece, fortuitously coincided with an earthquake of magnitude 5.4 on the Richter scale. A rise in seawater temperature was recorded prior to the quake, and gas bubbles were seen emerging from the pock marks for several days thereafter (Hasiotis *et al.*, 1996).

Pock marks are particularly well known from the petroliferous sedimentary basins of the northwest European continental shelf, and from northwest Australia (e.g., Hovland and Sommerville, 1985, and Hovland *et al.*, 1994, respectively). In the Gulf of Mexico, however, pock marks overlie gas-hydrate cemented sediments over mud diapirs and salt domes. Here the explosive event may be linked to the destabilization of gas hydrate, with a concomitant escape of methane gas through the water column to the atmosphere (Bagirov and Lerche, 1997). This phenomenon has been advocated as an explanation of unexplained maritime plane and ship losses, the so-called Bermuda Triangle effect (McIver, in Simmons and Jacobs, 1992).

Pock marks are also known on the floor of the Baltic Sea, where only a thin veneer of modern sediment overlies fractured igneous and metamorphic basement. Here it is more likely that the eruptions take place from mantle-derived gases (Gold, 1999).

5.4 PALEOCURRENT ANALYSIS

The preceding analysis of sedimentary structures shows that they can be used to determine depositional processes. Because depositional processes occur in several environments, few structures are immediately diagnostic of a specific environment; assemblages of structures are most useful, as for example in a tempestite (Section 5.3.3.4), a turbidite (Section 4.2.2), or a point bar (Section 6.3.2.2.3).

There is, however, one further use for sedimentary structures. They can indicate the direction of paleocurrent flow, paleoslope, paleogeography, and sand-body trend. Paleocurrent analysis, as this discipline is called, forms an integral part of facies analysis both at outcrop and, using the dipmeter, in subsurface studies. There is an extensive literature on this topic. Potter and Pettijohn (1977) is the definitive text. The methodology, interpretation, and applications of paleocurrent analysis are now described in turn.

5.4.1 Collection of Paleocurrent Data

A wide range of sedimentary structures can be used in paleocurrent analysis. Some structures yield only the sense of current flow, others yield both sense and direction. Examples of the first group include groove marks, channels, washouts and parting lineation. Examples of the second group include pebble imbrication (see Section 3.2.3.1.4), cross-lamination, cross-bedding, slump folds, flute marks, and the asymmetric profiles of ripples. The measurement of the orientation of sedimentary structures must be done with care. Ideally some kind of areal sampling grid should be used for regional paleocurrent mapping. In practice, this ideal approach is commonly restricted by limitations of access, exposure and time.

Each sample station will generally consist of a cliff, quarry, stream section, road cut, etc. If it is to be worth anything, paleocurrent analysis must be integrated with a full sedimentological study. Thus each sample station will also be the location of a measured section, or at least some detailed notes on stratigraphy, lithology, facies, and fauna. At each station it is necessary to record structural dip and strike. If it is excessive (greater than about 10°), each measurement must be corrected on a stereographic net. The orientation of the structures will be recorded, including both the azimuth and dip of planar structures that need correction. For linear structures and for planar structures in outcrops of low tectonic dip only the azimuth need be recorded. At the same time, it is necessary to note the type and scale of the structure and the lithology in which it occurs. Foreset dip directions from cross-bedding should always be measured in plan view. Dip directions seen in vertical sections should only be recorded as a last resort. There are two reasons for this. First, as pointed out earlier in this chapter, cross-beds do not always dip directly downcurrent. In troughs and laterally infilled channels, foresets are deposited obliquely or perpendicular to current flow. Examination of cross-bedding in plan view gives a clue to the structural arrangement of the foresets. If cross-bedding is measured from vertical facies generally only an apparent dip can be recorded. This may diverge considerably from the true dip direction, especially if there is well-developed jointing. The discrepancy will not be too erroneous, however, as foresets appear horizontal when viewed normal to their dip direction (Fig. 5.14).

The number of readings that need to be measured at a sample station is a matter for debate and may fortunately be dictated by the size of the exposure. Discussions of the statistics of sampling are given in Miller and Kahn (1962) and Krumbein and Graybill (1965). There is great scope here for statistical aerobics. As a rule of thumb in unipolar cross-bed systems, as in alluvium, 25 readings are generally sufficient to determine a vector mean with an accuracy of ±30°. This is sufficiently accurate for most purposes. Many more readings may be needed, however, to establish well-defined modes in a section of interbedded facies with different and often polymodal vectors. For example, in shoreline deposits, fluvial channel sands with unimodal downslope dipping foresets may be interbedded with marine sands with bipolar dipping foresets due to tidal currents unrelated to the paleoslope.

5.4.2 Presentation of Paleocurrent Data

Paleocurrent data may be entered in a field notebook and subsequently published in tabular form. The azimuths are, however, generally manipulated in some way to make their interpretation easier. The first step involves the removal of tectonic dip on a stereographic net where applicable (Schlumberger Ltd., 1970; Potter and Pettijohn, 1977, p. 371; Lewis and McConchie, 1994, p. 87). Then the azimuths are divided into class intervals from 0 to 360°. Class intervals of between 30 and 45° are about typical. The data may then be presented on a histogram. More usually, however, a compass rose is used

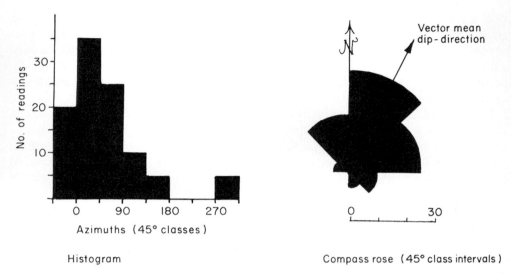

Histogram   Compass rose (45° class intervals)

Fig. 5.40. Presentation of paleocurrent data, by histogram (left) and by azimuth plot (right). Both methods display the same set of readings.

(Fig. 5.40). This is a histogram converted to a circular distribution. It is conventional to indicate the direction of dip of foresets. This is contrary to the convention for wind roses, which indicate the direction from which the wind blew. The dominant dip directions, termed modes, are at once apparent from a compass rose. Several types of azimuthal pattern are recognizable in paleocurrent data (Fig. 5.41). A vector mean may then be statistically determined for unimodal data. This may be done mathematically or graphically.

A graphic method was been developed by Reiche (1938) and Raup and Miesch (1957). Starting at a point, a unit length is drawn (i.e., 1 cm) on the azimuth of the first reading. A unit length is then drawn along the azimuth of the second reading starting at the end of the first. This process is continued until all the readings have been plotted. A line connecting the point of origin to the distal end of the last reading records the graphic vector mean. This is a quick simple method for innumerate geologists. It particularly useful for highlighting dip sections or dip meter runs of boreholes. A more sophisti-

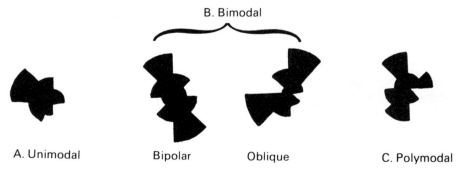

Fig. 5.41. Various characteristic azimuthal patterns of paleocurrent data.

cated way of calculating the vector mean is by the following formula (Harbaugh and Merriam, 1968, p. 42):

$$X_v = \arctan \left[\frac{\sum\limits_{i=1}^{n} n_i \sin X_i}{\sum\limits_{i=1}^{n} n_i \cos X_i} \right],$$

where X_v is the directional vector mean, n is the total number of observations, n_i is number of observations in each frequency class, and X_i is the midpoint azimuth in the ith class interval.

More simply, the humble arithmetic mean may be calculated by adding all azimuths and dividing by the total number of observations. This does not work if the azimuths are dispersed about the 360° point, because this is then likely to yield mean direction of about 180°, the exact opposite of the true mean. The arithmetic means of such sets of data may be calculated by using a false origin. Ninety degrees, for example, are added to all the data. The azimuths are summed and divided by the total number of readings, as before. Subtraction of 90° from the result then yields the true arithmetic mean.

Additional statistical methods are available for measuring the amount of dispersal of the data around the vector mean (Harbaugh and Merriam, 1968, p. 42; Potter and Pettijohn, 1977, p. 374).

These techniques are only applicable to unimodal distribution of azimuthal data. They may not be used on bimodal or polymodal data. In such instances it may be safer to present the data as compass roses (Tanner, 1959).

Considerable attention has been paid to the degree of scatter of paleocurrent data and to the calculation of statistical variance. This might give an insight into the sinuosity of fluvial channels and into the differentiation of unidirectional continental and polymodal marine current systems. Long and Young (1978) found that the statistical variance of fluvial paleocurrent data was less than 4000, and that of marine date tended to exceed that figure.

Paleocurrent data can be used as an element of regional facies mapping. Where there are sufficient sample points of unimodal data, their vector means may be plotted and contoured. The contours are dimensionless isolines, which record the regional paleostrike (e.g., Fig. 10.14). The azimuth vectors, hopefully aided by facies analysis, indicate the paleodip. Regional paleocurrent maps can be subjected to mathematical smoothing techniques such as trend surface analysis.

5.4.3 Interpretation of Paleocurrent Data

Paleocurrent analysis involves several stages before the data are actually interpreted:

1. Measurement of structures
2. Deduction of paleocurrent
3. Manipulation of paleocurrent data
4. Deduction of paleoslope.

The two deductive phases of the exercise deserve special attention. Considering the deduction of paleocurrent direction from sedimentary structures, it has already been pointed out how foresets are sometimes oblique or perpendicular to current flow; anti-dunes actually point upcurrent. Paleocurrents must thus be deduced carefully from the structures actually recorded. The sedimentary structures should be studied in the field and their genesis considered before measurement commences. Many published studies include compass roses of paleocurrents without making clear whether these are actual measured structural orientations or deduced flow directions.

A further point of deducing paleocurrents from structures concerns the weighting to be given to different structures. A ripple reflects a much more local and smaller current flow than a dune. A dune, in turn, reflects a smaller flow than a channel. A channel may itself meander and deviate from the regional topographic slope. The sedimentary structures are members of a hierarchy of the total flow system (Allen, 1966). Thus when measuring paleocurrent data a channel axis is immensely more significant than a few cross-bed orientations, and these should count for more than an equivalent number of cross-laminae. Few geologists have attempted to address the problem of weighting sedimentary structures of different rank (Iriondo, 1973). It is, however, reassuring to find that measurements of cross-beds in modern channels do give mean dip directions that correspond to the channel axis (e.g., Smith, 1972; Potter and Pettijohn, 1977, p. 103).

This leads to the second main problem of paleocurrent interpretation, namely, the relationship between paleocurrent and paleoslope. In certain environments the flow systems are slope controlled, in others they are not. In the first case, paleocurrent analysis can give valuable information on paleogeography and basin evolution. In the second case, it cannot. Paleocurrents are slope controlled in fluvial, deltaic, and (most) turbidite environments. Paleocurrents are not related to slope in eolian and marine shoreline environments. Klein (1967) has reviewed the relationship between sedimentary structures, paleocurrents, and paleoslopes in modern deposits. Selley (1968a) has defined a number of regional paleocurrent models that have been recognized in ancient sedimentary deposits.

Each major depositional environment is characterized by a particular paleocurrent model (Table 5.5). Figure 5.42 summarizes a typical example of paleocurrent analysis in a regional study of complex shoreline deposits of diverse facies.

Table 5.5
Classification of Some Paleocurrent Patterns

Environment		Local current vector	Regional pattern
Alluvial {	braided	Unimodal, low variability	Often fan-shaped
	meandering	Unimodal, high variability	Slope-controlled often centripetal basin fill
Eolian		Uni-, bi- or polymodal	May swing round over hundreds of kilometers around high-pressure systems
Deltaic		Unimodal	Regionally radiating
Shorelines and shelves		Bimodal (due to tidal currents), sometimes unipolar or polymodal	Generally consistently oriented onshore, off-shore, or long-shore
Marine turbidite		Unimodal (some exceptions)	Fan-shaped or, on a larger scale, trending into or along trough axes

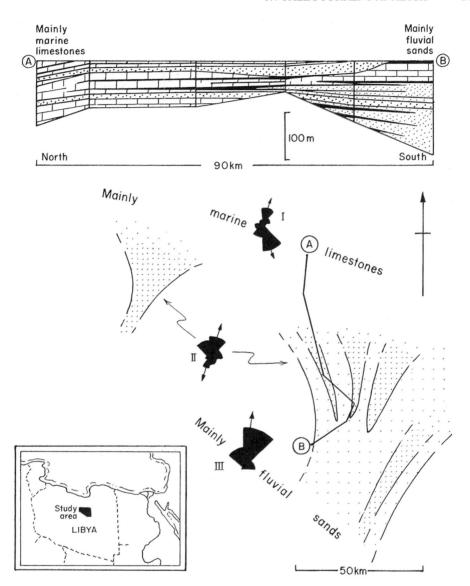

Fig. 5.42. An example of paleocurrent analysis integrated with a paleoenvironmental study of a Libyan Miocene shoreline. In the north of the area marine calcarenites show a bipolar azimuthal pattern of foreset dips (I).
This reflects tidal currents with a net onshore component. In the south fluvial sands show unipolar north
(seaward)-dipping foresets (III). Radiating estuarine channel complexes show bipolar cross-bedding dips (II),
suggesting tidal currents with the ebb current dominant. (From data in Selley, 1968b.)

On a still broader scale, paleocurrents can indicate the age of formation of structural
features. In particular, they may show whether a paleohigh was active during sedimentation or whether it rose after deposition of the sediments that drape it (Fig. 5.43). Similarly, paleocurrent analysis can distinguish syndepositional from postdepositional sedimentary basins. In the former, paleocurrents converge on the center of the basin. In
postdepositional (tectonic) basins, paleocurrents sweep across the basin with a more or

A. Post - depositional uplift

B. Syndepositional uplift

Fig. 5.43. Where paleocurrents reflect paleoslope, as in fluvial environments, they provide clues as to the age of formation of regional structural arches. Paleocurrents move in a constant direction across postdepositional arches (A), but diverge from axes of syndepositional uplift (B).

less uniform paleostrike (Fig. 5.44). The South Wales Pennsylvanian basin shows centripetal paleocurrents indicative of its syndepositional origin (Bluck and Kelling, 1963). The Illinois basin provides a good example of a postdepositional tectonic basin with a regionally consistent paleostrike (Potter *et al.,* 1958). Interpretive studies such as these can only be made from paleocurrents deduced from sediments deposited in slope-controlled environments. These examples do show how paleocurrent analysis can be an important and integral feature of regional facies analysis.

SELECTED BIBLIOGRAPHY

Collinson, J. D., and Thompson, D. B. (1988). "Sedimentary Structures," 2nd ed. Allen & Unwin, London. 194pp.

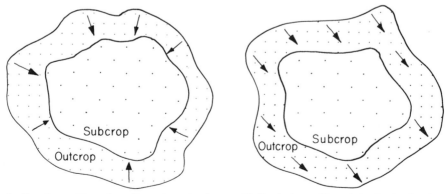

A. Syndepositional sedimentary basin   B. Post-depositional tectonic basin

Fig. 5.44. Where paleocurrents reflect paleoslope as in fluvial deposits, they provide clues as to the age of formation of sedimentary basins. (A) In syndepositional basins they converge on what was the topographic center of the basin. (B) In postdepositional basins, however, paleocurrents show no relation to present structural morphology. Examples of both types are cited in the text.

Ricci Lucchi, F. (1995). "Sedimentographica: A Photographic Atlas of Sedimentary Structures," 2nd ed. Columbia University Press, New York. 255pp.

REFERENCES

Ager, D. V. (1973). Storm deposits in the Jurassic of the Moroccan High Atlas. *Palaeogeogr., Palaeoclimatol., Palaeoecol.* **15,** 83–93.

Aigner, T. (1985). "Storm Depositional Systems: Dynamic Stratigraphy in Modern and Ancient Marine Sequences." Springer-Verlag, Berlin. 223pp.

Alexander, J., and Fielding, C. (1997). Gravel antidunes in the tropical Burdekin River, Queensland, Australia. *Sedimentology* **44,** 327–338.

Allen, J. R. L. (1963). The classification of cross-stratified units, with notes on their origin. *Sedimentology* **2,** 93–114.

Allen, J. R. L. (1964). Primary current lineation in the Lower Old Red Sandstone (Devonian) Anglo-Welsh basin. *Sedimentology* **3,** 89–108.

Allen, J. R. L. (1966). On bedforms and palaeocurrents. *Sedimentology* **6,** 153–190.

Allen, J. R. L. (1968a). On criteria for the continuance of flute marks, and their implications. *Geol. Mijnbouw* **47,** 3–16.

Allen, J. R. L. (1968b). "Current Ripples." North-Holland, Amsterdam. 433pp.

Allen, J. R. L. (1969). Some recent advances in the physics of sedimentation. *Proc. Geol. Assoc.* **80,** 1–42.

Allen, J. R. L. (1970). "Physical Processes of Sedimentation." Allen & Unwin, London. 248pp.

Allen, J. R. L. (1971). Transverse erosional marks of mud and rock: their physical basis and geological significance. *Sediment. Geol.* **5,** 167–385.

Allen, J. R. L. (1979). A model for the interpretation of wave ripple-marks using their wavelength textural composition, and shape. *J. Geol. Soc., London* **136,** 673–682.

Allen, J. R. L. (1980). Sandwaves: A model of origin and internal structure. *Sediment. Geol.* **26,** 281–328.

Allen, J. R. L. (1986a). On the curl of desiccation polygons. *Sediment. Geol.* **46,** 23–32.

Allen, J. R. L. (1986b). Earthquake magnitude-frequency, epicentral distance, and soft sediment deformation in sedimentary basins. *Sediment. Geol.* **46,** 67–76.

Allen, J. R. L., and Banks, N. L. (1972). An interpretation and analysis of recumbent-folded deformed cross-bedding. *Sedimentology* **19,** No. 3/4, 257–283.

Anketell, J. M., Gegla, J., and Dzulinski, S. (1970). On the deformational structures in system with reversed density gradients. *Ann. Soc. Geol. Pol.* **15,** 3–29.

Bagirov, E., and Lerche, I. (1997). Hydrates represent gas source, drilling hazard. *Oil & Gas J.,* December 1, pp. 99–104.

Bagnold, R. A. (1954). "The Physics of Blown Sand and Desert Dunes." Methuen, London. 265pp.

Barratt, P. J. (1966). Effects of the 1964 Alaskan earthquake on some shallow water sediments in Prince William Sound, S. E. Alaska. *J. Sediment. Petrol.* **36,** 992–1006.

Bigarella, J. J. (1972). Eolian environments — their characteristics, recognition and importance. *Spec. Publ. — Soc. Econ. Paleontol. Mineral.* **16,** 12–62.

Blanc, J. J. (1972). Slumpings et figures sedimentaires dans le Cretace supérieur du bassin de Beausset, France. *Sediment. Geol.* **7,** 47–64.

Bluck, B. H., and Kelling, G. (1963). Channels from the Upper Carboniferous coal measures of South Wales. *Sedimentology* **2,** 29–53.

Bouma, A. H. (1962). "Sedimentology of Some Flysch Deposits." Elsevier, Amsterdam. 168p.

Brenchley, P. J. (1990). Biofacies. *In* "Palaeobiology: A Synthesis" (D. E. G. Briggs and P. R. Crowther, eds.), pp. 395–400. Blackwell, Oxford.

Bristow, C. (1995). Internal Geometry of ancient tidal bedforms revealed by using ground penetrating radar. *In* "Tidal Signatures in Modern and Ancient Sediments" (B. W. Fleming and A. Bartholoma, eds.), pp. 313–328. Blackwell, Oxford.

Bromley, R. G. (1975). Trace fossils at omission surfaces. *In* "The Study of Trace Fossils" (R. W. Frey, ed.), pp. 97–120. Springer-Verlag, Berlin.

Bromley, R. G. (1990). "Trace Fossils, Biology and Taphonomy." Unwin Hyman, London. 304pp.

Bromley, R. G., and Asgaard, U. (1972). Freshwater *Cruziana* from the Upper Triassic of Jameson Land, East Greenland. *Groenl. Geol. Unders. Rapp.* **49,** 7–13.

Brooke, C. M., Trimble, T. J., and Mackay, T. A. (1995). Mounded shallow gas sands from the Quaternary of the North Sea: Analogues for the formation of sand mounds in deep water Tertiary sediments? *In* "Characteristics of Deep Marine Clastic Systems" (A. J. Harley and D. J. Prosser, eds.), Spec. Publ. No. 94, pp. 95–101. Geol. Soc. London, London.

Bucher, W. H. (1919). On ripples and related sedimentary surface forms and their palaeographic interpretation. *Am. J. Sci.* **47,** 149–210, 241–269.

Campbell, C. V. (1967). Lamina, laminaset, bed and bedset. *Sedimentology* **8,** 7–26.

Cita, M. B., Woodside, J. M., Ivanov, M. K., Kidd, R. B., Limonov, A. F., and Scientific Staff of Cruise TTR3 — Leg 2. (1995). Fluid venting from a mud volcano in the Mediterranean Ridge Diapiric Belt. *Terra Nova* **7,** 453–458.

Collinson, J. D., and Thompson, D. B. (1988). "Sedimentary Structures," 2nd ed. Allen & Unwin, London. 194pp.

Conybeare, C. E. B., and Crook, K. A. W. (1968). Manual of sedimentary structures. *Bull. — Bur. Miner. Resour. Geol. Geophys. (Aust.)* **12,** 1–327.

Crimes, T. P., and Harper, J. C., eds. (1970). "Trace Fossils." Liverpool Geol. Soc., Liverpool. 547pp.

Crimes, T. P., and Harper, J. C. (1977). "Trace Fossils," Vol. 2. Seel House Press, Liverpool. 351pp.

Darwin, G. H. (1883). On the formation of ripple-marks in sand. *Proc. R. Soc. London* **36,** 18–43.

Davies, H. G. (1965). Convolute lamination and other structures from the Lower Coal Measures of Yorkshire. *Sedimentology* **5,** 305–326.

Davies, S. N. (1971). Barbados: A major submarine gravity slide. *Geol. Soc. Am. Bull.* **82,** 2593–2602.

De la Beche, H. T. (1851). "The Geological Observer." John Murray, London. 846pp.

Dingle, R. V. (1980). Large allochthonous sediment masses and their role in the construction of the continental slope and rise of southwestern Africa. *Mar. Geol.* **37,** 333–354.

Dionne, J. C. (1973). Monroes: A type of so-called mud volcanoes on tidal flats. *J. Sediment. Petrol.* **43,** 848–856.

Doe, T. W., and Dott, R. J., Jr. (1980). Genetic significance of deformed crossbedding — with examples from the Navajo and Weber Sandstones of Utah. *J. Sediment. Petrol.* **50,** 793–812.

Donovan, R. N., and Foster, R. J. (1972). Subaqueous shrinkage cracks from the Caithness Flagstone Series (Middle Devonian) of Northeast Scotland. *J. Sediment. Petrol.* **42,** 309–317.

Dott, R. H., and Bourgeois, J. (1982). Hummocky stratification: Significance of its variable bedding sequences. *Geol. Soc. Am. Bull.* **93,** 663–680.

Dott, R. H., and Bourgeois, J. (1983). Hummocky stratification: Significance of its variable bedding sequences: Discussion and reply. *Geol. Soc. Am. Bull.* **94,** 1249–1251.

Dzulinski, S., and Sanders, J. E. (1962). Current marks on firm mud bottoms. *Trans. Conn. Acad. Arts Sci.* **42,** 57–96.

Dzulinski, S., and Slaczka, A. (1959). Directional structures and sedimentation of the Krosno beds *(Carpathian flysch)*. *Ann. Soc. Geol. Pol.* **28,** 205–260.

Dzulinski, S., and Walton, E. K. (1965). "Sedimentary Features of Flysch and Greywacke." Elsevier, Amsterdam. 274pp.

Fenies, H., De Resseguier, A., and Tastet, J. P. (1999). Intertidal clay-drape couplets (Gironde estuary, France). *Sedimentology* **46,** 1–15.

Flemming, B. W., and Bartholoma, A. (1995). "Tidal Signatures in Modern and Ancient Sediments." Blackwell, Oxford.

Frazier, D. E., and Osanik, A. (1961). Point-bar deposits. Old River Locksite, Louisiana. *Trans. — Gulf Coast Assoc. Geol. Soc.* **11,** 127–137.

Frey, R., ed. (1975). "The Study of Trace Fossils." Springer-Verlag, Berlin. 562pp.

Friend, P. F., and Moody-Stuart, M. (1972). Sedimentation of the Wood Bay Formation (Devonian) of Spitsbergen: Regional analysis of a late orogenic basin. *Skr., Nor. Polarinst.* **157,** 1–77.

Gall, J. C. (1983). "Ancient Sedimentary Environments and the Habitats of Living Organisms." Springer-Verlag, Berlin. 219pp.

Gill, W. D. (1979). Syndepositional sliding and slumping in the West Clare Namurian Basin, Ireland. *Geol. Surv. Irel., Spec. Pub.* **4,** 1–121.

Gill, W. D., and Kuenen, P. H. (1958). Sand volcanoes on slumps in the Carboniferous of County, Clare, Ireland. *Q. J. Geol. Soc. London* **113,** 441–460.

Glennie, K. W. (1987). Desert Sedimentary Environments, present and past — a summary. *Sediment. Geol.* **50,** 135–166.

Glennie, K. W. (1970). "Desert Sedimentary Environments." Elsevier, Amsterdam. 222pp.

Gold, T. (1999). "The Hot Deep Biosphere." Springer-Verlag, New York. 235pp.

Goldring, R. (1999). "Field Palaeontology," 2nd ed. Pearson Education, Harlow, England.

Gubler, Y., ed. (1966). "Essai de Nomenclature et de Caractérisation des Principales Structures Sedimentaires." Editions Technip, Paris. 291pp.

Hamblin, W. K. (1962). X-ray radiography in the study of structures in homogenous sediments. *J. Sediment. Petrol.* **32,** 201–210.

Hand, B. M., Middleton, G. V., and Skipper, K. (1972). Antidune cross-stratification in a turbidite sequence, Cloridorme Formation, Gaspe, Quebec. *Sedimentology* **18,** 135–138.

Hantzschel, W. (1975). Trace fossils and problematica. *In* "Treatise on Invertebrate Paleontology" (R. C. Moore, ed.), Part W. Geol. Soc. Am., New York. 269pp.

Harbaugh, J. W., and Merriam, D. F. (1968). "Computer Applications in Stratigraphic Analysis." Wiley, New York. 282pp.

Harms, J. C. (1975). Stratification and sequence in prograding shoreline deposits. *SEPM Short Course* **2,** 81–102.

Harms, J. C., and Fahnestock, R. K. (1965). Stratification, bedforms and flow phenomena (with an example from the Rio Grande). *Spec. Publ. — Soc. Econ. Palaeontol. Mineral.* **12,** 8–115.

Harms, J. C., MacKenzie, D. B., and McCubbin, D. G. (1963). Stratification in modern sands of the Red River, Louisiana. *J. Geol.* **71,** 566–580.

Harms, J. C., Southard, J. B., Spearing, D. R., and Walker, R. G. (1982). Depositional environments as interpreted from sedimentary structures and stratification sequences. *Econ. Palaeontol. Min. Short Course Notes* **9,** 1–166.

Hasiotis, T., Papatheodorou, G., Kastanos, N., and Ferentinos, G. (1996). A pockmark field in the Patras Gulf (Greece) and its activation during the 14/7/93 seismic event. *Mar. Geol.* **130,** 333–344.

Heckel, P. H. (1972). Recognition of ancient shallow marine environments. *Spec. Pub. — Econ. Palaeontol. Mineral.* **16,** 226–286.

Hovland, M., and Judd, A. G. (1988). "Sea Bed Pockmarks and Seepages." Kluwer Academic Publishing Group, Norwell, MA.

Hovland, M., and Somerville, J. H. (1985). Characteristics of two natural gas seepages in the North Sea. *Mar. Pet. Geol.* **2,** 319–326.

Hovland, M., Croker, P. F., and Martin, M. (1994). Fault-associated seabed mounds (carbonate knolls?) off western Ireland and north-west Australia. *Mar. Pet. Geol.* **11,** 232–246.

Hulsemann, J. (1955). Grossrippeln und Schragschichtungsgefuge im Nordsee-Watt und in Molasse. *Senckenbergiana Lethaea* **36,** 359–388.

Imbrie, J., and Buchanan, H. (1965). Sedimentary structures in Modern Carbonates in the Bahamas. *Spec. Pub. — Soc. Econ. Palaeontol. Mineral.* **12,** 149–172.

Iriondo, H. H. (1973). Volume factor in paleocurrent analysis. *Am. Assoc. Petrol. Geol. Bull.* **57,** 1341–1342.

Jopling, A. V. (1965). Hydraulic factors controlling the shape of laboratory deltas. *J. Sediment. Petrol.* **35,** 777–791.

Jopling, A. V., and Walker, R. G. (1968). Morphology and origin of ripple-drift cross-lamination with examples from the Pleistocene of Massachusetts. *J. Sediment. Petrol.* **38,** 971–984.

Kindle, E. M. (1917). Recent and fossil ripple marks. *Mus. Bull. Can. Geol. Surv.* **25,** 1–56.

Klein, G. de Vries (1967). Paleocurrent analysis in relation to modern marine sediment dispersal patterns. *Am. Assoc. Pet. Geol. Bull.* **51,** 366–382.

Klein, G. de Vries, de Melo, U., and Favera, J. C. D. (1972). Subaqueous gravity processes on the front of Cretaceous deltas, Reconcavo Basin, Brazil. *Geol. Soc. Am. Bull.* **83,** 1469–1492.

Krumbein, W. C., and Graybill, F. A. (1965). "An Introduction to Statistical Methods in Geology." McGraw-Hill, New York. 475pp.

Kuenen, P. H. (1948). Slumping in the Carboniferous rocks of Pembrokeshire. *Q. J. Geol. London* **104,** 365–385.

Kuenen, P. H. (1958). Experiments in geology. *Trans. Geol. Soc. Glasgow* **23,** 1–28.

Lewis, D. W., and McConchie, D. (1994). "Analytical Sedimentology." Chapman & Hall, London. 197pp.

Lewis, K. B. (1971). Slumping on a continental slope inclined at 1°–4°. *Sedimentology* **16,** 97.

Li, Y., Craven, J., Schweig, E. S., and Obermeir, S. F. (1996). Sand boils induced by the 1993 Mississippi River flood: Could they one day be misinterpreted as earthquake-induced liquifaction? *Geology* **24,** 171–174.

Lockley, M., and Hunt, A. P. (1996). "Dinosaur Tracks, and other Fossil Footprints of the Western United States." Columbia University Press, Columbia, NY. 388p.

Long, D. G. F., and Young, G. M. (1978). Dispersion of cross-stratification as a potential to the interpretation of Proterozoic arenites. *J. Sediment. Petrol.* **48,** 857–862.

Lowe, D. R. (1975). Water escape structures in coarse grained sediments. *Sedimentology* **22,** 204.

Lowe, D. R., and Lopiccolo, L. D. (1974). The characteristics and origins of dish and structures. *J. Sediment. Petrol.* **44,** 481–501.

Lyell, Sir C. (1865). "Elements of Geology," 6th ed. John Murray, London. 794pp.

Macar, P., and Antun, P. (1949). Pseudonodules et glissements sous-aquatiques dans l'Emsian inferior de l'Oesling. *Ann. Soc. Geol. Belg.* **73,** 121–150.

Martinsson, A. (1965). Aspects of a Middle Cambrian Thanatotope in Oland. *Geol. Foeren. Stockholm Foerh.* **87,** 181–230.

McBride, E. F., and Hayes, M. O. (1962). Dune cross-bedding on Mustang Island, Texas. *Bull. Am. Assoc. Pet. Geol.* **46,** 546–551.

McCalpin, J. P., ed. (1996). "Paleoseismology." Academic Press, San Diego, CA. 588pp.

McKee, E. D., and Weir, G. W. (1953). Terminology for stratification and cross-stratification. *Geol. Soc. Am. Bull.* **64,** 381–390.

McKee, E. D., Crosby, E. J., and Berryhill, H. L. (1967). Flood deposits, Bijou Creek, Colorado June 1965. *J. Sediment. Petrol.* **37,** 829–851.

Miller, R. L., and Kahn, J. S. (1962). "Statistical Analysis in the Geological Sciences." Wiley, New York. 483pp.

Mills, P. C. (1983). Genesis and diagnostic value of soft sediment deformation structures — a review. *Sediment. Geol.* **35,** 83–104.

Obermeir, S. F. (1996). Using liquefaction-induced features for paleoseismic analysis. *In* "Paleoseismology" (J. P. McCalpin, ed.), pp. 331–396. Academic Press, San Diego, CA.

Otto, G. H. (1935). The sedimentation unit and its use in field sampling. *J. Geol.* **46,** 569–575.

Owen, G. (1996). Experimental soft-sediment deformation: Structures formed by the liquefaction of unconsolidated sands and some ancient examples. *Sedimentology* **43,** 279–293.

Payne, T. G. (1942). Stratigraphical analysis and environmental reconstruction. *Bull. Am. Assoc. Pet. Geol.* **26,** 1697–1770.

Pettijohn, F. J., and Potter, P. E. (1964). "Atlas and Glossary of Primary Sedimentary Structures." Springer-Verlag, Berlin. 370pp.

Picard, M. D. (1966). Oriented, linear-shrinkage cracks in Green River Formation (Eocene), Raven Ridge area, Uinta basin, Utah. *J. Sediment. Petrol.* **36,** 1050–1057.

Plummer, P. S., and Gostin, V. A. (1986). Shrinkage cracks: desiccation or synaeresis. *J. Sediment. Petrol.* **51,** 1147–1156.

Potter, P. E., and Pettijohn, F. J. (1977). "Paleocurrents and Basin Analysis," 2nd ed. Springer-Verlag, Berlin. 425pp.

Potter, P. E., Nosow, E., Smith, N. W., Swann, D. H., and Walker, F. H. (1958). Chester cross-bedding and sandstone trends in Illinois basin. *Bull. Am. Assoc. Pet. Geol.* **42,** 1013–1046.

Prior, D. B., Doyle, E. H., and Kaluza, M. J. (1989). Evidence for Sediment eruption on Deep Sea Floor, Gulf of Mexico. *Science* **243,** 517–518.

Raup, O. B., and Miesch, A. T. (1957). A new method for obtaining significant average directional measurements in cross-stratification studies. *J. Sediment. Petrol.* **27,** 313–321.

Reiche, P. (1938). An analysis of cross-lamination of the Coconino sandstone. *J. Geol.* **44,** 905–932.

Reineck, H. E. (1961). Sediment bewegungen an kleinrippeln im watt. *Senckenbergiana Lethaea* **42,** 51–67.

Reineck, H. E., and Wunderlich, F. (1968). Classification and origin of flaser and lenticular bedding. *Sedimentology* **11,** 99–104.

Ricci Lucchi, F. (1995). "Sedimentographica: A Photographic Atlas of Sedimentary Structures," 2nd ed. Columbia University Press, New York. 255pp.

Rittenhouse, G. (1972). Cross-bedding dip as a measure of sandstone compaction. *J. Sediment. Petrol.* **42,** 682–683.

Roberts, D. G. (1972). Slumping on the eastern margin of the Rockall Bank, North Atlantic Ocean. *Mar. Geol.* **13,** 225–237.

Rodriguez, J., and Gutschick, R. C. (1970). Late Devonian–early Mississippian ichnofossils from Western Montana and Northern Utah. *In* "Trace Fossils" (T. P. Crimes and J. C. Harper, eds.), pp. 407–438. Liverpool Geol. Soc., Liverpool.

Rust, B. R. (1978). A classification of alluvial channel systems. *In* "Fluvial Sedimentology" (A. D. Miall, ed.), pp. 187–198. Can. Soc. Pet. Geol., Calgary.

Schlumberger Ltd. (1970). "Fundamentals of Dipmeter Interpretation." Schlumberger, New York. 145pp.

Schumm, S. A. (1969). River metamorphosis. *J. Hydraul. Div., Am. Soc. Civ. Eng.* **95,** 255–273.

Seilacher, A. (1964). Biogenic sedimentary structures. *In* "Approaches to Paleoecology" (J. Imbrie and N. D. Newell, eds.), pp. 296–315. Wiley, New York.

Seilacher, A. (1967). Bathymetry of trace fossils. *Mar. Geol.* **5,** 413–428.

Selley, R. C. (1965). Diagnostic characters of fluviatile sediments in the Pre-Cambrian rocks of Scotland. *J. Sediment. Petrol.* **35,** 366–380.

Selley, R. C. (1968a). A classification of paleocurrent models. *J. Geol.* **76,** 99–110.

Selley, R. C. (1968b). Nearshore marine and continental sediments of the Sirte basin, Libya. *Q. J. Geol. Soc. London* **124,** 419–460.

Selley, R. C. (1969). Torridonian alluvium and quicksands. *Scott. J. Geol.* **5,** 328–346.

Selley, R. C. (1970). Ichnology of Palaeozoic sandstones in the southern desert of Jordan: A study of trace fossils in their sedimentologic context. *In* "Trace Fossils" (T. P. Crimes and J. C. Harper, eds.), pp. 477–488. Liverpool Geol. Soc., Liverpool.

Selley, R. C. (1996). "Ancient Sedimentary Environments and their Subsurface Diagnosis," 4th ed. Chapman & Hall, London. 300pp.

Selley, R. C., Sutton, J., Shearman, D. J., and Watson, J. (1963). Some underwater disturbances in the Torridonian of Skye and Raasay. *Geol. Mag.* **100,** 224–243.

Simmons, J., and Jacobs, S. (1992). "The Bermuda Triangle. A Mystery Solved by Science?" Channel 4 Television, London. 33pp.

Skipper, K. (1971). Antidune cross-stratification in a turbidite sequence, Cloridorme Formation, Gaspe, Quebec. *Sedimentology* **17,** 51–68.

Smith, N. D. (1972). Some sedimentological aspects of planar cross-stratification in a sandy braided river. *J. Sediment. Petrol.* **42,** 624–634.

Smyers, N. B., and Peterson, G. L. (1971). Sandstone dikes and sills in the Moreno Shale, Panoche Hills, California. *Geol. Soc. Am. Bull.* **82,** 3201.

Sorby, H. C. (1859). On the structures produced by the currents present during the deposition of stratified rocks. *Geologist* **2,** 137–149.

Southard, J. B., and Dingler, J. R. (1971). Flume study of ripple propagation behind mounds on flat sandbeds. *Sedimentology* **16,** 251–263.

Swift, D. J. P., Figueiredo, A. G., Freeland, G. L., and Oertel, G. F. (1983). Hummocky cross-stratification and megaripples: a geological double standard. *J. Sediment. Petrol.* **53,** 1295–1317.

Tanner, W. F. (1959). The importance of modes in cross-bedding data. *J. Sediment. Petrol.* **29,** 211–226.

Tanner, W. F. (1967). Ripple mark indices and their uses. *Sedimentology* **9,** 89–104.

Tanner, W. F. (1971). Numerical estimates of ancient waves, water depth and fetch. *Sedimentology* **16,** 71–88.

Terwindt, J. H. J., and Breusers, H. N. C. (1972). Experiments on the origin of flaser, lenticular and sand-clay alternating bedding. *Sedimentology* **19,** 85–98.

van Loon, A. J., and Brodzikowski, K. (1987). Problems and progress in the research on soft sedimentation deformations. *Sediment. Geol.* **50,** 167–194.

Visher, G. S., and Cunningham, R. D. (1981). Convolute laminations — a theoretical analysis: Examples of a Pennsylvanian sandstone. *Sediment. Geol.* **28,** 175–188.

Visser, M. J. (1980). Neap-spring cycles reflected in Holocene sub-tidal large-scale bedform deposition: A preliminary note. *Geology* **8,** 543–546.

White, G. (1961). Colloid phenomena in the sedimentation of argillaceous rocks. *J. Sediment. Petrol.* **31** (4), 560–565.

Williams, G. E. (1970). Origin of disturbed bedding in Torridon Group Sandstones. *Scott. J. Geol.* **6,** 409–410.

Williams, P. B., and Kemp, P. H. (1971). Initiation of ripples on flat sediment beds. *J. Hydraul. Div., Am. Soc. Civ. Eng.* **97** (HY4) Proc. Pap. 8042, 502–522.

6 Depositional Systems

This chapter begins by describing the concepts of sedimentary environments, sedimentary facies, and sedimentary models. These concepts linked together provide the depositional systems approach to understanding the origins of sedimentary rocks. In this method the ancient depositional environment (process) of a sedimentary rock (the product) is deduced by comparison with modern depositional environments. The main part of the chapter provides a concise review of the various depositional systems in terms of their environments (process), facies (product), and economic importance. The chapter concludes by describing the spatial organization of sedimentary rocks, in terms of sequences and cyclicity.

6.1 SEDIMENTARY ENVIRONMENTS

6.1.1 Environments Defined

The modern earth's surface is the geologist's laboratory. Here can be studied the processes that generate sediments and their resultant deposits. By applying the principle of uniformitarianism ("the key to the past is in the present"), it is possible to deduce the origin of ancient sedimentary rocks. For this reason many geologists, and scientists of other disciplines, have studied modern sedimentary processes and products intensively.

The surface of the earth can be classified by geomorphologists into distinctive physiographic units, such as mountain ranges, sand deserts, and deltas. Similarly, oceanographers define morphologic types of sea floors, such as continental shelves, submarine fans, and abyssal plains. The most cursory of such studies shows that the number of physiographic types is finite. For example, lakes and deltas occur on most of the continents; submarine fans and coral reefs are spread wide across the oceans. From this observation it follows that the surface of the earth may be classified into different sedimentary realms or environments. A **sedimentary environment** has been defined as a "part of the earth's surface which is physically, chemically and biologically distinct from adjacent areas" (Selley, 1970, p. 1). As already pointed out, sand deserts, deltas, and submarine fans are examples of these different sedimentary environments.

The physical parameters of a sedimentary environment include the velocity, direction, and variation of wind, waves, and flowing water; they include the climate and weather of the environment in all their subtle variations of temperature, rainfall, snowfall, and humidity. The chemical parameters of an environment include the composition of the waters that cover a subaqueous sedimentary environment; they include the geochemistry of the rocks in the catchment area of a terrestrial environment. The biological parameters of an environment comprise both fauna and flora. On land these may have major effects on sedimentary processes. Overgrazing, defoliation, deforestation, and overcultivation of soils by animals can cause catastrophic increases in rates of erosion in one area, accompanied by accelerated rates of deposition elsewhere. Conversely, the colonization of deserts by a new flora has a moderating effect on sedimentary processes. In marine environments many of the lowliest forms of life are important, both because their skeletons can contribute to sediment formation, and because their presence in water can change its equilibrium, causing minerals to precipitate (see Section 9.2.3.2). The morphology and history of carbonates in general, and of organic reefs in particular, are inextricably related to the ecology of their biota.

This brief review of the physical, chemical, and biological parameters that define a sedimentary environment should be sufficient to demonstrate how numerous and complex these variables are.

6.1.2 Environments of Erosion, Equilibrium, and Deposition

The classification of sedimentary environments by their physical, chemical, and biological parameters will be returned to shortly. Now consider environments from a slightly different viewpoint. Examination of the modern earth shows that there are sedimentary environments of net erosion, environments of equilibrium (or nondeposition), and environments of net deposition. Sedimentary environments of net erosion are typically terrestrial and consist largely of the mountainous areas of the world. In such erosional environments weathering is often intensive and erosion is rapid. Locally sedimentation may take place from glacial, mud flow, and flash flood processes. Due to renewed erosion, however, such deposits are ephemeral and soil profiles have little time to develop on either bedrock or sediment. Erosional sedimentary environments also occur on cliffed coastlines and, under the sea, in submarine canyons and on current-scoured shelves. It is, however, in these shoreline and submarine situations that the products of deposition dominate the process of erosion. Subaqueous deposits must take up some 90% of the world's sedimentary cover. Probably some 60% of this total volume is composed of submarine and shoreline deposits. It appears that depositional sedimentary environments are predominantly subaqueous.

To environments of erosion and of deposition must be added a third category: environments of equilibrium or nondeposition. These are surfaces of the earth, both on the land and under the sea, which for long periods of time are neither sites of erosion nor of deposition. Because of their stability such environments often experience intense chemical alteration of the substrate. On land, environments of equilibrium are represented by the great peneplanes of the continental interiors (King, 1962). These, in the case of central Africa at least, are parts of the earth's surface that have been open to the sky for millions of years. Prolonged exposure to the elements is responsible for the development of weathering profiles and soil formation in the rocks which immediately underlie an environment of equilibrium (Thiry and Simon-Coincon, 1999). Laterite and bauxite horizons are the products of certain specific climatic conditions in conjunction with suitable rock substrates (see Section 2.3.3). They may be regarded as the products of sedimentary environments of equilibrium.

Environments of equilibrium can also be recognized beneath the sea. Extensive areas, both of the continental shelves and of the abyssal plains, are subjected to currents that are powerful enough to remove any sediment which may have settled by suspension, yet are too weak to erode the substrate. These scoured surfaces are susceptible to chemical reactions with seawater leading to the formation of manganese crusts and to phosphatization and other diagenetic changes (Cronan, 1980). In the geological column submarine sedimentary environments of equilibrium are represented by hardgrounds (see Section 9.2.5.3). These are mineralized surfaces, generally within limestones, often intensely bored and overlain by a thin conglomerate composed of clasts of the substrate. Examples occur in ancient shelf deposits such as the Cretaceous chalk of northern Europe (Jefferies, 1963) and from Jurassic pelagic deposits of the Alps (Fischer and Garrison, 1967). Table 6.1 summarizes these concepts of sedimentary environments of erosion, equilibrium, and deposition.

It is readily apparent that sedimentary geology is concerned primarily with depositional environments. These must be regarded as a particular type of sedimentary environment. It is important to make this distinction when interpreting the depositional environment of an ancient sedimentary rock. What is preserved is not just the result of the depositional process that deposited the rock. This will be shown unequivocally by the sedimentary structures. The sediment itself and its fossils may have originated to a large extent in erosional or equilibrial environments contiguous to that in which the sediment was actually deposited. Consider, for example, an ancient braided river channel. Its cross-bedding will indicate the direction, force, and nature of the depositing current. The composition and texture of the sand itself, however, may largely be inherited from an erosional sedimentary environment in the source area. A fossilized tree trunk enclosed in the channel sand testifies to the coeval existence of an, albeit temporary, environment of equilibrium.

Table 6.1
Illustration of the Concept of Sedimentary Environments
of Erosion, Deposition, and Equilibrium

		Erosional	Equilibrial	Depositional
Land	Subaerial	Dominant	Development of peneplanes, soils, laterites, and bauxites	Rare (eolian and glacial)
	Subaqueous	Localized	Unknown?	Localized (fluvial and lacustrine)
Sea		Rare	Development of "hardgrounds," often nodular and mineralized	Dominant

Table 6.2
Example of the Classical Type of Classification of Sedimentary Environments

Continental	Terrestrial {	Desert Glacial
	Aqueous {	Fluvial Paludal (swamp) Lacustrine Cave (spelean)
Transitional		Deltaic Estuarine Lagoonal Littoral (intertidal)
Marine		Reef Neritic (between low tide and 200 m) Bathyal (between 200 and 2000 m) Abyssal (deeper than 2000 m)

6.1.3 Environments Classified

Since the earliest days of geological studies it has been found convenient to classify sedimentary environments into various groups and subgroups. Such a classification provides a useful formal framework on which to base more detailed analyses of specific environmental types. Table 6.2 is an example of the classical classification of environments. Variants of this type will be found in many textbooks (e.g., Twenhofel, 1926, p. 784; Pettijohn, 1956, p. 633; Dunbar and Rodgers, 1957; Krumbein and Sloss, 1959, p. 196; Blatt et al., 1978).

This sort of all-embracing scheme, while excellent for modern environmental studies, is limited in its application to ancient sedimentary rocks. There are two reasons for this. First, this is a classification of sedimentary environments and, as already shown, only a limited number of these are quantitatively significant depositional environments. Thus rocks of spelean, glacial, and abyssal origin are very rare in the geological column. A second reason why this classification is difficult to apply to ancient sediments is because it is extremely hard to determine the depth of water in which ancient marine deposits originated. It is generally possible to recognize the relative position of a sequence of facies with respect to a shoreline, but it is often extremely hard to equate these with absolute depths (Hallam, 1967, p. 330). For this reason a classification of marine environments into depth-defined neritic, bathyal, and abyssal realms is hard to apply to ancient sediments.

In a review of the problems of environmental classifications, Crosby (1972) cites a version compiled by Shepard and McKee. This is particularly useful because its hierarchical structure allows the inclusions of many minor subenvironments which have been identified by studies of modern sediments (Table 6.3). Crosby (1972) has also prepared a useful classification of marine environments based on water depth, water circulation, and energy level. These three parameters define the limits of a number of depositional environments that have been commonly recognized in sedimentary rocks. Finally, Table 6.4 is a classification of major depositional sedimentary environments. It tabulates only those environments that generate quantitatively significant deposits and

Terrestrial	Landslide Talus Alluvial fans and plains River channels Floodplains Glacial moraine Outwash plains Dunes	Unidirectional wind types Bidirectional wind types Multidirectional wind types
Lacustrine	Playas Salt lakes Deep lakes	
Delta	Onshore	Distributary channel Levée Marsh and swamp Interdistributary Beach
	Offshore	Channel and levée extensions Distributary mouth bar Delta front platform Prodelta slope
Beach	Back-shore Berm Fore-shore	
Near-shore zone Offshore zone		
Barrier	Beach Dunefield Barrier flat Washover fan Inlet	
Bar (submerged)	Longshore bar Bay bar	
Tidal-flat area	Salt marsh Tidal flat Tidal channel	
Lagoon	Hypersaline Brackish Fresh	
Estuary	Shallow Deep	
Continental shelf Epicontinental sea		
Deep intracontinental depression	Trough Basin	
Continental borderland	Basin Trough	

continues

Table 6.3
Continued

Continental slope		
Deep sea	$\left\{\begin{array}{l}\text{Deep-sea fan}\\\text{Abyssal plain}\\\text{Marine areas marginal to}\\\quad\text{glaciers}\end{array}\right.$	
Reef	$\left\{\begin{array}{l}\text{Linear}\\\text{Patch}\\\text{Fringing}\end{array}\right.$	

can be identified confidently in ancient sediments. This scheme recognizes the three major environmental types of continental, transitional (shoreline), and marine. Within the continental environments no room is found for the rare glacial and cave deposits. Swamp deposits are omitted; they can generally be considered to be subenvironments of fluvial, lacustrine, and shoreline deposits. The classification of shorelines lays stress on whether they are linear-barrier shorelines or lobate deltaic ones. Though estuaries are clearly recognizable at the present day they are extremely hard to recognize in ancient sediments and their parameters are ill defined. Cliffed coastlines are omitted because they are essentially environments of nondeposition. Tidal flats, tidal channels, lagoons, salt marshes, and barrier bars are all considered as subenvironments of deltaic or linear shorelines.

The classification of marine environments shuns all attempts to recognize depth zones. Shelf deposits can be recognized by a combination of lithologic, paleontologic, and sedimentary features, as well as their structural setting. Terrigenous and carbonate shelves are easily differentiated. Reefs, in the broadest sense of the word, are recogniz-

Table 6.4
Classification of Depositional Sedimentary Environments

Continental	$\left\{\begin{array}{l}\text{Fanglomerate}\\\text{Fluviatile}\\\text{Lacustrine}\\\text{Eolian}\end{array}\right.$		$\left\{\begin{array}{l}\text{Braided}\\\text{Meandering}\end{array}\right.$
Shorelines	$\left\{\begin{array}{l}\text{Lobate (deltaic)}\\\text{Linear (barrier)}\end{array}\right.$		$\left\{\begin{array}{l}\text{Terrigenous}\\\text{Mixed carbonate-}\\\quad\text{terrigenous}\\\\\text{Carbonate}\end{array}\right.$
Marine	$\left\{\begin{array}{l}\text{Reef}\\\text{Shelf}\\\text{Submarine}\\\quad\text{channel and fan}\\\text{Pelagic}\end{array}\right.$	$\left\{\begin{array}{l}\text{Terrigenous}\\\\\text{Carbonate}\end{array}\right.$	

This table tabulates only those environments that have generated large volumes of ancient sediments. From Selley (1996, Table 1.2).

able to most geologists (saving the geosemanticists). Logically, most reefs are a variety of the carbonate shelf environment, but their abundance and geologic importance justifies giving them independent status. Submarine channel and fan environments result from turbidity currents and associated processes. (see Section 4.2.2). Pelagic deposits are those literally "of the open sea." These are largely fine-grained chemical deposits, argillaceous, calcareous, or siliceous. Pelagic sediments were deposited away from terrestrial influence, but this may have been due either to great water depth or distance from land. The designation "pelagic" evades this dilemma.

The classification of depositional environments shown in Table 6.4 has certain limitations. Nevertheless it forms a useful framework for the discussion of the main sedimentary models that are described later in this chapter.

6.2 SEDIMENTARY FACIES

Having examined the concept of sedimentary environments and their classification, it is now appropriate to discuss their ancient products. Geologists first began to study sedimentary rocks in some detail in the early part of the penultimate century. In 1815, William Smith published his "Geological Map of England and Wales, with Part of Scotland." The principles on which this map was based were that strata occurred in sequences of diverse rock types, that these strata had a lateral continuity which could be mapped, and that they were characterized by different assemblages of fossils. As the discipline of stratigraphic paleontology progressed, it became apparent that some fossils appeared to be restricted to certain geological time spans, while others were long ranging but appeared to occur in certain rock types. The implications of these facts were realized by Prevost in 1838. He proposed the name "formation" for lithostratigraphic units. By using biostratigraphy he demonstrated that different "formations" were formed in the same geological "epoch," and that similar formations could occur in different epochs. Almost simultaneously, Gressly (1838), working in the Alps, reached similar conclusions. He coined the name "facies" for units of rock which were characterized by similar lithological and paleontological criteria. His original definition (as translated by Teichert, 1958b) reads:

> To begin with, two principal facts characterize the sum total of the modifications which I call facies or aspects of a stratigraphic unit: one is that a certain lithological aspect of a stratigraphic unit is linked everywhere with the same paleontological assemblage; the other is that from such an assemblage fossil genera and species common in other facies are invariably excluded.

Gressly took the word "facies" from the writings of Steno (1667) and, though he was not aware of the distinction of zonal and facies fossils, he formulated a number of laws that govern the vertical and lateral transitions of facies. By the mid-nineteenth century it was established that bodies of rock could be defined and mapped by a distinctive combination of lithological and paleontological criteria. These rock units were called "formations" or "facies" by different geologists. Over the years, however, these two terms assumed different meanings. "Formation" became ever more rigorously defined, while "facies" became used in ever broader meanings.

In 1865, Lyell was still writing:

> The term "formation" expresses in geology any assemblage of rocks which have some character in common, whether of origin, age, or composition. Thus we speak of stratified and unstratified, freshwater and marine, aqueous and volcanic, ancient and modern, metalliferous and non-metalliferous formations.

In modern usage this definition better fits facies than formation.

Over the years, however, the term "formation" became more and more restricted to the description of discrete mappable rock units. This concept was finally embalmed in Article 6 of the Code of Stratigraphic Nomenclature of the American Commission on Stratigraphic Nomenclature, namely:

> The **formation** is the fundamental unit in rock stratigraphic classification. A formation is a body of rock characterized by lithologic homogeneity. It is prevailingly, but not necessarily tabular and is mappable at the earth's surface or traceable in the subsurface.

Similarly, the Geological Society of London states:

> The formation is the basic practical division in lithostratigraphical classification. It should possess some degree of internal lithological homogeneity, or distinctive lithological features that constitute a form of unity in comparison with adjacent strata. (Geol. Soc. London, *Recommendations on Stratigraphical Usage,* 1969)

Thus Prevost's original term used for describing a rock unit by lithology and paleontology has been restricted to a lithostratigraphic unit. Meanwhile, Gressly's term "facies" became used in ever vaguer ways. In particular, it ceased to be used to define rocks purely by lithology and paleontology. The paleogeographic and tectonic situation of a rock also became one of the defining parameters (e.g., Stille, 1924). Thus the terms "geosynclinal," "orogenic," and "shelf facies" came into use. The concept of facies was also applied to metamorphic rocks (Eskola, 1915). Recognition of the breadth of usage of facies is shown by Krumbein and Sloss's terms "lithofacies," "biofacies," and "tectofacies" (1959). "The expressions of variation in lithologic aspect are lithofacies and the expressions of variation in biologic aspect as biofacies" (1959, p. 268). "Tectofacies are defined as the laterally varying tectonic aspect of a stratigraphic unit" (1959, p. 383). Further extensive use of the term "facies" came when environmental connotations were added. As early as 1879, von Mojsisovics wrote: "Following Gressly and Oppel, one now customarily applies the term facies to deposits formed under different environmental conditions" (translation in Teichert, 1958b). As the analysis of ancient sedimentary environments progressed this approach to facies became increasingly common until the same rock formation could be termed "flysch," "geosynclinal," or "turbidite facies." Only the first of these three usages is descriptive and consistent with the original definition. The terms "geosynclinal" and "turbidite" give a genetic connotation to the term, describing its interpreted tectonic situation and depositional process.

More recently still the term "facies" has been applied to describe the character of seismic data. Terms like "mounded," "even," "wavy," and "contorted" are applied to seismic character, and used to interpret the depositional environment (e.g., Mitchum *et al.,* 1977; Neal *et al.,* 1993). Further reviews of the origin of the facies concept and its sub-

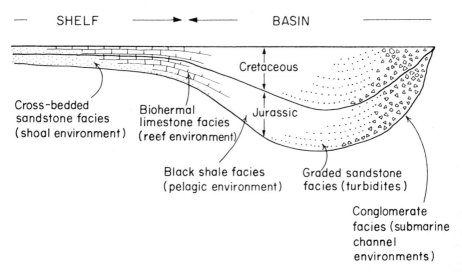

Fig. 6.1. Cross-section of a sedimentary basin showing the relationship between facies, environments, and time.

sequent growth and diversification are found in Teichert (1958b), Erben (1964), Anderton (1985), Plint (1995) and Reading (1996).

It seems most useful to restrict the term "facies" to its original usage. This text uses facies as a descriptive term as stated by Moore (1949): "**Sedimentary facies** is defined an any areally restricted part of a designated stratigraphic unit which exhibits characters significantly different from those of other parts of the unit." A further refinement of this definition is that a facies has five defining parameters: geometry, lithology, paleontology, sedimentary structures, and paleocurrent pattern (Selley, 1970, p. 1). Thus one may talk of red bed facies, an evaporite facies, or a flysch facies. Terms like "shelf facies" or "back-arc facies" are unnecessary and should not be used. Similarly, terms such as "fluvial facies" or "turbidite facies" are naughty and inadmissible; they are genetic and conflict with the original definition. It may be correct, however, to make statements to the effect that "this pebbly sand facies was deposited in a fluvial environment, and this flysch facies was deposited by turbidity currents." Figure 6.1 illustrates the differences between facies and environments.

6.3 SEDIMENTARY MODELS

6.3.1 The Model Concept

The academic geologist has always been concerned with interpreting geologic data. The industrial geologist has to go one step further and make predictions based on these interpretations. One of the most useful tools in both these exercises is the concept of the sedimentary model. This is due to the concatenation of the concepts of environments and facies which were discussed in the two preceding sections.

The concept of the sedimentary model is based on two main observations and one major interpretation. These may be dogmatically stated as follows:

Observation 1: There are on the earth's surface today a finite number of sedimentary environments. (Qualifying remarks: Detailed examination shows no two similar environments to be identical. Environments show both abrupt and gradational lateral transitions.)

Observation 2: There are a finite number of sedimentary facies which recur in time and space in the geological record. (Qualifying remarks: Detailed examination shows no two similar facies to be identical. Facies show both abrupt and gradational lateral and vertical transitions.)

Interpretation: The parameters of ancient sedimentary facies of unknown origin can be matched with modern deposits whose environments are known. Thus may the depositional environments of ancient sedimentary facies be discovered.

Conclusion: There are, and always have been, a finite number of sedimentary environments which deposit characteristic sedimentary facies. These can be classified into various ideal systems or models.

These ideas have been implicit, though seldom so baldly stated, in most, if not all facies analyses. Potter in particular has discussed sedimentary and facies models (apparently as interchangeable terms). Potter and Pettijohn (1977, p. 314) write: "A sedimentary model in essence describes a recurring pattern of sedimentation" and again "the fundamental assumption of the model concept is that there is a close relationship between the arrangement of major sedimentation in a basin and direction structures in as much as both are a product of a common dispersal pattern" (p. 228).

Additional discussion of models, sedimentary, and facies has been given by Potter (1959), Visher (1965), Pettijohn *et al.* (1972, p. 523), Blatt *et al.* (1978), Curtis (1978), Walker (1984a,b), Reading and Levell (1996), and Selley (1996).

The next part of this chapter is devoted to the description of the major sedimentary models that can be defined. The classification of these models is based on the classification of depositional environments given in Table 6.4. Each major depositional environment can be correlated with a sedimentary model. More detailed accounts of these are found in books by Laporte (1979), Klein (1980), Reineck and Singh (1980), Scholle and Spearing (1982), Scholle *et al.* (1983a,b), Anderton (1985), Reading (1996), and Selley (1996).

6.3.2 Some Models Described

6.3.2.1 Piedmont Fanglomerates

Between mountains and adjacent lowlands it is often possible to define a distinct belt known as the **piedmont zone** (literally from the French: mountain foot). At the feet of mountains, valleys debouch their sediments into the plains. The piedmont zone is characterized by small alluvial cones which are fed by gullies and build out onto pediment surfaces cut subhorizontally into the bedrock (Fig. 6.2). Larger alluvial fans are fed by complex valley systems and grade out into the deposits of the alluvial plain (Fig. 6.3).

Depositional gradients in the piedmont zone are steep, up to 30° in marginal screes, but diminish radially down-fan. This change in gradient correlates with changes in process and sediment type. Alluvial cones and the heads of alluvial valleys are characterized by boulder beds and conglomerates deposited by gravity slides from the adjacent mountain sides. These grade down the fan into conglomerates and massive or crudely

Fig. 6.2. Escarpment with minor marginal alluvial fans (background) and pediment equilibrial surface (foreground), southern Libya.

bedded argillaceous pebbly sandstones and siltstones (Fig. 6.3). Deposits of this type, termed "diamictites," originate from mud flows. (see Section 4.5.1). They in turn grade down-fan into poorly sorted massive or flat-bedded pebbly sandstones. They sometimes show irregular scours and silt laminae. These beds are deposited from flash floods, and pass downslope into the alluvium of braided channel systems. Deposits of the piedmont zone are characterized, therefore, by extremely coarse grain size and poor sorting, by

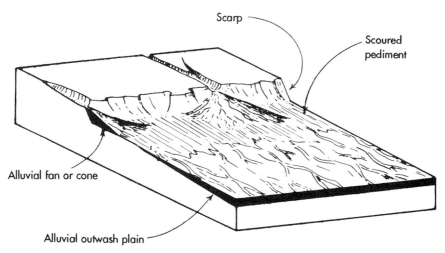

Fig. 6.3. Sketch illustrating the morphology of the piedmont zone. Coarse sands and gravels are deposited by landslides, mud flows, and flash floods on the alluvial fans. Sands with minor silts are deposited in braided channel systems by ephemeral floods on the outwash plains.

Fig. 6.4. Old Red Sandstone (Devonian) fanglomerate outcrop. Hornelen basin, near Nordfjord, western Norway. (Courtesy of J. M. Cohen.)

massive or subhorizontal bedding, and by an absence of fossils. The term **fanglomerate** is applied to deposits of the piedmont zone, aptly denoting their lithology and geometry (Lawson, 1913). The previous account has been drawn largely from studies of modern piedmont zones, especially in the Rocky Mountains, by Trowbridge (1911), Black-welder (1928), Blissenbach (1954), Bluck (1964), Denny (1965), and Van Houten (1977).

Modern piedmont zone deposits are widespread around mountain chains from the Arctic to the Equator. Piedmont zone deposits have been frequently identified in ancient sediments. They are especially characteristic of fault-bounded intracratonic rifts. Spectacular examples rim the margins of the Devonian Old Red Sandstone basins of Scotland and Norway (Bluck, 1967) (Fig. 6.4). In such situations great thicknesses of fanglomerates can form due to episodic synsedimentary movement along fault scarps. Repeated fault scarp rejuvenation generates upward coarsening conglomerate cycles as fans prograde toward the basin depocenter.

A second common setting for piedmont zone deposits is as thin laterally extensive veneers above major basal unconformities of thick continental sequences. Examples of this type have been documented from the Torridon Group (Pre-Cambrian) of northwest Scotland by Williams (1969) and from the sub-Cambrian unconformity of Jordan (Selley, 1972). There are many other instances of ancient fanglomerate piedmont deposits.

6.3.2.2 Fluvial Processes and Models

The processes and deposits of modern rivers have been intensively studied for a number of very good reasons. Many of the largest concentrations of population of the modern world lie on major alluvial valleys such as the Ganges, the Indus, the Nile, and the Mississippi. It is important, therefore, to study alluvial processes and deposits because of the way in which they influence settlement patterns, farming, irrigation, water supply, communications, and pollution. Important accounts of modern river systems have been given by Marzo and Puigdefabregas (1993), Collinson (1996), and Smith and Rogers (1999).

6.3.2.2.1 Fluvial processes and modern alluvium

Piedmont fans pass gradationally downslope into the alluvial environment. Alluvial systems may be arbitrarily divided into two main types according to channel morphology, though there are gradations between the two. Alluvial fans are characterized by ephemeral flood discharge that generates low sinuosity braided channels.

Alluvial fans often grade into, or occasionally flank, alluvial floodplains. These normally contain a single sinuous river channel with continuous discharge. The lateral changes from piedmont to alluvial fan to floodplain are associated with changes in gradient, grain size, and sorting (Fig. 6.5). The factors that determine whether a channel meanders or braids appear to include gradient, grain size, and discharge. Braided channels are normally associated with steep slopes, coarse, often gravelly, sediment, and ephemeral discharge. Meandering channels are normally associated with fine-grained sediment, gentle gradient, and steady discharge. There are, however, exceptions to these generalizations, and the relationships between these factors and channel morphology are complex. Analyses of these problems will be found in Allen (1964), Leopold *et al.* (1964), Miall (1978), Collinson and Lewin (1983), and Collinson (1996). Many different fluvial models have been defined; see, for example, Cant (1982) and Miall (1981). All pre-Silurian alluvial deposits appear to be of braided channel type. Alluvium attributable to meandering channels is absent. This may be because the land was not colonized by plants, whose roots and stems would slow down runoff, until this time. Thus in Precambrian and Early Paleozoic time rainwater would have drained quickly off the land

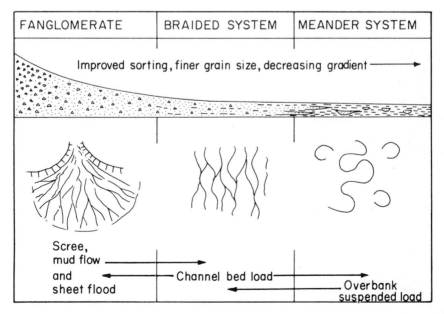

Fig. 6.5. Diagram illustrating the transitional relationship between piedmont fans, braided and meandering alluvial systems, and facies.

via ephemeral, braided river channels. The steady perennial discharge, characteristic of meander channels, did not occur until the colonization of the land by vegetation. Braided and meandering alluvial depositional systems are now described in turn.

6.3.2.2.2 Alluvium of braided rivers

Braided channel systems are characterized by a network of constantly shifting low sinuosity anastomozing courses. Modern braided river systems occur on alluvial fans in semi-arid and arid climates, along many mountain fronts, the edges of ice caps, and the "snouts" of glaciers. They are characterized by relatively higher gradients and coarser sediments than meandering river systems. Discharge is seasonal and ephemeral in glacial and mountainous realms, and still more sporadic in desert climates. With their coarse grain size and erratic flow, braided channels are generally overloaded with sediment (Fig. 6.6). Continuous formation of channel bars causes thalwegs to diverge until they meet up with another channel course.

Braided alluvial plains thus consist essentially of a network of channels with no clearly defined overbank terrain. Fine suspended load sediment can only come to rest in rare abandoned channels and in the pools of active channels when a flood abates. Channel abandonment occurs by channel bars enlarging until they block a course and cause it to divert. Alternatively, the headward erosion of channel gullies in soft detritus results in capture of a previously active channel course. The alluvium of braided river systems consists, therefore, largely of channel lag gravels and of cross-bedded channel bar and braid bar sands. Laminated, cross-laminated, and desiccation-cracked fine sands and silts occur in rare shoe-string forms infilling abandoned channels (Fig. 6.7).

Accounts of mountain front, glacial, and arid braided alluvial deposits, ancient and

Fig. 6.6. Modern river, Tunsbergdabru in northern Norway, showing braided channels separated by longitudinal gravel bar. (Courtesy of J. M. Cohen.)

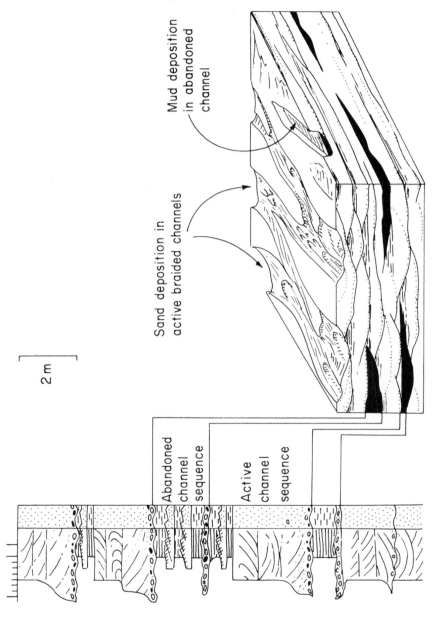

Mud deposition in abandoned channel

Sand deposition in active braided channels

Abandoned channel sequence

Active channel sequence

2 m

Fig. 6.7. Physiography and deposits of a braided alluvial channel system. Sedimentation occurs almost entirely in the rapidly shifting complex of channels. Silts are rarely deposited in abandoned channels. A floodplain is absent.

modern, have been given by Doeglas (1962), Rust (1972), Williams (1971), and Best and Bristow (1993). Miall (1977), Rust (1978), and Cant (1982) have proposed several depositional models for braided channel systems. Miall (1977) defined four types of braided alluvial deposit, named after the recent Scott, Donjek, and Platte rivers and Bijou Creek. The Scott River type is composed almost entirely of flood gravels, with minor cross-bedded sand bars. The Donjek type consists of upward fining channel lag gravels and braid bar sands. The Platte River type is predominantly sandy. It contains thin channel lag gravels but is mainly composed of cross-bedded sands deposited from migrating longitudinal and linguoid bars. Finally, the Bijou Creek type is the finest grained of all. There are no gravels, apart from rare intraformational clay pebble conglomerates. Cross-bedded braid bar sands predominate, with minor abandoned channel muds.

Examples of alluvium that was deposited from braided channel systems are found worldwide in rocks of diverse ages from the Pre-Cambrian to the recent. One noted instance occurs in the great Cambro-Ordovician sand blanket that stretches from the Atlantic Ocean, across the Sahara and Arabia as far as the Arabian Gulf (Bennacef *et al.,* 1971; Selley, 1972, 1996; al-Laboun, 1986). These sands were derived from the Pre-Cambrian shields, transported northward across vast braided outwash plains, some as periglacial outwash, to interfinger with shallow marine shelf sands. With excellent porosity and permeability, and great lateral continuity, these formations are important aquifers and petroleum reservoirs in Algeria, Libya, and Arabia.

6.3.2.2.3 Alluvium of meandering rivers

With increasing distance from source, the gradient of river profiles lessens, grain size decreases, and channels diminish in number on the floodplain and increase in sinuosity. Thus braided alluvial plains can change downstream into broad floodplains traversed by meandering rivers. The resultant alluvium shows the whole suite of active channel, abandoned channel, and overbank deposits (Table 6.5). There is, therefore, a much higher proportion of silt and clay and far less sand and gravel than in the alluvium of braided rivers.

Figure 6.8 illustrates the physiography and mutual relationships of these subenvironments. The characteristic feature of the subfacies of each subenvironment is now

Table 6.5
Hierarchical Classification of Alluvial Subenvironments

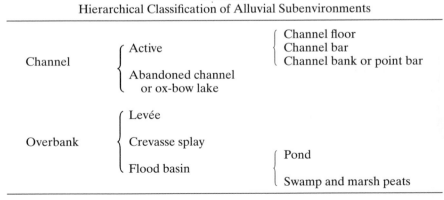

Based on Shantser (1951) and Allen (1965).

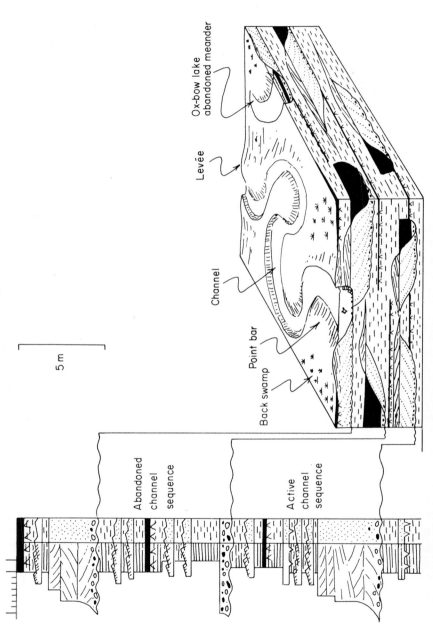

Fig. 6.8. Physiography and deposits of an alluvial flood plain cut by meandering channels. This illustration shows how the lateral migration of a channel generates an upward-fining-grain size profile on its inner convex bank.

Fig. 6.9. Photograph of the alluvial valley of the River Cuckmere, Sussex, to show the floodplain and meander belt. (Compare with the braided channel system of Fig. 6.6.)

described, and processes that form them explained. The banks of a river are inherently unstable due to the erosive power of the current. This is particularly so where rivers flow through their own detritus. This instability shows itself both by sudden switching of channels from place to place, and by the gentle lateral erosion of channel walls. This process deserves discussion in some detail. In certain circumstances it is an inherent property of river channels to meander sinuously across their floodplain (Fig. 6.9). As the water flows round a bend the current velocity increases on the outer bank of the curve and decreases on the inner bank. This leads to erosion of the outer bank, to form a sub-vertical cliff. On the inner bend of a meander a slackening of current velocity allows sedimentation of bed load, and the formation of a gently sloping point bar profile. On the point bar of major rivers subaqueous dunes are present which migrate downcurrent round the bend, depositing cross-bedded sands. On the river bed in the center of the channel the cross-sectional profile remains about constant. A lag gravel of intraformational and extraformational clasts may be present together with abraded bones, teeth, shells, and waterlogged driftwood.

As these processes continue through time, the channel migrates sideways to deposit a characteristic sequence of grain size and sedimentary structures. At the base of the sequence is a scoured intraformational erosion surface bevelled across older alluvium or bedrock. This is overlain by a channel lag conglomerate whose composition has already been described. This may be a veneer only one clast thick, or it may occur in crudely bedded or cross-bedded sequences measurable in tens of meters or more. Above this unit come the cross-bedded sandbar deposits. Some studies of modern and ancient point bar sequences record a vertical decline in grain size and set height. This reflects the progressive lateral decline in current velocity from the channel floor across the point bar up to the inner bank of the meander. Vertical fining of grain size will not be present if the source of detritus does not contain a broad enough spectrum of grain sizes.

The rate of discharge in a river channel is seldom constant. Diminishing discharge will result in the river shrinking within its own major channel, to find its way in a braided pattern through the bars which it deposited at full flood. An increase in discharge, by contrast, causes the river to rise until it bursts its banks. On flowing over the channel lip, current velocity may diminish; thus depositing layers of sediment, termed **levées,** which decrease in grain size away from the lip. The levées may build the banks up higher and higher on either side of the channel. They separate the channel from low-lying flood basins on either side of the alluvial plain. Flood basin deposits are generally fine-grained sands, clays, and silts. They are interlaminated, cross laminated, and characteristically desiccation-cracked. Flood basin deposits are often burrowed, frequently pierced by plant roots, and, under suitably waterlogged conditions, may become peat-forming swamps and marshes. This assemblage of levée, flood basin, and swamp sediments is collectively referred to as an **overbank deposit,** to distinguish it from the assemblage of channel deposits.

One further type of overbank deposit to mention is the **crevasse-splay.** Rivers at bank full scour channels through the levées termed "crevasses"; lobes of fine sand are deposited where these debouch into the flood basins. These crevasse-splays are similar to, though smaller than, the lobes of deltas.

Returning now to processes of channel sedimentation, recall that the classification in Table 6.5 differentiates active and abandoned channel types. The sedimentary processes of channels just described may be terminated abruptly by switching or avulsion of the channel course. This can happen in several ways. A sinuous river channel can ultimately meander back on itself to short circuit its flow by necking, as it is termed. Another type of channel diversion occurs when a river has raised itself up above the floodplain by levée building to the extent that its floor is above the level of the floodplain. Channel diversion of this type can often occur on a huge scale with resultant catastrophic flooding and loss of life. Both types of channel switching are characteristic of meandering river systems. In some situations the lower end of the abandoned channel may still open out as a backwater into the main channel. After a time, however, this may become blocked to form an isolated ox-bow lake, as it is picturesquely termed, lying within the floodplain away from the present active channel. Abandoned channel deposits are similar to those of flood basins. They are laminated, cross-laminated, desiccation-cracked, burrowed fine sands, silts, and clays, which are occasionally pierced by rootlets and interbedded with peats. Abandoned channel deposits are distinguishable from flood basin deposits, however, by their channel-shaped geometry, and by the fact that they abruptly pass down into channel-floor lag gravels instead of passing down gradually into cross-bedded point bar sands.

This review of the processes and products of modern alluvial sedimentation is based on a wide range of publications. Particularly important studies have been documented by Leopold *et al.* (1964), Allen (1965), Miall (1978, 1981), Cant (1982), Collinson and Lewin (1983), and Collinson (1996).

The upward-fining sequences of meandering alluvial deposits tend to be repeated rhythmically. Classic examples of this phenomenon are described from the Devonian Old Red Sandstone of the North Atlantic margins of the Appalachians, South Wales, and Spitzbergen (Allen and Friend, 1968; Allen, 1964; Friend, 1965, respectively). The cycles vary in average thickness from 1 or 2 m, up to 20 m. The genesis of the fining-upward sequence is implicit in the sedimentary model of the laterally migrating channel. The cause of its repetition has been a subject for debate (Miall, 1981).

Four fining-up cycles in the Quaternary alluvium of the Mississippi River Valley are correlative with eustatic sea level changes during the ice age (Turnbull *et al.,* 1950). There is also a strong presumption that erratic subsidence of basins bounded by active faults may cause repeated regrading of floodplains, thus generating thick piles of alluvium with upward-fining cyclothems. Cyclic climatic changes, especially precipitation, may effect runoff and river discharge patterns. These can cause pulsating fluctuations in sediment

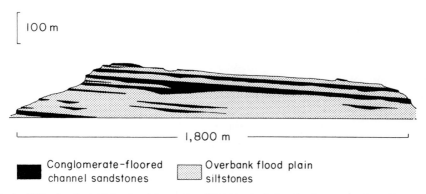

100 m

1,800 m

■ Conglomerate-floored   ▢ Overbank flood plain
  channel sandstones   siltstones

Fig. 6.10. A cliff in the fluvial Morrison Formation, Slick Rock, Colorado. A measured section at any point in that cliff would record a series of upward-fining cyclothems. Individual channel sequences have only limited lateral extent, however, suggesting an absence of any external (allocyclic) controlling mechanism. (From Shawe *et al.,* 1968.)

input which, again, may be responsible for the repetition of the upward-fining channel motif.

Beerbower (1964) suggested, however, that it is unnecessary to invoke such external processes as the cause of repeated alluvial cycles. They may be simply explained by the to-and-fro lateral migration of a river channel across its floodplain, coupled with gradual isostatic adjustment of the basin floor in response to the weight of sediment. Individual cyclothems should be basin wide if due to external climatic or tectonic causes (termed **allocyclic** by Beerbower). By contrast, cyclothems due to the lateral meandering of channels, interrupted by abandonment, superimposed on gradual subsidence (termed **autocyclic** by Beerbower) may be expected to be of only local extent.

It is interesting to note that studies of the small-scale facies changes of ancient alluvium sometimes show uncorrelatable sections spaced only a few hundred meters apart (Kazmi, 1964; Friend and Moody-Stuart, 1972, fig. 21). Extensive paleostrike trending cliffs in the Morrison Formation of Colorado also show that upward-fining cycles occur repeatedly at any one point, but are of extremely local extension (Fig. 6.10).

Examples of alluvium that was deposited from meandering channel systems are found worldwide in rocks of diverse ages from the Silurian to the Recent. Well-documented examples occur in the Old Red Sandstone (Devonian) deposits that ring the North Atlantic from England, Wales, Ireland, Norway, Spitzbergen, Greenland, and North America. The Old Red Sandstone includes deposits of many continental environments, including fanglomerate, braided outwash, eolian dune, and lake fill. Reference has already been made, however, to many accounts of repeated upward-fining cycles attributed to the deposition of point bar sands from channels that meandered across muddy floodplains.

6.3.2.2.4 Economic aspects of alluvial deposits

Alluvial deposits are of great economic importance for many reasons. They serve as aquifers, as petroleum reservoirs, and as the hosts for deposits of coal, uranium, and placer minerals.

Alluvium may serve as an aquifer or a petroleum reservoir because of its porosity and permeability. Obviously the storage capacity and flow characteristics of alluvial deposits will vary according to the sorting and cementation of the sands, and to the overall sand:shale ratio. Thus the blanket sands of braided outwash plains normally have better continuity than the often isolated point bar sands of meander channel alluvium. Alluvial aquifers range from small Holocene valley fill deposits to the vast alluvial blankets of the Early Cretaceous "Nubian Sandstones" that extend across whole Saharan sedimentary basins (Pallas, 1980). The same generalizations apply to petroleum reservoirs. Braided alluvial reservoirs in structural traps host such giant oil and gas fields as Hassi Messaoud, Algeria (Balducci and Pommier, 1970), Prudhoe Bay, Alaska (Melvin and Knight, 1984; Atkinson *et al.*, 1990), and Sarir and Messla, Libya (Clifford *et al.*, 1980). These fields have reservoirs over 100 m thick and areal extents of hundreds of square kilometers (Fig. 6.11, upper).

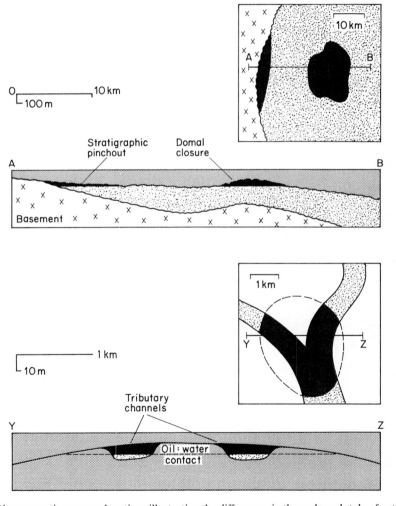

Fig. 6.11. Diagrammatic maps and sections illustrating the differences in the scale and style of petroleum entrapment in the alluvium of braided (upper) and meandering (lower) channel systems. Giant structural traps occur in the thick reservoir sands of braided channel systems. Small stratigraphic traps occur in the discrete point bar sands of meandering fluvial channels. For examples, see text.

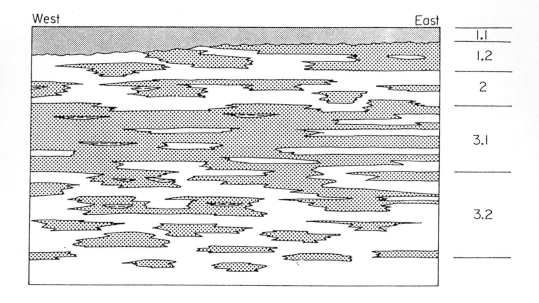

West   East

1.1
1.2
2
3.1
3.2

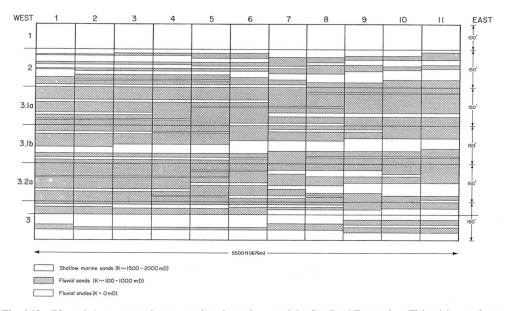

WEST   1   2   3   4   5   6   7   8   9   10   11   EAST

1   100'
2   150'
3.1a   150'
3.1b   150'
3.2a   150'
3   150'

5500 ft (1676m)

Shallow marine sands (K~1500 - 2000 mD)

Fluvial sands (K~100 - 1000 mD)

Fluvial shales (K = 0 mD)

Fig. 6.12. (Upper) A conceptual cross-section through part of the Statfjord Formation (Triassic) petroleum reservoir in the northern North Sea showing the complex arrangement of reservoir sands, and nonreservoir shales. A blanket shallow marine sand at the top of the reservoir presents no problem, but note the problems of correlating and predicting sand continuity in the lower fluvial part of the section. In such reservoirs a probabilistic approach must be used. (Lower) Stochastic and other statistical methods are employed to produce a reservoir simulation model to compute the way in which the reservoir will produce petroleum (From Johnson and Stewart, 1985, by courtesy of the Geological Society of London).

Such petroleum accumulations are in marked contrast to those found in the alluvium of meandering channels. These are normally small stratigraphic traps where the reservoir may be only several meters thick and extend for a few square kilometers (Fig. 6.11, lower). The Little Creek field in the Tuscaloosa trend (Upper Cretaceous) of Mississippi is an excellently documented example of a small field in a fluvial point bar stratigraphic trap (Werren *et al.*, 1990).

Finally, once a fluvial petroleum reservoir has been found, the production of petroleum from it can be quite challenging. Attention must be paid to the size and frequency of channels, to establish their intercommunication (refer back to Fig. 6.10 to see the nature of the problem). Ideally one would hope to be able to use the depositional model to correlate channels, but in many cases the problem is too complex, and necessitates a shift from deterministic to probabilistic methods, such as stochastic modeling (e.g., Martin, 1993). Reservoir units must be defined and characterized for computer simulation (Fig. 6.12). On a smaller scale, recall that in Chapter 3, it was shown how channels normally exhibit an upward-fining grain-size profile, concomitant with an upward decrease in permeability. Crevasse-splays, by contrast, are the reverse. These variations must be considered in small-scale reservoir modeling of fluvial systems (Fig. 6.13).

In certain circumstances permeability barriers within alluvium are actually economically advantageous because they favor the precipitation of valuable minerals, such as the

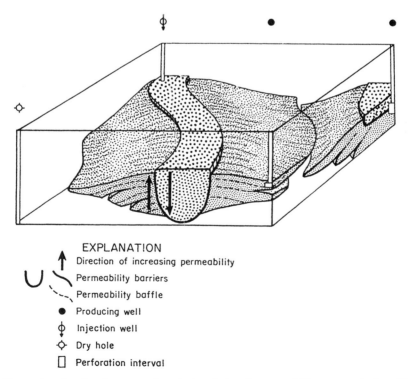

EXPLANATION
↑ Direction of increasing permeability
Permeability barriers
Permeability baffle
● Producing well
φ Injection well
◇ Dry hole
☐ Perforation interval

Fig. 6.13. Illustration showing the opposed vertical variations in permeability commonly found in channels and their associated levée sands. This situation is encountered in both fluvial and deltaic petroleum reservoirs. (From Tyler and Finley, 1991, by courtesy of the Society for Sedimentary Geology.)

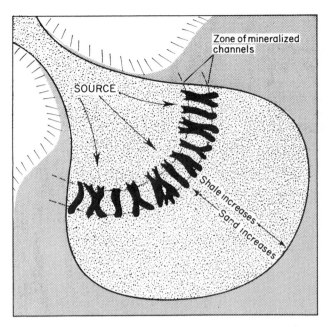

Fig. 6.14. Diagram illustrating the facies control on uranium mineralization that is seen in some alluvial fans. "SOURCE" refers both to the origin of the fan detritus and of the mineralizing fluids. The latter generally originate from the leaching of volcanic rocks. For examples, see text.

uranium ore carnotite. Reference has already been made to the relationship of this mineral to recent meteoric water flow (see Section 1.3.2.3). There is, however, often a strong facies control on the actual locus of mineral emplacement. On a regional scale mineralization sometimes occurs halfway down alluvial fans (Fig. 6.14). This phenomenon has been noted both in Tertiary alluvium in Wyoming, and in Triassic alluvium in Utah (Galloway and Hobday, 1983). Precipitation of carnotite occurs where uranium charged meteoric water meets connate water. The zone of mixing occurs where the permeability reaches a critical sand:shale ratio. Considered on a smaller scale, uranium deposits are restricted to channel sands, as shown previously in Fig. 2.14.

Alluvial deposits are also important hosts for placer ores of detrital heavy minerals notably gold (Bache, 1987). Placers are variously termed eluvial, coluvial, and alluvial. Eluvial concentrations occur in place above the parent ore (mother lode). Eluvial placers are those transported a short distance downslope, while alluvial placers have undergone fluvial transportation. The processes that govern placer formation still seem to be little understood, but see Macdonald (1983). Placers are often developed associated with lag gravels on channel floors, and may even be preserved in fissures and potholes on the river bed. Placers occur both on the floor of the modern channel, and also along the basal erosion surfaces of river terraces and buried river vallies. Paleoplacers are found in analogous settings in pre-Holocene alluvium. Heavy mineral segregation occurs on the deeper downstream parts of point bars, on the upstream slopes of braid bars, and adjacent to stream confluences (Fig. 6.15).

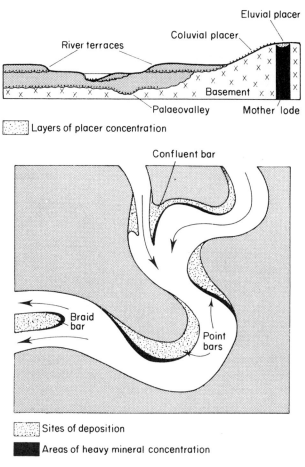

Fig. 6.15. Cross-section and plan showing the loci of alluvial and related placer mineral deposits.

There are several Proterozoic alluvial deposits that host both gold and uranium. These are scattered around the world, but are broadly similar in sedimentology and age. They consist of conglomerates that were deposited in fluvial channel systems. Detailed studies show that these range from tributary valley fill, via braided alluvial outwash plains, to fan deltas that debouched into large standing bodies of water (Fig. 6.16).

The best known example is the reef of the Witwatersrand Group in the basin of the same name in South Africa (Fig. 6.17). Other examples occur at Blind River and Montgomery Lake, Canada, the Moeda and Jacobina deposits of Brazil, and Deep Lake, USA (Armstrong, 1981; Cox, 1986). Radiometric dates show that these deposits were formed between 2 and 2.8 billion years ago. Uraninite, the common ore mineral, is unstable on the surface of the earth today. It presumably formed before algal photosynthesis depleted the atmosphere of carbon dioxide to generate the modern oxygenated atmosphere. The deposits themselves are quartz and quartzite conglomerates and pebbly

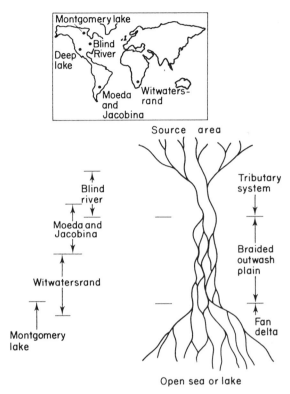

Fig. 6.16. Diagram illustrating the various settings of the Pre-Cambrian gold- and uranium-bearing conglomerates that occur around the world. (From Skinner, 1981.)

sands (Fig. 6.18). Paleocurrent analysis commonly delineates subparallel or radiating orientations of the ore-bearing conglomeratic channels (Fig. 6.19).

The matrix of the conglomerates is a poorly sorted carbonaceous wacke that includes detrital grains of gold and uraninite. The carbonaceous material is believed to be of bacterial or algal origin. The presence of diagenetic pyrite indicates reducing conditions. Some of the uraninite, however, appears to be authigenic. There has, therefore, been great debate as to what extent the gold is a placer deposits, to what extent the uranium was concentrated by bacterial action, and if both may have been subsequently remobilized. The occurrence of rounded gold particles on foresets show that the gold in the Witwatersrand basin began its life as a placer ore, but isotopic and fluid inclusion studies demonstrate that it subsequently underwent two phases of hydrothermal remobilization (Frimmel, 1997).

This brief review shows that alluvial deposits are of economic importance as aquifers, as petroleum reservoirs and as hosts to certain mineral deposits. Peat, the precursor of coal, often forms in association with fluvial sediments. For convenience, however, peat formation will be discussed in the context of deltaic sedimentation (see Section 6.3.2.5).

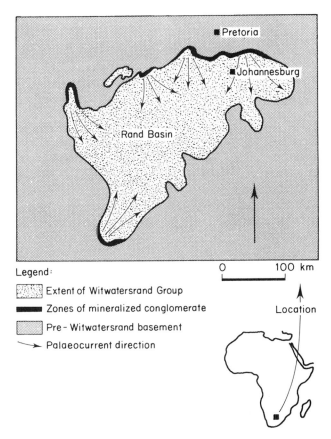

Fig. 6.17. Map of the Pre-Cambrian Witwatersrand basin of South Africa, to show the location of gold- and uranium-bearing alluvial fans. (Based on Pretorius, 1979, and Brock and Pretorius, 1964.)

6.3.2.3 Eolian Depositional Systems

6.3.2.3.1 Modern deserts and eolian environments

Significant volumes of eolian sediment are only deposited in deserts, though minor quantities may develop along beaches and barrier islands. Deserts are regions characterized by aridity, and often by extremes of temperature (Williams and Balling, 1996). Modern deserts occur in the centers of continents away from marine influence, and along the western seaboards of continents where prevailing winds blow offshore (Fig. 6.20). Not all deserts are characterized by high temperatures. In some arid periglacial environments wind action forms ventifacts and dust clouds that may ultimately settle as loess (see Section 4.3.2).

Eolian sands are only one of a wide range of deposits that occur in modern deserts (Fig. 6.21). Many deserts are equilibrial environments which range from sand-blasted rock pediments, via boulder or gravel strewn surfaces, to flat sand sheets. Vast areas of

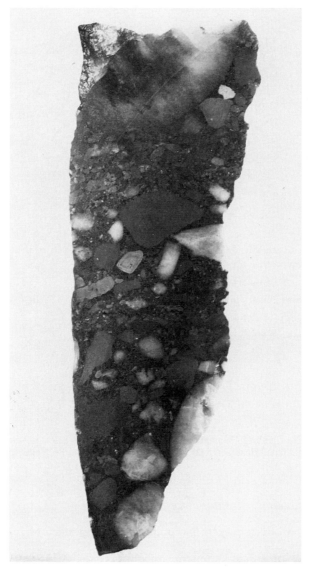

Fig. 6.18. A 4-cm-diameter core of gold- and uranium-bearing conglomerate from the Witwatersrand Group, Pre-Cambrian, of the Witwatersrand basin, South Africa.

modern deserts are in fact old Pleistocene alluvial fans formed during pluvial periods. During the subsequent arid phase the wind has deflated sand, silt, and clay until the fans are protected by a lag gravel. This is the "sarir" surface (Fig. 6.22). Many modern eolian sands are well-rounded, well-sorted chemically mature orthoquartzites (Fig. 3.9D). Roundness may reflect prolonged abrasion; sorting may reflect prolonged transportation. Note, however, that many modern desert sands are polycyclic in origin, having

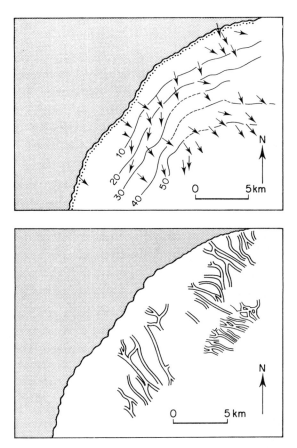

Fig. 6.19. (Upper) Isopach and paleocurrent map of the Vaal Reef on the northwestern side of the Witwatersrand basin (C.I. = 10 m). (Lower) Ore-bearing conglomerate channels of the same. (Based on data in Minter, 1981, and Minter and Loen, 1991.)

undergone multiple cycles of weathering, erosion, transportation, and deposition. They would be texturally and mineralogically mature irrespective of environment. Some modern eolian sands are arkosic and volcaniclastic, though well rounded and well sorted. This reflects the absence of chemical weathering.

Eolian sedimentary processes and their resultant bedforms and structures have already been described. Further accounts of modern deserts are found in Glennie (1970, 1987), McKee (1979), Collinson (1996), Nickling (1986), Pye and Lancaster (1993), Lancaster (1995), and Kokurek (1996).

6.3.2.3.2 Ancient eolian deposits

The foregoing brief review of modern deserts suggests that ancient eolian sands will be associated with other arid environment deposits, such as braided alluvium, sabkha

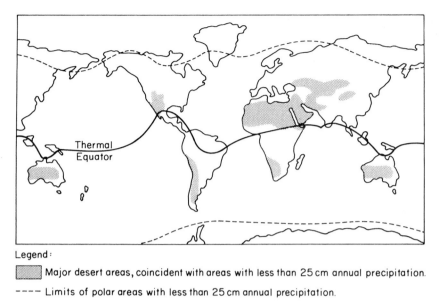

Legend:

▨ Major desert areas, coincident with areas with less than 25 cm annual precipitation.

---- Limits of polar areas with less than 25 cm annual precipitation.

Fig. 6.20. World map showing the distribution of modern deserts. Note that they occur in the centers of continents, and on the western seaboards of continents where the prevailing winds blow offshore.

evaporites, and playa lake shales. Ancient eolian sandstones have been identified in three main areas:

1. Pennsylvanian–Jurassic deposits of Arizona, Colorado, Utah, and New Mexico (Ahlbrandt and Fryberger, 1982)
2. The Late Jurassic–Early Cretaceous Botucatu Sandstone of Brazil (Bigarella and Salamuni, 1961; Bigarella, 1979)
3. Permian–Triassic "New Red Sandstone" of northwest Europe (Laming, 1966, Thompson, 1969; Glennie, 1972, 1982, 1990)

In all of these examples the presumed eolian sands occur in red bed facies, and are often associated with fanglomerates, alluvial sands, red shales, and evaporites. The sands are texturally mature, and exhibit large-scale cross-bedding in sets often several meters in height (Fig. 6.23). Detailed facies analysis has been used to map out the paleogeographies of these past desert terrains (Fig. 6.24). Despite these wide-ranging studies no viable eolian sedimentary model has emerged. There are perhaps two main reasons for this. First of all, eolian deposits are perhaps one of the hardest of all types to recognize in ancient sediments. This is because no criteria so far proposed seem to be exclusive to eolian deposits. An eolian origin for a facies can only be postulated when one can prove an absence of criteria suggesting aqueous sedimentation. This is not easy and the wind-blown origin of the classic "eolian" deposits has not gone unchallenged. Visher (1971), Freeman and Visher (1975), Stanley *et al.* (1971), and Jordan (1969) all questioned an eolian origin for many of the Colorado Plateau sandstone formations. Pryor (1971) questioned the generally accepted eolian origin of the Permian Yellow Sands of north-

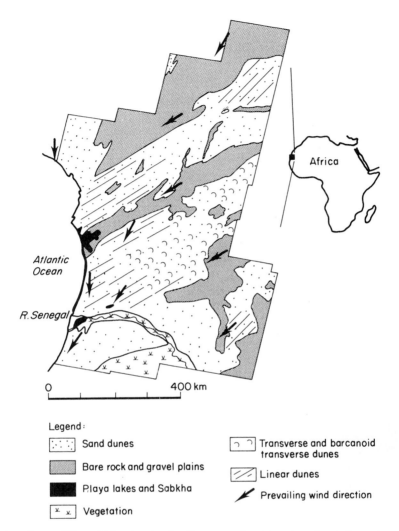

Fig. 6.21. Map of part of West Africa showing the diverse terrain of a modern desert. Note that much of the area is composed of planar equilibrial surface, with no significant erosion or deposition. Areas of net eolian deposition are relatively rare. (Modified from interpretations of LANDSAT by Breed *et al.*, 1979.)

eastern England. In both cases it has been suggested that the maturity, good sorting, and thick foresets of these facies are due to sedimentation from subaqueous dunes in a shelf sea. In this context it is significant that these groups of rock lie in similar settings. They both occur associated with red beds, evaporites, and carbonate facies. This assemblage suggests a paleogeography in which fluvial deposits passed laterally into sabkha and marine-shelf environments. By analogy with modern shorelines, eolian dunes, and submarine sand shoals would both be expected and, with repeated marine advances and retreats, could be so reworked as to make it difficult to detect the last process that actually deposited them.

These arguments suggest that an eolian sedimentary model has failed to emerge

Fig. 6.22. Photograph of a section in a wadi in the Sarir Calanscio, eastern Libya. This shows the lag gravel of a sarir surface developed on a poorly sorted Pleistocene alluvial fan deposit. The gravel surface prevents further eolian deflation.

Fig. 6.23. Large-scale cross-bedding in Permian red sands of presumed eolian origin. Dawlish, southwest England.

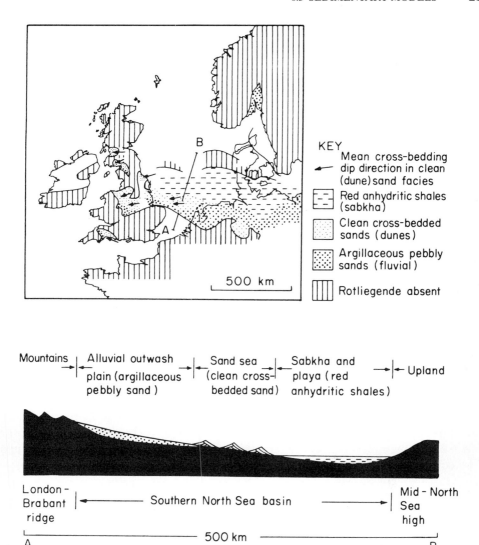

Fig. 6.24. Map and paleogeographic cross-section showing distribution, facies and environment of the Rotliegende (Permian) Sandstones of the North Sea basin. (After Glennie, 1972. American Association of Petroleum Geologists Bulletin, AAPG © 1972, reprinted by permission of the American Association of Petroleum Geologists whose permission is required for further use.)

because ancient eolian deposits are hard to identify. A second reason for this lack of a viable model is that modern deserts are largely environments of erosion or of equilibrium. Areas of actual net eolian sand sedimentation are rare and difficult to study. Most published work describes isolated dunes that migrate across deflation surfaces. This raises questions: Do eolian deposits have a low preservation potential? Eolian processes may play a large part in determining the texture, shape, and sorting of sediments, but are these sands actually laid rest in the geological column by sporadic catastrophic floods, rather than by the process that shapes them? These may be some of the reasons why a viable eolian sedimentary model has not been defined.

6.3.2.3.3 Economic aspects of eolian sands

Because of their good sorting and absence of clay matrix, eolian sands have excellent primary porosity and permeability. They may thus serve as aquifers and petroleum reservoirs if they have not undergone extensive diagenesis. Like braided alluvium they normally lack shale interbeds that may serve as permeability barriers. It has been noted, however, that permeability varies vertically through cross-bedded dune units (see Section 3.2.3.1.5). The finer-grained toesets are often poorly permeable, especially if they have undergone early evaporitic cementation where they grade into interdune sabkhas.

Eolian petroleum reservoirs have been identified in the Triassic–Jurassic red beds of the Rocky Mountain thrust belt in Colorado and Utah (Lundquist, 1983). Here the Norphlet (Jurassic) sands are particularly well documented (see papers in Barwis *et al.*, 1990). Eolian sands in the Lower Permian Rotliegendes of northwest Europe contain many giant gas fields, both in the southern North Sea and onshore. Detailed paleogeographic mapping has been essential to locate favorable reservoirs. Only the eolian sands are permeable enough to be commercially productive. The fluvial and sabkha deposits with which they are interbedded are too tight (Glennie, 1987; Sweet, 1999).

Modern eolian deposits are also hosts to placer minerals. Wind action is not as effective as aqueous action in segregating minerals of different density. This is because the difference in their specific gravities in air is not as large as in water. Heavy mineral segregation takes place in two settings. Fine-grained heavy minerals settle out on the stoss side of dunes, while large grains of heavy minerals segregate in small deflation hollows (Fig. 6.25). Eolian placers occur in many deserts, but are seldom sufficiently abundant to be commercial. Diamond placers of Namibia are an exception (Smirnov, 1976).

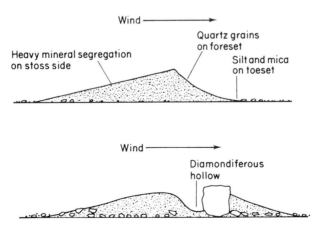

Fig. 6.25. Modes of eolian placer formation. (Modified from Smirnov, 1976.)

6.3.2.4 Lakes and Lacustrine Models

Lakes are bodies of water that are restricted from communication with the open sea. They vary widely in water depth, size, and salinity, ranging from freshwater to hypersaline. Accounts of modern lakes have been given by Hawarth and Lund (1984) and Burgis and Morris (1987). Ancient lake sediments have been documented by Fouch and Dean (1982), Dean and Fouch (1983), Anadon *et al.* (1991), Gierlowski-Kordesch and Kelts (1994), and Talbot and Allen (1996). These studies show that lakes contain a similar range of subenvironments to those found in the marine realm. They range from deltas and barrier island coasts, to subaqueous fans deposited by turbidity currents. Lakes deposit terrigenous and carbonate sediment, including oolite shoals and even what may be loosely termed "reefs" (of algal origin). In arid climates hypersaline lakes deposit evaporites in both marginal and median sabkhas. Lake deposits only differ from marine sediments in their paleontology, and perhaps in some aspects of their geochemistry.

Attempts to define lacustrine sedimentary models have been made by Visher (1965), Kukal (1971), and Picard and High (1972a, 1979). These attempts reach only a broad consensus. Visher, and Picard and High worked essentially from the evidence of ancient lacustrine deposits. They defined an ideal model showing an upward-coarsening grain-size profile. Picard and High write (1972b, p. 115): "Inasmuch as all lakes are ultimately filled, regression dominates the history of a lake." Visher reached a similar conclusion comparing the lacustrine sedimentary sequence to that produced by a regressive marine shoreline. According to these concepts an ideal lake sedimentary model generates a sequence that commences with laminated fine sediments deposited in the deep lake center. As the lake is infilled, marginal fluvial, deltaic, and swamp environments encroach inward. The laminated fines may grade up through turbidite sands, such as are known from Lake Zug and Lake Meade, into beach sands, cross-bedded fluviodeltaic channel sands, and marsh peats.

By contrast to this simple model, Kukal (1971) defined four different lake types, based on the areal distribution of different kinds of sediment in modern lakes. By combining the classic geological approach of looking at lakes vertically with the bird's-eye view of a recent sedimentologist, it is possible to define the sedimentary models shown in Fig. 6.26. In addition to Kukal's four classes of lakes, this scheme introduces two models for the ephemeral lakes of arid regions. The six models have been arranged according to the degree of aridity and of topographic relief that may be expected to generate the various lacustrine types.

6.3.2.4.1 Terrigenous permanent lakes

Kukal's first type of lake is the permanent variety which is infilled by terrigenous sediment (Fig. 6.26, 1). These are commonly found today in mountainous terrain where high precipitation and sediment runoff are combined. Numerous examples may be cited, such as Lake Constance and Lake Zug in the Alps, and Lake Titicaca in the Andes. Such lakes correspond to the typical lake model of Visher in that marginal fluviodeltaic sands prograde into the lake to bury finer sediment settled out from suspension. Because of their location in mountainous terrains, the preservation of such lakes in the

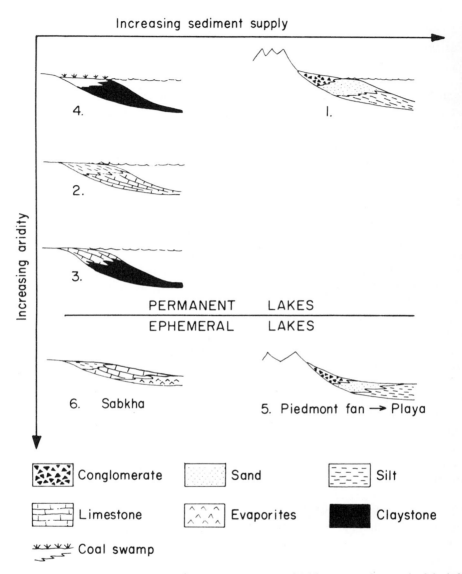

Fig. 6.26. Lacustrine sedimentary models. This combines Visher's (1965) concept of regressive lake infilling with Kukal's (1971) classification of lakes based on centripetal lithological variation. Two additional models are included for playa lakes and sabkhas in arid ephemeral lake basins.

geological record must be slim. These, and other types of lake, often show a remarkable regular lamination in their muds (Fig. 6.27). These laminae are termed "varves." Varves appear to be annual in origin, and have been subjected to extensive statistical analysis to detect sun spot and other cycles (e.g., Fischer and Roberts, 1991). In Pleistocene periglacial lakes it has been possible to use varves to establish a geochronology extending far back in time. This can be calibrated both with dendrochronology, based on growth rings in trees, and longer term climatic cycles. In periglacial lakes each varve is com-

Fig. 6.27. Photograph of varved layering from an Irish Pleistocene periglacial lake. (Courtesy of J. M. Cohen.)

posed of a couplet silt and clay. The silt was deposited from summer glacial melt water. The clay settles from suspension during winter when there was no runoff into the lake. Lakes in arid and/or low-lying areas may lack terrigenous input, but their deposits are still often varved. In these cases the varves consist of couplets of organic mud and lime.

6.3.2.4.2 Autochthonous permanent lakes

Kukal's second class of lake occurs in low-lying terrain in temperate and warm humid climates. Minor amounts of fines are brought in by rivers. Carbonate sedimentation occurs away from the river mouths, both around the lake shores and in the deeper parts of the lake. This takes the form of charophyte calcareous algal marls in the center of the lake. Around the shores wave action may disaggregate freshwater mollusk shells to form calcarenites. This type of sedimentation occurs in the north German lakes such as Lake Schonau and the Great Ploner Lake, and in the lakes of southern Canada (Fig. 6.26, 2).

In Kukal's third type of lake, sapropelites in the center are ringed by carbonate shoreline sediments of algal and molluscan origin (Fig. 6.26, 3). Good examples of permanent lakes with extensive autochthonous sedimentation occurred in the Tertiary basins of the Rocky Mountain foothills. One of the best known is Lake Uinta, which, during the Eocene, covered some 23,000 km^2 of Utah and Colorado, depositing over 2000 m of diverse sediment types. These included marginal deltaic sands and coal swamps, which prograded basinward over varved oil shales of the Green River Formation. These regressive phases alternated cyclically with transgressions when carbonate marls in the basin center passed into skeletal sands and algal oolites around the margin. Intensively studied, because of oil fields in the marginal facies, key papers on Lake Uinta include those by Bradley (1948), Picard (1967), Picard and High (1968, 1972b), Eugster and Surdam (1973), Dean and Fouch (1983), and Fischer and Roberts (1991).

Oil shales, such as occur in the Green River Formation, are found also in Permo-Carboniferous lakes of South Africa, South America, and New South Wales and in the Carboniferous Lake Cadell in the Midland Valley of Scotland (Greensmith, 1968).

Kukal's fourth type of lake consists of marginal marshes which prograde centripetally to overlie organic muds (sapropelites) deposited in the central part of the lake basin (Fig. 6.26, 4). Examples of this type of sedimentation occur in the cold wet gytta lake of northern Canada and northern Europe and Asia. Although sapropelites and oil shales are common lacustrine deposits, they seldom occur unassociated with other rock types in ancient lake sediments.

6.3.2.4.3 Ephemeral lakes

Modern lakes that only exist for intermittent spans of time include the "playas" of the North American desert, the Kavirs of Iran, and the great inland drainage basins of the Sahara, Arabia, and the Australian interior. In such desert climates precipitation is very erratic. Though rain may only come once every few years, when it does fall it pours down. In the 1967 rainstorms of the Australian Lake Eyre basin, overnight falls of 150 mm were recorded (Williams, 1971). Downpours such as these lead to the formation of ephemeral inland lakes (Fig. 6.26, 5).

Intermontane playa lakes are rimmed by piedmont alluvial fans, whose sedimentology has already been described (Fig. 6.28). The width of this marginal facies can be very variable. In playas such as Qa Saleb and Qa Disi in the Southern desert of Jordan, the mud flats locally impinge on sheer cliffs hundreds of meters high.

The typical deposit of modern playa lakes is a red-brown mudstone containing varying amounts of clay, silt, and disseminated carbonate. Scattered windblown sand grains are common, and are often trapped in desiccation cracks (Fig. 6.29). Eolian mud rocks can be diagnosed from the presence of scattered well-rounded sand grains. Because of their tendency to evaporation, ephemeral lakes can also deposit evaporite minerals. Hence the soda flats of the North American desert and the inland sabkhas of the Saharan and Arabian deserts (Fig. 6.26, 6).

Ancient lakes analogous to the playas of modern arid climates have been well documented, notably from Triassic rocks of Europe and North America. Specific examples include the Popo Agie member of the Chugwater Formation in Wyoming (Picard and

Fig. 6.28. Ephemeral playa lake, with braided alluvial outwash fans at the feet of distant jebels, Sinai, Egypt.

High, 1968), the Lockatong Formation of the Newark Group of New Jersey and Pennsylvania (Van Houten, 1964), and the red Keuper marls and Mercia mudstones and associated evaporites of northwest Europe.

In summary, ancient lacustrine deposits can generally be recognized with some confidence by the absence of a marine fauna. A single well-defined lacustrine sedimentary model has not been established, because of the variability of climate and sediment

Fig. 6.29. View of playa lake mud surface with desiccation cracks, Sinai, Egypt. Note the scattered granules and sand grains. These were carried out on to the playa surface by flash floods. The presence of scattered well-rounded sand grains in a red mudstone is a diagnostic feature of a playa lake origin.

supply rate. Lake sediments tend to reflect an upward-coarsening regressive sequence. This model will not bear up under close scrutiny because the diverse environmental parameters of lakes can generate a wide range of allochthonous and autochthonous lithologies. Furthermore, the susceptibility of lakes to climatic changes can cause fluctuating lake levels. These are reflected in transgressive:regressive cycles that disrupt any overall regressive lacustrine sequence.

6.3.2.4.4 Economic aspects of lake deposits

Lacustrine deposits are of great economic importance. They contain oil shales and other organic-rich petroleum source beds. Temperate lakes may be infilled with peat to form coal deposits on burial. Hypersaline lakes may form commercial quantities of evaporites.

Many lake waters become stratified. Commonly a low-density upper layer, the epilimnion, overlies the cooler, denser hypolimnion. Algae photosynthesize in the sunny upper layers, generating oxygen. The combination of plant food and oxygen allows abundant life to thrive in the epilimnion. Oxygen is soon used up in the hypolimnion, however, and due to the lack of sunlight, may not be replenished by photosynthesis. Thus lake beds often become anoxic and stagnant. Organic detritus drifting down from the hypolimnion may therefore be preserved from the normal processes of decay. In this way organic-rich muds may be deposited (Fig. 6.30). After burial the organic component may evolve into kerogen from which petroleum can be generated.

The Green River Shale Formation (Eocene) was deposited in Tertiary lake basins in Utah and Wyoming. Not only does this contain commercial oil shales (Yen and Chilingarian, 1976), but it has generated petroleum that has migrated into reservoirs in peripheral sands and also into older formations (Ray, 1982). Most of China's onshore oil production is from lacustrine basins and indeed lacustrine source beds characterize the Tertiary petroleum provinces of the Far East (Grunau and Gruner, 1978). Aside from petroleum, oil shale, and coal, lakes often contain evaporites, diatomaceous muds (diatomite or kieselguhr), iron ores, and china clay (see Section 8.3.2.1).

6.3.2.5 Deltaic Models

The term "delta," the Greek character Δ, was used to describe the mouth of the Nile by Herodotus nearly 2500 years ago. This term is still used by geographers and geologists

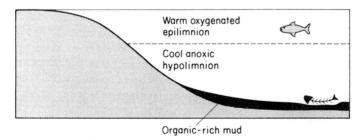

Fig. 6.30. Naivogram showing how lake waters may become stratified as shallow water, warmed by the sun, overlies denser cooler waters (these layers are known as the epilimnion and hypolimnion, respectively). Photosynthesis replenishes the oxygen supply of the upper layer. In the cold, dark hypolimnion, however, oxygen is rapidly depleted, anoxic conditions prevail, and organic-rich muds may be deposited.

alike. A modern definition cites a delta as "the subaerial and submerged contiguous sediment mass deposited in a body of water (ocean or lake) primarily by the action of a river" (Moore and Asquith, 1971, p. 2563). This definition, though broadly correlative with the original meaning of Herodotus, lays no stress on a triangular geometry. Not all deltas, as presently defined, possess this feature.

6.3.2.5.1 Processes in a model delta

Reduced to its simplest elements, a delta forms where a jet of sediment-laden water intrudes a body of standing water (Fig. 6.31). Current velocity diminishes radially from the jet mouth, depositing sediment whose settling velocities allow grain size to diminish radially from the jet mouth. Sedimentation around the jet mouth builds up to the air/water interface, but the force of the jet maintains a scoured channel out through the sediment. Ridges on either side of the distributary channel are termed "levées." As

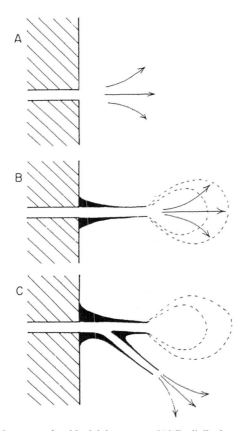

Fig. 6.31. Stages in the development of an ideal delta system. (A) Radially decreasing current velocities from jet mouth deposit concentric arcs of sand, silt, and clay. (B) Delta progrades, forcing a channel through marginal levées. (C) Channel mouth chokes, levée ruptures, and a new delta builds out from the crevasse. Old abandoned lobe compacts and subsides. Its top is reworked by marine processes. Thus a transgression and a regression occur simultaneously adjacent to one another. In sequence stratigraphic terminology, "upper low stand" and "transgressive" system tracts form synchronously, but without any change in sea level.

sedimentation continues, the delta progrades out into the standing body of water. Three main morphological units appear. The delta platform is the subhorizontal surface nearest the jet mouth. It is basically composed of sand and is traversed by the distributary channel and its flanking levées.

The delta platform grades away from the source into the delta slope on which finer sands and silts come to rest. This in turn passes down into the prodelta area on which clay settles out of suspension. A vertical section through the apex of a delta thus reveals a gradual vertical increase in grain size. At the base the prodelta clays grade up through delta slope silts into sands of the delta platform. Classically, these three elements have been termed the bottomset, foreset, and topset, respectively (Fig. 6.32).

Eventually a distributary channel extends so far that its mouth becomes choked with sediment. At a point of weakness the levée bursts and a new distributary system is established. The abandoned distributary is choked by suspended sediment, and the whole abandoned lobe sinks beneath the water as it compacts.

Theoretically, this process may continue indefinitely as the distributaries switch from side to side from the original point of sediment input. This ideal delta model consists of a series of interdigitating lobes, each one showing a gradual upward increase in grain size, and a decrease in grain size from its point of origin. This ideal model is next contrasted with modern deltas.

6.3.2.5.2 Modern delta systems

Without doubt the Mississippi is one of the most intensely studied modern deltas. Some of the key papers include those of Fisk (1955), Coleman and Gagliano (1965), Kolb and Van Lopik (1966), and Gould (1970).

The Mississippi delta bears a close relation to the ideal model (Fig. 6.33). A series of seven separate Quaternary delta lobes can be mapped (Fig. 6.34). Topset, foreset, and bottomset sedimentary facies can be recognized, each with a characteristic suite of lithologies, biota, and sedimentary structures. Though the modern Mississippi compares well with the previously derived "ideal model," it is in fact a very dangerous analog to use to interpret ancient deltas. It is unusual for two reasons. It has a far higher ratio of mud to sand than most deltas, ancient or modern. Indeed, as Mark Twain remarked, the water of the Mississippi is too thick to drink, too thin to plough. Further-

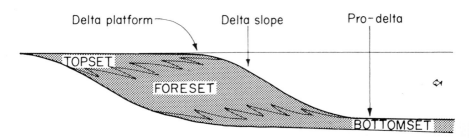

Fig. 6.32. Terminology of a delta profile. The platform, slope, and prodelta environments have also been termed "undaform," "clinoform," and "fundoform" (Rich, 1950, 1951). The topset, foreset, and bottomset deposits have been termed "undathem," "clinothem," and "fundothem" (ibid.). Only "clinoform" has survived, normally used to describe prograding reflectors seen on seismic sections.

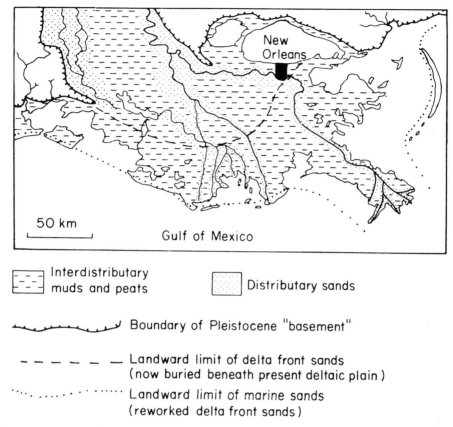

Interdistributary muds and peats

Distributary sands

Boundary of Pleistocene "basement"

Landward limit of delta front sands
(now buried beneath present deltaic plain)

Landward limit of marine sands
(reworked delta front sands)

Fig. 6.33. Map of the modern Mississippi delta showing distribution of major sand facies. The present active delta lobe is prograding to the southeast, causing a local regression. In the northeast, the sea transgresses over the subsiding earlier Chandeleur delta lobe. Marine processes rework the subsiding delta lobe depositing the arcuate sands of the Chandeleur barrier islands. In sequence stratigraphic terminology, "upper low stand" and "transgressive" system tracts form synchronously, but without any change in sea level.

more, the Mississippi builds into a sheltered marine embayment of low tidal range. Thus the marine processes that redistribute the alluvial sediments of most deltas are largely absent from the Mississippi. Only about 25% of the modern Mississippi's load is sand, the rest is silt and clay. This means that a very small amount of each delta lobe is sand. Almost all the sand load is deposited at the mouths of the distributary channels in what are called "bar-finger sands." As the distributary extends seaward, the bar-finger sands take on a linear geometry. Switching of the delta means that these bar-finger sands generate an overall pattern of radiating shoe-strings, analogous to the fingers of a hand.

In the older lobes of the Mississippi the main locus of sand deposition was the seaward edge of the delta platform, where it is possible to define an arcuate belt of delta-front sheet sands which were deposited at the mouths of distributary channels (Fig. 6.35). The delta-front sheet sands pass shoreward into, and are overlain by, silts, clays, and peats deposited in levée, interdistributary, and swamp environments.

Seaward progradation of the delta has thus generated an upward-coarsening sequence ranging from marine clays of the prodelta up into the delta-front sheet sands.

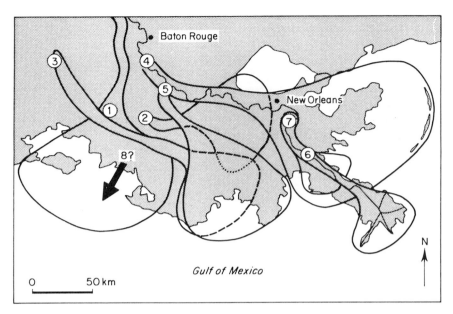

Fig. 6.34. Map showing the location of past, present and probable future lobes of the Mississippi delta. Six lobes have been identified that predate the present one. The next, eighth, lobe may develop in Atchafalaya Bay, unless the modern environment is grossly abused. (Based on Kolb and Van Lopik, 1966, and Shlemon and Gagliano, 1972.)

These are in turn overlain by a suite of brackish and nonmarine fine-grained facies dissected by radiating distributary channel sands.

To the east of the present active mouth of the Mississippi lies the arcuate archipelago of the Chandeleur islands. These mark the edge of the abandoned lobe of the old St Bernard delta. Once the sediment supply of a delta is cut off, it is extremely suscep-

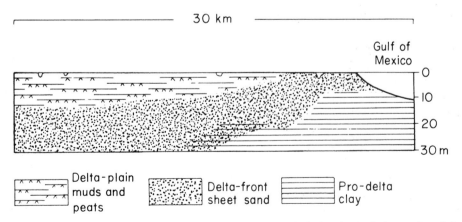

Fig. 6.35. Section of the Lafourche subdelta of the modern Mississippi, showing main locus of sand deposition as a sheet at the delta front. This is gradually buried beneath a prograding sheet of delta-plain muds and peats deposited in interdistributary bays and swamps, with minor distributary channel sand bodies. (After Gould, 1970. Courtesy of the Society for Sedimentary Geology.)

tible to reworking by marine influences, both tides and waves. As the delta lobe compacts and subsides, the sea transgresses it and reworks its upper part. Fine sediment tends to be winnowed out and settles in deeper quieter water. The sands are reworked and redeposited as a transgressive marine blanket which may thus unconformably overstep the various facies of the delta.

Instead of consisting of a simple upward-coarsening sequence, examination of the modern Mississippi shows an upper fine-grained nonmarine unit. Furthermore, while a delta is, of its very nature, a regressive prograding prism, each lobe may contain both constructive regressive and destructive transgressive phases. The bird-foot Mississippi type of delta, with radiating distributary channel networks, is rare worldwide, and is seen more commonly in lakes rather than seas, such as, for example, the St. Clair River delta of Canada (Pezzetta, 1973). There may be two reasons for this. First, not all rivers carry as much fine material as the Mississippi. Gravel fan deltas such as those of the Arctic, and of steep desert coasts, such as the eastern side of the Gulf of Aqaba, retain the basic form of alluvial cones (Fig. 6.36). Fan deltas pass from mountain front to seabed with no differentiation into the diverse subenvironments described from the muddy Mississippi.

The second main qualifying factor of deltaic geometry, other than sediment type, is the relative importance of marine and fluvial influences. The Mississippi maintains a bird-foot geometry because of its relatively sheltered position in the Gulf of Mexico. The tidal range of the northern part of the Gulf of Mexico is low (less than 1 m) so tidal currents are relatively insignificant. Prevailing winds are from the north and east so the fetch of waves is short, apart from those due to the occasional hurricane. More exposed

Fig. 6.36. Modern fan delta of undifferentiated boulders and gravel debouching into Lake Tunsberjvatnet, Jostedalen, northern Norway. (Courtesy of J. M. Cohen.)

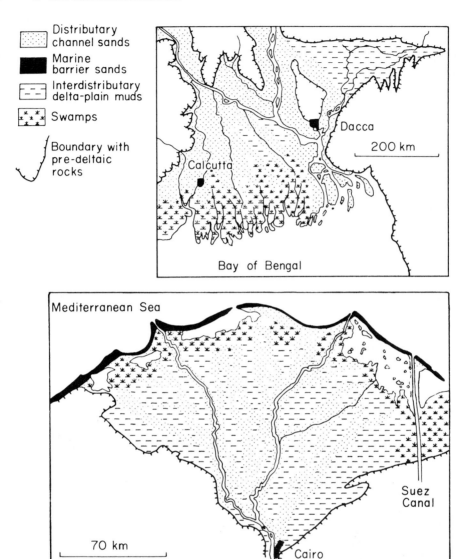

Fig. 6.37. Map showing the distribution of modern deltaic deposits of the Ganges-Brahmaputra delta (upper) and the Nile delta (lower). The first is an example of a tide-dominated delta with braided seaward-trending channel sands. The Nile is an example of a wave-dominated delta with extensive fringing barrier sand bodies. (Based on Coleman *et al.*, 1970; Morgan, 1970; Wright and Coleman, 1973. Courtesy of the Society for Sedimentary Geology.)

deltas, such as those of the Nile and Niger, show smoothed arcuate coastlines (Fig. 6.37). This is because sand is no sooner deposited at a distributary mouth than it is reworked by the sea and redeposited along the delta front, often in the form of barrier islands. Marine influence, however, may take the form of wave action on exposed coasts, or of tidal scour on coasts with high tidal ranges. In the Bay of Bengal tidal ranges vary from 3 to 5 m. Largely because of this, deltas of the Bay of Bengal are quite different from

Table 6.6
Classification of Deltas Based on the Dominant Processes

Dominant process		Environments	Sand facies	Example
Fluvial		Radiating bird-foot distributary/ levée systems	Radiating mouth-bar sands	Mississippi
Marine	Waves	Distributaries truncated by barrier sands	Arcuate delta-front barrier sands	Nile and Niger
	Tides	Extensive tidal flats and scoured braided estuaries	Delta-front sheet sand	Mekong and Ganges–Brahmaputra

Based on Fisher *et al.* (1969), Morgan (1970), and Wright and Coleman (1973).

that of the Mississippi. Tidally dominated deltas occur at the mouths of the Ganges-Brahmaputra, Klang, Langat, and Mekong rivers (Coleman *et al.,* 1970; Morgan, 1970). The scouring effect of strong tidal currents redistributes fluvial sediment in broad tidal flats where extensive mangrove swamp development acts as an additional sediment trap. The distributary channels themselves are wide, deep, straight, braided estuaries (Fig. 6.37).

The preceding analysis shows that deltas are composed of a series of upward- and landward-coarsening clastic lobes. These lobes, essentially regressive, may contain fine-grained topsets, and embody both constructive and destructive phases. Furthermore the environments and facies of deltas vary widely according to the relative importance of fluvial, tidal, and wave processes. Table 6.6 attempts to classify the different delta types that are recognizable.

6.3.2.5.3 Ancient deltaic deposits

Ancient deltas are common in the geological record. Accounts of ancient deltas are found in Coleman and Prior (1982), Barwis *et al.* (1990), Oti and Postma (1995), and Reading and Collinson (1996).

In a general way it is possible to distinguish fluvial-dominant from marine-dominated deltas. Notable examples of cyclic deltaic sediments have been described from the Upper Carboniferous (Pennsylvanian) in many parts of the world, notably northwest Europe, North America, and Australia (e.g., Potter, 1962; Wanless *et al.,* 1970). These deltas include the coal measures that have been of such economic importance to these regions. Detailed mapping of sediment increments between regionally widespread coal and thin marine shale marker horizons shows radiating shoe-string sand bodies analogous to those of the modern Mississippi. The sedimentology and field relationships of these sands suggest that they are channel sands, rather than bar-finger sands. Thus, though the Pennsylvanian sands may be different in genesis from those of the modern Mississippi, it is reasonable to suppose that they were laid down in fluvially dominated deltas.

A notable example of a marine-dominated delta is provided by the Brent Group (Middle Jurassic) of the northern North Sea. This is an important petroleum reservoir (Morton *et al.*, 1992). The Brent delta prograded northward down the axis of the Viking graben, debouching its sediments into the opening ocean. It provides a classic example of Walther's law (see Section 6.4.1). The delta slope deposits of the Rannoch Formation are overlain by the delta-front sands of the Etive Formation, which are overlain in turn by the delta plain coal-bearing Ness Formation. Rising sea level reworked the top of the delta to deposit the shallow marine sands of the Tarbert Formation (Fig. 6.38).

The Brent delta, just described, graded seaward into open marine muds. Others, such as the Pennsylvanian deltas of the Appalachian plateau, and those of the Yoredale series of northern England, prograded across carbonate platforms (e.g., Ferm, 1970; Moore, 1959). There is, however, yet another important type of delta that generates turbidite sands at its foot. The development of slumps and slides is known from modern deltas, and there is evidence that these transport sand by turbidity currents onto the basin floor. This has been described from the Mississippi, Fraser, and Niger deltas (Shepard, 1963, pp. 494 and 500; Burke, 1972). Great depth of both water and sediment make it hard to study modern delta-front turbidites. Many ancient examples have been described however. Notable case histories have been documented from the Carboniferous of England (Walker, 1966; De Raaf *et al.*, 1964), from the Ordovician rocks of the Appalachians (Horowitz, 1966), from the Coaledo Formation of Oregon (Dott, 1966), and from the Tertiary-Recent wedge of the Niger delta (Fig. 6.39).

To conclude, at its simplest the delta process generates upward-coarsening lobes of sediment that grade from marine muds, upward and shoreward, into diverse nonmarine sands, muds, and often coals. This simple model may be modified by marine destructive influences. Furthermore, if the delta slope was sufficiently unstable to slide and slump, then redeposited turbidite sands may be present at the delta foot.

Recognition of these diverse deltaic models is critical to the effective exploitation of hydrocarbons from ancient deltas. Sand reservoirs in fluvial-dominated deltas are radiating shoe-strings on the delta platform. Marine-dominated deltas tend to have arcuate motifs of shoal sands. Additional reservoir sands may be present in the submarine canyons and fans of high-slope deltas (see Section 6.3.2.9.1).

6.3.2.5.4 Economic aspects of deltaic deposits

Ancient deltaic deposits are extremely important economically. They host most of the world's coal, and many major petroleum provinces. Environments of coal formation are not discussed now. They are dealt with in Section 9.3.3. Deltas make excellent petroleum provinces because they fulfil all the conditions necessary for petroleum source bed formation, petroleum generation, and entrapment (Selley, 1977).

The deltaic process is a way of depositing lobes of sand (potential reservoirs) into envelopes of organic-rich marine muds (potential source beds). Deltaic environments deposit many potential stratigraphic traps, including mouth bars, barrier bars, and channels. Rapid deposition often leads to overpressuring. This may generate diapiric traps and roll-over anticlines. Deltas need a basin, or at least some subsidence, before they may form. Subsidence implies crustal stretching and thus increased heat flow. This expedites the maturation of source beds. No wonder then that ancient deltas are major

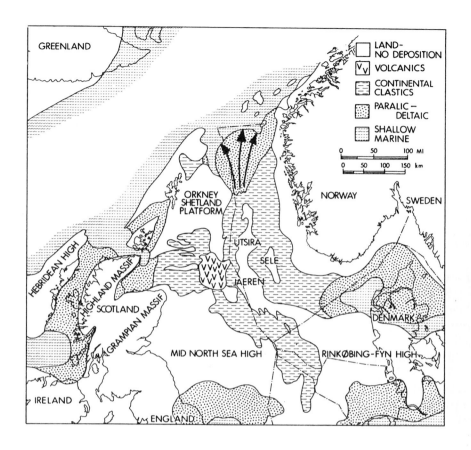

LAND-
NO DEPOSITION

Vᵥ VOLCANICS

CONTINENTAL
CLASTICS

PARALIC–
DELTAIC

SHALLOW
MARINE

0   50   100 MI

0   50   100   150 km

GREENLAND

ORKNEY
SHETLAND
PLATFORM

NORWAY

SWEDEN

UTSIRA

SELE

JAEREN

HEBRIDEAN HIGH

HIGHLAND MASSIF

SCOTLAND

GRAMPIAN MASSIF

MID NORTH SEA HIGH

DENMARK

RINKØBING-FYN HIGH

IRELAND

ENGLAND

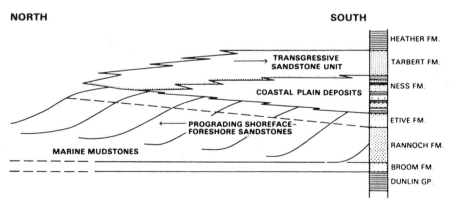

NORTH   SOUTH

HEATHER FM.

TRANSGRESSIVE
SANDSTONE UNIT

TARBERT FM.

COASTAL PLAIN DEPOSITS

NESS FM.

ETIVE FM.

PROGRADING SHOREFACE–
FORESHORE SANDSTONES

RANNOCH FM.

MARINE MUDSTONES

BROOM FM.

DUNLIN GP.

Fig. 6.38. (Upper) Middle Jurassic paleogeographic map of the North Sea basin. [From Eynon, G. 1981. Basin development and sedimentation in the middle Jurassic of the northern North Sea. *In* "Petroleum Geology of the Continental Shelf of Northwest Europe" (L. V. Iling, and D. G. Hobson, eds.), pp. 196–204, Fig, 3. Courtesy of the Institute of Petroleum.] Arrows indicate the location and progradational direction of the Brent delta. (Lower) Stratigraphic section of the Brent Group showing how its constituent formations demonstrate the progradation, transgressive reworking, and drowning of the delta. In sequence stratigraphic terminology, upper low stand systems tract overlain by a transgressive systems tract. [From Brown, S. 1990. The Jurassic. *In* "Introduction to the Petroleum Geology of the North Sea," 3rd. Ed. (K. W. Glennie, ed.), pp. 219–254, Fig. 8.14. Courtesy of Blackwell Scientific.]

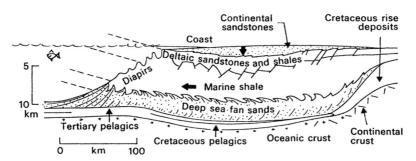

Fig. 6.39. Cross-section of the Tertiary–Recent sediment prism of the Niger turbidite fan sands at the foot of the delta slope. (From Burke, 1972. American Association of Petroleum Geologists Bulletin, AAPG © 1972, reprinted by permission of the American Association of Petroleum Geologists whose permission is required for further use.)

petroleum provinces. The Tertiary Niger Delta and the Tertiary Gulf Coast province of the USA are two classic examples (see Reijers *et al.*, 1997, and Dow, 1978, respectively).

Figure 6.40 illustrates some of the ways in which petroleum may be trapped in deltaic deposits. The up-dip alluvial pinchout (Fig. 6.40A) is illustrated by the Clareton and Fiddler Creek fields of the Powder River basin, Wyoming (Woncik, 1972), and distributary channel entrapment (Fig. 6.40B) by the Pennsylvanian Booch delta of Oklahoma (Busch, 1971). Mouth bar entrapment (Fig. 6.40C) is illustrated by the West Tuscola field of Texas (Shannon and Dahl, 1971). Roll-over anticlines (Fig. 6.40D) are major productive traps in the Niger and Mississippi provinces, and deep-sea sands domed over diapiric mud lumps (Fig. 6.40E) occur in the Beaufort Sea of Arctic Canada (Hubbard *et al.*, 1985).

Once a deltaic petrolum accumulation has been found, however, sedimentology must be applied to develop it efficiently. The earlier discussion of fluvial reservoirs introduced the problems of mapping channels, first trying to establish their continuity deterministically, but then often having to resort to modelling the reservoir statistically. Similar situations are encountered in deltaic reservoirs. Here the problems are complicated by the fact that, not only may there be downslope trending channels, but there may also be shallow marine sands elongated perpendicular to the distributaries. This situation is encountered in the Brent deltaic reservoir of the northern North Sea fields mentioned earlier (Fig. 6.41).

6.3.2.6 Linear Barrier Coasts

With increasing marine influence deltas pass gradationally into linear barrier coasts. All stages of the transition are present in modern shorelines and in ancient rocks; yet the two end members are sufficiently distinctive to be clearly definable models. Modern deltas that pass along the shore into linear barrier island coastlines can be seen in the Gulf of Mexico on either side of the Mississippi delta, on the Dutch coast to the northeast of the Rhine delta, on the Sinai coast to the east of the Nile delta, and on the coast to the west of the Niger delta.

In addition, many modern barrier coasts are unrelated to any major delta. Examples include much of the east coast of North America between New Jersey and Florida, and

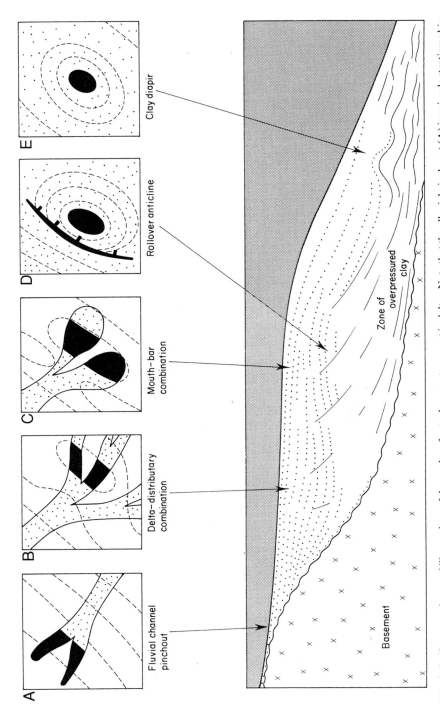

Fig. 6.40. Cross-sections and illustrations of the modes of petroleum entrapment in deltas. Note that the alluvial pinch-out (A) is a simple stratigraphic trap. Types (B) and (C) are combined stratigraphic and structural traps. Types (D) and (E) are both contingent on the presence of overpressured muds in the distal part of the delta. Examples of each type of trap are cited in the text.

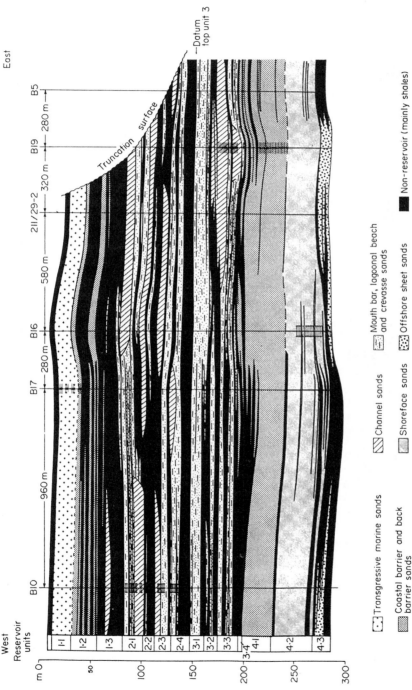

Fig. 6.41. West–east (i.e., paleostrike) cross-section through part of the Brent Group correlated with wells between 960 and 280 m apart. Note the good west–east continuity of the transgressive marine sands of the Tarbert Formation, at the top. Compare this with the lenticular nature of the channel sands in the Ness Formation beneath. (From Johnson and Stewart, 1985. Courtesy of the Geological Society of London.)

parts of the north German and Polish coasts and the Younghusband peninsula of South Australia. These coasts do show, nevertheless, considerable input of sediment from rivers draining their hinterlands. Coastlines have been classified according to their tidal range. Microtidal coasts have a tidal range of less than 2 m. Mesotidal coasts have tidal ranges between 2 and 4 m. Macrotidal coasts have tidal ranges in excess of 4 m (Davies, 1973). Barrier islands are generally best developed where tidal range is relatively low.

6.3.2.6.1 Modern barrier coasts

Recent barrier islands and the processes that form them have been intensely studied and debated (Davis, 1978; Swift and Palmer, 1978; Carter,1995). There is considerable controversy over the genesis of modern barrier islands. The two major mechanisms proposed are the progressive up-building of offshore bars and the submergence of coastal beaches and dune belts. One of the problems of studying modern bars is that we live in an ice age, albeit an interglacial, and eustatic shoreline changes have occurred several times in the last million years. Modern shorelines show both raised and drowned beach features and many modern barrier islands contain cores of older sediments. This is, therefore, not the best moment in geological time to study barrier island formation. It seems reasonable to postulate a polygenetic origin for barrier islands (Schwarz, 1971); additional data and discussions are found in Hoyt (1967), Guilcher (1970), Steers (1971), and King (1972).

Some of the best studied of modern barrier island complexes include the Gulf Coast of west Texas in general, and Padre and Galveston islands in particular (Shepard, 1960; Bernard et al., 1962). Of the barrier coastlines of the eastern coast of North America, Sapelo Island is one of the most studied (Hoyt et al., 1964; Hoyt and Henry, 1967). Outside North America, the Dutch and German barrier coasts are some of the best known (Horn, 1965; Van Straaten, 1965).

Coupling studies of these modern coasts with their ancient analogs, a well-defined barrier island sedimentary model has been put together by Visher (1965), Potter (1967), Shelton (1967), Davies et al. (1971), Davis (1978), McCubbin (1982), and Reading and Collinson (1996). Essentially, a barrier island is a linear sand body exposed at high tide that runs parallel to the coast, separating the open sea from sheltered bays, lagoons, and tidal flats (Fig. 6.42). In most modern barrier coasts, two high-energy environments alternate laterally with two low-energy environments. On the landward side is a fluvial coastal plain of sands, silts, clay, and peats. This grades seaward, generally through salt marsh deposits, into quiet-water tidal flats and lagoons (Black et al., 1998). The sediments of these environments consist of laminated, cross-laminated, and flaser-bedded fine sands, silts, and clays. This zone is characterized by intense bioturbation, by shell beds, often of oysters and mussels, and by upward-fining tidal creek sequences (e.g., Evans, 1965; Ginsburg, 1975).

The barrier island itself consists of a number of distinctive physiographic units. On its landward side there may be a complex of wash-over fans and barrier flats formed from sand washed over the barrier island during storms. The crest of the barrier island is often marked by eolian sand dunes, sometimes stabilized by soil horizons and vegetation. This crestal area passes seaward through a beach zone to the open sea.

Fig. 6.42. Chesil beach, Dorset, southern England. This barrier island, or more strictly tombolo, joins Portland Island to the mainland. It is some 13 km in length and is separated from the land by a brackish lagoon.

Barrier island sands are generally mature and well sorted (Fig. 3.9c). Attempts to distinguish beach, dune, and river sands by granulometry are described elsewhere (see Section 3.1.3). Internally the beach deposits are horizontally or subhorizontally bedded with gentle seaward dips. Trough and planar foresets are also present in subordinate amounts. Barrier islands are sometimes cross-cut by tidal channels. These may possess flood and ebb tide deltas on their landward and seaward sides, respectively (Armstrong-Price, 1963). Cross-bedded sand sequences are formed from the lateral accretion of these channels; such deposits may comprise a considerable part of the barrier sand body that is actually preserved in the geological record (Hoyt and Henry, 1967).

Seaward, the beach deposits of the barrier island pass out into open marine environments. Many modern barrier sands, particularly those of the North Sea, pass out into scoured marine shelves, which are essentially environments of erosion or equilibrium. In the coasts of Nigeria and of the Gulf of Mexico, by contrast, barrier islands pass offshore into deeper water. Here fine sediment may settle out of suspension. Detailed seaward traverses from such barrier beaches reveal a gradual decrease in grain size from the high-energy surf zone of the shoreface to the lower energy environment beneath the limit of wave and tidal current action. Bioturbated muddy flaser-bedded sands and silts below low water pass seaward into the laminated muds of the open marine environment. The subenvironments and the facies generated by an ideal sedimentary model for a prograding barrier island complex are shown in Fig. 6.43. As one would expect, the resultant sedimentary sequence is very similar to that generated by a wave-dominated delta. The main differences are that the deltaic sequence contains distributary channel sands within the lagoon and tidal flat sequence, and in the upper part of the sand sheet.

6.3.2.6.2 Ancient barrier coasts

A large number of ancient sedimentary facies have been attributed to deposition in bar environments. Examples are given in the papers cited in the previous section. The Cretaceous rocks of the Rocky Mountain interior basins provide excellent examples of barrier coasts and wave-dominated deltas. In the Late Cretaceous a broad sea-way stretched along the eastern edge of the rising Rocky Mountains from the Canadian

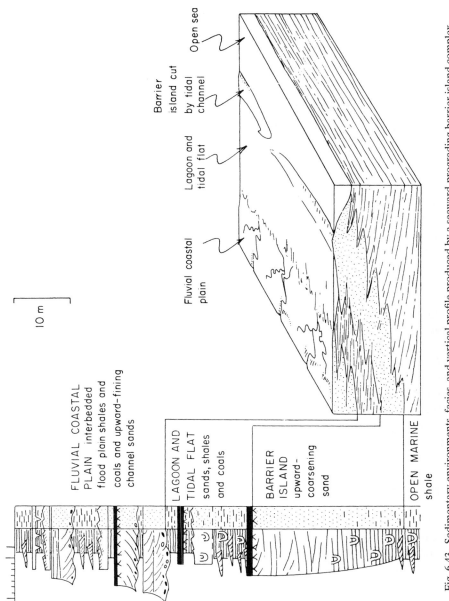

Fig. 6.43. Sedimentary environments, facies, and vertical profile produced by a seaward-prograding barrier island complex.

Arctic to the Gulf of Mexico. In response to the uplift of the Laramide orogeny, vast quantities of detritus were brought into this sea-way from the west. This was deposited in a wide range of environments ranging from the piedmont fanglomerates of the Mesaverde Group to the offshore marine shales of the Pierre and Lewis formations. Transitional shoreline deposits were laid down in a great diversity of environments. Both linear shoreline and lobate shoreline deposits have been recognized. The deltaic deposits seem to have been deposited in marine dominant deltas analogous to the Niger and the Nile. Notable accounts of these beds have been given by Weimer (1970, 1992), Asquith (1970), van de Graaff (1972), McCubbin (1982), Devine (1991), and Martinsen *et al.* (1993).

Specific examples of Rocky Mountain Cretaceous barrier island sand bodies include the Eagle and Muddy sandstones of Montana (Shelton, 1965; Berg and Davies, 1968), and the Bisti bar oil field of New Mexico (Sabins, 1963, 1972).

These examples, taken from both surface and subsurface studies, have a number of common features. They all tend to show an upward-coarsening grain-size profile. The bases of the sand bodies are transitional with open marine shales. The tops of the sands are abrupt, often marked by a break in sedimentation, and overlain by marine or nonmarine shales. Internally these sand bodies are generally bioturbated in their lower parts and massive or subhorizontally bedded toward the top. The most diagnostic feature of all, however, is that these sand bodies are linear shoe-strings running parallel to the local shoreline.

Modern coastal geomorphologists rigorously define coastal sand bodies into barrier islands, offshore bars, spits, and tombolos. It is easy to make these distinctions of modern sand bodies. They are not easy to make of ancient sand bodies because their spatial relationships are seldom sufficiently clear. Most geologists are content to label a particular unit as a bar sand. Additional refinements of its degree of exposure or geomorphology are largely academic.

Lateral progradation of barrier bars can generate not just a shoe-string, but a sheet sand body. This requires a critical balance between sediment input and subsidence. Nevertheless, examples of barrier sheet sands have been described, notably from the Cretaceous of the Rocky Mountain foothills and from the Gulf Coast of Mexico Tertiary. Descriptions of Rocky Mountain bar sheet sands have been given by Hollenshead and Pritchard (1961), Weimer (1961), and Asquith (1970). Descriptions of the Gulf Coast sand bodies have been given by Burke (1958) and Boyd and Dyer (1966). These studies show that the sand sheets occur not just as regressive sand sheets, which pass down into marine and up into nonmarine facies, but also as trangressive sands which pass up into marine facies. The transgressive sand bodies tend to be less well developed, however, and the transgressions are more often represented by surfaces of erosion than by depositional units. Close examination of the transgressive sand bodies shows that they are actually composed of a series of upward-coarsening shoe-string units, which are vertically arranged *en echelon* (e.g., Asquith, 1970, Fig. 34). This implies that deposition of sand during a marine trangression actually takes place only during still-stands of the sea when bars may prograde seaward. In sequence stratigraphic terminology (discussed later in this chapter) barrier bar sands are part of the transgressive systems tract in a third-order sequence cycle.

Barrier bar sands are often stacked in a series of transgressive and regressive cycles (Fig. 6.44). Unlike the deltaic sedimentary model, the barrier model lacks a built-in

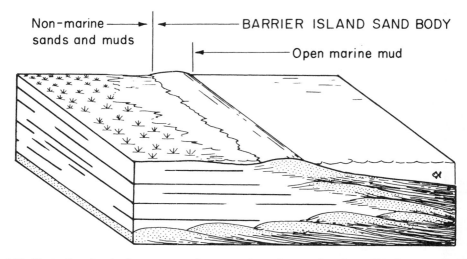

Non-marine ——→ | ←——————— BARRIER ISLAND SAND BODY
sands and muds   ←————————— Open marine mud

Fig. 6.44. Illustration showing how transgressive–regressive cycles may deposit sand blankets composed of a multitude of discrete sandbars. Note how individual sand increments are upward coarsening and progradational, not only during regressions, but also during transgressions. The transgressive sands are themselves composed of upward-coarsening increments. In modern sequence stratigraphic terminology, these are the deposits of a transgressive systems tract in a third-order sequence cycle. They are arranged in fourth-order cycle parasequences. (Hopefully that makes it clearer.)

cycle generator. Such cycles are generally attributed to external causes such as local tectonic movement or to global climate eustatic changes (Haq *et al.*, 1988).

To conclude, a clearly defined barrier island sedimentary model can be recognized in modern shorelines. Ancient bar sands are often recognizable by their sequence of grain size and sedimentary structures. There can be little doubt of the bar sand origin of a shoe-string enclosed in shale, trending parallel to the paleoshoreline. It is hard, and often irrelevant, to prove whether such a sand body was an offshore bar or a barrier island. Sand sheets that vertically separate marine shales from nonmarine sediments can be formed by both prograding marine-dominated deltas and by barrier island coasts. Detailed studies of such sand bodies show that bar sands form an integral part of the overall prograding sediment prism, and that a division of the shoreline into deltaic or barrier models is sometimes irrelevant and may actually be misleading. It is important to make accurate environmental diagnoses of discrete sand lenses to better predict their geometry. Weber's (1971) classic unravelling of the barrier and tidal channel sands on the Tertiary topset of the Niger delta illustrates this point.

6.3.2.6.3 Economic aspects of barrier bar and beach sands

It is obvious that, because of their well-sorted texture, barrier bar and beach sands have excellent porosity and permeability. Ancient barrier bar sands may therefore be important petroleum reservoirs, if they have not lost their porosity through diagenesis. In the subsurface there are many blanket sands that are obviously of high-energy shallow marine origin. It is, however, often very difficult to prove whether these were deposited from open marine shelf sand waves, or by the lateral accretion of beaches or barrier islands. It is only where the lagoonal sediments are preserved that a barrier island origin

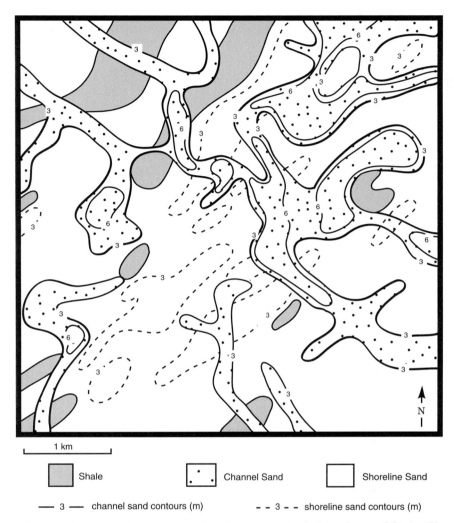

Fig. 6.45. An example of a complex petroleum reservoir in coastal sands. Isopach map of the Aux Vases Formation (Upper Mississipian) sands in the Rural Hill field, Illinois, USA. Well locations are not shown, but the
field has been drilled up with a well-spacing of some 300 m, so the contours are tightly constrained. Shoreline
sands trending NE–SW along the paleostrike are cross-cut at right angles by channel sands aligned along the
paleodip. (After Weimer *et al.,* 1982, American Association of Petroleum Geologists Bulletin, AAPG © 1982,
reprinted by permission of the American Association of Petroleum Geologists whose permission is required
for further use.)

for a sand may be proved. Reference has already been made to instances of petroleum
trapped in "shoe-string" barrier island sands in the Cretaceous basins of the Rocky
Mountains of Canada and the USA.

Figure 6.45 illustrates the problems of producing petroleum from a reservoir composed of paleostrike trending barrier bar sands into which are cut paleodip aligned channel sands. These two types of reservoir are interbedded with impermeable nonreservoir
shales.

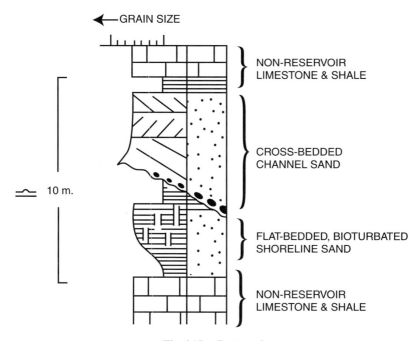

Fig. 6.45 — *Continued*

Shoreline processes also favor the segregation of heavy minerals to form placer ores (Macdonald, 1983, pp. 139–156). Modern beach placers consist of gold, monazite, sphene, ilmenite, and cassiterite. Figure 6.46 shows the general location of strandline placer segregation. Commercial beach placers occur in Sri Lanka, India, New Zealand, New South Wales, and along the coasts of Florida and New Jersey, USA (Force, 1986). As with alluvial placers, beach placers require the intense weathering of a basement source terrain, followed by extensive reworking of the detrital residue. Figure 6.47 illustrates the setting of ilmenite and monazite coastal placers in Sri Lanka. These sands were derived from Pre-Cambrian igneous and metamorphic rocks. They have been deeply weathered in the tropical climate, and the residue transported down to the sea by rivers. Coastal processes have concentrated the heavy minerals along the beaches and, to

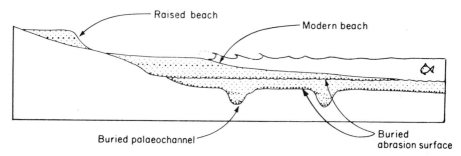

Fig. 6.46. Sketch section illustrating the zones of heavy mineral concentration in coastal and near-shore marine sediments. Placers shown by heavy stipple.

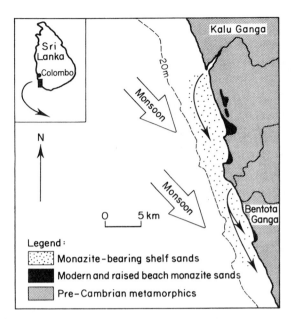

Fig. 6.47. Map showing the mode of occurrence of monazite and ilmenite placers on the western coast of Sri Lanka. Note how major heavy mineral concentrations occur in bays that open toward the prevailing direction of longshore drift. (Simplified from Wickremeratne, W. S. 1986. Preliminary studies on the offshore occurrences of monazite bearing heavy mineral placers, southwestern Sri Lanka. *Mar. Geol.* **72**, 1–10. Copyright 1986, with permission from Elsevier Science.)

a lesser extent, out on the shelf. Major concentrations occur in coastal embayments that face the direction of longshore drift. Here, up to 14% of the sand is composed of ilmenite, monazite, rutile, and garnet. They have been worked since 1918 (Wickremeratne, 1986). Beach paleoplacers also occur, but they are seldom commercial unless the sands are still near the surface and unconsolidated. Thus most commercial paleoplacers are found in Tertiary and Pleistocene formations (Peterson *et al.,* 1987). Paleoplacer gold deposits, for example, occur in Pleistocene raised beaches in the Yukon, Alaska.

6.3.2.7 A Shelf Sea Model

6.3.2.7.1 Introduction

The continental shelf is defined as the area between low-water mark and the continental shelf margin, normally considered to be about 200 m. Shelves are affected by many sedimentary processes. These include wave action and wave-generated currents, storms, tidal currents, geostrophic contour currents (see Section 4.2.2) and debouching fluvial waters and sediment (Walsh, 1987). Most shelves are thus exposed to high-energy current conditions and may variously be described as wave dominated, storm dominated, or tide dominated as appropriate. These processes either constantly scour the seabed or deposit sand, and often form large wave-like bedforms. Wave, tide, and storm processes

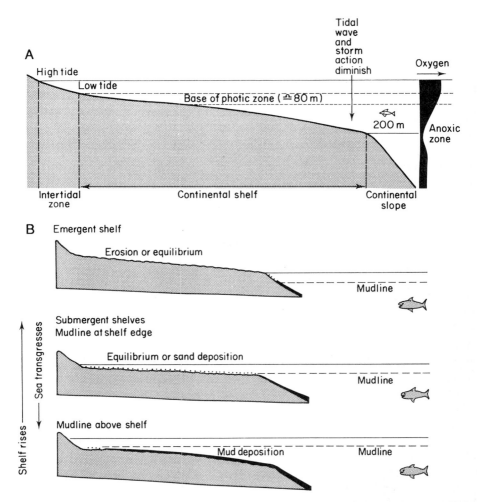

Fig. 6.48. (A) Cross-section to indicate the critical parameters of continental shelves. (B) Cross-sections showing how fluctuating sea level and the mudline control shelf deposition. Note that the depth to the mudline is controlled by tidal, storm, and wave processes. It also varies seasonally and geographically.

diminish with depth until a critical surface is reached below which mud may settle from suspension. This surface has been termed the **effective wave base,** but this is not wholly appropriate because other processes are also involved. The term **mudline** is to be preferred because it is purely descriptive (Stanley and Weir, 1978). As global sea level varies with time continental shelves may be emergent or submergent. During submergent phases the mudline may move up from the continental margin to encroach across the shelf itself (Fig. 6.48).

   Other important continental shelf parameters include vertical variations in light and oxygen. Several studies show that in modern oceans oxygen content is high in the shallow zone where planktonic algae photosynthesize, diminishes rapidly to almost zero at some 200 m, and then increases slightly right down to the ocean deeps (e.g., Dietrich,

1963; Emery, 1963). The relationship between the anoxic zone and the mudline obviously has important implications for the deposition of organic-rich muds that may become petroleum source beds. The sunlight that planktonic algae require for photosynthesis decreases with depth. The lower limit of the photic zone varies according to the turbidity of the water, but is normally taken as about 80 m. The relationship between the oxygenated zone and the photic zone is important because most carbonate sediment is organically precipitated. The abundance of life is closely related to sunlight and oxygen (Fig. 6.48).

It is therefore appropriate to consider terrigenous and carbonate shelves separately. In the same way that modern coasts are in some ways unsuitable for studying the evolution of barrier deposits, so are modern shelves not ideal for study as a key to their ancient counterparts. Modern continental shelves include environments of erosion, equilibrium, and deposition. These have inherited features gained when the shelves were largely exposed due to low sea levels during glacial maxima. The present-day sediments of the continental shelves were largely laid down in fluvial, periglacial and lagoonal environments. They have subsequently been reworked since the last postglacial rise in sea level. It has been estimated that some 70% of the continental shelves of the world are covered by these "relict" sediments (Emery, 1968).

Modern shelf deposits are of three kinds: those in which the Pleistocene sediments have been reworked so as to be in equilibrium with the present hydrodynamic environment; those which are truly "relict" and have been essentially unaltered by the present regime; and those which are now being buried beneath a mud blanket (Curray, 1965; Klein, 1977; Johnson and Baldwin, 1986). Continental shelf deposits in equilibrium with the present hydrodynamic environment tend to be shallow and inshore, whereas the unmodified relict deposits are largely preserved in the deeper waters near the continental margins. Burial of relict Pleistocene deposits beneath mud occurs near the mouths of major rivers in areas of low current velocities. These facies types contrast with the concept of a graded shelf, that is, a shelf whose sedimentary cover is in equilibrium with its hydrodynamic regime and, ideally, one in which sediment grade fines seaward (Swift, 1969). The Bering Sea has been cited as one example of such a modern graded shelf (Sharma, 1972). The erratic behavior of sea level in the last million years makes it difficult to use modern continental shelves to study the evolution of deposits through time. Nevertheless, the processes and structures of individual environments can be studied and applied to ancient sedimentary facies.

There are three main environments on modern shelves: an inshore zone of sandbars and tidal flats, an open marine high-energy environment, and an open marine low-energy environment. In the ideal situation of a gentle uniformly dipping shelf, these three zones would trend parallel to the shore in linear belts. In practice, because of the complex tectonic and glacial histories of many modern shelves, these zones are erratically arranged. The inshore zone tidal flats deposit an overall upward-fining prograding sequence. The thickness of the individual units of this sequence is controlled by the tidal range. It has been argued that the paleotidal range of ancient tidal flat sequences may be measured from them (Klein, 1971; Fleming and Bartholoma, 1995). The relatively sheltered inshore environments of tidal flats are generally protected from the open sea by a high-energy zone. This may be a barrier island, as already described. Alternatively rock reefs and subaqueous sand shoals may provide protection. The high-energy shelf environment may be one of erosion, where a marine terrace is cut across bedrock; it may be an environment of equilibrium, where sand shoals migrate to and fro by tidal scour; rarely is it an environment of deposition (Stride, 1982). Carbonate and terrigenous continental shelves are described separately.

6.3.2.7.2 Terrigenous shelves

Modern terrigenous shelves have been intensively studied (Swift *et al.*, 1972, 1973, 1992; Stanley and Swift, 1976; Bouma *et al.*, 1982; McCave, 1985). Continental shelves can be classified into tide, ocean, storm, and wave-dominated models (Johnson and Baldwin, 1986). The continental shelf of northwestern Europe is an example of a tidally dominated shelf. The southeast African coast is an example of a shelf dominated by oceanic currents. The Oregon–Washington coast is an example of a storm-dominated shelf, and the Baltic Sea is an example of a wave-dominated shelf.

The tide-dominated shelf of northwest Europe is particularly well known (e.g., Stride, 1963, 1982; Kenyon and Stride, 1970; Belderson *et al.*, 1971; Banner *et al.*, 1979). Much of this shelf is floored by Pleistocene glacial and fluvioglacial deposits. Three major zones can be recognized, one predominantly gravel floored, the second sand, and the third mud (Fig. 6.49). The gravel floored parts of the shelf are those subjected to the strongest tidal currents. They are essentially areas of erosion from which sand and mud have been winnowed to leave a lag gravel deposit. These gravel seabeds are traversed by ephemeral sand ribbons up to 2.5 km long and 100 m wide, which are aligned parallel to the axis of tidal flow.

The sand floored parts of the shelf are essentially environments of equilibrium. Much sediment is moved to and fro, but there is little net sand deposition. The dominant bed form of these areas is sand waves (Fig. 6.50). These are large underwater dunes with heights of up to 20 m and wavelengths of up to 1 km. The surface of these sand bodies is modified by smaller dunes and ripples. Acoustic and sparker surveys show low-angle bedding within these sand bodies. Coring reveals cross-bedded sets of shelly sand. Detailed studies of these sand bodies show a complex relationship between external morphology and internal structure (Houbolt, 1968). There is no correlation between dune height and water depth (Stride, 1970). The complex morphology and structure reflect the response of the sand waves to the ever-changing tidal flow regime.

Studies of analogous smaller scale inshore sand bodies have been carried out where they are exposed at low tide. Bipolar cross-bedding has been observed dipping in the two opposing tidal current directions (Hulsemann, 1955; Reineck, 1963, 1971). Ancient analogs of these tidal sand bodies have been recognized by Narayan (1970), De Raaf and Boersma (1971), Reineck (1971), and Swett *et al.* (1971).

Mud is the third sediment type to be deposited on the modern shelf of western Europe. This takes place in two settings. Tidal mudflats deposit upward-fining sequences in which subtidal sands grade up through flaser-bedded and bioturbated muds into salt marsh peats (Fig. 6.51). These intertidal deposits have been described from the Wadden Zee of Holland, and from the Wash embayment of England by Van Straaten (1965) and Evans (1965), respectively. Finally, mud settles out on the continental shelf below the

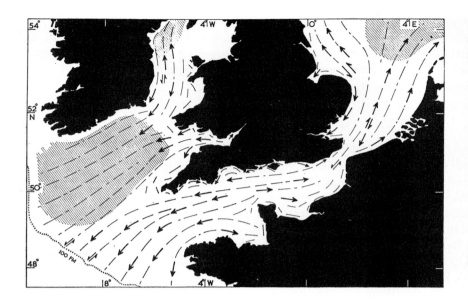

Fig. 6.49. Maps showing the alignment of tidal currents (upper) and of sediment (lower) for part of the north-west European continental shelf. A = mud, B = sand sheet and ridges, C = gravel and scoured sea floor. (From Stride, 1963, by courtesy of the Geological Society of London.)

Fig. 6.50. A 200-m-wide swath from a high-resolution side-scan sonar image of sand waves on the Rockall bank (North Atlantic Ocean) at a depth of about 500 m. There are two wavelengths present, at about 100 and 15 m. They are formed by strong along-slope thermohaline currents. (Courtesy of N. Kenyon and the UNESCO-IOC Floating University Programme.)

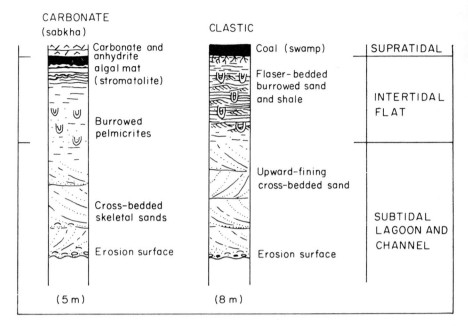

Fig. 6.51. Comparative profiles of tidal flat sequences in arid sabkha carbonate and humid clastic sediments, based on the coasts of Abu Dhabi and the Dutch Wadden Sea. The thickness of the sequences is related to tidal range. (After Evans, 1970. Courtesy of the Geologists' Association.)

"mudline." Notice that these areas are not just in deep water, but also in shallow embayments sheltered from tidal currents (Fig. 6.49).

6.3.2.7.3 Economic aspects of terrigenous shelves

Shelf deposits, both modern and ancient, are of considerable economic importance. Modern shelves have been studied intensively, not just because of their importance to navigation, but also because they contain economic deposits of phosphates and heavy minerals. The latter include cassiterite off the Malaysian shelf, iron from sand waves off the northwest coast of New Zealand, and monazite from the shelves of New South Wales and South Australia (Macdonald, 1983; Glasby, 1986). Many of these placers are relict deposits, reworked from drowned Pleistocene beaches and alluvium.

Shelf sands are normally well sorted and free of clay. Therefore they have excellent porosity and permeability. Sometimes individual tidal sand waves are enveloped in mud, but commonly continental shelf sands occur in laterally extensive blankets. The lateral permeability of such formations may make them excellent aquifers and petroleum reservoirs. Examples include the Simpson Sand of Oklahoma, and the Haouaz and Um Sahm formations of the Sahara and Arabia, respectively (Plate 3A; Bennacef *et al.*, 1971; Selley, 1972; al-Laboun, 1986). These are all of Ordovician age. This is not a coincidence. The worldwide distribution of hypermature orthoquartzites of Ordovician–Silurian age reflects a global rise in sea level that drowned many continental shelves.

6.3.2.7.4 Carbonate shelves

A general theory of carbonate shelf sea sedimentation was put forward, based largely on the study of Paleozoic deposits of the Williston basin, North America (Shaw, 1964; Irwin, 1965; Heckel, 1972). The thesis on which this model is based states that in quiescent tectonic epochs of the past there were broad stable subhorizontal shelves with gradients of less than one in a thousand. These gently sloping surfaces were intersected by two horizontal surfaces of great significance: sea level and effective wave base (what might now be called the "mudline"). The intersections of these surfaces with the sea bed define three sedimentary environments. In the deepest part of the shelf, below effective wave base, fine-grained mud settles out of suspension. Resultant sedimentary facies are laminated shales and calcilutites, sometimes with chert bands, and a biota of sparse well-preserved macrofossils and pelagic foraminifera. Upslope of the point at which effective wave base impinges on the seabed is a high-energy environment. Because of the gentle gradient of the shelf, this belt may be tens of kilometers wide. This is a zone of shoals and bars. The resultant sedimentary facies include biogenic reefs, cross-bedded oolites, and skeletal and mature quartz sands. To the lee of this high-energy belt is a sheltered zone which may stretch for hundreds of kilometers to the shoreline. This low-energy environment generates pelmicrites, micrites, dolomicrites, and evaporites in the lagoons, tidal flats, and sabkhas of arid carbonate realms (Warren, 1989; Kendall and Harwood, 1996). Clays, sands, and peats form in the analogous environments of humid terrigenous realms. Regressions and transgressions cause the three facies belts to migrate to and fro over each other in a cyclic manner.

The X-Y-Z zone model was based on the study of ancient limestones. Subsequent research into recent carbonate environments has allowed much more detailed carbonate facies models to be devised, with up to eight separate facies (Wilson, 1975; Enos, 1983; Scholle *et al.*, 1983b; Wilson and Jordan, 1983; Tucker and Wright, 1990; Wright and Burchette, 1996, 1999). A major distinction is made between **carbonate ramps** and **rimmed carbonate platforms** (Ahr, 1973; Read, 1985). Ramps are gently sloping surfaces, broadly comparable to the X-Y-Z zone model. Ramps tend to accrete across a shelf depositing a sequence of pelagic muds, overlain by shallow water high-energy skeletal and/or oolitic sand, succeeded in turn by lagoonal and intertidal muds. Rimmed carbonate platforms, in contrast, drop sharply off from shallow to deep water, and are thus also referred to as the carbonate **drop-off** model. There is a close correlation between the grain type and texture of carbonates and their depositional environment, for reasons explained in some detail in Chapter 9. Figure 6.52 illustrates this correlation for carbonate accretionary ramps and rimmed platforms. Modern examples of these two models are briefly described and illustrated next.

The modern northeastern coast of Arabia is an example of a modern carbonate ramp (Fig. 6.53). In the deeper water of the Gulf, below about 30 m, lime mud is being deposited. As water depth gradually shallows toward the Arabian Shield, skeletal wackestones pass shoreward, via skeletal packstones, into shallow water oolite grainstones and reefs that accrete around Pleistocene limestone islands (Purser, 1973). Carbonate muds, algal stromatolites, and evaporites form in **sabkhas** (Arabic = salt marsh) in sheltered coastal lagoons and embayments (Evans *et al.*, 1969). These form upward-fining sequences analogous to those of temperate terrigenous intertidal flats (Fig. 6.51).

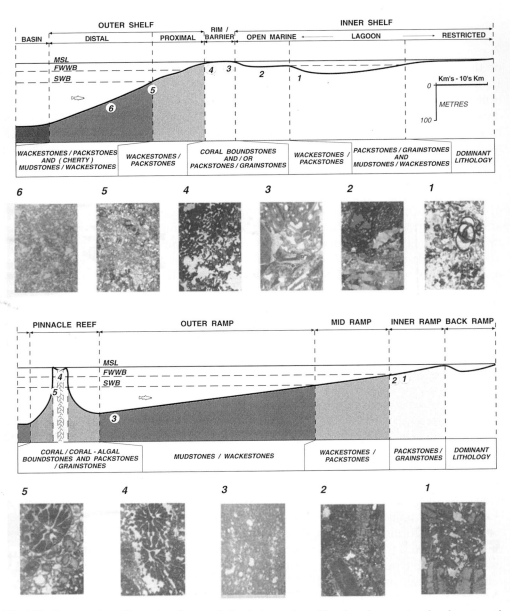

Fig. 6.52. Cross-sections illustrating the correlation between depositional environment and carbonate rock types for a rimmed carbonate platform (upper) and a carbonate ramp (lower). Details of limestone grains and rock type names are given in Chapter 9. (From Spring and Hansen, 1998, by courtesy of the Geological Society of London.)

The Bahama Bank is a modern example of a rimmed carbonate platform (Fig. 6.54). The bank margins are marked by minor reefs and extensive areas of oolite shoals. Morphologically these are analogous to the sand waves of the northwestern shelf of Europe previously described (Fig. 6.55). The sand waves migrate across a scoured surface of

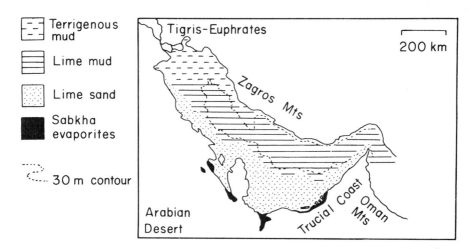

Terrigenous mud

Lime mud

Lime sand

Sabkha evaporites

30 m contour

Tigris-Euphrates

Zagros Mts

200 km

Arabian Desert

Trucial Coast

Oman Mts

Fig. 6.53. Map showing the present-day sediment distribution of the gulf between Arabia and Iran. The gently sloping Arabian shelf is a modern example of a carbonate ramp. (From Emery, 1956. American Association of Petroleum Geologists Bulletin, AAPG © 1956, reprinted by permission of the American Association of Petroleum Geologists whose permission is required for further use.)

Pleistocene limestone (Newell and Rigby, 1957; Purdy, 1961, 1963; Imbrie and Buchanan, 1965; Ball, 1967). The central part of the Great Bahama Bank is a site of fecal pellet mud deposition. Tidal flats are present around land areas, like Andros Island. These deposit algal stromatolites and dolomitic muds, analogous to those of the Arabian coast. Evaporites are largely absent, however, because of the higher precipitation.

Rimmed carbonate platforms may often be controlled tectonically. The platform margin is often a fault and may mark a change from continental to oceanic crust. However, this is not always the case. Benthonic carbonate organic growth is most prolific where the photic zone impinges on the seabed. This is because nutrients from the open ocean will be removed by the beasties along this zone, so beasties on the landward side will suffer malnutrition. Thus carbonate shoals or biogenic reefs may prograde basinward. A drop in sea level will cause the erosion of a sea cliff into newly cemented limestone. When sea level rises, optimum organic carbonate growth will take place at the crest of the drowned sea cliff. If this process is repeated a rimmed carbonate platform margin of atectonic origin may form (Fig. 6.56). This evolutionary pattern has been noted on modern carbonate shelf margins where sedimentation takes place over a paleokarst topography of drowned Pleistocene and earlier limestones (Purdy, 1974).

Carbonate shelf sediments are of great economic importance, but it may be more appropriate to discuss this topic after looking at "reef" models. Reefs are a particular constituent of carbonate shelf facies of great economic interest.

6.3.2.8 Reefs

The reef is one of a spectrum of marine shelf environments. Such is its importance, however, that it merits special attention. A "reef" has been defined as "a chain of rocks at, or near the surface of water: a shoal or bank" (Chambers Dictionary, 1972 edition).

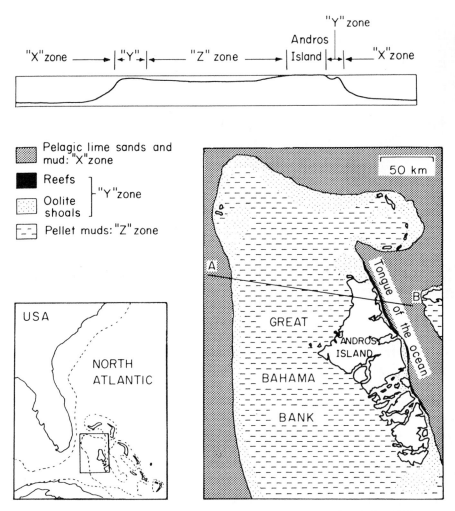

Fig. 6.54. Cross-section and map of part of the Great Bahama Bank, showing the relationship between physiography and facies. This is a modern example of a rimmed carbonate platform. (After Purdy, E. G. 1963. Recent calcium carbonate facies of the Great Bahama bank. *J. Geol.* **71**, 472–497. By permission of the University of Chicago Press.)

This lay definition places no stress on the organic origin or on the wave-resistant potential of a reef. Strictly speaking, therefore, a reef is thus a rocky protuberance on the seabed with the potential to wreck a ship or disembowel a careless cruising ichthyosaur. To geologists, however, reefs have, in a vague sort of way, always meant wave-resistant organic structures, built largely of corals (Fig. 6.57).

6.3.2.8.1 Present-day reefs

Modern organic reefs have been intensely studied by biologists and also by geologists interested in carbonate sedimentation (e.g., Jones and Endean, 1973). The morphology,

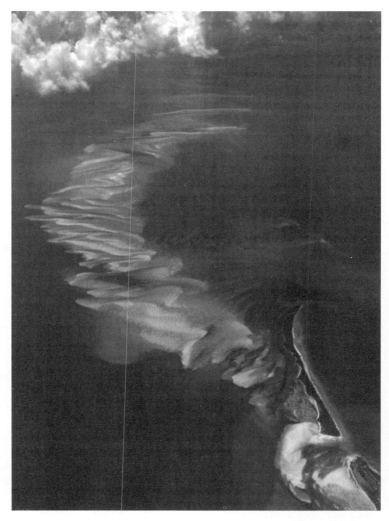

Fig. 6.55. Air photograph of oolite shoal megaripples on the Great Bahama Banks. Flood tides are stronger than ebb tides in the Bahamas so the shoals are migrating to the left over muddy lagoonal sediments. The Bahama Bank drops off into deeper water to the right. (Courtesy of E. Purdy.)

structure, and ecology of modern reefs are well known. It is accepted that organic reefs can form in a wide range of water depths, temperatures, and salinities. Reefs of calcareous algae and mollusks can grow in lakes. Marine reefs can be made of almost any of the sedentary lime-secreting invertebrates. Deep-water carbonate reefs and mounds do occur rarely around the globe. It has been known for many years that some corals can grow in deep, cold water (Teichert, 1958a; Maksimova, 1972). In recent years, however, deepwater carbonate mounds have been found that are composed of carbonate-skeleton-secreting tube worms and assorted mollusks. The basis of the food chain is not photosynthesising algae, as with shallow water reefs, but methanogenic bacteria feeding off petroleum seepages (e.g., Hovland and Somerville, 1985). Nevertheless, the majority of

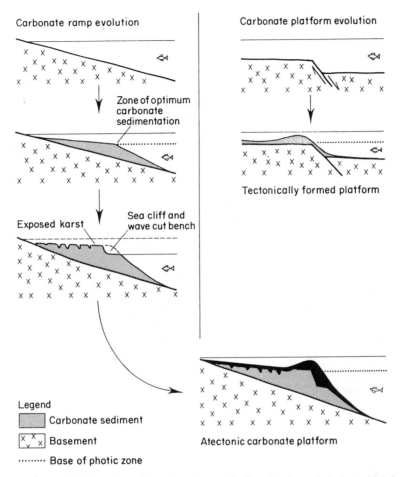

Carbonate ramp evolution

Zone of optimum
carbonate
sedimentation

Exposed karst

Sea cliff and
wave cut bench

Carbonate platform evolution

Tectonically formed platform

Legend

Carbonate sediment

Basement

········ Base of photic zone

Atectonic carbonate platform

Fig. 6.56. Illustrations of the formation of a carbonate ramp (left) and a rimmed platform (right). Carbonate sedimentation is most rapid where the base of the photic zone impinges on the sea floor. Note how rimmed platforms may be controlled tectonically or may develop by alternate phases of erosion and deposition when sea level fluctuates across a carbonate ramp. (For examples, see Purdy, 1974.)

modern reefs occur in warm, clear, shallow seawater and they contain coralline frame-building organisms in significant amounts.

Morphologically reefs have been traditionally grouped into four main types: fringing reefs, barrier reefs, atolls, and pinnacle or patch reefs (Fig. 6.58). There are obviously transitions between these various types and, as this figure shows, there are three ways in which an atoll may form (Fig. 6.59).

The classic theory of atoll evolution was advanced by Charles Darwin in the light of his observations during the voyage of the *Beagle* (Darwin, 1837). He postulated that as volcanic islands subsided, due to isostatic adjustment, a fringing reef could evolve into a barrier, and then, as the tip of the volcano submerged, into an atoll. On some modern shelves, however, a complete transition from isolated patch reefs to atolls can be observed. Patches evolve into atolls by coral prograding over talus deposited on the lee

Fig. 6.57. Colonial corals. Lasaga reef, south of Lae, Papua New Guinea. The high primary porosity of corals is often infilled by detritus and early cementation. Later, secondary porosity may develop by solution or do-lomitization. (Courtesy of M. C. Cooper.)

side of the reef (Fairbridge, 1950). The third mode of atoll formation was expounded when the origin of carbonate platforms was discussed. This is where a carbonate mound, of whatever origin, has undergone emergence, erosion, submergence, and encrusting carbonate deposition on the crests of the drowned sea cliffs (Purdy, 1974). The evidence suggests that most modern atolls owe their shape to more than one of these processes. The three modes of atoll formation are illustrated in Fig. 6.58.

Modern coral reefs are divisible into a number of distinct physiographic units. Each unit is characterized by distinctive sediment types and biota. There are considerable variations in the species and genera of reefs, yet a well-defined ecological zonation characterizes the subenvironments of any specific example (Fig. 6.60). As Fig. 6.61 shows there are three main morphological elements to a reef: the fore-reef, the reef flat, and the back-reef. Additional terms are recognized for lower order features. The fore-reef, or reef talus, grades seaward into deeper water toward the open sea. This zone slopes away from the reef front with a decreasing gradient. It is composed largely of detritus broken off the reef and is, therefore, a kind of subaqueous scree. Boulders of reef rock may be present at the foot of the reef front, but grain size decreases from calcirudite, through calcarenite to calcilutite downslope. The talus is itself colonized by corals, calcareous algae, and other invertebrates that tend to bind the scree together. Progradation of the scree means that seaward-dipping depositional bedding may be preserved.

The reef itself is composed largely of sessile colonial calcareous frame-builders. The reef front of modern reefs is characteristically encrusted by highly wave-resistant Litho-thamnion algae. Other organisms grow behind this in ecological zones. The reef itself is flat, because its biota cannot withstand prolonged subaerial exposure, and it is generally emergent at low tide. The reef flat is locally cross-cut by tidal channels. Modern

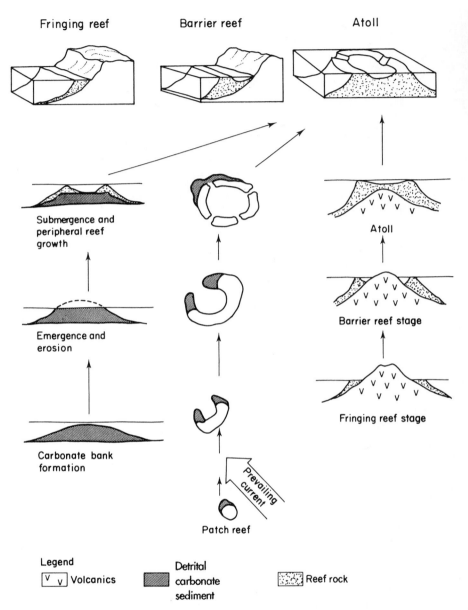

Fig. 6.58. Illustrations of the morphology and evolution of the different types of reefs. Note that atolls may form by three processes: isostatically subsiding volcanoes (Darwin, 1837), leeward detrital accretion and progradation (Fairbridge, 1950), or by bank emergence, erosion, subsidence, and peripheral reef growth (Purdy, 1974).

reefs have a high primary porosity present between the skeletal framework. This gradually diminishes through time due to infiltration by lime mud and to cementation (see Section 9.2.5.2).

Behind the reef flat is a third zone, the back-reef. Immediately behind the growing

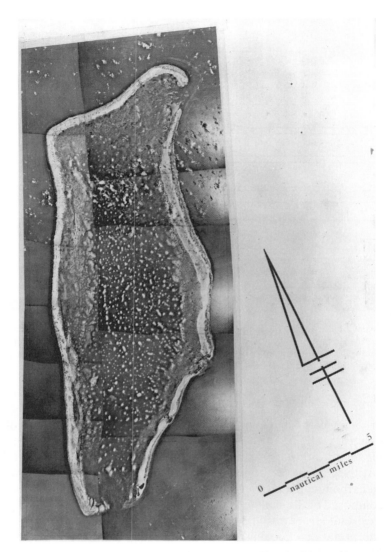

Fig. 6.59. Air photograph of a modern coral atoll, Glovers reef, Belize, Caribbean. (Courtesy of E. Purdy.)

reef there is an area of smashed-up reef debris washed over by storms. The grain size of the back-reef deposits grades away from the reef into micrite and fecal pellet muds of a lagoon. Back-reef lagoons sometimes contain small patch reefs.

It is important to remember that reefs do not grow in isolation, but form an integral part of carbonate ramp and platform deposits (Figs. 6.53 and 6.54). Present-day reefs show complex cross-sectional geometries (where they can be studied), reflecting Pleistocene sea level changes. Likewise, their diagenetic histories have been complex. It is evident, more from the study of ancient reefs than modern ones, that reefs may prograde seaward over their own fore-reef talus, that they may grow vertically, or that they may migrate landward in response to a marine transgression.

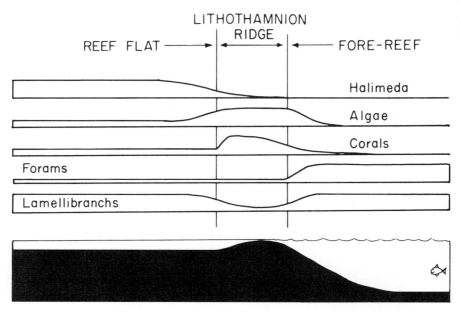

Fig. 6.60. Cross-section illustrating the morphology and ecological zonation of a modern Florida reef. (After Ginsburg, 1956, American Association of Petroleum Geologists Bulletin, AAPG © 1956, reprinted by permission of the American Association of Petroleum Geologists whose permission is required for further use.)

6.3.2.8.2 Ancient "reefs"

Lenses of carbonate contained in other sediments are found in many parts of the world in Phanerozoic rocks. Some of these have been identified as ancient reefs by careful study of their fossils, lithology, and facies relationships. Others have been labeled as reefal without the benefit of detailed study. A reef origin is impossible to prove for many ancient carbonate lenses if a rigid definition is used. For example "reefs are bodies of rock composed of the skeletons of organisms which had the ecologic potential to build wave-resistant structures" (Lowenstam, 1950). To avoid such precise and often untenable interpretations, geologists have approached the problem under a blizzard of names (build-up, biostrome, bioherm, mound, organic reef, stratigraphic reef, ecologic reef, and so on).

There are two fundamental reasons for the problems of defining a viable ancient reef model. First, there is the confusion of facies and environment; second, many of these rock masses are so extensively modified by diagenesis that their original nature is undetectable. In no other group of ancient sediments are environments and facies nomenclature so confused (Braithwaite, 1973). Terms like "reef limestone," "intrareef detritus," and "fore-reef facies" are widely used. Many problems could be avoided by describing ancient carbonates in lithofacies terms only. In particular the term **carbonate buildup** though ugly, has much to commend it for describing a lenticular organic carbonate rock unit of uncertain origin (Stanton, 1967).

The second factor that has complicated the study of ancient reefs and banks is their

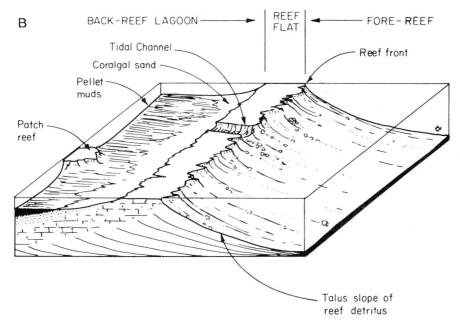

Fig. 6.61. (A) Air photograph of the Columbus barrier reef, off the Yucatan peninsula, Belize. Open sea to the right, tidal channel cross-cuts reef flat in the foreground, back-reef lagoon to the left. (Courtesy of E. Purdy.) (B) Illustration of the physiography and facies of a modern barrier reef. Note that the steep slope of a modern fore-reef is often a drowned Pleistocene sea cliff.

extensive diagenesis. This is due to a variety of factors, including high primary porosity, unstable mineralogy, and invasion by hypersaline fluids when sea levels drop and lagoons evaporate (see Section 9.2.5.2). Because of these factors, reefs and banks are frequently recrystallized and dolomitized, so that their primary fabric is undetectable. Many of the Devonian "reefs" of Canada are of this type. Because of this diagenesis, it may be impossible to prove whether a lenticular carbonate buildup was actually once a reef in the ecologic sense.

There is a second important point related to the diagenesis of carbonate buildups. It has already been mentioned that a drop in sea level exposes the crests, not only of orthodox reefs, but also of shell banks and carbonate sand shoals. These may all be cemented by early diagenesis. When sea level rises again, all of these features, regardless of their genesis, now form wave-resistant masses complete with talus slopes and lagoons. Repetition of this process can build up wide shelves with margins formed of early cemented carbonates of diverse facies with aprons of slope deposits that grade down into a basin. The Capitan Formation (Permian) of west Texas rims the Delaware basin, separating shallow water evaporites and continental red beds on the northwestern shelf from deep water shales and turbidites (Fig. 6.62). This was once interpreted as a classic example of a barrier reef (e.g., Newell *et al.,* 1953). The tendency now is to interpret this as a feature that had a wave-resistant scarp, but that this was composed of organic bioclastic shoals and banks which gained their rigidity, not from the syndepositional activity of encrusting algae, but from cementation during marine low stands (Achauer, 1969; Dunham, 1969; Kendall, 1969).

The story of the Capitan "reef" suggests that it may be helpful to recognize and dis-

Fig. 6.62. Photograph of the jebel El Capitan, type locality of the Permian Capitan "reef," Guadalupe Mountains, west Texas. The massive reef carbonates overlie basinal black shales and turbidites of the Delaware Mountain Group that crop out in the foreground. (Courtesy of E. Purdy).

tinguish between stratigraphic reefs and ecologic reefs (Dunham, 1970). A **stratigraphic reef** is a carbonate buildup that is inorganically (e.g., sparite) cemented. An **ecologic reef** is one that is actually composed of frame-building skeletons bound by organic means, such as encrusting algae. In the words of Dunham (1970, p. 1931):

> The stratigraphic reef is an objective concept concerning three dimensional geometric masses. The ecologic reef is a subjective concept concerning inferred interactions of organisms and topographic evolution. One involves geometry, the other topography.

In conclusion an organic reef sedimentary model is definable based on the study of modern tropical coral reefs. Application of this model to ancient sediments is difficult (Elloy, 1972). Lenticular carbonate masses are common, but it is hard to prove that these were always wave-resistant topographic features composed of organic skeletons in growth positions. This is especially hard because many buildups have lost their primary depositional features by obliterative diagenesis. There may be sufficient data, however, to distinguish whether the buildup originated as a soft sediment bank of shells or ooliths, or whether it was an ecologic or a stratigraphic reef. Though it may be impossible to be specific about the origin of a buildup, or of a shelf margin, models for these can be defined based on their lithology and gross morphology (Fig. 6.63).

Ancient analogs of the modern deep-water reefs are the curious features termed **mud mounds** (Monty *et al.,* 1995). These are lenses of micritic limestone, with a sparse fossil assemblage, that occur enclosed in marine shales. Ancient mud mounds are known in Devonian rocks of Belgium and Triassic rocks of Spain. They are especially common in Carboniferous rocks, where they are termed "Waulsortian banks" (Lees and Miller, 1995). These mud mounds commonly occur at platform margins, where faults extend down into the "devil's kitchen" where petroleum generation takes place.

This situation is analogous to the modern deep-water carbonate "reefs" described earlier, where the escape of methane and oil to the seabed along faults, gives rise to

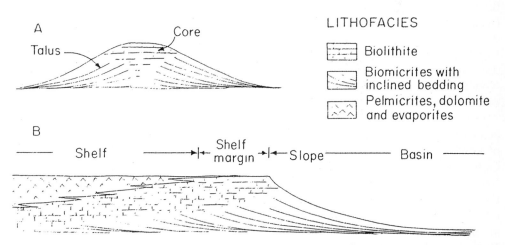

Fig. 6.63. Nonemotive terminology for (A) carbonate buildups and (B) shelf/basin transition facies models. This uses lithologic and physiographic terms, avoiding genetic terms with reefal connotations.

pock marks and cold seeps, as described elsewhere (see Section 5.3.5.7). The petroleum may have fed methanogenic bacteria, which formed the base of a higher food chain of invertebrates. The occurrence of stromatactis structure (see Section 3.2.2.3.3) in mud mounds suggests that gas bubbles were trapped in rapidly cemented lime mud, and the resultant voids then infilled by carbonate cement.

6.3.2.8.3 Economic aspects of carbonates in general and reefs in particular

Carbonate rocks, especially reefs, are of great economic importance. They may be quarried for building stone or aggregate, they may serve as aquifers, and they host many ores and nearly 50% of the world's oil. In carbonates, like sandstones, the amount and distribution of porosity and permeability control their suitability as aquifers, petroleum reservoirs, and sites of mineralization. There is, however, a major difference between the distribution of porosity and permeability in carbonates, when compared with sandstones. In sandstones the effects of diagenesis are normally not too extreme. Porosity is largely primary and related to facies. As Chapter 9 shows, however, porosity in carbonates is largely secondary due to extensive diagenesis. The porosity and permeability of ancient carbonates is therefore often unrelated to their original facies-related primary porosity distribution. Thus many limestone aquifers owe their capacity to tectonically controlled fracture pore systems, or to leached paleokarst horizons where unconformities cross-cut facies and stratigraphy.

Limestones and dolomites often make excellent petroleum reservoirs. This is because they are commonly developed along basin margins, acting as natural traps for petroleum migrating up from basinal source beds. Reefs have been aptly termed "sedimentary anticlines." They are often transgressed by organic-rich muds that serve both as source and seal (Fig. 6.64). There are problems, however. The eccentric distribution of porosity and permeability in carbonate petroleum reservoirs has mystified geologists and engineers for many years. Numerous excellent studies, mostly unpublished, describe the diagenesis and porosity evolution of carbonate reservoirs in the most intimate detail (e.g., Reeckman and Friedman, 1982). Sadly, most of these studies are of limited predictive value, and are unable to guide the drilling and development program of a field. In many carbonate provinces the distribution of petroleum often appears to be random, and within the fields themselves porosity is unrelated to depositional environment. Nonetheless, Wilson (1975, 1981) has described a series of characteristic petroleum reservoir models (Fig. 6.65).

Carbonates in general, and reefs in particular, are hosts to ore bodies. The most characteristic variety is lead–zinc sulfide mineralization. This is often referred to as being of **Mississippi Valley** type, because that is one region noted for such ores (Brown, 1968; Ohle, 1980; Gustafson and Williams, 1981; Wolf, 1981; Anderson and McQueen, 1982; Briskey, 1982; Clemmey, 1985). The Mississippi Valley mining district extends through Oklahoma, Kansas, and Missouri. Other similar North American examples are known in Wisconsin, and at Pine Point in the Northwest Territories of Canada (Fig. 6.66). There are many examples in the Carboniferous limestones of Europe, ranging from Poland to Yugoslavia, England and Ireland (Hitzman and Large, 1986). Fluid inclusion studies of

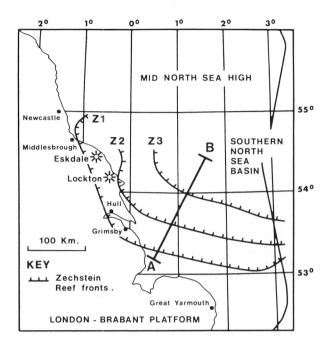

Progradation of Zechstein reef fronts ➡

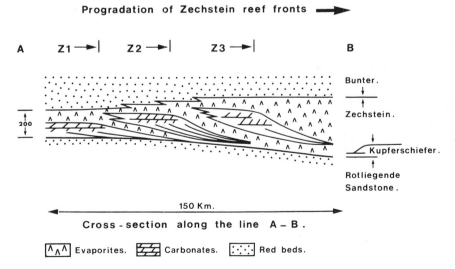

Cross-section along the line A – B.

Fig. 6.64. Map and cross-section showing the cyclic progradation of the Zechstein (Upper Permian) barrier "reef" of the southern North Sea basin. This is coeval with, and similar in many ways to, the Capitan "reef" illustrated in Fig. 6.62. (After Taylor, J. C. M. and Colter, V. S. 1975. Zechstein in the English Sector of the southern North Sea Basin. *In* "Petroleum and the Continental Shelf of Northwest Europe" (A. E. Woodland, ed.), pp. 249–263, Figs. 2 & 3. Courtesy of the Institute of Petroleum.)

I Primary porosity preserved in shoal grainstones
   (Murban field, UAE; Arab D, Saudi Arabia)

II Primary porosity dolomitization and fracturing in reefs
    (Golden Lane, Mexico; Alberta, Devonian)

III Porosity in fore-reef talus
     (Poza Rica and Reforma pools, Mexico)

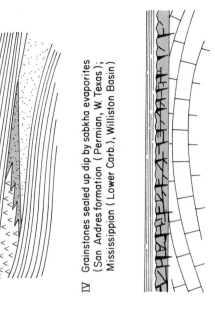

IV Grainstones sealed up dip by sabkha evaporites
    (San Andres formation (Permian, W. Texas);
    Mississippian (Lower Carb.), Williston Basin)

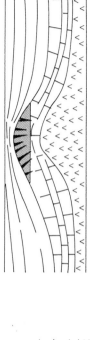

V Epidiagenetic solution porosity and dolomitization
   (Trenton Limestone, Cincinnati Arch; Fahud field, Oman)

VI Fractured chalk over diapirs
    (Ekofisk (Cretaceous) fields, North Sea)

Fig. 6.65. Illustrations of the different carbonate petroleum reservoir models defined by Wilson (1975, 1981). References to examples are cited in the text.

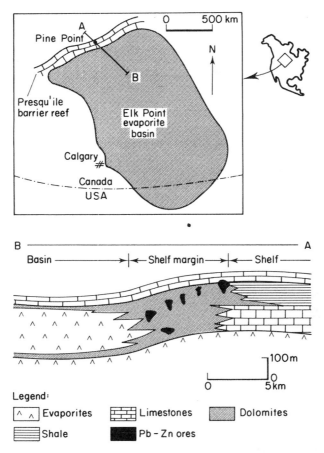

Fig. 6.66. Map and cross-section illustrating the location of the Devonian lead–zinc sulfide ore belt of Pine Point, northern Canada. Note that mineralization occurs in the dolomitized shelf margin of an evaporite basin. Irrespective of the precise mode of mineralization, the location of the ore is obviously related to facies. (Based on Skall, 1975, and Kyle, 1981.)

these lead–zinc sulfides indicate mineral precipitation from hypersaline fluids with temperatures up to 225°C. An association of lead–zinc carbonate-hosted ores with evaporites and petroleum (frequently degraded to bitumen) has often been noted. Seven main mechanisms for lead–zinc stratiform carbonate ores have been proposed, including syngenetic, diagenetic, and epigenetic origins (Guilbert and Park, 1986, pp. 888–910). Currently it is fashionable to believe that mineralization was due to connate fluids enriched in metallic ions migrating from basinal muds. Regional trends of mineralization have been noted but, as with petroleum, the specific locus of ore emplacement is eccentric, occurring variously in fore-reef, reef core, and back-reef deposits.

6.3.2.9 Deep-Water Sand Models

Deep-water sands may be defined as those that are deposited below the mudline. Normally, but not invariably, this is below the continental margin at some 200 m. At these depths sand deposition by turbidity currents and other gravity processes is the norm, with minimal reworking by traction currents, except for relatively low-velocity geostrophic ones. Deep-water sands occur principally in marine basins, but also in large freshwater lake basins. Geomorphologically, deep-water sand deposition occurs in channels cut into a slope and on fan-shaped lobes at the foot of the slope. Sedimentary models for terrigenous and carbonate deep-sea sands are rather different, so they are discussed separately.

6.3.2.9.1 Terrigenous deep-sea sands

Some of the best known modern continental margin turbidite systems occur off the Californian coast (Fig. 6.67). The submarine valley and fan systems of Monterey and La Jolla are particularly well documented (Gorsline and Emery, 1959; Hand and Emery, 1964; Shepard et al., 1969; Wolf, 1970; Horn et al., 1971). The seaward progradation of this type of submarine slope produces the following sequence: At the top is a facies of sands, tending to be well sorted, often glauconitic and with a fraction of skeletal sand. These sediments are cross bedded, cross laminated, and often burrowed. This facies is deposited by traction currents on the continental shelf. It passes down, generally abruptly, into the second facies composed of varying amounts of two distinct subfacies. One consists of laminated clays and silts which were deposited out of suspension in the quieter deeper water of the slope zone. These slope deposits tend to slump and slide because of their high water saturation and unstable situation. The Canary Island slide was described and illustrated earlier (see Section 4.5). The slope shales are cut into by varying amounts of the second slope subfacies. These are the submarine canyon or valley-fill deposits. They consist principally of debris flow, fluidized flow, and grain flow deposits. Where the submarine channels reach the base of the slope they radiate like the distributaries of a bird foot delta, the analogy even extends to the existence of raised levées on the channel sides. This channel:levée system overlies fan-shaped lobes of interbedded turbidite sands and pelagic shales. Sand content and grain size increase toward the top of the fan and toward the apex.

Though the submarine channels of the Californian fans radiate, this pattern is by no means universal. The submarine channels on the Bengal fan are braided, while those of the Amazon and Indus fans are highly meandriform (Fig. 6.68).

Ancient terrigenous deep-sea sands are commonly found in linear troughs, sometimes of rifted origin, sometimes associated with zones of subduction. The latter areas, once termed **geosynclines,** are now referred to as **foredeep basins** (see Section 10.2.3). These sedimentary sequences consist of thick formations of interbedded shales and graded beds, often interpreted as turbidites (Fig. 6.69). This type of facies has often been termed "flysch" (Dzulinski and Walton, 1965; Hsu, 1970).

Terrigenous deep sea sands and their sedimentary models have been described by Howell and Normark (1982), Stow (1985, 1991), Hartley and Prosser (1995), Pickering et al. (1995), and Stow et al. (1996). There are two common settings. Deep-water sediments are found on continental margins where the slopes are cut in bedrock, and they are found at the edges of many prograding deltas. The latter arrangement has already been described as an integral part of the high-slope delta model and modern and ancient examples have been cited.

Reference has already been made to the studies of the La Jolla, Monterey, and asso-

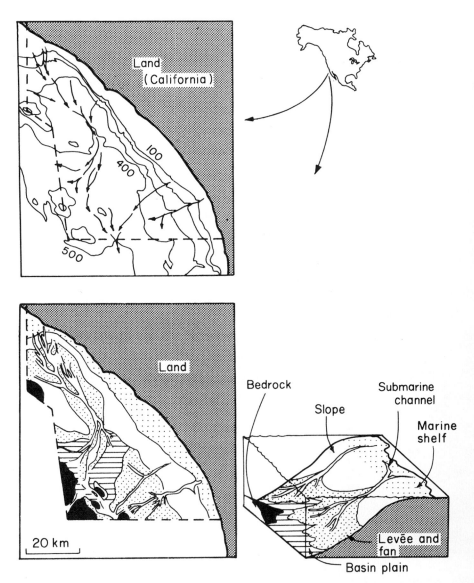

Fig. 6.67. Physiography of the continental margin off California. (From Hand, B. M., and Emery, K. O. 1964. Turbidites and topography of north end of San Diego Trough, California. *J. Geol.* **72,** 526–552. By permission of the University of Chicago Press.) (Upper) Bathymetric chart showing main routes of sediment transport; contours in fathoms. (Lower) Physiographic units showing the mutual relationship between the slope channels and their associated submarine levees and basin floor fans.

ciated channels of the Californian continental margin. Stanley and Unrug (1972) synthesized many data on modern submarine channel deposits and used them to identify ancient analogs in Tertiary flysch basins of the Alps and of the Carpathians. Submarine channel deposits consist largely of sands with varying amounts of conglomerate and traces of shale. The sands occur in thick units with erosional bases. They are seldom

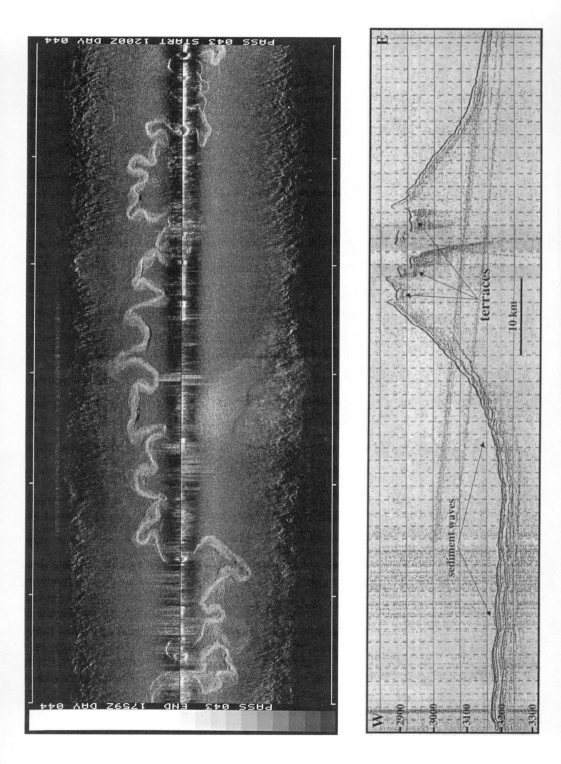

PASS 043 START 1200Z DAY 044

PASS 043 END 1759Z DAY 044

E

W

terraces

sediment waves

10 km

2900

3000

3100

3200

3300

Fig. 6.69. Graded greywacke turbidite sands and shales of the Aberystwyth Grits (Silurian), Wales; an example of deep-sea sands. (Courtesy of J. M. Cohen.)

conspicuously graded, often structureless or slump-bedded, and contain scattered clasts and slump blocks. These sands are attributed to deposition by grain flow, slumping and sliding down the valleys approaching the state of a true water-saturated turbid flow. They are sometimes referred to as fluxoturbidites (see Section 4.5.1).

This facies passes downward into the true turbidite deposits of the submarine fans at the feet of the channels (Fig. 6.69). This transition is shown by a decline in grain size and bed thickness, and by an increase in graded bedding and intervening shale units; simultaneously channels become shallower and wider. Examples of this transition down from channel to fan deposits have been described from the Appennine flysch of Italy by Mutti and Lucchi (1972). The sands and shales of the fan facies grade down into thinner bedded distal turbidites with increasing amounts of fine-grained pelagic muds, which settled out in the basin away from the prograding slope. A characteristic feature of these sediments is that the shales may contain a deep-water (*Nereites*) trace fossils suite, radiolaria and pelagic forams. The turbidite sands themselves may contain glauconite, carbonaceous detritus, and an assemblage of abraded shallow water fossils (Fig. 6.70).

Though there are many documented examples of at least one of the facies transitions of the turbidite slope sedimentary model, it is unusual to be able to find the whole suite present in any one vertical section. An example of this has been described, however, from Eocene sediments of the Santa Ynez Mountains, California (Stauffer, 1967; Van de Kamp *et al.*, 1974). In this particular case the slope deposits are represented by extensive

Fig. 6.68. (Upper) Sinuous submarine channel on the Indus fan, Arabian Sea. This channel is approximately 10 km wide and 30 m deep. (Lower) High-resolution seismic profile through the above channel showing flanking levées and intrachannel terraces. [From Kenyon, N. H., Amir, A., and Cramp, A. 1995. Geometry of the younger sediment bodies of the Indus fan. *In* "An Atlas of Deep Water Environments." (K. T. Pickering, R. Hiscott, N. H. Kenyon, F. Ricci Lucchi, and R. D. A. Smith, eds.), pp. 89–93. Copyright 1995 Kluwer Academic, Dordrecht. Figs. 15.4 & 15.5, p. 92. with kind permission of Kluwer Academic Publishers.]

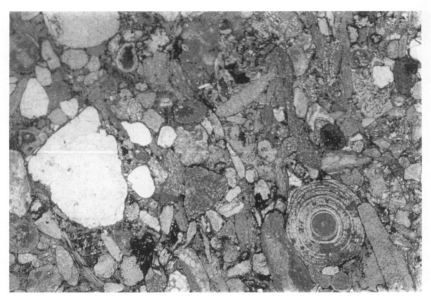

Fig. 6.70. Thin section of basinal turbidite sandstone showing transported skeletal grains and ooliths. Conway Castle Grits (Lower Paleozoic), North Wales (20×).

grain flow sandstones. Slope muds and silts are rare and channels are not extensively observed (Fig. 6.71). To conclude, two models for terrigenous deep-sea sand sedimentation are recognized. In one the channels are cut into a submarine fault scarp, in the other the slope is produced by a prograding delta (Fig. 6.72).

6.3.2.9.2 Carbonate deep-water sands

Most carbonate sediment forms in shallow water, but some of this detritus finally comes to rest in deep water. These deposits are referred to as **allodapic limestones.** The sedimentary processes are similar for terrigenous and carbonate sands, but it is sometimes surprising to encounter graded oolitic turbidites and allochthonous blocks of biolithite in basinal settings. Excellent accounts of allodapic limestones have been given by Cook and Mullins (1983) and Enos and Moore (1983).

These authors document modern deep-sea carbonates from the foot of the Bahamas Platform, and review ancient examples from the slopes of carbonate shelf margins and reefs in the Devonian of Canada, the Canning basin of Australia, the Permian basin of west Texas, and elsewhere. These studies point not only to the abundance of deep-sea carbonates, but also to their diversity of grain size. Because of early lithifaction, partly due to beach rock processes, and partly by organic encrustation, allodapic limestones are often much coarser than terrigenous deep-sea sands. Not only are transported boulder beds quite common, but large slumps of biolithite have been transported far down the slope of the platform rim on which they grew. At the other extreme allodapic carbonate turbidite muds have been described from the Cretaceous Chalk of the North Sea (Kennedy, 1987).

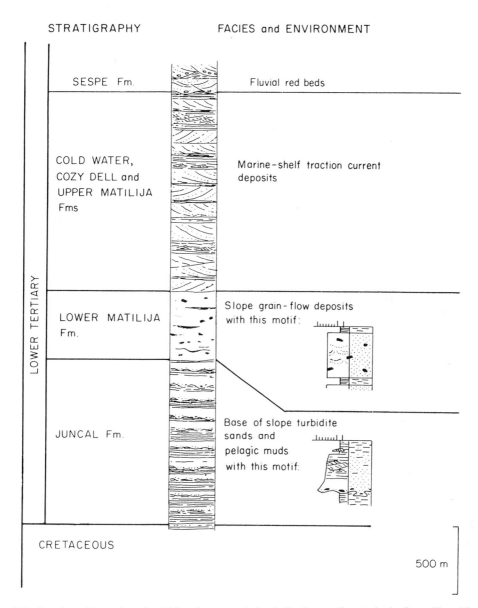

STRATIGRAPHY   FACIES and ENVIRONMENT

SESPE Fm.   Fluvial red beds

COLD WATER,   Marine-shelf traction current
COZY DELL and   deposits
UPPER MATILIJA
Fms

LOWER TERTIARY

LOWER MATILIJA   Slope grain-flow deposits
Fm.   with this motif:

JUNCAL Fm.   Base of slope turbidite
                                sands and
                                pelagic muds
                                with this motif:

CRETACEOUS

                                                    500 m

Fig. 6.71. Stratigraphic section of turbidite slope association in Tertiary sediments in the Santa Ynez Mountains, California. This sequence reflects the basinward progradation of pelagic muds, slope base turbidite sands, slope grain flow sands, marine shelf sands, and alluvium. (After Stauffer, 1967. Courtesy of the Society for Sedimentary Geology.)

Carbonate deep-sea sand models also appear to differ from terrigenous ones (Collacicchi and Baldanza, 1986; Mullins and Cook, 1986). Deep-sea carbonates seldom debouch from a prograding micrite slope, because there is no carbonate analog to a delta. It is more usual for allodapic carbonates to be shed from a carbonate shelf margin or,

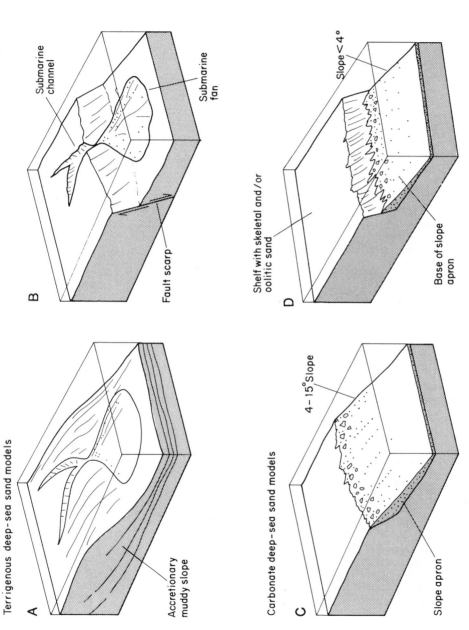

Terrigenous deep-sea sand models

A

Accretionary muddy slope

B

Submarine channel

Submarine fan

Fault scarp

Carbonate deep-sea sand models

C

4–15°Slope

Slope apron

D

Slope < 4°

Shelf with skeletal and/or oolitic sand

Base of slope apron

Fig. 6.72. Geophantasmograms illustrating terrigenous and carbonate deep-sea sedimentary models. Note that terrigenous sands are normally deposited from channel point sources. The slope may either be (A) a prograding delta or (B) a submarine fault scarp. Deep-sea carbonates develop from linear shelf edges of tectonic or atectonic origin. (C) Carbonate slope apron and (D) base of slope apron models proposed by Mullins and Cook (1986).

as in the Cretaceous Chalk of the North Sea, from the crest of a rising salt dome. Fan lobes are normally absent. A continuous apron of detritus is more common. The shelf from which the detritus is derived is normally a scoured hardground (see Section 9.2.5.2) covered by a veneer of skeletal or oolitic sands. As the sands are carried by currents over the platform rim they may be mixed with biolithite, or contemporaneously cemented limestone boulders dislodged from the shelf edge by storms or earthquakes. Mullins and Cook (1986) recognize two allodapic carbonate models: slope aprons, with detritus sloping at angles of < 4° that rise from the basin floor to the shelf rim, and base of slope aprons, with slopes of 4–15° that rise from the basin floor to the base of the shelf rim (Fig. 6.72).

6.3.2.9.3 Economic aspects of deep-sea sands

Deep-sea sands are of particular interest as petroleum reservoirs. This is because distal deep-sea sands may interfinger with organic-rich basinal muds that may serve as petroleum source beds. Petroleum may migrate from the source shales into the interlaminated distal fan sands, and migrate up-dip toward the basin margin, moving into progressively thicker, coarser (and thus more permeable) sands. Reservoir continuity may diminish as reservoir quality increases, however, because the basin plain sheet sands pass up-dip into channelized units.

Petroleum can be trapped in deep-sea sand reservoirs in many ways, some structural, some stratigraphic. Conventional anticlinal entrapment is illustrated by the Long Beach–Wilmington field of California, and the Forties field of the North Sea. In the former the anticline results from compression, in the latter from drape and compaction over a deep seated horst (see Mayuga, 1970, and Walmsley, 1975, respectively). Closure may be caused by diapirs, either mud lumps, such as those mentioned earlier from the Beaufort Sea (see Section 6.3.2.5.4), or salt domes, like the Cod field of the North Sea (Kessler et al., 1980).

In all of the examples just cited the reservoirs are principally laterally extensive basin plain deposits of interbedded outer fan turbidites and shales. Reservoir heterogeneities due to channels are relatively minor. Petroleum entrapment in deep-sea sands along basin margins is somewhat different, and is sometimes stratigraphic. Closure may be provided by the preserved paleotopography of a submarine fan, as in the Frigg field of the North Sea (Heritier et al., 1981). Bizarre stratigraphic traps may occur associated with the submarine channels cut into the slope. Sometimes channel sands may serve as a reservoir. There are fields, however, in which turbidite reservoirs are sealed up-dip by a clay-plugged channel (e.g., the Rosedale field of California, described by Martin, 1963). Figure 6.73 illustrates the diversity of petroleum entrapment in terrigenous deep-sea sands.

Allodapic carbonates also serve as petroleum reservoirs. As discussed earlier (see Section 2.2.3) reservoir quality in limestones is often unrelated to facies-controlled primary porosity, but corresponds to diagenetically induced secondary porosity. One of the most spectacular allodapic limestone reservoirs is the Tamabra Formation of Mexico (Guzman, 1967; Enos, 1977). This consists of grainstones, packstones, and carbonate debris flows that were shed off the Golden Lane atoll — a carbonate platform made of the Cretaceous El Abra limestones. A ring of fields has been discovered along the platform

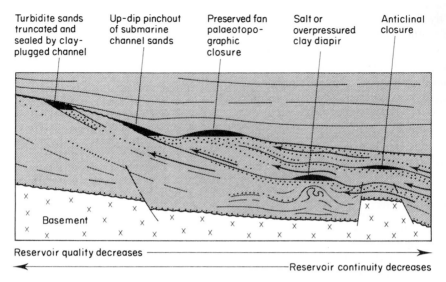

Fig. 6.73. Regional cross-section illustrating styles of petroleum entrapment in deep-sea sands. Arrows indicate direction of petroleum migration. Examples of the various plays are cited in the text.

edge, but a separate trend occurs in the allodapic Tamabra limestones. This fairway is apparently restricted to the southwestern flank of the atoll (Fig. 6.74). It has been suggested that allodapic limestones were only shed off the leeward side of the platform, and that they are absent from the windward flank, where currents would be moving toward the bank (Mullins and Cook, 1986).

Moving to the other extreme, the Cretaceous limestones of the North Sea include both pelagic and resedimented chalks. These contain oil fields where the chalk has been domed and fractured over salt domes (see Fig. 6.65, VI). Permeability is thus largely related to fracture distribution. Porosity, however, is related to facies. The resedimented chalks have higher porosities than the pelagic ones (Kennedy, 1987).

6.3.2.10 Pelagic Deposits

Pelagic sediments are the last major facies to consider. It was pointed out earlier that it is hard to detect the absolute depth of ancient sediments; for this reason a classification into neritic, bathyal, and abyssal deposits was avoided (see Section 6.1.3). The term "pelagic" is applied to those sediments that were deposited in the sea away from terrestrial influence (Jenkyns, 1986). This term has no depth connotation, and pelagic deposits may be found on the distal extremes of broad continental shelves as well as on the abyssal plains. Terrigenous relict sediments of modern shelves and abyssal turbidite wedges are not deemed pelagic.

6.3.2.10.1 Modern pelagic environments

Modern oceanic sediments have been studied from the *Challenger* cruise of 1872, to the more recent Deep Sea Drilling Project (DSDP) and its progeny the International Proj-

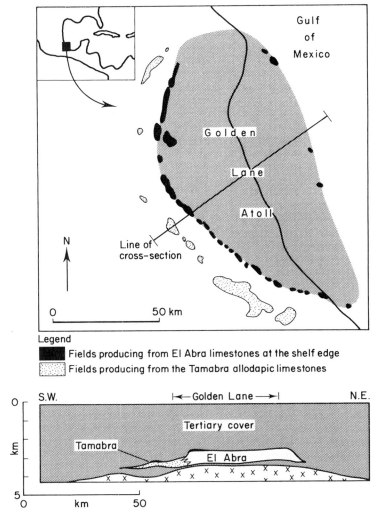

Fig. 6.74. Map and cross-section through the Golden Lane (Cretaceous) atoll of Mexico. Oil fields occur in Rudist reefs around the platform rim and also in allodapic limestones at the foot of the leeward edge. (Based on Guzman, 1967, and Enos, 1977.) The cross-section has been restored to a late Cretaceous datum. The Golden Lane atoll currently dips to the northeast.

ect for Ocean Drilling (IPOD) (Arrhenius, 1963; Riedel, 1963; Kukal, 1971; Hsu and Jenkyns, 1974; Stow and Piper, 1984; Apel, 1987).

The main sediment types of modern ocean floors are the oozes and the clays (Table 6.7). Oozes are defined as pelagic sediments with over 30% organic skeletal detritus. There are three major types of ooze. Pteropod ooze is composed largely of the tests of the microscopic gastropod of that name. Globigerina ooze is made up largely of the foraminifera of that name, together with other foram species and coccolithic nannoplankton. Radiolarian ooze is dominantly composed of the siliceous tests of radiolaria. Inorganic detritus also occurs in pelagic sediments. This includes distal turbidites from

Table 6.7
Simplified Classification of Pelagic Deposits (from Berger, 1974)
and Commonly Found Sediment Types

Classification	Common sediment type

I. Pelagic
    <25% of fraction > 5 µm of terrigenous,
    volcanic, or neritic origin
    A. Pelagic clays
        <30% bioclasts
	Red Clays

B. Oozes
        >30% bioclasts
        (i) Calcareous oozes
	Pteropod ooze (aragonitic)
	Globigerina ooze (calcitic)

(ii) Siliceous oozes	Radiolarian ooze Diatom ooze

II. Hemipelagic deposits
    >25% of fraction > 5 µm of terrigenous,
    neritic, or volcanic origin
	Terrigenous, calcareous,
	   and volcanogenic muds

continental slopes, silt and clay transported out to sea by desert sandstorms, cosmic dust, and volcanic ash. When the inorganic component of pelagic sediment is dominant the deposit is referred to as Red Clay.

Not all ocean floors are environments of deposition. In some areas currents scour the seabed maintaining exposed lithified surfaces ("hardgrounds," see Section 9.2.5.2) and areas encrusted with manganese nodules.

The distribution of the various modern deep-sea sediments is controlled by several factors. Distance from land controls the terrigenous fraction, and proximity to volcanoes controls the volcanogenic source. Both of these factors are also related to prevailing wind direction. The presence or absence of ocean floor currents affects sedimentation rates, the distribution of hardgrounds and areas of manganese encrustation. The organic oozes are closely related to depth. The CO_2 concentration of modern oceans increases with depth and this lowers the pH. The solubility of calcium carbonate thus increases with depth. The aragonitic pteropod oozes are seldom found below 3000 m, while the calcitic Globigerina oozes die out at about 5000 m, below which depth radiolarian ooze or Red Clay prevail. Attempts to apply the concept of the **carbonate compensation depth** (often referred to as just CCD) to ancient rocks are fraught with difficulty because the CCD is not depth dependent so much as temperature related. Thus assumptions must be made of ancient oceanic temperatures before trying to use the CCD concept to diagnose the depth of ancient pelagic sediments (Van Andel, 1975).

A depositional model for pelagic sediments has been defined by combining observations of modern ocean floors with plate tectonic setting (see Section 10.1.3). The mid-ocean ridges are axes of sea floor spreading composed largely of volcanic bedrock. This is often starved of pelagic sediment, and intensely altered with iron and manganese encrustations. Moving away from the mid-ocean ridge this volcanic layer is progressively blanketed by carbonate ooze. Traced further away, the sea floor slopes below the CCD so the carbonate ooze is replaced by siliceous ooze or red clay (Davies and Gorsline,

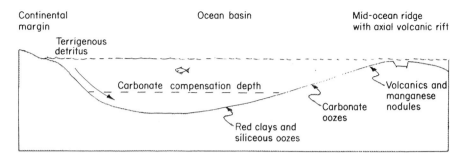

Fig. 6.75. Generalized cross-section through an ocean basin showing distribution of pelagic sediment.

1976). The clays may become gradually replaced by terrigenous sediment as the continental rise is approached or may be digested within a zone of subduction (Fig. 6.75).

6.3.2.10.2 Ancient pelagic deposits

A distinct sedimentary facies can be attributed to a pelagic environment with some confidence. It consists of interbedded laminated radiolarian cherts, micrites, and red manganiferous shales. It overlies pillow lavas of the old ocean floor and in turn is overlain by flysch facies. Characteristically this facies is generally highly tectonized. Ancient pelagic deposits have been described by Aubouin (1965), Garrison and Fischer (1969), Wilson (1969), and Leggett (1985), to mention but a few. Specific examples of this facies assemblage include the late Mesozoic "ammonitico rosso" of the Alps, the red cephalopod limestones of the Variscan trough in northern Europe, and the Franciscan cherts of California (Fig. 6.76). This type of facies assemblage is directly comparable to modern

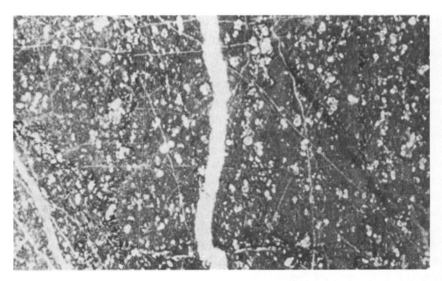

Fig. 6.76. Thin section of Franciscan (Jurassic) chert with pelagic microfossils, California. This is believed to be a deep-water deposit that underwent later intense structural deformation as shown by extensive fracturing. Extensive stylolites with black insoluble residue can also be seen (30×).

carbonate and siliceous oozes and to red manganiferous clays. Though the ancient facies can be interpreted as pelagic, their actual depositional environment is a matter for debate.

Another important ancient type of pelagic deposit that occurs throughout the stratigraphic record from the Pre-Cambrian to Recent is black shales (Wignall, 1994). These contain pelagic micro- and macrofossils. Benthonic body fossils and bioturbation are generally absent. The black color is in part due to the presence of organic carbon, but largely due to abundant disseminated pyrites (FeS_2). Noted black shales include the Lower Paleozoic Alum shales of the Baltic, the Liassic and Kimmeridgian shales of northwest Europe, the Cretaceous shales of the Atlantic Ocean margin basins, and the Devonian shales of North America. The fauna and lithology of these black shales clearly indicate deposition in anoxic conditions. These may occur in various water depths and environments that include lakes, lagoons, continental shelves, and deep oceans (Tyson and Pearson, 1991). Research has particularly concentrated on establishing the global and stratigraphic distribution of black shales. These data have been integrated with the movements of tectonic plates and computer-generated paleoclimatic models (Huc, 1995).

6.3.2.10.3 Economic aspects of pelagic deposits

Pelagic deposits are of considerable economic interest as petroleum source beds, and because they often contain phosphates, barytes, and a variety of metallic ores. Phosphate deposits will not be described here because they are dealt with elsewhere (see Section 9.5). When discussing shelf deposits it was mentioned that modern ocean beds are not anoxic, but that there is an anoxic zone between some 180 and 220 m that corresponds to the modern continental shelf margin. Modern deep oceans are moderately oxygenated because of a constant downward flow of oxygenated cold polar waters. Thus one would not normally expect ancient pelagic deposits to be organic-rich potential petroleum source beds. There appear to be two ways in which pelagic environments generate petroleum source beds (Katz, 1995). Some geologists have noted that source rock formation was a worldwide event in late Jurassic and early Cretaceous times. They have correlated this with an equable global climate. With no polar ice caps there would be no cold water descending into the deep ocean basins to keep them oxygenated. Anoxic conditions might thus favor source rock formation (Schlanger and Cita, 1982; Stein, 1986). This idea has not received universal support. In particular it is noted that most Upper Jurassic and Lower Cretaceous source beds occur around the modern Atlantic Ocean. They may therefore reflect restricted oceanic circulation as the American plates separated from Europe and Africa (see Section 10.2.4).

There is another way in which deep-water organic-rich source beds may form. In the Miocene the Mediterranean dried out to deposit basin-wide evaporites. This is termed the Messinian Salinity Crisis. Today organic-rich muds are found in deep parts of the Mediterranean, such as the Strabo trench, and the Bacino and Bannock basins. These occur because of the presence of anoxic brines at depths greater than about 3000–3200 m. Messinian evaporites crop out on the adjacent sea floor, and are leached out by the seawater. The resultant dense brines flow down into the basinal lows, where there is insufficient current activity to rework them with waters of normal salinity. Organic matter is thus preserved in these troughs, like pickled herring in a barrel.

Acounts of the diverse pelagic mineral deposits have been given by Cronan (1980, 1986). Of the nonmetallic minerals, barytes $(BaSO_4)$ is one of the most common associated with pelagic deposits, though it occurs in several geological settings, both igneous and sediment related. Bedded barite deposits are found associated with pelagic shales, limestones, and cherts in Ireland, Canada, the USA, Australia, Thailand, and India (Orris, 1986). It has been argued that the barium is derived from the solution of biogenic detritus (Scholle *et al.*, 1983b), but some bedded barite deposits are associated with Pb–Zn exhalative centers.

Studies of modern mid-oceanic ridges have led to the discovery of brine springs, whose waters are enriched in many metallic ions, notably Cu, Mn, Fe, and Pb. The emergent fluids are reducing, acid (pH about 4), and have temperatures of up to 350°C. These brine springs were first encountered in the Red Sea (Degens and Ross, 1969), but have subsequently been found on many other mid-ocean ridges. Reactions between the brines and seawater cause mineral precipitation. Amorphous mineral particles impart a cloudy appearance to the emergent spring waters, and are deposited as a chimney, sometimes several meters high, around the spring vent. These are termed **smokers.** White smokers and black smokers have been distinguished. Black smokers emit black clouds of finely disseminated pyrrhotite. They begin with the precipitation of an anhydrite chimney. This is followed by a phase of replacement of the original anhydrite and direct sulfide precipitation. Bornite, chalcopyrite, and chalcocite are the principal minerals that form (Haymon, 1983). White smokers are rather cooler, with temperatures of less than 300°C. They emit white clouds of amorphous silica, barytes, and pyrite. The resultant chimneys are encrusted with polychete worms and have a peripheral ecosystem adapted to high temperatures and salinities. Mn–oxyhydroxite crusts develop on the sea bed within a few meters of the vents (Haymon and Kastner, 1981).

These modern smokers are of great interest because they are analogs of an important group of mineral deposits, variously termed "volcanogenic," "volcano exhalative," "sedimentary-exhalative," or simply **sedex** deposits (Briskey, 1986; Guilbert and Park, 1986, pp. 572–603). These are massive sulfide ore bodies, principally of lead, zinc, copper, and iron, sometimes associated with bedded baryte deposits. They commonly occur interbedded with volcanic and volcaniclastic rocks, together with pelagic shales, limestones, and cherts. Noted examples include Rammelsberg (Germany), Mount Isa (Australia), Broken Hill (Australia and South Africa), and Silver mines (Ireland). Andrews *et al.* (1986) give a fascinating account of the analogy of modern "smokers" with the massive sulfide ores of Ireland. The similarities even include the discovery of paleosmokers complete with an encrusted fossilized worm tube (Fig. 6.77).

Moving away from the axis of mid-ocean ridges, vast areas of their flanks and of major ocean basins have been found covered by manganese nodules (Fig. 6.78). There is a gradual change in the ferromagnesian ratio of these nodules away from the ridge axis (Figure 6.79). Manganese nodules form very slowly from oxygenated seawater of normal salinity (Cronan, 1980). The foregoing brief review shows that pelagic deposits are of great commercial interest.

6.4 SEDIMENTARY MODELS, INCREMENTS, AND CYCLES

The preceding pages have described the sedimentary models that are recognized as major facies generators. It is now appropriate to review the principles behind this approach

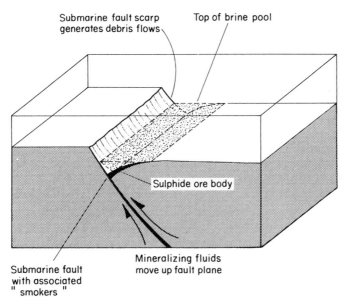

Submarine fault scarp
generates debris flows

Top of brine pool

Sulphide ore body

Submarine fault
with associated
" smokers "

Mineralizing fluids
move up fault plane

Fig. 6.77. Cartoon illustrating the formation of "sedex" (sedimentary-exhalative) massive sulfide ore bodies. The regional setting is a mid-oceanic ridge, with crustal thinning, rifting, and high heat flow. Metal-enriched brines emerge from vents along fault-planes forming "smokers." Metallic-sulfide precipitation occurs in the adjacent brine pools entrapped in the rollover hollow on the downthrown side of the fault. For modern and ancient analogs see text. (Distilled from Andrews *et al.,* 1986.)

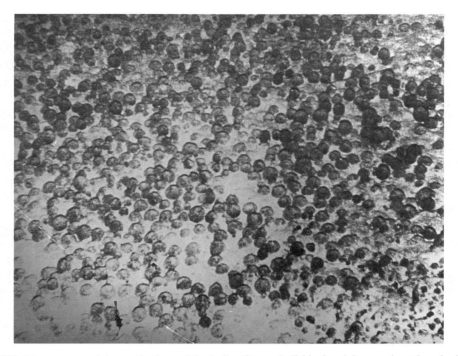

Fig. 6.78. Manganese nodules on the floor of the Indian Ocean. Individual nodules are about 8 cm in diameter. (From Cronan, 1980, Fig. 88. Courtesy of Academic Press.)

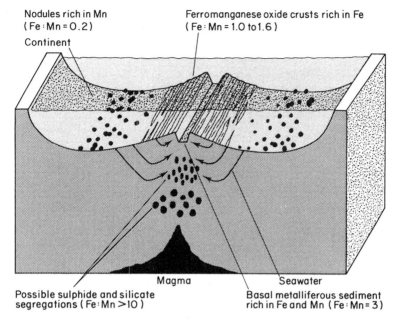

Nodules rich in Mn
(Fe:Mn = 0.2)

Continent

Ferromanganese oxide crusts rich in Fe
(Fe:Mn = 1.0 to 1.6)

Magma

Seawater

Possible sulphide and silicate
segregations (Fe:Mn >10)

Basal metalliferous sediment
rich in Fe and Mn (Fe:Mn= 3)

Fig. 6.79. Diagram showing the association of manganese nodules with mid-ocean ridges. (From Cronan, 1980, Fig. 67. Courtesy of Academic Press.)

before examining these sedimentary models in their broader stratigraphic context. The fundamental principle behind this chapter is that a finite number of sedimentary environments generate a finite number of distinctive facies. This concept is a good servant, but a bad master. It is useful as a means of communicating ideas, so long as one geologist's submarine valley/fan model is similar to their colleagues'. It is dangerous if one person's reef is another's bioherm. Sedimentary models are useful for demonstrating the relationship between physiography, process, and resultant facies. As students gain experience with real rocks, they must accommodate the model to fit the rocks, not the other way round. One must not be afraid to discard the model altogether if it proves untenable in the face of hard data.

One very important principle can be learned from the sedimentary models described in this chapter. Example after example shows a pattern of environments that prograde sideways to deposit a series of facies arranged in a predictable vertical sequence. This concept is the key to modern dynamic stratigraphic analysis and is now discussed in some detail.

6.4.1 Walther's Law

Walther (1893, 1894) continued the development of the facies concept begun by Prevost and Gressly (Middleton, 1973). He recognized that facies are seldom randomly arranged. Environmental analysis shows that vertical sections of strata originated in sequences of environments that are seen side by side on the earth's surface today. Walther coined the term "faciesbezirk" (facies tract) for a conformable vertical sequence of genetically related facies. Walther distinguished environments of erosion, equilibrium, and deposition and included intraformational erosional intervals as an integral part of a facies tract. Walther realized the significance of this arrangement, writing (1894, p. 979):

> The various deposits of the same facies area and similarly, the sum of the rocks of different facies areas, were formed beside each other in space, but in a crustal profile we see them lying on top of each other . . . it is a basic statement of far-reaching significance that only those facies and facies areas can be superimposed, primarily, that can be observed beside each other at the present time. (translation from Blatt *et al.,* 1980, p. 187)

This principle is termed **Walther's law,** and may be succinctly stated as "a conformable vertical sequence of facies was generated by a lateral sequence of environments." This principle is a vital one in environmental analysis. One of the ways of identifying the environment of a facies is by analyzing the environments of the facies above and below to come up with the most logical sequence of events.

6.4.2 Genetic Increments and Sequences of Sedimentation

Walther's concept of "faciesbezirk" has been widely used in facies analysis, though it has been renamed **systems tract** (Brown and Fisher, 1977). This idea can be usefully extended to consider not just a vertical sequence, but a whole body of rock; what has been termed a genetic increment of strata (Busch, 1971).

   A **genetic increment** of strata is a mass of sedimentary rock in which the facies or subfacies are genetically related to one another. A typical genetic increment of strata would consist of a single prograding delta sequence containing delta platform, delta-front, and prodelta deposits (Fig. 6.80).

   A **genetic sequence** of strata includes more than one increment of the same genetic type. These terms were developed to aid the subsurface mapping of deltaic deposits. The isopach mapping of genetic increments of strata define individual delta lobes. Iso-

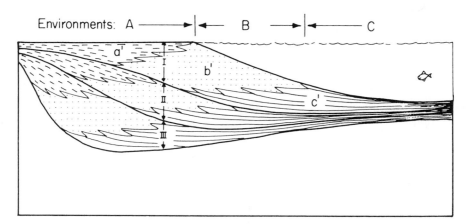

Fig. 6.80. Cross-section illustrating the relationship between environments, facies, and increments of sedimentation, as defined by Busch (1971). Environments A, B, and C deposited facies a', b', and c', respectively, in the increment of sedimentation designated I. The increments I, II, and III together constitute a genetic sequence of strata. In sequence stratigraphic terminology these would be referred to as parasequences, and a parasequence set, respectively.

pach maps of genetic sequences of strata define larger morphologic units of shelf, hinge line, and basin.

It is apparent though, from the data in this chapter, that increments of sedimentation are generated by many sedimentary models. Therefore, incremental mapping is actually feasible with many sedimentary facies. The concept of incremental mapping, though seldom the formal terminology, is widely applied in facies analysis. It forms an integral part of environmental and paleogeographic analyses to locate favorable hydrocarbon reservoir facies.

6.4.3 Sequence Stratigraphy

The depositional environments of ancient sedimentary rocks is normally carried out simultaneously with a stratigraphic analysis of the formations studied. It is necessary, therefore, to consider the relationship between facies analysis and stratigraphy. For over a century fossils provided a vital link between these two pursuits, paleoecology aiding environmental interpretation, and biostratigraphy establishing the chronology. Now, however, paleontology is joined by seismic surveying as a linking tool. The advent of modern high-quality seismic data has lead to a new discipline, known by its advocates as seismic sequence stratigraphy (Payton, 1977; Berg and Wolverton, 1985; Wilgus *et al.*, 1988; Steele *et al.*, 1995; Emery and Myers, 1996; Miall, 1997).

Sequences of sedimentary rocks may be differentiated into units, and the units correlated from section to section in many different ways. These include facies, lithostratigraphy, and chronostratigraphy. **Facies** are defined by their geometry, lithology, sedimentary structures, paleocurrent pattern, and paleontology. In the subsurface facies may also be characterized by their geophysical properties, such as log profile and seismic character (Selley, 1996). Facies may be subdivided into subfacies, and genetic increments and sequences may be recognized.

The **lithostratigraphy** delineates mappable rock units arranged in a hierarchy of groups, formations, and members. The **chronostratigraphy** is commonly based on biostratigraphy, and can be used to attribute the rocks to geological systems, stages, or series. These units can be divided in to upper, middle, and lower, as appropriate.

The foregoing are essentially observational activities. They form the basis for subsequent interpretations. The facies may be studied to interpret their environments of deposition. The chronostratigraphy may be interpreted to establish geochronology. Geochronologic units are the intervals of time during which a particular chronostratigraphic unit was deposited. They are arranged in the hierarchy of periods, epochs, or ages. Geochronologic units may be subdivided into early, middle, and late, as appropriate. A useful way of distinguishing between chronostratigraphic and geochronologic units is to consider an hourglass, or its miniaturized version, the egg timer. The sand that flows through the hourglass is the chronostratigraphic unit. The time taken is the geochronologic unit.

Sequence stratigraphy is basically "a geologic approach to the stratigraphic interpretation of seismic data" (Vail *et al.,* 1977, p. 51). Of fundamental importance is the seismic delineation of depositional sequences. A depositional sequence is defined as a "stratigraphic unit composed of a relatively conformable succession of genetically related strata, bounded at top and bottom by unconformities or their correlative conformities" (Vail *et al.,* 1977, p. 53).

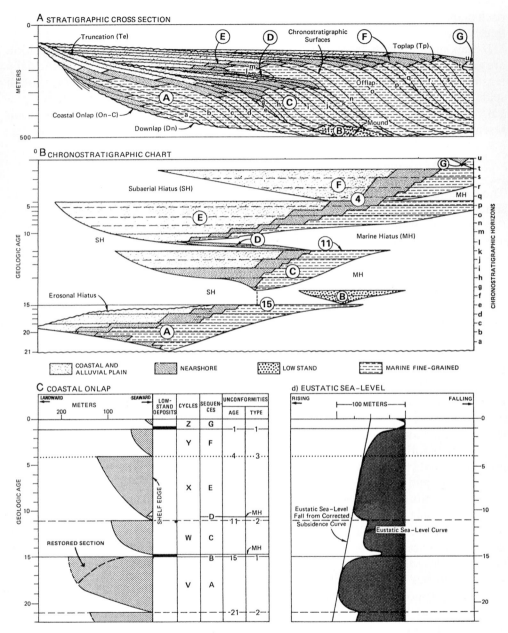

Fig. 6.81. Diagrammatic cross-section illustrating the procedure for constructing regional coastal onlap and eustatic curves from unconformity characteristics, stratal patterns, and facies relations. (A) Stratigraphic cross-section, (B) chronostratigraphic chart, and (C) coastal onlap. (From Vail, P. R., and Todd, R. G. 1981. Northern North Sea Jurassic unconformities, chronostratigraphies, and sea level changes from seismic stratigraphy. *In* "Petroleum Geology of the Continental Shelf of Northwest Europe" (L. V. Iling and D. G. Hobson, eds.), pp. 216–235, Fig. 2. Courtesy of the Institute of Petroleum.)

The sedimentary facies of a depositional sequence are grouped into three systems tracts (a translation of Walther's *faciesbezirk*). These are defined according to relative sea level at the time of deposition, namely, **low stand, transgressive,** and **high stand** tracts, corresponding to low, rising, and highest sea level (Fig. 6.81). The highest sea level is termed a **maximum flooding surface.** This normally results in the deposition of a laterally extensive **condensed sequence.** High stands correspond to what used to be termed transgressions, and low stands, to regressions. Similarly Busch's terms "genetic increment" and "genetic sequence," defined earlier, are renamed **parasequence** and **parasequence set,** respectively (Van Wagoner *et al.,* 1988, 1990). One depositional sequence may be composed of the deposits of high and low stand systems tracts (transgressions and regressions) arranged in many parasequences (increments). Deposition will only take place where there is **accommodation space,** the vertical interval between sea level and the sea floor.

Depositional sequences are associated with two types of chronostratigraphic surface. Bedding, or stratal, surfaces occur within a sequence. Unconformities, and their relative correlative conformities, occur at, and define, **sequence boundaries.** Unconformities may be recognized on seismic sections in their own right. Their presence may also be inferred where stratal surfaces onlap, toplap, offlap, downlap, or are truncated by an unconformity (these terms are illustrated in Fig. 6.81).

Unconformities are of three types, illustrated in Fig. 6.82, and are related to high stands and low stands of sea level:

- *Type 1 unconformities* occur when sea level is below the shelf edge, leading to the extensive erosion of the shelf, with incision by fluvial channels.
- *Type 2 unconformities* occur when sea level is at or landward of the shelf edge, with overlying transgressive marine sediments or deeper water deposits and no evidence of a low stand systems tract.
- *Type 3 unconformities* occur when sea level is at or landward of the shelf edge, with coastal deposits above and below the unconformity, either with coastal deposits above and below the unconformity, or with deeper water deposits overlain by shallower.

It is argued that these three types of unconformity can be recognized and correlated in sedimentary basins around the world. Vail *et al.* (1977) and Haq *et al.* (1988) believe that these unconformities correspond to global changes of sea level, and that it is possible to construct a curve of global changing sea level (Fig. 6.83). Type 1, 2, and 3 unconformities are used to define a hierarchy of cycles with periodicities of 15, 11, and 4 my, according to (Vail and Todd, 1981), or 50, 3–50, and 0.5–3 my, according to Neal *et al.* (1993). The first-order cycle is attributed to continental breakup, the second to changes in the rate of subsidence and uplift due to plate movements, the third is the sequence cycle, described more fully later. Fourth and higher order cycles are attributed to climatic changes.

The third-order sequence cycle is illustrated in Plate 2. Conventionally the cycle starts with the lowest sea level at which time a low stand systems tract overlies a type 1 unconformity (Plate 2A). Sandy lobate submarine fans are deposited at the foot of the continental slope. As sea level begins to rise turbidite sands and shales are deposited on the continental slope (Plate 2B). Continuing submergence increases the accommodation space between basin floor and sea level, allowing upward-coarsening deltas to prograde out across the basin (Plate 2C). Once sea level has risen over the shelf edge it will migrate rapidly across the shelf, cutting off sediment supply to the basin, and deposit a

STRATIGRAPHIC AND FACIES PATTERNS

INDICATED EUSTATIC SEA–LEVEL CHANGE

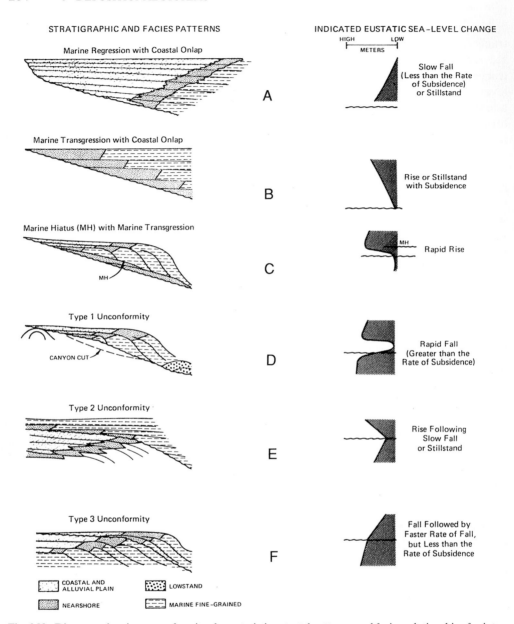

Fig. 6.82. Diagrams showing unconformity characteristics, stratal patterns, and facies relationships for interpreting eustatic sea level changes. [From Vail, P. R., and Todd, R. G. 1981. Northern North Sea Jurassic unconformities, chronostratigraphies, and sea level changes from seismic stratigraphy. *In* "Petroleum Geology of the Continental Shelf of Northwest Europe" (L. V. Iling and D. G. Hobson, eds.), pp. 216–235, Fig. 3. Courtesy of the Institute of Petroleum.]

thin transgressive systems tract of shallow marine sands (Plate 2D). The top of the transgressive systems tract, termed the maximum flooding surface, is marked by a condensed sequence. As the rate of sea level rise slows down, a thin high stand systems tract is deposited, with fluvial sands on land, sandy shoreline sands, and open marine shales (Plate 2E).

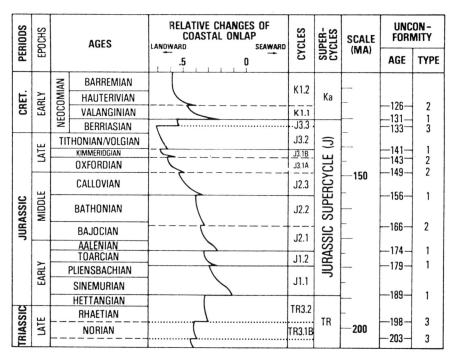

Fig. 6.83. Global sea level curve for the Late Triassic to Early Cretaceous. [From Vail, P. R., and Todd, R. G. 1981. Northern North Sea Jurassic unconformities, chronostratigraphies, and sea level changes from seismic stratigraphy. *In* "Petroleum Geology of the Continental Shelf of Northwest Europe" (L. V. Iling and D. G. Hobson, eds.), pp. 216–235, Fig. 3. Courtesy of the Institute of Petroleum.]

It is important to note that the sequence stratigraphic concepts outlined here were developed from the study of terrigenous sediments, principally in continental margins where there was little local tectonic disturbance. They have also been extended, with some modifications, to inland continental basins, where an equilibrial surface, other than sea level must be established (Jervey, 1988).

Sequence stratigraphic concepts have also been extended to carbonate sequences (Loucks and Sarg, 1993; Tucker *et al.*, 1990). In carbonate environments, as with terrigenous shelf margins, the upper limit of accommodation space is defined by sea level, but the source of sediment supply is intrabasinal, rather than extrabasinal. Spring and Hansen (1998) have developed sequence stratigraphic variants for both carbonate rimmed platforms and ramps (Plate 8).

Sequence stratigraphy, as briefly outlined earlier, is just one way of splitting up and correlating rock sequences, on a par with lithostratigraphy, biostratigraphy, magneto-stratigraphy, and so forth. It has been a stimulating approach for geologists in general and for sedimentologists in particular. Nonetheless the methodology and conclusions have been criticized.

There are two points that deserve particular scrutiny: that local changes in shorelines reflect global changes in sea level, and that there is a periodicity to sea level change.

Many geologists have commented on the global uniformity of sedimentary formations (e.g., Ager, 1993). Concomitantly many geologists have noted global episodes of sea level change (e.g., Stille, 1924, 1936; Sloss, 1963; Dott, 1994). Not everyone is prepared to accept, however, that sea level changes are global and eustatic. There is plenty of evidence for global climatic change, and evidence for global changes in sea level. Every geologist knows, however, that sea level varies at a point in response to:

1. Global changes in ocean water volume — probably driven by climatic change
2. Local tectonic uplift and subsidence of the crust — probably driven by plate movements
3. Local changes in the availability of sediment supply — probably driven by climatic change, changing rates of sediment runoff, and the initiation and extinction of drainage systems.

Sadly it is normally impossible to determine which of these mechanisms, singly or together, may be responsible for a particular sea level change in a given area. There is obviously no universal benchmark or Plimsoll line against which global changes in sea level may be calibrated, and which may be used therefore to differentiate the various processes of sea level change. This point is demonstrated by many modern deltas. The modern Mississippi (Fig. 6.33) and the Arctic fan deltas of Spitzbergen (Martinsen, 1995) show that regressions (high stand systems tract) occur at the active river mouth. Simultaneously, a few kilometers away, the sea advances across a subsiding compacting abandoned delta lobe depositing a transgressive systems tract.

6.4.4 Cyclostratigraphy

Cyclicity, the repetition of genetic increments of strata, is a common feature in many stratigraphic sections. This is a topic that has been widely studied and written about for many years (e.g., Merriam, 1965; Duff *et al.,* 1967; Schwarzacher, 1975). Twenhofel remarked in 1926 "The finding of rhythms by geologists seems at the present time to be a pleasant diversion" (p. 612).

This chapter has decribed many sedimentary models that produce predictable facies sequences. The formation of genetically related sequences is an obvious and integral part of sedimentation. The origin of cyclicity, the phenomenon of repeated sequences, has aroused lively debate. Beerbower (1964) made an important distinction between **autocyclic** mechanisms, which are an integral part of the sedimentary model, and **allocyclic** mechanisms, which originate outside the model. The to-and-fro migration of a river channel and the switching of a delta distributary can generate cyclic sedimentation in a steadily subsiding sedimentary basin. These are examples of autocyclicity. Not all sedimentary models have this ability, however, and external mechanisms must be sought to explain cyclicity where it is present (Frostick and Steel, 1994).

Mechanisms capable of causing allocyclic sedimentation can be seen operating in the recent past. The worldwide distribution of raised beach deposits, and relict terrestrial sediments on continental shelves, provide evidence of sea level changes correlatable with the waxing and waning of the ice caps. Individual raised beaches are not always of constant elevation when traced along the coast. These testify to tectonic tilting of the earth's crust, sometimes in isostatic response to the vanishing load of ice caps. Cyclic climatic variations are known both in historic times and, from pollen analyses, from prehistoric time. These facts show that transgressions and regressions capable of generating sedimentary cycles may occur due to climatic, eustatic, and tectonic changes. It is not always possible, however, to detect which of these mechanisms caused any specific instance of cyclicity. Furthermore, as at the present time, several cycle mechanisms may operate simultaneously. In deltaic deposits there will be local autocyclic motifs due to delta switching, on which may be superimposed more extensive cycles due to allocyclic mechanisms.

One of the simplest tests for cyclicity is to construct a transition matrix array that records the number of times beds of one type overlie the other types of bed (e.g., sandstone, limestone, shale). Then a transition probability matrix is constructed to show the number of expected transitions, given the numbers of each type of bed in the section. Finally, a matrix is prepared by subtracting predicted from observed transitions (Fig. 6.84).

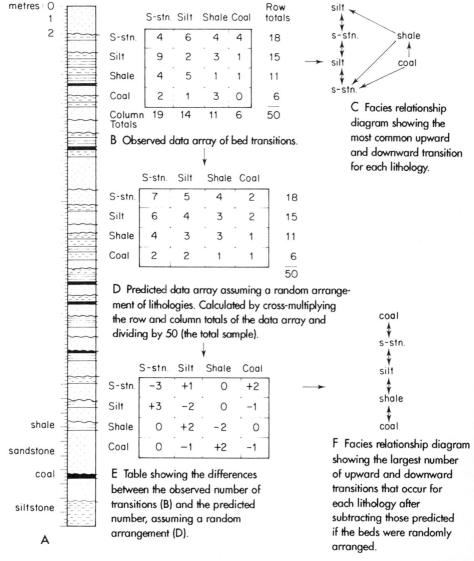

Fig. 6.84. Diagram illustrating how to test for cyclicity in sedimentary sequences using a simple mathematical device. For explanation and qualifications see text. (From Selley, 1970. Courtesy of the Geological Society of London.)

This reveals the number of times that beds overlie other bed types more often than would be expected if sedimentation took place randomly (Selley, 1970). Thus can any regular sedimentary cycle be identified. This technique has both stimulated and irritated geologists for many years (Harper, 1984; Walker, 1984a). One of the main problems is that, when measuring sedimentary sections, few geologists record the same facies overlying itself, except perhaps multistory channel sands. Thus the diagonals in the matrix array record zeros, something that disturbs the mathematical mind. This problem is simply overcome by recording bed types at regular vertical intervals down the sequence, and recording the resultant transitions accordingly.

In recent years studies of cyclicity in sedimentary successions has led to the establishment of what is termed **cyclostratigraphy** (Schwarzacher, 1993; de Boer and Smith,

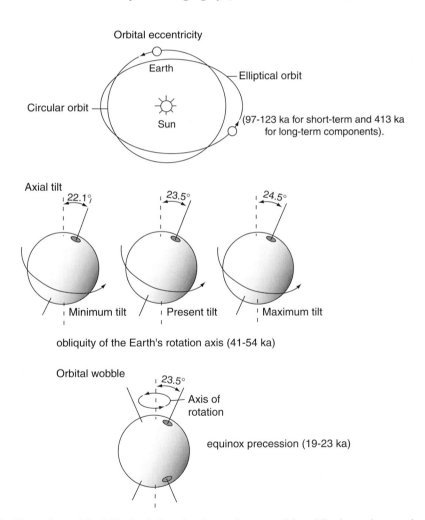

Fig. 6.85. Illustrations of the Milankovitch cycles due to the eccentricity, obliquity, and precessional cycles of the earth. These in combination may affect the earth's sedimentary successions. (From Kroon, D., Norris, R. D., Klaus, A., and the ODP Leg 171B Shipboard Scientific Party, 1998. Drilling Blake Nose: The search for evidence of extreme Paleogene-Cretaceous climates and extraterrestrial events. *Geol. Today* **14,** 222–225, Fig. 4. Courtesy of Blackwell Science.)

1994; House and Gale, 1995). Since the work of Milankovitch (1941) it has been recognized that extraterrestrial processes impact sedimentation. Three types of Milankovitch cycles are recognized (Fig. 6.85). The precession of the earth's axis has a cycle with two peaks of 19–23 ka, the obliquity, or tilt of the earth's axis, has a cycle of 41 ka, and the eccentricity of the earth's orbit changes shape in 100- and 400-ka cycles (House and Gale, 1995). Each one of these impacts the earth system, the various signals combining to produce what may appear random to the human eye. Many studies of cyclic sequences have distinguished hierarchical systems of "mega cyclothems," "cyclothems," and "microcyclothems." Not all of these are imaginary. There are a number of mathematical techniques for statistically testing a section for cyclicity and for determining cycle wavelength (Krumbein, 1967; Merriam, 1967). Gale (1998) has reviewed the diverse statistical gymnastics that may be employed to separate the various astronomical cycles from the earthly residual signal (Fig. 6.86).

In conclusion it is commonplace to find predictable sequences within genetic increments of strata. Cyclicity is the repetition of these sequences. Autocyclic sedimentation is an integral part of several sedimentary models. Allocyclic sedimentation occurs from time to time in almost all deposits irrespective of environment; it is caused by a variety of mechanisms. More than one cyclic process may control the deposition of a sedimentary section, including autocyclic, terrestrial, and extraterrestrial mechanisms.

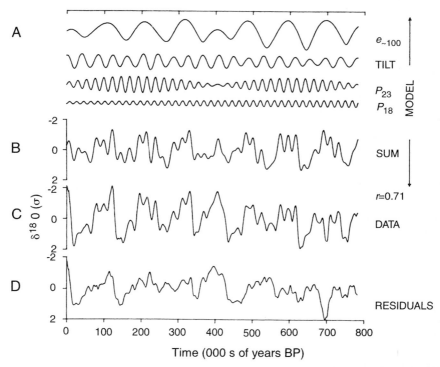

Fig. 6.86. Variations in the earth–sun system. (A) Oscillations caused by eccentricity, obliquity, and precession. (B) The sum of the four signals, a measure of solar energy received by the earth. (C) The oxygen isotope record, a manifestation of the Milankovitch signal. (D) The residual products after deducting B from C. [From Gale, A. S. 1998. Cyclostratigraphy. *In* "Unlocking the Stratigraphic Record" (P. Doyle and M. R. Bennett, eds.), pp. 195–220, Fig, 7.3. Copyright John Wiley & Sons Limited. Reproduced with permission.]

SELECTED BIBLIOGRAPHY

For an introduction to stratigraphy see:

Doyle, P., and Bennett, M. R., eds. (1998). "Unlocking the Stratigraphic Record." Wiley, Chichester. 532pp.

For sequence stratigraphy see:

Emery, D., and Myers, K. (1995). "Sequence Stratigraphy." Blackwell Science, Oxford. 320pp.
Miall, A. D. (1997). "The Geology of Stratigraphic Sequences." Springer-Verlag, Berlin. 433pp.

For modern sedimentary environments and the diagnosis of ancient ones see:

Reading, H. G., ed. (1996). "Sedimentary Environments: Processes, Facies and Stratigraphy," 3rd ed. Blackwell, Oxford. 688pp.

And to do it underground see:

Selley, R. C. (1996). "Ancient Sedimentary Environments: And their Subsurface Diagnosis," 4th ed. Chapman & Hall, London. 300pp.

REFERENCES

Achauer, C. W. (1969). Origin of Capitan Formation, Guadalupe Mountains, New Mexico and Texas. *Am. Assoc. Pet. Geol. Bull.* **53,** 231–2323.
Ager, D. V. (1993). "The Nature of the Stratigraphic Record." 3rd ed. Wiley, Chichester. 151pp.
Ahlbrandt, T. S., and Fryberger, S. G. (1982). Eolian deposits. *Mem. — Am. Assoc. Pet. Geol.* **31,** 11–48.
Ahr, W. M. (1973). The carbonate ramp: An alternative to the shelf model. *Trans. Gulf Coast Assoc. Geol. Soc.* **23,** 221–225.
al-Laboun, A. A. (1986). Stratigraphy and hydrocarbon potential of the Paleozoic succession in both the Tabuk and Windayan basins, Arabia. *Mem. — Am. Assoc. Pet. Geol.* **40,** 373–398.
Allen, J. R. L. (1964). Studies in fluviatile sedimentation: Six cyclothems from the Lower Old Red Sandstone, Anglo-Welsh Basin. *Sedimentology* **3,** 163–198.
Allen, J. R. L. (1965). A review of the origin and characteristics of Recent Alluvial sediments. *Sedimentology, Spec. Issue* **5,** No. 2.
Allen, J. R. L., and Friend, P. F. (1968). Deposition of the Catskill facies, Appalachian region: with notes on some other Old Red Sandstone basins. *Spec. Pap. — Geol. Soc. Am.* **206,** 21–74.
Anadon, P., Cabrera, L. L., and Kelts, K., eds. (1991). "Lacustrine Facies Analysis." Blackwell, Oxford. 328pp.
Anderson, G. M., and McQueen, R. W. (1982). Ore deposit models 6. Mississippi Valley-Type lead-zinc deposits. *Geosci. Can.* **9,** 108–117.
Anderton, R. (1985). Clastic facies models and facies analysis. *In* "Sedimentology: Recent Advances and Applied Aspects" (P. J. Brenchley and B. P. J. Williams, eds.), pp. 31–48. Blackwell, Oxford.
Andrews, C. J., Crowe, R. W., Finlay, S., Pennell, W. M., and Payne, J. F., eds. (1986). "Geology and Genesis of Mineral Deposits in Ireland." Irish Association for Economic Geology, Dublin. 704pp.
Apel, J. R. (1987). "Principles of Ocean Physics." Academic Press, London. 500pp.
Armstrong, F. C., ed. (1981). "Genesis of Uranium and Gold-bearing Precambrian Quartz-pebble Conglomerates," Prof. Pap. 1161-A-BB. U.S. Geol. Surv., Washington, DC.
Armstrong-Price, W. (1963). Patterns of flow and channelling in tidal inlets. *J. Sediment. Petrol.* **33,** 279–290.

Arrhenius, G. (1963). Pelagic sediments. *In* "The Sea" (M. N. Hill, ed.), Vol. 3, pp. 655–727. Interscience, New York.

Asquith, D. O. (1970). Depositional topography and major marine environments, late Cretaceous, Wyoming. *Am. Assoc. Pet. Geol. Bull.* **54,** 1184–1224.

Atkinson, C. D., McGowen, J. H., Bloch, S., Lundell, L. L., and Trumbly, P. N. (1990). Braidplain and Deltaic reservoir, Prudhoe bay Field, Alaska. *In* "Sandstone Petroleum Reservoirs" (J. H. Barwis, J. G. McPherson, and J. R. J. Studlick, eds.), pp. 7–30. Springer-Verlag, Berlin.

Aubouin, J. (1965). "Geosynclines." Elsevier, Amsterdam. 335pp.

Bache, J. J. (1987). "World Gold Deposits." North Oxford Academic, London. 192pp.

Balducci, A., and Pommier, A. (1970). Cambrian oil field of Hassi Messaoud, Algeria. *Mem.— Am. Assoc. Pet. Geol.* **14,** 477–488.

Ball, M. M. (1967). Carbonate sand bodies of Florida and the Bahamas. *J. Sediment. Petrol.* **37,** 556.

Banner, F. T., Dollins, M. B., and Massie, K. S., eds. (1979). "The North-West European Shelf Seas: The Sea Bed and the Sea in Motion. I. Geology and Sedimentology." Elsevier, Amsterdam. 300pp.

Barwis, J. H., McPherson, J. G., and Studlick, J. R. J., eds. (1990). "Sandstone Petroleum Reservoirs." Springer-Verlag, Berlin. 583pp.

Beerbower, J. R. (1964). Cyclothems and cyclic depositional mechanisms in alluvial plain sedimentation. *Bull. Kans. Univ. Geol. Surv.* **169,** 35–42.

Belderson, R. H., Kenyon, N. H., and Stride, A. H. (1971). Holocene sediments on the continental shelf west of the British Isles. *In* "The Geology of the East Atlantic Continental Margin," Vol. 2, No. 70/14, pp. 160–170. Eur. Inst. Geol. Sci., Berlin.

Bennacef, A., Beuf, S., Biju-Duval, B., de Charpal, O., Gariel, O., and Rognon, P. (1971). Example of cratonic sedimentation: Lower Paleozoic of Algerian Sahara. *Am. Assoc. Pet. Geol. Bull.* **55,** 2225–2245.

Berg, R. R., and Davies, D. K. (1968). Origin of Lower Cretaceous Muddy sandstone at Bell Creek Field, Montana. *Am. Assoc. Pet. Geol. Bull.* **52,** 1888–1898.

Berg, O. R., and Wolverton, D. G., eds. (1985). Seismic Stratigraphy II. An Integrated Approach. *Mem.— Am. Assoc. Pet. Geol.* **39,** 276pp.

Berger, W. H. (1974). Deep-sea sedimentation. *In* "The Geology of Continental Margins" (C. A. Burke and C. L. Drake, eds.), pp. 213–241. Springer-Verlag, Berlin.

Bernard, H. A., Le Blanc, R. J., and Major, C. F. (1962). Recent and Pleistocene geology southwest Texas. *In* "Geology of Gulf Coast and Central Texas," pp. 175–224. Houston Geol. Soc., Houston, TX.

Best, J. L., and Bristow, C. S. (1993). "Braided Rivers," Spec. Publ. 75. Geol. Soc. London, London. 528pp.

Bigarella, J. J. (1979). Botucatu and Sambaiba Sandstones of South America (Jurassic and Cretaceous). *Geol. Surv. Prof. Pap. (U.S.)* **1052,** 233–236.

Bigarella, J. J., and Salamuni, R. (1961). Early Mesozoic wind patterns as suggested by dune bedding in the Botucatu Sandstone of Brazil and Uruguay. *Geol. Soc. Am. Bull.* **72,** 1089–1106.

Black, K. S., Paterson, D. M., and Cramp, A., eds. (1998). "Sedimentary Processes in the Intertidal Zone," Spec. Publ. No. 139. Geol. Soc. London, London.

Blackwelder, E. (1928). Mudflow as a geologic agent in semi-arid mountains. *Geol. Soc. Am. Bull.* **39,** 465–483.

Blatt, H., Middleton, G., and Murray, R. (1980). "Origin of Sedimentary Rocks," 2nd ed. Prentice-Hall, Englewood Cliffs, NJ. 782pp.

Blissenbach, E. (1954). Geology of alluvial fan in Southern Nevada. *Geol. Soc. Am. Bull.* **65,** 175–190.

Bluck, B. J. (1964). Sedimentation of an alluvial fan in southern Nevada. *J. Sediment. Petrol.* **34,** 395–400.

Bluck, B. J. (1967). Deposition of some Upper Old Red Sandstone conglomerates in the Clyde area: A study in the significance of bedding. *Scott. J. Geol.* **3**(2), 139–167.

Bouma, A. H., Berryhill, H. L., Knebel, H. J., and Brenner, R. L. (1982). Continental shelves. *Mem.— Am. Assoc. Pet. Geol.* **31,** 281–328.

Boyd, D. R., and Dyer, B. F. (1966). Frio Barrier Bar System of S. Texas. *Bull. Am. Assoc. Pet. Geol.* **50,** 170–178.

Bradley, W. H. (1948). Limnology and the Eocene lakes of the Rocky Mountain region. *Geol. Soc. Am. Bull.* **59,** 635–648.

Braithwaite, C. J. R. (1973). Reefs: Just a problem of semantics? *Am. Assoc. Pet. Geol. Bull.* **57,** 1100–1116.

Breed, C. S., Fryberger, S. C., Andrews, S., McCauley, C., Lennartz, F., Gebel, D., and Horstman, K. (1979). Regional studies of sand seas using Landsat (ERTS) imagery. *Geol. Surv. Prof. Pap. (U.S.)* **1052,** 305–398.

Briskey, J. A. (1982). Descriptive model of southeast Missouri Pb-Zn. *Geol. Surv. Bull. (U.S.)* **1693,** 220–221.

Briskey, J. A. (1986). Descriptive model of sedimentary exhalative Zn-Pb. *Geol. Surv. Bull. (U.S.)* **1693,** 211–212.

Brock, B. B., and Pretorius, D. A. (1964). Rand basin sedimentation and tectonics. *In* "The Geology of some Ore Deposits of Southern Africa," (S.H. Houghton, ed.), Vol. 1, pp. 549–559. Geol. Soc. South Africa, Johannesburg.

Brown, J. S. (1968). "Genesis of Stratiform Lead-Zinc-Barite Fluorite Deposits," Monogr. No. 3. Econ. Geol., Blacksberg, VA. 443pp.

Brown, S. (1990). The Jurassic. *In* "Introduction to the Petroleum Geology of the North Sea." Third ed. (K.W. Glennie, ed.), 219–254. Blackwell Scientific. Oxford.

Brown, L. F., and Fisher, W. L. (1977). Seismic-stratigraphic interpretation of depositional systems: Examples from Brazilian rift and pull-apart basins. *In* "Seismic Stratigraphy Applications to Hydrocarbon Exploration" (C. E. Payton, ed.), *Am. Assoc. Petrol. Geol.,* **26,** 213–248.

Burgis, M. J., and Morris, P. (1987). "The Natural History of Lakes." Cambridge University Press, Cambridge, UK. 232pp.

Burke, K. (1972). Longshore drift, submarine canyons, and submarine fans in development of Niger delta. *Bull. Am. Assoc. Pet. Geol.* **56,** 1975–1983.

Burke R. A. (1958). Summary of oil occurrence in Anahuac and Frio Formations of Texas and Louisiana. *Am. Assoc. Pet. Geol. Bull.* **42,** 2935–2950.

Busch, D. A. (1971). Genetic units in delta prospecting. *Am. Assoc. Pet. Geol. Bull.* **55,** 1137–1154.

Cant, D. J. (1982). Fluvial facies models. *Mem.—Am. Assoc. Pet. Geol.* **31,** 115–138.

Carter, R. W. G., ed. (1995). "Coastal Evolution." Cambridge University Press, Cambridge, UK. 490pp.

Chambers (1972). "Twentieth Century Dictionary." T & A Constable. Edinburgh, Scotland. 1649pp.

Clemmey, H. (1985). Sedimentary ore deposits. *In* "Sedimentology: Recent Developments and Applied Aspects" (P. J. Brenchley and B. P. J. Williams, eds.), pp. 229–248. Blackwell, Oxford.

Clifford, H. G., Grund, R., and Musrati, H. (1980). Geology of a stratigraphic giant: Messla oil field, Libya. *Mem.—Am. Assoc. Pet. Geol.* **30,** 507–522.

Coleman, J. M., and Gagliano, S. M. (1965). Sedimentary structure: Mississippi River deltaic plain. *Spec. Publ.—Soc. Econ. Palaeontol. Mineral.* **12,** 133–148.

Coleman, J. M., and Prior, D. B. (1982). Deltaic environments. *Mem.—Am. Assoc. Pet. Geol.* **31,** 139–178.

Coleman, J. M., Gagliano, S. M., and Smith, W. C. (1970). Sedimentation in a Malaysian high tide tropical delta. *Spec. Publ.—Soc. Econ. Paleontol. Mineral.* **15,** 185–197.

Collacicchi, R., and Baldanza, A. (1986). Carbonate turbidites in a Mesozoic pelagic basin: Scaglia Formation, Appenines-comparison with siliciclastic depositional models. *Sediment. Geol.* **48,** 81–106.

Collinson, J. D. (1996). Alluvial sediments. *In* "Sedimentary Environments: Processes, Facies and Stratigraphy" (H. G. Reading, ed., 3rd ed.), pp. 37–82. Blackwell, Oxford.

Collinson, J. D., and Lewin, J., eds. (1983). "Modern and Ancient Fluvial Systems," Spec. Pub. No. 6. Int. Assoc. Sedol.

Cook, H. E., and Mullins, H. T. (1983). Basin margin. *Mem.—Am. Assoc. Pet. Geol.* **33,** 539–618.

Cox, D. P. (1986). Descriptive model of quartz-pebble conglomerate Au-U. *Geol. Surv. Bull. (U.S.)* **1693,** 1–379.

Cronan, D. S. (1980). "Underwater Minerals." Academic Press, London. 362pp.

Cronan, D. S., ed. (1986). "Sedimentation and Mineral Deposits in the Southwestern Pacific Ocean." Academic Press, London. 344pp.

Crosby, E. J. (1972). Classification of sedimentary environments. *Spec. Publ.— Soc. Econ. Paleontol. Mineral.* **16,** 1–11.

Curray, J. R. (1965). Late Quaternary history, continental shelves of the United States. *In* "Quaternary of the United States" (H. E. Write and D. C. Frey, eds.). pp. 723–735. University Press, Princeton, NJ.

Curtis, D. M., ed. (1978). "Environmental Models in Ancient Sediments," Soc. Econ. Paleontol. Mineral. Repr. Ser. No. 6. 240pp.

Darwin, C. (1837). On certain areas of elevation and subsidence in the Pacific and Indian Oceans as deduced from the study of coral formations. *Proc. Geol. Soc. London* **2,** 552–554.

Davies, D. K., Ethridge, F. G., and Berg, R. R. (1971). Recognition of barrier environments. *Am. Assoc. Pet. Geol. Bull.* **55,** 550–565.

Davies, J. L. (1973). "Geographical Variation in Coastal Development." Oliver & Boyd, Edinburgh. 435pp.

Davies, T. A., and Gorsline, D. S. (1976). Oceanic sediments and sedimentary processes. *In* "Chemical Oceanography" (J. P. Riley and G. Skirrow, eds.), 2nd ed., vol. 5, pp. 1–80. Academic Press, London.

Davis, R. A. (1978). "Coastal Sedimentary Environments." Springer-Verlag, Berlin. 420pp.

Dean, W. E., and Fouch, T. D. (1983). Lacustrine environment. *Mem.— Am. Assoc. Pet. Geol.* **33,** 97–130.

de Boer, P. L., and Smith, D. G. (1994). "Orbital Forcing and Cyclic Sequences." Blackwell, Oxford. 576pp.

Degens, E. T., and Ross, D. A., eds. (1969). "Hot Brines and Recent Heavy Metal Deposits in the Red Sea." Springer-Verlag, Berlin. 600pp.

Denny, C. S. (1965). Alluvial fans in the Death Valley Region, California and Nevada. *Geol. Surv. Prof. Pap. (U.S.)* **466,** 1–62.

De Raaf, J. F. M., and Boersma, J. R. (1971). Tidal deposits and their sedimentary structures. *Geol. Mijnbouw* **50,** 479–504.

De Raaf, J. F. M., Reading, H. G., and Walker, R. G. (1964). Cyclic sedimentation in the Lower Westphalian of North Devon. *Sedimentology* **74,** 373–420.

Devine, P.E. (1991). Transgressive origin of channeled estuarine deposits in the Point Lookout Sandstone, Northwestern New Mexico: A model for Upper Cretaceous, cyclic regressive parasequences of the U.S. Western interior. *AAPG Bull.* **75,** 1039–1063.

Dietrich, G. (1963). "General Oceanography: An Introduction." Wiley, Chichester. 588pp.

Doeglas, D. J. (1962). The structure of sedimentary deposits of braided rivers. *Sedimentology* **1,** 167–190.

Dott, R. H. (1966). Eocene deltaic sedimentation at Coos Bay, Oregon. *J. Geol.* **74,** 373–420.

Dott, R. H., ed. (1994). "Eustasy: The Historical Ups and Downs of a Major Geological Concept," Mem No. 180. Geol. Soc. Am. Boulder, CO. 120pp.

Dow, W. C. (1978). Petroleum source beds on continental slopes and rises. *AAPG Bull.* **62,** 1584–1606.

Duff, P. McL. D., Hallam, A., and Walton, E. K. (1967). "Cyclic Sedimentation." Elsevier, Amsterdam. 280pp.

Dunbar, C. O., and Rodgers, J. (1957). "Principles of Stratigraphy." Wiley, New York. 356pp.

Dunham, R. J. (1969). Vadose pisolite in the Capitan Reef (Permian), New Mexico and Texas. *Spec. Publ.— Soc. Econ. Paleontol. Mineral.* **14,** 182–191.

Dunham, R. J. (1970). Stratigraphic reefs versus ecologic reefs. *Am. Assoc. Pet. Geol. Bull.* **54,** 1931–1932.

Dzulinski, S., and Walton, E. K. (1965). "Sedimentary Features of Flysch and Greywackes." Elsevier, Amsterdam. 274pp.

Elloy, R. (1972). Reflexions sur quelques environments recifaux de Paleozoique. *Bull. Cent. Rech. Pau* **6,** 1–106.

Emery, D., and Myers, K. (1996). "Sequence Stratigraphy." Blackwell Science, Oxford. 320pp.

Emery, K. O. (1956). Sediments and water of Persian Gulf. *Am. Assoc. Pet. Geol. Bull.* **40,** 235–2383.

Emery, K. O. (1963). Oceanic factors in accumulation of petroleum. *Proc.—World Pet. Congr.* **6,** 483–491.

Emery, K. O. (1968). Relict structures on continental shelves of world. *Am. Assoc. Pet. Geol. Bull.* **52,** 445–464.

Enos, P. (1977). Tamabra limestone of the Poza Rica trend, Cretaceous, Mexico. *Spec. Publ.—Soc. Econ. Paleontol. Mineral.* **25,** 273–314.

Enos, P. (1983). Shelf. *Mem.—Am. Assoc. Pet. Geol.* **33,** 267–296.

Enos, P., and Moore, C. H. (1983). Fore-reef slope. *Mem.—Am. Assoc. Pet. Geol.* **33,** 507–538.

Erben, H. K. (1964). Facies developments in the Marine Devonian of the old world. *Proc. Ussher Soc.* **1** (Pt. 3), 92–118.

Eskola, P. (1915). Om sambandet mellan kemisk och mineralogisk sammansattning hos Orijarvitraktens metamorfa bergarter. *Bull. Commun. Geol. Finl.* **44,** 1–145.

Eugster, H. P., and Surdam, R. C. (1973). Depositional environment of the Green River Formation of Wyoming. A preliminary report. *Geol. Soc. Am. Bull.* **84,** 1115–1120.

Evans, G. (1965). Intertidal flat sediments and their environments of deposition in the Wash. *Q. J. Geol. Soc. London* **121,** 209–245.

Evans, G. (1970). Coastal and nearshore sedimentation: A comparison of clastic and carbonate deposition. *Proc. Geol. Assoc.* **81,** 493–508.

Evans, G., Schmidt, V., Bush, P., and Nelson, H. (1969). Stratigraphy and geologic history of the sabkha, Abu Dhabi, Persian Gulf. *Sedimentology* **12,** 145–159.

Eynon, G. (1981). Basin development and sedimentation in the Middle Jurassic of the Northern North Sea. *In* "Petroleum Geology of the Continental Shelf of Northwest Europe" (L. V. Iling and D. G. Hobson, eds.), pp. 196–204. Inst. Pet., London.

Fairbridge, R. W. (1950). Recent and Pleistocene coral reefs of Australia. *J. Geol.* **58,** 330–401.

Ferm, J. C. (1970). Allegheny deltaic deposits. *Spec. Publ.—Soc. Econ. Paleontol. Mineral.* **15,** 246–255.

Fischer, A. G., and Garrison, R. E. (1967). Carbonate lithification on the sea floor. *J. Geol.* **75,** 488–496.

Fischer, A. G., and Roberts, L.T. (1991). Cyclicity in the Green River Formation (lacustrine, Eocene) of Wyoming. *J. Sediment. Petrol.* **61,** 1146–1154.

Fisher, W. L., Brown, L. F., Scott, A. J., and McGowen, J. H. (1969). Delta systems in the exploration for oil and gas. *Bureau Econ. Geol. Univ. Texas.* 78pp.

Fisk, H. N. (1955). Sand facies of Recent Mississippi delta deposits. *Proc.—World Pet. Congr.* **4** (Sect. 1), 377–398.

Fleming, B. W., and Bartholoma, A. (1995). "Tidal Signatures in Modern and Ancient Sediments." Blackwell, Oxford. 368pp.

Force, E. R. (1986). Descriptive model of shoreline placer Ti. *Geol. Surv. Bull. (U.S.)* **1693,** 270.

Fouch, T. D., and Dean, W. E. (1982). Lacustrine deposits. *Mem.—Am. Assoc. Pet. Geol.* **31,** 87–114.

Freeman, W. E., and Visher, G. S. (1975). Stratigraphic analysis of the Navajo Sandstone. *J. Sediment. Petrol.* **45,** 651–658.

Friend, P. F. (1965). Fluviatile sedimentary structures in the Wood Bay Series (Devonian) of Spitsbergen. *Sedimentology* **5,** 39–68.

Friend, P. F., and Moody-Stuart, M. (1972). Sedimentation of the Wood Bay Formation (Devonian) of Spitsbergen: Regional analysis of a late orogenic basin. *Skr., Nor. Polarinst.* **157,** 1–77.

Frimmel, H. E. (1997). Detrital origin of the Witwatersrand gold—a review. *Terra Nova* **9,** 192–197.

Frostick, L. E., and Steel, R. J., eds. (1994). "Tectonic Controls and Signatures in Sedimentary Successions." Blackwell, Oxford. 528pp.

Gale, A. S. (1998). Cyclostratigraphy. *In* "Unlocking the Stratigraphic Record" (P. Doyle and M. R. Bennett, eds.), pp. 195–220. Wiley, Chichester.

Galloway, E. E., and Hobday, D. K. (1983). "Terrigenous Clastic Depositional Systems." Springer-Verlag, Berlin. 433pp.

Garrison, R. E., and Fischer, A. G. (1969). Deepwater limestones and radiolarites of the Alpine Jurassic. *Spec. Publ.— Soc. Econ. Paleontol. Mineral.* **14,** 20–56.

Geological Society of London (1969). Recommendations on Stratigraphic Usage.

Gierlowski-Kordesch, E., and Kelts, K., eds. (1994). "Global Geological Record of Lake Basins." Cambridge University Press, Cambridge, UK. 440pp.

Ginsburg, R. N. (1956). Environmental relationships of grain size and constituent particles in some S. Florida sediments. *Am. Assoc. Pet. Geol. Bull.* **40,** 2384–2427.

Ginsburg, R. N. (1975). "Tidal Deposits: A Casebook of Recent Examples and Fossil Counterparts." Springer-Verlag, Berlin. 428pp.

Glasby, G. P. (1986). Nearshore mineral deposits in the SW Pacific. *In* "Sedimentation and Mineral Deposits in the Southwestern Pacific Ocean" (D. S. Cronan, ed.), pp. 149–182. Academic Press, London.

Glennie, K. W. (1970). "Desert Sedimentary Environments." Elsevier, Amsterdam. 222pp.

Glennie, K. W. (1972). Permian Rotliegendes of northwest Europe interpreted in light of modern sediment studies. *Am. Assoc. Pet. Geol. Bull.* **56,** 1048–1071.

Glennie, K. W. (1982). Early Permian (Rotliegendes) palaeowinds of the North Sea. *Sediment. Geol.* **34,** 245–265.

Glennie, K. W. (1987). Desert sedimentary environments, present and past — A summary. *Earth Sci. Rev.* **50,** 135–166.

Glennie, K. W. (1990). Early Permian-Rotliegendes. *In* "Introduction to the Petroleum Geology of the North Sea" (E. K. Glennie, ed.), 3rd ed., pp. 120–152. Blackwell, Oxford.

Gorsline, D. S., and Emery, K. O. (1959). Turbidity current deposits in San Pedro and Santa Monica basins off Southern California. *Geol. Soc. Am. Bull.* **70,** 279–290.

Gould, H. R. (1970). The Mississippi delta complex. *Spec. Publ.— Soc. Econ. Paleontol. Mineral.* **15,** 3–30.

Greensmith, J. T. (1968). Paleogeography and rhythmic deposition in the Scottish Oil-Shale group. *Proc. U.N. Symp. Dev. Util. Oil Shales Res.,* Tallin, pp. 1–16.

Gressly, A. (1838). Observations geologiques sur le Jura Soleurois. *Neue Denkschr. Allg. Schweiz. Ges. Gesamten Naturwiss.* **2,** 1–112.

Grunau, H. R., and Gruner, U. (1978). "Source rock and origin of natural gas in the Far East." *J. Pet. Geol.* **1,** 3–56.

Guilbert, J. M., and Park, C. F. (1986). "The Geology of Ore Deposits." Freeman, New York. 985pp.

Guilcher, A. (1970). Symposium on the evolution of shorelines and continental shelves in their mutual relations during the Quaternary. *Quaternaria* **12,** 1–229.

Gustafson, L. B., and Williams, N. (1981). Sediment-hosted stratiform deposits of copper, lead and zinc. *Econ. Geol., 75th Anniv. Vol.,* pp. 179–213.

Guzman, E. J. (1967). Reef type stratigraphic traps in Mexico. *Proc.— World Petrol. Congr.* **7** (2), 461–470.

Hallam, A. (1967). Editorial comment. *Mar. Geol. Spec. Issue* **5** (5/6), 329–332.

Hand, B. M., and Emery, K. O. (1964). Turbidites and topography of north end of San Diego Trough, California. *J. Geol.* **72,** 526–552.

Haq, B. U., Hardenbohl, J., and Vail, P. (1988). Mesozoic and Cenozoic chronostratigraphy and eustatic cycles. *In* "Sea-Level Changes: An Integrated Approach" (C. K. Wilgus, H. Posamentier, C. A. Ross, and C. G. St. C. Kendall, eds.). *Soc. Econ. Pal. Min. Spec. Publ.* **42,** pp. 71–108.

Harper, C. W. (1984). Improved methods of facies sequence analysis. *In* "Facies Models" (R. G. Walker, ed.), 2nd ed., Repr. Ser. No. 1, pp. 11–14. Geoscience Canada.

Hartley, A. J., and Prosser, D. J., eds. (1995). "Characterization of Deep Marine Clastic Systems," Spec. Publ. No. 94 Geol. Soc. London, London. 247pp.

Hawarth, E. Y., and Lund, J. W. G., eds. (1984). "Lake Sediments and Environmental History." Leicester University Press, Leicester. 411pp.

Haymon R. M. (1983). Growth history of hydrothermal black smoker chimneys. *Nature (London)* **301,** 695–698.

Haymon, R. M., and Kastner, M. (1981). Hot spring deposits on the East Pacific Rise at 21°N: preliminary description of mineralogy and genesis. *Earth Planet. Sci. Lett.* **53,** 363–381.

Heckel, P. H. (1972). Recognition of Ancient shallow marine environments. *Spec. Publ.— Soc. Econ. Paleontol. Mineral.* **16,** 226–286.

Heritier, F. E., Lossel, P., and Wathne, E. (1981). The Frigg Gas Field. *In* "Petroleum Geology on the Continental Shelf of Northwest Europe" (L. V. Illing and G. D. Hobson, eds.), pp. 380–394. Heydon Press, London.

Hitzman, M. W., and Large, D. (1986). A review of Irish carbonate-hosted base metal deposits. *In* "Geology and Genesis of Mineral Deposits in Ireland" (C. J. Andrew, ed.), pp. 217–238. Irish Association for Economic Geology, Dublin.

Hollenshead, C. T., and Pritchard, R. L. (1961). Geometry of producing Mesaverde Sandstones, San Juan basin. *In* "Geometry of Sandstone Bodies" (J. A. Peterson and J. C. Osmond, eds.), pp. 98–118. Am. Assoc. Pet. Geol., Tulsa, OK.

Horn, D. (1965). Zur geologischen Entwicklung der sudlichen Schleimundung im Holozan. *Meyniana* **15,** 42–58.

Horn, D. R., Ewing, M., Delach, M. N., and Horn, B. M. (1971). Turbidites of the northeast Pacific. *Sedimentology* **16,** 55–69.

Horowitz, D. H. (1966). Evidence for deltaic origin of an Upper Ordovician sequence in the Central Appalachians. *In* "Deltas in their Geologic Framework" (M. L. Shirley and J. A. Ragsdale, eds.), pp. 159–169. Geol. Soc., Houston, TX.

Houbolt, J. J. H. C. (1968). Recent sediments in the southern bight of the North Sea. *Geol. Mijnbouw* **47,** 245–273.

House, M. R., and Gale, A. S., eds. (1995). "Orbital Forcing Timescales and Cyclostratigraphy," Spec. Publ. 85. Geol. Soc. London, London. 204pp.

Hovland, M., and Somerville, J. H. (1985). Characteristics of two natural gas seepages in the North Sea. *Mar. Pet. Geol.* **2,** 319–326.

Howell, D. G., and Normark, W. R. (1982). Submarine fans. *Mem.— Am. Assoc. Pet. Geol.* **31,** 365–404.

Hoyt, J. H. (1967). Barrier Island formation. *Geol. Soc. Am. Bull.* **78,** 1125–1136.

Hoyt, J. H., and Henry, V. J. J. (1967). Influence of island migration on Barrier Island sedimentation. *Geol. Soc. Am. Bull.* **78,** 77–86.

Hoyt, J. H., Weimer, R. J., and Vernon, J. H. (1964). Late Pleistocene and Recent sedimentation, central Georgia coast. *In* "Deltaic and Shallow Marine Deposits" (L. M. J. U. Van Straaten, ed.), pp. 170–176. Elsevier, Amsterdam.

Hsu, K. (1970). The meaning of the word 'Flysch', a short historical search. *Spec. Pap.— Geol. Assoc. Can.* **7,** 1–11.

Hsu, K. J., and Jenkyns, H. C., eds. (1974). "Pelagic Sediments: On Land and Under the Sea," Spec. Publ. 1. Int. Assoc. Sediment. Oxford. 447pp.

Hubbard, R. J., Pape, J., and Roberts, D. G. (1985). Depositional sequence mapping to illustrate the evolution of a passive continental margin. *Mem.— Am. Assoc. Pet. Geol.* **39,** 79–92.

Huc., A. Y. (1995). Paleogeography, paleoclimate and source rocks. *Am. Assoc. Pet. Geol. Stud. Geol.* **10,** 1–347.

Hulsemann, J. (1955). Grossrippeln und Schragschichtungs-Gefuge im Nordsee-Watt und in der Molasse. *Senckenbergiana Lethaea* **36,** 359–388.

Imbrie, J., and Buchanan, H. (1965). Sedimentary structures in modern carbonate sands of the Bahamas. *Spec. Publ.— Soc. Econ. Paleontol. Mineral.* **12,** 149–172.

Irwin, M. L. (1965). General theory of epeiric clear water sedimentation. *Bull. Am. Assoc. Pet. Geol.* **49,** 445–459.

Jefferies, R. (1963). The stratigraphy of the *Actinocamax plenus* subzone (Turonian) in the Anglo-Paris basin. *Proc. Geol. Assoc.* **74,** 1–34.

Jenkyns, H. C. (1986). Pelagic environments. *In* "Sedimentary Environments and Facies" (H. G. Reading, ed.), 2nd ed., pp. 343–397. Blackwell, Oxford.

Jervey, M. T. (1988). Quantitative geological modelling of siliciclastc rocks sequences and their seismic expression. *Spec. Publ.— Soc. Econ. Paleontol. Mineral.* **42,** 47–69.

Johnson, H. D., and Stewart, D. J. (1985). Role of clastic sedimentology in the exploration and production of oil and gas in the North Sea. *In* "Sedimentology. Recent Developments and Applied Aspects" (P. J. Brenchley and B. P. J. Williams, eds.), pp. 249–310. Blackwell, Oxford.

Jones, O. A., and Endean, R. (1973). "Biology and Geology of Coral Reefs," 3 vols. Academic Press, London. 12,045pp.

Jordan, W. M. (1969).The enigma of Colorado Plateau eolian sandstones. *Am. Assoc. Petrol. Geol. Bull.* **53,** 725 (abstr.).

Katz, B., ed. (1995). "Petroleum Source Rocks." Springer-Verlag, Berlin. 327pp.

Kazmi, A. H. (1964). Report on the geology and ground water investigations in Rechna Doab, West Pakistan. *Rec. Geol. Surv. Pak.* **10,** 1–26.

Kendall, C. G. St. C. (1969). An environmental reinterpretation of the Permian evaporite/carbonate shelf sediments of the Guadalupe Mountains. *Geol. Soc. Am. Bull.* **80,** 2503–2525.

Kendall, C. G. St. C., and Harwood, G.M. (1996). Marine evaporites: Arid shorelines and basins. *In* "Sedimentary Environments: Processes, Facies and Stratigraphy" (H. G. Reading, ed.), 3rd ed., pp. 281–324. Blackwell, Oxford.

Kennedy, W. J. (1987). Late Cretaceous and Early Palaeocene sedimentation in the Greater Ekofisk Area, North Sea Central Graben. *Bull. Cent. Rech. Explor.-Prod. Elf-Aquitaine* **II,** 91–130.

Kenyon, N. H., and Stride, A. H. (1970). The tide-swept continental shelf sediments between the Shetland Isles and France. *Sedimentology* **14,** 159–173.

Kenyon, N. H., Amir, A., and Cramp, A. (1995) Geometry of the younger sediment bodies of the Indus fan. *In* "An Atlas of Deep Water Environments" (K. T. Pickering, R. Hiscott, N. H. Kenyon, F. Ricci Lucchi, and R. D. A. Smith, eds.), pp. 89–93. Kluwer Academic Press, Dordrecht, The Netherlands.

Kessler, L. G., Zang, R. D., Englehorn, J. A., and Eger, J. D. (1980). Stratigraphy and sedimentology of a Palaeocene submarine fan complex, Cod Field, Norwegian North Sea. *In* "The Sedimentation of North Sea Reservoir Rocks" (R. Hardman, ed.), Pap. VII, pp. 1–19. Norw. Petrol. Soc., Oslo.

King, C. A. M. (1972). "Beaches and Coasts," 2nd ed. Edward Arnold, London. 570pp.

King, L. C. (1962). "Morphology of the Earth." Oliver & Boyd, Edinburgh. 699pp.

Klein, G. de Vries (1971). A sedimentary model for determining paleotidal range. *Bull. Geol. Soc. Am.* **82,** 2585–2592.

Klein, G. de Vries (1977). "Clastic Tidal Facies." CEPCO, Champaign, IL. 149pp.

Klein, G. de Vries (1980). "Sandstone Depositional Models for Exploration for Fossil Fuels." CEPCO, Burgess, MN. 149pp.

Kokurek, G. A. (1996). Desert aeolian systems. *In* "Sedimentary Environments: Processes, Facies and Stratigraphy" (H. G. Reading, ed.), 3rd ed., pp. 125–153. Blackwell, Oxford.

Kolb, C. R., and Van Lopik, J. R. (1966). Depositional environments of the Mississippi River deltaic plain — southeastern Louisiana. *In* "Deltas in their Geologic Framework" (M. L. Shirley and J. A. Ragsdale, eds.), pp. 17–62. Geol. Soc., Houston, TX.

Kroon, D., Norris, R. D., Klaus, A., and the ODP Leg 171B Shipboard Scientific Party (1998). Drilling Blake Nose: The search for evidence of extreme Palaeogene-Cretaceous climates and extraterrestrial events. *Geol. Today* **14,** 222–225.

Krumbein, W. C. (1967). Fortran IV computer programs for Markov chain experiments in geology. *Comput. Contrib., Kans. Univ. Geol. Surv.* No. 13.

Krumbein, W. C., and Sloss, L. L. (1959). "Stratigraphy and Sedimentation." Freeman, San Francisco. 497pp.

Kukal, Z. (1971). "Geology of Recent Sediments." Academic Press, London and New York. 490pp.

Kyle, J. R. (1981). Geology of the Pine Point lead-zinc district. *In* "Handbook of Strata-bound and Stratiform Ore Deposits" (K. H. Wolf, ed.), Vol. 9, pp. 643–741. Elsevier, Amsterdam.

Laming, D. J. C. (1966). Imbrication, paleocurrents and other sedimentary features in the Lower New Red Sandstone, Devon, England. *J. Sediment. Petrol.* **36,** 940–957.

Lancaster, N. (1995). "Geomorphology of Desert Dunes." Routledge, London. 290pp.

Laporte, L. F. (1979). "Ancient Environments," 2nd ed. Prentice-Hall, Englewood Cliffs, NJ. 163pp.

Lawson, A. C. (1913). The petrographic designation of alluvial-fan formations. *Univ. Calif., Berkeley, Publ. Geol. Sci.* **7,** 325–334.

Lees, A., and Miller, J. (1995). Waulsortian banks. *In* "Carbonate Mud Mounts" (C. L. V. Monty, D. W. J. Bosence, P. H. Bridges, and B. R. Pratt, eds.), pp. 191–171. Blackwell, Oxford.

Leggett, J. K. (1985). Deep sea pelagic sediments and palaeo-oceanography: A review of recent progress. *In* "Sedimentology: Recent Advances and Applied Aspects" (P. J. Brenchley and B. P. J. Williams, eds.), pp. 95–122. Blackwell, Oxford.

Leopold, L. B., Solman, M. G., and Miller, J. P. (1964). "Fluvial Processes in Geomorphology." Freeman, San Francisco. 522pp.

Loucks, R. G., and Sarg, J. F., eds. (1993). "Carbonate Sequence Stratigraphy," Mem. No. 57. Am. Assoc. Pet. Geol., Tulsa. OK. 545pp.

Lowenstam, H. A. (1950). Niagaran reefs of the Great Lakes area. *J. Geol.* **58**, 430–487.

Lundquist, S. J. (1983). Nugget formation: Reservoir characteristics affecting production in the overthrust belt of Southwestern Wyoming. *JPT, J. Pet. Technol.* **83**, 1355–1365.

Lyell, C. (1865). "Elements of Geology." John Murray, London. 794pp.

Macdonald, E. H. (1983). "Alluvial Mining." Chapman & Hall, London. 508pp.

Maksimova, S. V. (1972). Coral reefs in the Arctic and their paleogeographic interpretation. *Int. Geol. Rev.* **14**, 764–769.

Martin, D. B. (1963) Rosedale Channel: Evidence for late Miocene submarine erosion in the Great Valley of California. *Bull. Am. Assoc. Pet. Geol.* **47**, 441–456.

Martin, J. H. (1993). A review of braided fluvial hydrocarbon reservoirs: The petroleum engineer's perspective. *In* "Braided Rivers," Spec. Publ. No. 75, pp. 33–367. Geol. Soc. London, London.

Martinsen, O. J. (1995). Sequence stratigraphy, three dimensions and philosophy. *In* "Sequence Stratigraphy of the Northwest European Margin" (Steel, R. J., Felt, V. L., Johannsen, E. P., and Mathieu, C., eds.), pp. 23–30. Elsevier, Amsterdam.

Martinsen, O. J., Martinsen, R. I., and Steidmann, J. A. (1993). Mesaverde Group (Upper Cretaceous), southeastern Wyoming: Allostratigraphy versus sequence stratigraphy in a tectonically active area. *AAPG Bull.* **77**, 1351–1373.

Marzo, M., and Puigdefabregas, C., eds. (1993). "Alluvial Sedimentation." Blackwell, Oxford. 600pp.

Mayuga, M. N. (1970). California's giant — Wilmington Oil Field. *Mem. — Am. Assoc. Pet. Geol.* **14**, 158–184.

McCave, I. N. (1985). Recent shelf clastic sediments. *In* "Sedimentology: Recent Advances and Applied Aspects" (P. J. Brenchley and B. P. J. Williams, eds.), pp. 49–66. Blackwell, Oxford.

McCubbin, D. G. (1982). Barrier-island and strand plain facies. *Mem. — Am. Assoc. Pet. Geol.* **31**, 247–280.

McKee, E. D., ed. (1979). "A Study of Global Sand Seas," Prof. Pap. 1052. U. S. Geol. Surv., Washington, DC. 431pp.

Melvin, J., and Knight, A. S. (1984). Lithofacies, diagenesis and porosity of the Ivishak Formation. Prudhoe Bay Area, Alaska. *Mem. — Am. Assoc. Pet. Geol.* **37**, 347–366.

Merriam, D. F., ed. (1965). "Symposium on Cyclic Sedimentation," Bull. No. 169, Kans. Univ. Geol. Surv., Lawrence. 636pp.

Merriam, D. F., ed. (1967). "Computer Applications in the Earth Sciences," Compu. Contrib. No. 18. Kans. Geol. Surv., Lawrence.

Miall, A. D. (1977). A review of the braided river depositional environment. *Earth Sci. Rev.* **13**, 1–62.

Miall, A. D., ed. (1978). "Fluvial Sedimentology," Spec. Publ. No. 5. Can. Soc. Pet. Geol. Calgary, 859pp.

Miall, A. D., ed. (1981). "Sedimentation and Tectonics in Alluvial Basins," Spec. Pap. No. 23. Geol. Soc. Can. 272pp.

Miall, A. D. (1997). "The Geology of Stratigraphic Sequences." Springer-Verlag, Berlin. 433pp.

Middleton G. V. (1973). Johannes Walther's law of correlation of facies. *Geol. Soc. Am. Bull.* **84**, 979–988.

Milankovitch, M. (1941). "Kanton der Erdbestrahlung und seine Anwendung auf das Eiszeiten-problem." Serbian Academy of Science. Belgrade. 133pp.

Minter, W. E. L. (1981). The cross-bedded nature of Proterozoic Witwatersrand placers in distal environments and a paleocurrent analysis of the Vaal Reef placer. *Geol. Surv. Prof. Pap. (U.S.)* **1161-A-BB,** Pap. G, 1–17.

Minter, W. E. L., and Loen, J. S. (1991). Palaeocurrent dispersal pattens of Witwatersrand gold placers. *S. Afr. J. Geol.* **99,** 339–449.

Mitchum, R. M., Vail, P. R., and Thompson, S. (1977). Part Two: The depositional Sequence as a Basic Unit for Stratigraphic Analysis. *In* "Seismic Stratigraphy — Application to Hydrocarbon Exploration" (Payton, C. E., ed.), pp. 53–62. *Am. Assoc. Petrol. Geol. Mem. No. 2.*

Monty, C. L. V, Bosence, D. W. J. Bridges, P. H., and Pratt, B. R., eds. (1995). "Carbonate Mud Mounds." Blackwell, Oxford. 544pp.

Moore, D. (1959). Role of deltas in the formation of some British Lower Carboniferous cyclothems. *J. Geol.* **67,** 522–539.

Moore, G. T., and Asquith, D. O. (1971). Delta: Term and concept. *Geol. Soc. Am. Bull.* **82,** 2563–2568.

Moore, R. C. (1949). Meaning of facies. *Mem. — Geol. Soc. Am.* **39,** 1–34.

Morgan, J. P. (1970). Depositional processes and products in the deltaic environment. *Spec. Publ. — Soc. Econ. Paleontol. Mineral.* **15,** 31–47.

Morton, A. C., Haszeldine, R. S., Giles, M. R., and Brown, S., eds. (1992). "Geology of the Brent Group," Spec. Publ. No. 61. Geol. Soc. London, London. 506pp.

Mullins, H. T., and Cook, H. E. (1986). Carbonate apron models: Alternatives to the submarine fan model for paleoenvironmental analysis and hydrocarbon exploration. *Sediment. Geol.* **48,** 37–80.

Mutti, E., and Lucchi, F. R. (1972). Le torbiditii dell'Appennino settentrionale: Introduzione all'analisi di facies. *Mem. Soc. Geol. Ital.* **11,** 161–199.

Narayan, J. (1970). Sedimentary structures in the Lower Greensand of the Weald, England, and Bas-Boulonnais, France. *Sediment. Geol.* **6,** 73–109.

Neal, J., Risch, D., and Vail, P. (1993). Sequence Stratigraphy — A global theory for local success. *Schlumberger Oilfield Review* **5** (1), 51–62.

Newell, N. D., and Rigby, J. K. (1957). Geological studies of the Great Bahama bank. *Spec. Publ. — Soc. Econ. Paleontol. Mineral.* **5,** 15–72.

Newell, N. D., Rigby, J. K., Fischer, A. G., Whiteman, A. J., Hickox, J. E., and Bradbury, J. S. (1953). "The Permian Reef Complex of the Guadalupe Mountains Region, Texas and New Mexico." Freeman, San Francisco. 236pp.

Nickling, W. G., ed. (1986). "Aeolian Geomorphology." Allen & Unwin, London. 307pp.

Ohle, E. L. (1980). Some considerations in determining the origin of ore deposits of the Mississippi Valley type. *Econ. Geol.* **75,** 161–172.

Orris, G. J. (1986). Descriptive model of bedded barite. *Geol. Surv. Bull. (U.S.)* **1693,** 216–217.

Oti, M. N., and Postma, G. (1995). "Geology of Deltas." A.A. Balkema, Rotterdam.

Pallas, P. (1980). Water resources of the Socialist People's Libyan Arab Jamahariya. *In* "The Geology of Libya" (M. J. Salem and M. T. Busrewil, eds.), pp. 539–554. Academic Press, London.

Payton, C. E. (ed.) (1977). Seismic Stratigraphy — Application to hydrocarbon exploration. *Am. Assoc. Petrol. Geol. Mem. No. 2.* 516pp.

Peterson, C. D., Gleeson, G. W., and Wetzel, N. (1987). Stratigraphic development, mineral — sources and preservation of marine placers from Pleistocene terraces in southern Oregon U.S.A. *Sediment. Geol.* **53,** 203–209.

Pettijohn, F. J. (1956). "Sedimentary Rocks." Harper Bros, New York. 718pp.

Pettijohn, F. J., Potter, P. E., and Siever, R. (1972). "Sand and Sandstone." Springer-Verlag, Berlin. 618pp.

Pezzetta, J. M. (1973). The St. Clair River delta: sedimentary characteristics and depositional environments. *J. Sediment. Petrol.* **43,** 168–187.

Picard, M. D. (1967). Paleocurrents and shoreline orientations in Green River Formation (Eocene), Raven Ridge and Red Wash areas, north-eastern Uinta basin, Utah. *Am. Assoc. Pet. Geol. Bull.* **5,** 383–392.

Picard, M. D., and High, L. R., Jr. (1968). Sedimentary cycles in the Green River Formation (Eocene) Uinta basin, Utah. *J. Sediment. Petrol.* **38,** 378–383.

Picard, M. D., and High, L. R., Jr (1972a). Criteria for recognizing lacustrine rocks. *Spec. Publ.—Soc. Econ. Paleontol. Mineral.* **16**, 108–145.

Picard, M. D., and High, L. R., Jr. (1972b). Paleoenvironmental reconstructions in an area of rapid facies change, Parachute Creek member of Green River Formation (Eocene), Uinta basin, Utah. *Geol. Soc. Am. Bull.* **83**, 2689–2708.

Picard, M. D., and High, L. R., Jr. (1979). Lacustrine Stratigraphic Relations. *Spec. Publ.— Geol. Soc. S. Afr.* **6**, 122.

Pickering, K. T., Hiscott, R. N., Kenyon, N. H., Ricci-Lucchi, F., and Smith, R. D. A., eds. (1995). "Atlas of Deep Water Environments: Architectural Styles in Turbidite Systems." Chapman & Hall, London.

Plint, A. G., ed. (1995). "Sedimentary Facies Analysis." Blackwell, Oxford. 400pp.

Potter, P. E. (1959). Facies model conference. *Science* **129**, 1292–1294.

Potter, P. E. (1962). Late Mississippian sandstones of Illinois. *Circ.—Ill. State Geol. Surv.* **340**, 1–36.

Potter, P. E. (1967). Sand bodies and sedimentary environments: A review. *Am. Assoc. Pet. Geol. Bull.* **51**, 337–365.

Potter, P. E., and Pettijohn, F. J. (1977). "Paleocurrents and Basin Analysis," 2nd ed. Springer-Verlag, Berlin. 425pp.

Pretorius, D. A. (1979). The depositional environment of the Witwatersrand goldfields: a chronological review of speculations and observations. *Spec. Pub.— Geol. Soc. S. Afr.* **6**, 33–56.

Prevost, C. (1838). Quelque vieux roc que j'ai connu. *Bull. Soc. Geol. Fr.* **9**, 90–95.

Pryor, W. A. (1971). Petrology of the Permian Yellow Sands of northeast England and their North Sea basin equivalents. *Sediment. Geol.* **6**, 221–254.

Purdy, E. G. (1961). Bahamian oolite shoals. *In* "The Geometry of Sandstone Bodies: A Symposium" (J. A. Peterson and J. C. Osmond, eds.), pp. 53–62. Am. Assoc. Pet. Geol., Tulsa, OK.

Purdy, E. G. (1963). Recent calcium carbonate facies of the Great Bahama bank. *J. Geol.* **71**, 334–355, 472–497.

Purdy, E. G. (1974). Karst-determined facies patterns in British Honduras: Holocene carbonate sedimentation model. *AAPG Bull.* **58**, 825–855.

Purser, B. H., ed. (1973). "The Persian Gulf." Springer-Verlag, Berlin. 471pp.

Pye, K., and Lancaster, N., eds. (1993). "Aeolian Sediments," Spec. Publ. No. 16. Int. Assoc. Sedol. Oxford. 176pp.

Ray, R. R. (1982). Seismic stratigraphic interpretation of the Fort Union Formation, Western Wind River Basin: Example of subtle trap exploration in a nonmarine sequence. *Mem.— Am. Assoc. Pet. Geol.* **32**, 169–180.

Read, J. F. (1985). Carbonate platform facies models. *AAPG Bull.* **69**, 1–21.

Reading, H. G., ed. (1996). "Sedimentary Environments: Processes, Facies and Stratigraphy," 3rd ed. Blackwell, Oxford. 688pp.

Reading, H. G., and Collinson, J. D. (1996). Clastic coasts. *In* "Sedimentary Environments: Processes, Facies and Stratigraphy" (H. G. Reading, ed.), 3rd ed., pp. 154–231. Blackwell, Oxford.

Reading, H. G., and Levell, B. K. (1996). Controls on the sedimentary rock record. *In* "Sedimentary Environments and Facies" (H. G. Reading, ed.), 3rd ed., pp. 5–36. Blackwell, Oxford.

Reeckman, A., and Friedman, G. M. (1982). "Exploration for Carbonate Petroleum Reservoirs." Wiley, Chichester, 213pp.

Reijers, T. J. A., Petters, S. W., and Nwajde, C. S. (1997). The Niger Delta Basin. *In* "African Basins" (R. C. Selley, ed.), pp. 151–172. Elsevier, Amsterdam.

Reineck, H. E. (1963). Sedimentgefuge im Bereich der sudlichen Nordsee. *Abh. Senckenb. Naturforsch. Ges.* **505**, 1–138.

Reineck, H. E. (1971). Marine sandkorper, rezent und fossil (Marine sand bodies recent and fossils). *Geol. Rundsch.* **60**, 302–321.

Reineck, H. E., and Singh, I. B. (1980). "Depositional Sedimentary Environments," 2nd ed. Springer-Verlag, Berlin. 549pp.

Rich, J. L. (1950). Flow markings, groovings, and intrastratal crumplings as criteria for recognition of slope deposits with illustrations from Silurian rocks of Wales. *Bull. Am. Assoc. Petrol.* **34**, 717–741.

Rich, J. L. (1951). Three critical environments of deposition and criteria for recognition of rocks deposited in each of them. *Bull. Geol. Soc. Am.* **62,** 1–20.

Riedel, W. R. (1963). The preserved record: paleontology of Pelagic sediments. *In* "The Sea" (M. N. Hill, ed.), vol. 3, pp. 866–887. Interscience, New York.

Rust, B. R. (1972). Structure and process in a braided river. *Sedimentology* **18,** 221–246.

Rust, B. R. (1978). A classification of alluvial channel systems. *Mem.— Can. Soc. Pet. Geol.* **5,** 187–198.

Sabins, F. F. (1963). Anatomy of stratigraphic traps, Bisti field, New Mexico. *Bull. Am. Assoc. Pet. Geol.* **47,** 193–228.

Sabins, F. F. (1972) Comparison of Bisti and Horseshoe Canyon stratigraphic traps, San Juan basin, New Mexico. *Mem.— Am. Assoc. Pet. Geol.* **16,** 610–622.

Schlanger, S. O., and Cita, M. B. (1982). "Nature and Origin of Carbon-Rich Facies." Academic Press, London. 224pp.

Scholle, P. A., and Spearing, D., eds. (1982). "Sandstone Depositional Environments," Mem. No. 37. Am. Assoc. Pet. Geol. Tulsa, OK. 410pp.

Scholle, P. A., Bebout, D. G., and Moore, C. H., eds. (1983a). "Carbonate Depositional Environments," Mem. No. 33. Am. Assoc. Pet. Geol., Tulsa, OK. 708pp.

Scholle, P. A., Arther, M. A., and Ekdale, A. A. (1983b). Pelagic. *Mem.— Am. Assoc. Pet. Geol.* **33,** 619–692.

Schwarz, M. L. (1971). The multiple causality of barrier islands. *J. Geol.* **79,** 91–94.

Schwarzacher, W. (1975). "Sedimentation Models and Quantitative Stratigraphy." Elsevier, Amsterdam. 396pp.

Schwarzacher, W. (1993). "Cyclostratigraphy and the Milankovitch Theory." Elsevier, Amsterdam. 238pp.

Selley, R. C. (1970). Studies of sequence in sediments using a simple mathematical device. *Q. J. Geol. Soc. London* **125,** 557–581.

Selley, R. C. (1972). Diagnosis of marine and non-marine environments from the Cambro-Ordovician sandstones of Jordan. *Q. J. Geol. Soc. London* **128,** 135–150.

Selley, R. C. (1977). Deltaic facies and petroleum geology. *In* "Developments in Petroleum Geology" (G. D. Hobson, ed.), Vol. 1, pp. 197–224. Applied Science, London.

Selley, R. C. (1996). "Ancient Sedimentary Environments: And Their Subsurface Diagnosis," 4th ed. Chapman & Hall, London. 300pp.

Shannon, J. P., and Dahl, A. R. (1971). Deltaic stratigraphic traps in west Tuscola field, Taylor County, Texas. *Am. Assoc. Pet. Geol. Bull.* **55,** 1194–1205.

Shantzer, E. V. (1951). Alluvium of river plains in a temperate zone and its significance for understanding the laws governing the structure and formation of alluvial suites. *Akad. Nauk. S.S.S.R. Geol. Ser.* **135,** 1–271.

Sharma, G. D. (1972). Graded sedimentation on Bering shelf. *Int. Geol. Congr., Rep. Sess., 24th,* Montreal, Sect. 8, pp. 262–271.

Shaw, A. B. (1964). "Time in Stratigraphy." McGraw-Hill, New York. 365pp.

Shawe, D. R., Simmons, G. C., and Archbold, N. L. (1968). Stratigraphy of Slick Rock district and vicinity, San Miguel and Dolores counties, Colorado. *Geol. Surv. Prof. Pap. (U.S.)* **576-A.** 157pp.

Shelton, J. W. (1965). Trend and genesis of Lowermost sandstone unit of Eagle Sandstone at Billings, Montana. *Bull. Am. Assoc. Pet. Geol.* **49,** 1385–1397.

Shelton, J. W. (1967). Stratigraphic models and general criteria for recognition of alluvial, barrier bar and turbidity current sand deposits. *Am. Assoc. Pet. Geol. Bull.* **51,** 2441–2460.

Shepard, F. P., ed. (1960). "Recent Sediments, North-western Gulf of Mexico," Am. Assoc. Pet. Geol., Tulsa, OK. 394pp.

Shepard, F. P. (1963). Submarine canyons. *In* "The Sea" (M. N. Hill, ed.), Vol. 3, pp. 480–506. Wiley, New York.

Shepard, F. P., Dill, R. F., and von Rad, U. (1969). Physiography and sedimentary processes of La Jolla submarine fan and Fan-Valley, California. *Am. Assoc. Pet. Geol. Bull.* **53,** 390 420.

Shlemon, R. J., and Gagliano, S. M. (1972). Birth of a delta — Atchafalaya Bay, Louisiana. *Proc.— Int. Geol. Congr.* **24** (6), 437–441.

Skall, H. (1975). The paleoenvironment of the Pine Point lead-zinc district. *Econ. Geol.* **70,** 22–45.

Skinner, B. J. (1981). Thoughts about uranium-bearing quartz-pebble conglomerates: A summary of ideas presented at the workshop. *Geol. Surv. Prof. Pap. (U.S.)* **1161-A-BB,** Pap. BB, 1–5.

Sloss, L. L. (1963). Sequence in a cratonic interior of North America. *Geol. Soc. Am. Bull.* **74,** 93–114.

Smirnov, V. I. (1976). "Geology of Mineral Deposits." Mir Publishers, Moscow. 520pp.

Smith, N., and Rogers, J., eds. (1999). "Fluvial Sedimentology VI." Blackwell, Oxford. 438pp.

Smith, W. (1816). Geological Map of England and Wales, with Part of Scotland.

Spring, D., and Hansen, O. P. (1998). The influence of platform morphology and sea level on the development of a carbonate sequence: The Harash Formation, Eastern Sirt Basin, Libya. *In* "Petroleum Geology of North Africa" (D. S. MacGregor, R. T. J. Moody, and D. D. Clark-Lowes, eds.), Spec. Publ. No. 132. pp. 335–353. Geol. Soc. London, London.

Stanley D. J., and Swift, D. J. P. (1976). "Marine Sediment Transport and Environment Management." Wiley, New York. 602pp.

Stanley D. J., and Unrug, R. (1972). Submarine channel deposits, fluxoturbidites and other indicators of slope and base of slope environments in modern and ancient marine basins. *Spec. Publ. — Soc. Econ. Paleontol. Mineral.* **16,** 787–340.

Stanley, D. J., and Weir, C. M. (1978). The "mudline": An erosion-deposition boundary on the upper continental slope. *Mar. Geol.* **28,** 19–29.

Stanley, K. O., Jordan, W. M., and Dott, R. H. (1971). New hypothesis of early Jurassic paleogeography and sediment dispersal for Western United States. *Am. Assoc. Pet. Geol. Bull.* **55,** 10–19.

Stanton, R. J. (1967). Factors controlling shape and interval facies distribution of organic build ups. *Am. Assoc. Pet. Geol. Bull.* **51,** 2462–2467.

Stauffer, P. H. (1967). Grainflow deposits and their implications, Santa Ynez Mountains, California. *J. Sediment. Petrol.* **37,** 487–508.

Steel, R. J., Felt, V. L., Johannsen, E. P., and Mathieu, C. (eds.) (1995). "Sequence Stratigraphy of the Northwest European Margin." Elsevier, Amsterdam. 608pp.

Steers, J. A. (1971). "Introduction to Coastline Development." Macmillan, London. 229pp.

Stein, R. (1986). Organic carbon and sedimentation rate — further evidence for anoxic deep-water conditions in the Cenomanian Turonian Atlantic Ocean. *Mar. Geol.* **72,** 199–210.

Stille, H. (1924). "Grundfragen der Vergleichenden Tektonik." Borntraeger, Berlin. 443pp.

Stille, H. (1936). The present tectonic state of the Earth. *Am. Assoc. Petrol. Geol. Bull.* **20,** 848–880.

Steno, N. (1667). "Elementorum myologiae specimen se musculi descripti geometrica." Firenze, Italy. Stellae. 123pp.

Stow, D. A. V. (1985). Deep-sea clastics: where are we and where are we going? *In* "Sedimentology: Recent Advances and Applied Aspects" (P. J. Branchley and B. P. J. Williams, eds.), pp. 67–94. Blackwell, Oxford.

Stow, D. A. V., ed. (1991). "Deep-water Turbidite Systems." Blackwell, Oxford. 480pp.

Stow, D. A. V., and Piper, D. J. W., eds. (1984). "Fine-grained Sediments: Deep Water Processes and Facies," Spec. Publ. No. 15. Geol. Soc. London. Blackwell, Oxford. 660pp.

Stow, D. A. V., Reading, H. G., and Collinson, J. D. (1996). Deep seas. *In* "Sedimentary Environments: Processes, Facies and Stratigraphy" (H.G. Reading, ed.), 3rd ed., pp. 395–453. Blackwell, Oxford.

Stride, A. H. (1963). Current-swept sea floors near the southern half of Great Britain. *Q. J. Geol. Soc. London* **119,** 175–200.

Stride, A. H. (1970). Shape and size trends for sandwaves in a depositional zone of the North Sea. *Geol. Mag.* **107,** 469–478.

Stride, A. H. (1982). "Offshore Tidal Sands: Process and Deposits." Chapman & Hall, London. 213pp.

Sweet, M. L. (1999). Interaction between eolian, fluvial and playa environments in the Permian Upper Rotliegende Group, UK southern North Sea. *Sedimentology* **46,** 171–187.

Swett, K., Klein, G. de Vries, and Smit, D. E. (1971). A Cambrian tidal sand body — the Eriboll Sandstone of Northwest Scotland: An ancient: recent analogue. *J. Geol.* **79,** 400–415.

Swift, D. J. P. (1969). Evolution of the shelf surface, and the relevance of the modern shelf studies to the rock record. *In* "The New Concepts of Continental Margin Sedimentation" (D. J. Stanley, ed.), pp. 37–49. Am. Geol. Inst. Washington, DC.

Swift, D. J. P., and Palmer, H. D., eds. (1978). "Coastal Sedimentation," Benchmark Papers in Geology, Vol. 42. Dowden, Hutchinson & Ross, Stroudsburg, PA. 339pp.

Swift, D. J. P., Duane, D. B., and Pilkey, O. H. (1972). "Shelf Sediment Transport: Process and Pattern." Dowden, Hutchinson & Ross, Stroudsburg, PA. 405pp.

Swift, D. J. P., Duane, D. B., and Orrin, H. P. (1973). "Shelf Sediment Transport: Process and Pattern." Wiley, Chichester. 670pp.

Swift, D. J. P., Oertel, G. F., Tilman, R. W., and Thorne, J. A. (1992). "Shelf Sand and Sandstone Bodies," Spec. Publ. No. 14. Int. Assoc. Oxford. 540pp.

Talbot, M. R., and Allen, P. A. (1996). Lakes. *In* "Sedimentary Environments: Processes, Facies and Stratigraphy" (H. G. Reading, ed.), 3rd ed., pp. 83–123. Blackwell, Oxford.

Tanner, W. F. (1968). Tertiary sea-level fluctuations. *Paleogeogr., Paleoclimatol. Paleoecol., Spec. Issue*, pp. 1–178.

Taylor, J. C. M., and Colter, V. S. (1975). Zechstein in the English Sector of the Southern North Sea Basin. *In* "Petroleum and the Continental Shelf of North West Europe" (A. E. Woodland, ed.), pp. 249–263. Inst. Pet., London.

Teichert, C. (1958a). Cold and deep-water coral banks. *Bull. Am. Assoc. Pet. Geol.* **42,** 106–1082

Teichert, C. (1958b). Concept of facies. *Bull. Am. Assoc. Pet. Geol.* **42,** 2718–2744.

Thiry, M., and Simon-Coincon, R., eds. (1999). "Palaeoweathering, Palaeosurfaces and Related Continental Deposits." Blackwell, Oxford. 408pp.

Thompson, D. B. (1969). Dome-shaped aeolian dunes in the Frodsham Member of the so called "Keuper" sandstone formation (Scythian; ?Anisian: Triassic) at Frodsham, Cheshire (England). *Sediment. Geol.* **3,** 263–289.

Trowbridge, A. C. (1911). The terrestrial deposits of Owens Valley, California. *J. Geol.* **19,** 707–747.

Tucker, M. E., and Wright, V. P. (1990). "Carbonate Sedimentology." Blackwell, Oxford. 496pp.

Tucker, M. E., Wilson, J. L., Crevello, P. D., Sarg, J. F., and Read, J. F. (1990). "Carbonate Platforms, Facies, Sequences and Evolution," Spec. Publ. No. 9. Int. Assoc. Sedimentol. Oxford.

Turnbull, W. J., Krinitsky, E. S., and Johnson, L. J. (1950). Sedimentary geology of the Alluvial valley of the Mississippi River and its bearing on foundation problems. *In* "Applied Sedimentation" (P. D. Trask, ed.), pp. 210–226. Wiley, New York.

Twenhofel, W. H. (1926). "Treatise on Sedimentation." Dover, New York. 926pp.

Tyler, N., and Finley, R. J. (1991). Architectural controls on the recovery of hydrocarbons from sandstone reservoirs. *In* "The Three-Dimensional Facies Architecture of Terrigenous Clastic Sediments and its Implications for Hydrocarbon Discovery and Recovery" (A.D. Miall and N. Tyler, eds.), pp. 1–5. Soc. Econ. Paleontol. Mineral., Tulsa, OK.

Tyson, R. V., and Pearson, T. H., eds. (1991). "Modern and Ancient Continental Shelf Anoxia," Spec. Publ. No. 58. Geol. Soc. London, London. 467pp.

Vail, P. R., Mitchum, R. M., Todd, R. G., Widmier, J. M., Thompson, S., Sangree, J. B., Bubb, J. N., and Hatledid, W. G. (1977). Seismic stratigraphy and global changes in sea level. *In* "Seismic Stratigraphy — Application to Hydrocarbon Exploration" (C. E. Payton, ed.), pp. 49–212. *Am. Assoc. Petrol. Geol. Mem. No.* **26.**

Vail, P. R., and Todd, R. G. (1981). Northern North Sea Jurassic unconformities, chronostratigraphy and sea level changes from seismic stratigraphy. *In* "Petroleum Geology of the Continental Shelf of North-West Europe." (L. V. Illing and D. G. Hobson, eds.). Inst. Pet., London. 216–235.

Van Andel, Tj. H. (1975). Mesozoic/Cenozoic calcite compensation depth and the global distribution of calcareous sediments. *Earth Planet. Sci. Lett.* **26,** 187–194.

van de Graaff, F. R. (1972). Fluvial-deltaic facies of the Castlegate sandstone (Cretaceous), East Central Utah. *J. Sediment. Petrol.* **42,** 558–571.

Van de Kamp, P. C., Coniff, J. J., and Morris, D. A. (1974). Facies relations in the Eocene – Oligocene of the Santa Ynez Mountains, California. *Q. J. Geol. Soc. London* **130,** 554–566.

Van Houten, F. B. (1964). Cyclic lacustrine sedimentation, Upper Triassic Lockatong Formation, central New Jersey and adjacent Pennsylvania. *Bull. Kans. Univ. Geol. Surv.* **169,** 497–531.

Van Houten, F. B., ed. (1977). "Ancient Continental Deposits," Benchmark Papers in Geology, Vol. 43. Dowden, Hutchinson & Ross, Stroudsburg, PA. 367pp.

Van Straaten, L. M. J. U. (1965). Coastal barrier deposits in south and north Holland in particular in the areas around Scheveningen and Ijmuiden. *Meded. Geol. Sticht, Nieuwe Ser. (Neth.)* **17,** 41–75.

Van Wagoner, J. C., Posamentier, H. W., Mitchum, R. M., Vail, P. R., Sarg. J. F., Loutit, T. S., and J. Hardenbol. (1988). An overview of the fundamentals of sequence stratigraphy and key definitions. *In* "Sea-Level Changes: An Integrated Approach." (C. K. Wilgus, B. S. Hastings, H. Posamentier, C. A. Ross, and G. St. C. Kendall, eds.) *Soc. Econ. Min. Pal. Sp. Pub. No. 42.* pp. 39–46.

Van Wagoner, J. C., Mitchum, R. M., Campion, K. M., and Rahmanian, V. D. (1990). Siliciclastic sequence stratigraphy in well logs, cores and outcrops. Am. Assoc. Petrol. Geol. Tulsa, Ok. *Methods in Exploration Series 7.* 55pp.

Visher, G. S. (1965). Use of vertical profile in environmental reconstruction. *Bull. Am. Assoc. Pet. Geol.* **49,** 41–61.

Visher, G. S. (1971). Depositional processes of the Navajo sandstone. *Geol. Soc. Am. Bull.* **82,** 1421–1424.

von Mojsisovics, M. E. (1879). "Die Dolomit-Riffe Von Sud Tirol und Venetien." A. Holder, Vienna. 552pp.

Walker, R. G. (1966). Shale grit and Grindslow shales: transition from Turbidite to shallow water sediments in the Upper Carboniferous of Northern England. *J. Sediment. Petrol.* **36,** 90–114.

Walker, R. G., ed. (1984a). "Facies Models," 2nd ed., Repr. Ser. No. 1. Geoscience Canada. 317pp. Toronto.

Walker, R. G. (1984b). General introduction: Facies, facies sequences and facies models. *In* "Facies Models" (R. G. Walker, ed.), 2nd ed., Repr. Ser. No. 1, pp. 1–9. Geoscience Canada. Toronto.

Walmsley, P. J. (1975). The Forties Field. *In* "Petroleum and the Continental Shelf of North West Europe" (A. W. Woodland, ed.), pp. 477–486. Applied Science, London.

Walsh, J. J. (1987). "On the Nature of Continental Shelves." Academic Press, London. 515pp.

Walther, J. (1893). "Einleithung in die Geologie als Historische Wissenschaft," vol. I. Fischer, Jena. 196pp.

Walther, J. (1894). "Einleitung in die Geologie als Historische Wissenschaft," vol. 3, pp. 535–1055. Fischer, Jena.

Wanless, H. R., Baroffio, J. R., Gamble, J. C., Horne, J. C., Orlopp, D. R., Rocha-Campos, A., Souter, J. E., Trescott, P. C., Vail, R. S., and Wright, C. R. (1970). Late Paleozoic deltas in the central and eastern United States. *Spec. Publ.— Soc. Econ. Paleontol. Mineral.* **15,** 215–245.

Warren, J. K. (1989). "Evaporite Sedimentology." Prentice-Hall, Englewood Cliffs, NJ. 285pp.

Weber, K. J. (1971). Sedimentological aspects of oil fields in the Niger Delta. *Geol. Mijnbouw* **50,** 559–576.

Weimer, R. J. (1961). Spatial dimensions of Upper Cretaceous Sandstones, Rocky Mountain area. *In* "Geometry of Sandstone Bodies" (J. A. Peterson and J. C. Osmond, eds.), pp. 82–97. Am. Assoc. Pet. Geol., Tulsa, OK.

Weimer, R. J. (1970). Rates of deltaic sedimentation and intrabasin deformation, Upper Cretaceous of Rocky Mountain region. *Spec. Publ.— Soc. Econ. Paleontol. Mineral.* **15,** 270–293.

Weimer, R. J. (1992). Developments in sequence stratigraphy: foreland and cratonic basins. *AAPG Bull.* **76,** 965–982.

Weimer, R. J., Howard, J. D., and Lindsey, D. R. (1982). Tidal flats. *Mem.— Am. Assoc. Pet. Geol.* **37,** 191–246

Werren, E. G., Shew, R. D., Adams, E. R., and Sutcliffe, R. J. (1990). Meander-belt reservoir geology, Mid-Dip Tuscaloosa, Little Creek Field, Mississippi. *In* "Sandstone Petroleum Reservoirs" (J. H. Barwis, J. G. McPherson, and J. R. J. Studlick, eds.), pp. 85–108. Springer-Verlag, Berlin.

Wickremeratne, W. S. (1986). Preliminary studies on the offshore occurrences of monazite bearing heavy mineral placers, southwestern Sri Lanka. *Mar. Geol.* **72,** 1–10.

Wignall, P. B. (1994). "Black Shales." Oxford University Press, Oxford. 144pp.

Wilgus, C. K., Posamentier, H., Ross, C. A., and Kendall, G. St. C. (eds.) (1988). Sea-level changes: an integrated approach. *Soc. Econ. Pal. & Min. Sp. Pub. 42.*

Williams, G. E. (1969). Characteristics and origin of a PreCambrian pediment. *J. Geol.* **77,** 183–207.

Williams, G. E. (1971). Flood deposits of the sandbed ephemeral streams of Central Australia. *Sedimentology* **17,** 1–40.

Williams, M. A. J., and Balling, R. C. (1996). "Desertification and Climate." Arnold, London.

Wilson, H. H. (1969). Late Cretaceous and eugeosynclinal sedimentation, gravity tectonics and ophiolite emplacement in Oman Mountains, southeast Arabia. *Am. Assoc. Pet. Geol. Bull.* **53,** 626–671.

Wilson, J. E., and Jordan, C. (1983). Middle shelf. *Mem.—Am. Assoc. Pet. Geol.* **33,** 345–462.

Wilson, J. L. (1975). "Carbonate Facies in Geologic History." Springer-Verlag, Berlin. 471pp.

Wilson, J. L. (1981). A review of carbonate reservoirs. *Bull. Can. Pet. Geol.* **29,** 95–117.

Wolf, K. H. (1981). "Handbook of Strata-bound Ore Deposits," 9 vols. Elsevier, Amsterdam.

Wolf, S. C. (1970). Coastal currents and mass transport of surface sediments over the shelf regions of Monterey Bay, California. *Mar. Geol.* **8,** 321–336.

Woncik, J. (1972). Recluse field, Campbell County, Wyoming. *Mem.—Am. Assoc. Pet. Geol.* **16,** 376–382.

Wright, L. D., and Coleman, J. M. (1973). Variations in morphology of major river deltas as functions of ocean wave and river discharge regimes. *Am. Assoc. Pet. Geol. Bull.* **57,** 370–417.

Wright, V. P., and Burchette, T. P. (1996). Shallow-water carbonate environments. *In* "Sedimentary Environments: Processes, Facies and Stratigraphy" (H. G. Reading, ed.), 3rd ed., pp. 325–394. Blackwell, Oxford.

Wright, V. P., and Burchette, T. P., eds. (1999). "Carbonate Ramps," Spec. Publ. No. 149. Geol. Soc. London, London. 472pp.

Yen, T. F., and Chilingarian, G. V., eds. (1976). "Oil Shale." Elsevier, Amsterdam. 292pp.

Part III
Sediment to Rock

The action of heat at various depths in the earth is probably the most powerful of all causes in hardening sedimentary strata.

Charles Lyell (1838)

7 The Subsurface Environment

The previous chapters have traced the story of sediments, from weathering, erosion, and transportation, until their deposition in diverse sedimentary environments. The third and final part of the story is to tell how sediments are turned back into solid rock, a process termed **lithifaction.** Before doing this, it is essential to consider the subsurface environment in which these changes take place. First, it is necessary to consider the temperatures and pressures that prevail, then the chemistry of the various subsurface fluids in which the many chemical reactions take place, and finally the motions of those fluids through the sedimentary basin.

7.1 SUBSURFACE TEMPERATURES

It is abundantly clear to miners that temperature increases with depth beneath the surface of the earth. Detailed observations show, however, that the increase in temperature with depth varies regionally and vertically. The internal heat of the earth is believed to be derived from the radioactive breakdown of potassium, uranium, and thorium (Lillie, 1999). The heat flow of the earth's crust is the product of the geothermal gradient and the thermal conductivity of the rocks. The global average heat flow is some 1.5 μcal/(cm^2)(s), but it varies from place to place. Old Pre-Cambrian shield areas have rates of 1.0 μcal/(cm^2)(s), active zones of subduction have rates of about 1.5 μcal/(cm^2)(s), and mid-ocean ridges have rates in excess of 3 μcal/(cm^2)(s) (Press and Siever, 1982). These variations are directly related to the thickness of the earth's crust and to convection cells in the mantle, as discussed in Section 10.1.3. The geothermal gradient is normally expressed as the rate of temperature increase with increasing depth. The global average geothermal gradient is taken as about 22°C/km, but ranges from as low as 10 in old shield areas, to as much as 50°C/km in active zones of sea floor spreading (North, 1985). When measured in a borehole, however, there may be several zones with different geothermal gradients. These normally reflect variations in the thermal conductivity of the strata penetrated.

The thermal conductivity of rocks ranges from as high as 5.5 W/m/°C for evaporites such as halite and anhydrite, down to 0.3 W/m/°C for coal. Values of conductivity for sandstones range from 2.6 to 4.0, and for limestones from 2.8 to 3.5 W/m/°C (data from Evans, 1977). Variations in conductivity in sands, clays, and carbonates are largely due

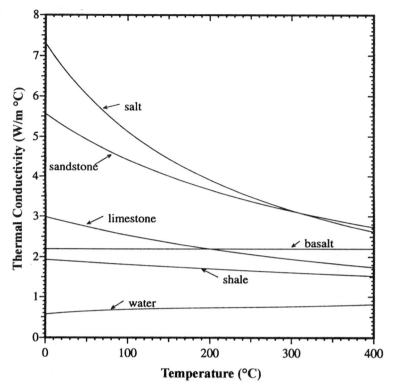

Fig. 7.1. Thermal conductivity of rock grains and water as a function of temperature. Rock grain thermal conductivity is equivalent to that of sediments with zero porosity. (From Mello, U. T., Karner, G. D., and Anderson, R. N. 1995. Role of salt in restraining the maturation of subsalt source rocks. *Mar. Pet. Geol.* **12**, 697–716. Copyright 1995, with permission from Elsevier Science.)

to porosity, and to some extent to mineralogy. Porous formations have lower conductivity than their lithified counterparts, due to the low conductivity of pore fluids. The thermal conductivity of minerals declines with increasing temperature, though water is hardly affected (Fig. 7.1). Thus as a sediment is buried, temperature increases and mineral conductivity decreases. Since porosity normally declines with increasing depth, there is a concomitant decrease in the amount of low conductive water. This counterbalances the previous effect, to some degree.

A sequence of strata of different interbedded lithologies will show an erratic vertical geothermal gradient profile (Fig. 7.2). Low gradients characterize highly conductive rocks, such as evaporites and cemented sandstones and carbonates. High gradients characterize rocks of low conductivity, such as overpressured clays. Note, however, that heat can be transferred through rocks in two ways. It can be conducted through minerals and also through fluids. In the case of igneous and metamorphic rocks, there are few fluid-filled pores, so heat is transferred through minerals alone. In sediments that are porous, but impermeable, the heat transfer will be by conduction through rock grains and pore fluids. In sediments that are porous and permeable, however, though heat may be transferred by conduction through grains and fluid, it may also be transferred by fluids migrating in response to thermal, pressure, or density gradients.

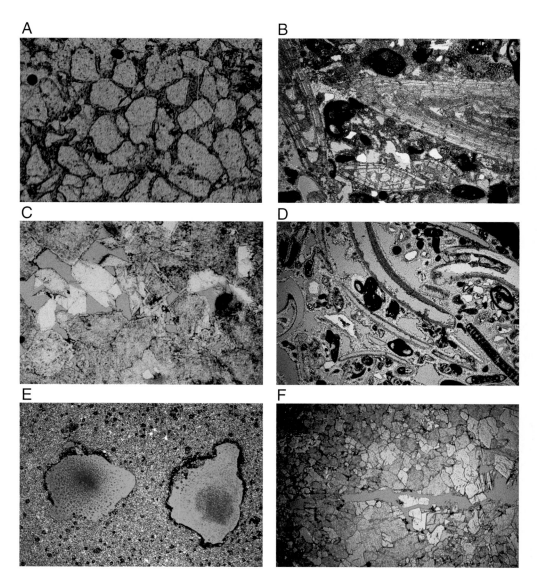

Plate 1 Photomicrographs illustrating the various types of pore systems. All thin sections have been impregnated with blue resin to display porosity, and are photographed in ordinary light. (A) Primary intergranular (interparticle) porosity in Folkestone Beds, Lower Cretaceous, Dorking, Surrey. (x40) (B) Skeletal limestone with primary intergranular and intragranular (within the bioclasts) porosity. (x10) (C) Intercrystalline porosity in Lower Carboniferous (Mississippian) dolomite, Forest of Dean, England. (x40) (D) Biomoldic (and primary intergranular) porosity in skeletal limestone. (x20) (E) Vuggy porosity, Zechstein (Upper Permian) dolomite, North Sea. Total porosity high, but effective porosity and permeability low because of the isolated nature of the pore system. (x10) (F) Fracture porosity in Zechstein (Upper Permian), North Sea. Fracture porosity normally gives high permeability even for low-porosity values. (x20)

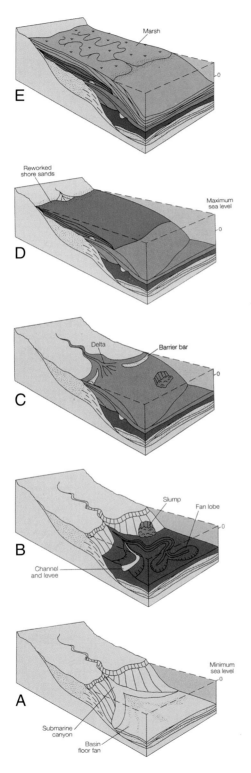

Plate 2 Illustrations of the components of a third-order sequence cycle. Read captions and admire figures in stratigraphic sequence from bottom to top of page. For full explanation see Section 6.4.3. (From Neal *et al.*, 1993, courtesy of *Schlumberger Oilfield Review*.)

(E) Slowly rising sea level permits the slow basinward progradation of thin high-stand systems tract (orange). Then the sea level begins to drop....

(D) Rapid flooding of the shelf permits deposition of transgressive systems tract of shallow marine shelf and coastal sands (green). The top of this tract is marked by a maximum flooding surface, indentified by a condensed sequence.

(C) Sea floods shelf providing accommodation space for prograding deltas (pink).

(B) Rising sea level floods the continental slope, permitting the deposition of slope fans (brown).

(A) Sea level at its lowest point. At the foot of the emergent continental slope a low-stand basin floor submarine fan systems tract is deposited (yellow).

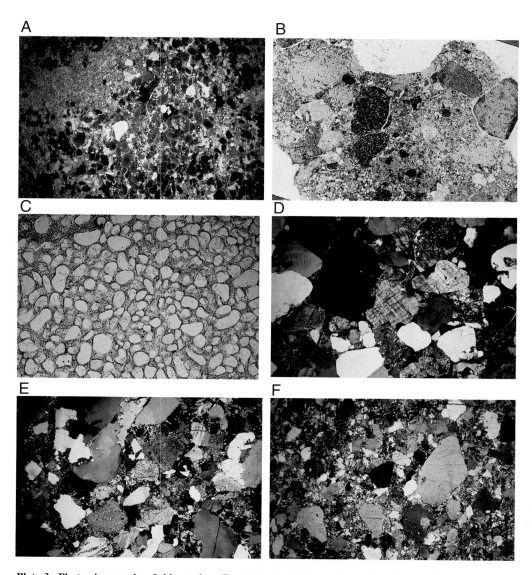

Plate 3 Photomicrographs of thin sections illustrating the different types of sandstone. (A) Sandstone with green glauconite pellets, Lower Cretaceous, North Sea. Ordinary light. (x20) (B) Volcaniclastic sandstone, Richmond Formation, a Tertiary fanglomerate, Jamaica. Composed largely of lithic grains of volcanic origin. Ordinary light. (x20) (C) Simpson Group (Ordovician) orthoquartzite, Arbuckle mountains, Oklahoma. Ordinary light. This is an example of a hypermature orthoquartzite. Note that, despite the age, this is a completely uncemented sand that retains excellent primary intergranular porosity. The pores are filled with a pale brown oil. (x20) (D) Arkosic sandstone from the Torridon Group (Pre-Cambrian), northwest Scotland. Though this sand was deposited in a braided fluvial environment, the well-rounded feldspar grains suggest eolian abrasion prior to deposition. Polarized light. (x40) (E) Jurassic greywacke from the North Sea. Unlike many sands from the same age and area, this sand lacks the properties of a petroleum reservoir. Polarized light. (x40) (F) Paleozoic quartz-wacke from the island of Chios, eastern Aegean. Though this is a poorly sorted sand with a "wacke" texture, the detrital grains are largely composed of quartz and lithic grains of quartzite and chert. Polarized light. (x40)

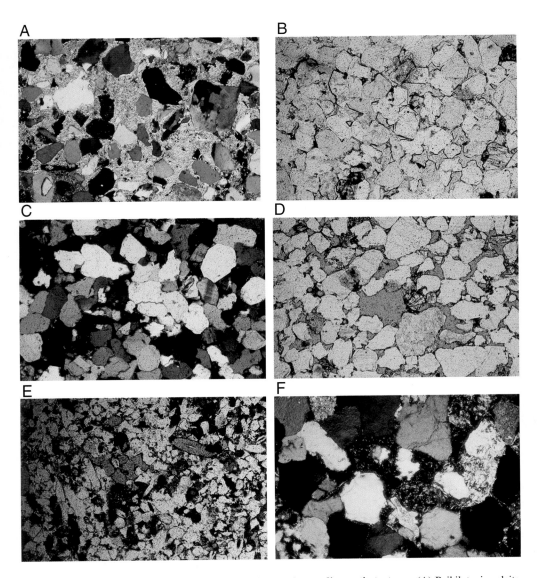

Plate 4 Photomicrographs of thin sections illustrating sandstone diagenetic textures. (A) Poikilotopic calcite cement in which the carbonate crystals completely enclose several detrital grains. At high magnification the grain boundaries are often seen to be corroded. In extreme cases this may have developed to such an extent that the detrital grains appear to "float" in the carbonate cement. Upper Jurassic, Helmsdale, Scotland. Polarized light. (x40) (B) Early-stage quartz cementation in Lossiemouth Sandstone (Permian), Scotland. Silica overgrowths can be seen developed in optical continuity on parent quartz grains. Euhedral crystal faces define boundaries of blue pore spaces. Ordinary light. (x40) (C) Terminal stage quartz cementation in Jurassic sandstone, North Sea. Porosity is negligible. At first glance the grains appear to be welded together by pressure solution. On looking closely, however, one can see that the original boundaries of some grains are defined by dust rims. This shows that porosity loss has been achieved almost entirely by cementation. Polarized light. (x40) (D) Secondary porosity in Brent Group sandstone (Middle Jurassic), North Sea. Note in particular the large pore in the center of the photo. This is too large to be a primary pore preserved during burial. It more probably formed from the dissolution of a feldspar grain. Ordinary light. (x40) (E) Biomoldic secondary porosity in a sandstone due to epidiagenesis in a modern weathering profile. Hythe Beds, Lower Cretaceous, Leith Hill, Surrey. Ordinary light. (x20) (F) Authigenic quartz and kaolinite cement (irregular gray patches of "pepper and salt" texture). Brent Group, Middle Jurassic, North Sea. Polarized light. (x50)

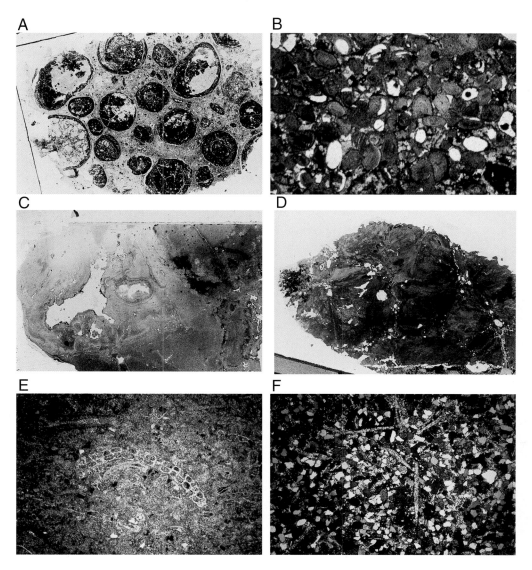

Plate 5 Photomicrographs of miscellaneous sedimentary rocks. (A) Lateritic bauxite with well-developed orbicular texture. The individual orbs are about 0.5 cm in diameter, Le Baux, France. For discussion of origin, see Section 2.3.3.1. (x10) (B) Sedimentary ironstone, Northampton, Middle Jurassic, England. Note the concentric structure of the chamosite ooids. For discussion of origin, see Section 9.4.1. (x20) (C) Guano phosphate deposit, St. Helena, South Atlantic. For discussion of origin, see Section 9.5.1. (x5) (D) Nodular anhydrite from the modern sabkha of Abu Dhabi, United Arab Emirates. For discussion of origin, see Section 9.6.3. (x20) (E) Chert formed by the silicification of skeletal lime mudstone. Chalk, Upper Cretaceous, Dover, England. Polarized light. (x60) (F) Silicified sandstone with chalcedonic sponge spicules and intergranular cement. Sandgate Beds, Lower Cretaceous, Leith Hill, Surrey. Polarized light. (x20)

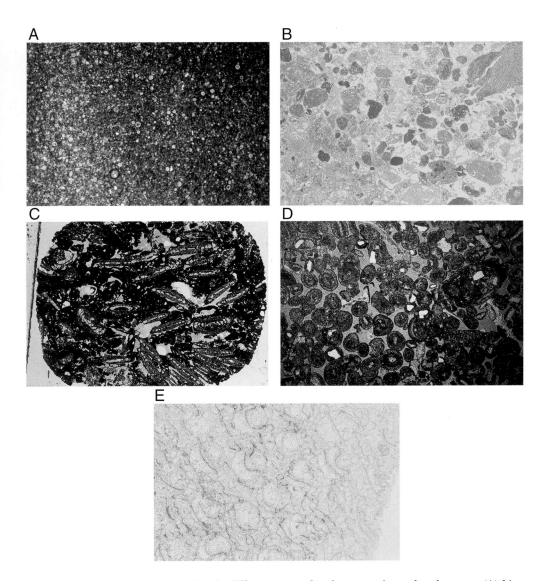

Plate 6 Photomicrographs illustrating the different types of carbonate grains and rock names. (A) Lime mudstone. Chalk, Upper Cretaceous, Beer, Devon. (x60) (B) Wackestone, a lime mudstone with over 10% grains, including both rounded peloids and angular micrite intraclasts. Marada Formation (Miocene), Jebel Zelten, Libya. (x40) (C) Foraminiferal packstone, a grain-supported rock, with over 5% lime mud. Oligocene, West of Marada Oasis, Libya. (x20) (D) Oolitic grainstone, Portland Group, Upper Jurassic, Dorset. This is a grain-supported rock with less than 5% micrite matrix. Note the preserved primary intergranular porosity. (x40) (E) Biolithite, boundstone, or reef rock. A colonial coral, *Halysites*, in which the primary pores of the coral calyces are now infilled by sparite cement. Silurian, Wenlock, England. (x10)

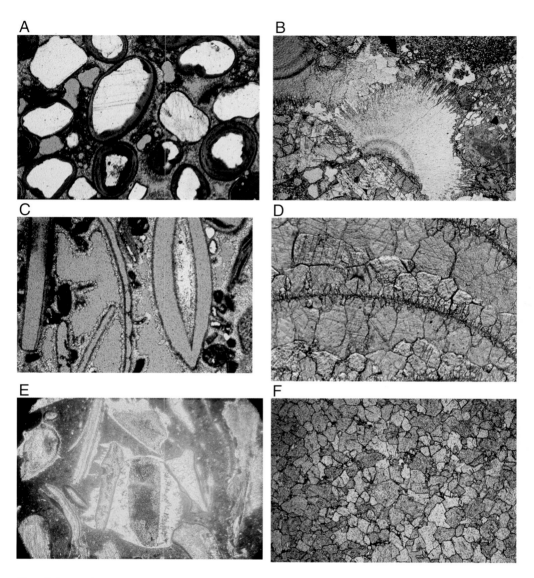

Plate 7 Photomicrographs illustrating carbonate diagenetic textures. (A) Pleistocene beach rock, Kuwait. Note that the fibrous cement rimming the grains is best developed at the pore throats. This "meniscus" effect indicates subaerial exposure and dehydration of the pore fluid in the vadose zone. (x40) (B) Botryoidal aragonite cement in Pleistocene volcaniclastic raised beach. Isle of Raasay, Scotland. (x60) (C) Skeletal limestone still in early diagenesis. Primary intergranular porosity is preserved. Aragonitic skeletal fragments have leached out to form biomoldic pores having boundaries marked by a granular cement that has grown into the primary pores. (x40) (D) Skeletal fragment with advanced calcite cementation. Sparry crystals have grown, with increasing size, both into the intergranular pores, and into the biomoldic pore caused by the solution of the orginal aragonite shell. (x60) (E) Skeletal wackestone with calcite stained pink. Note the syntaxial rim cement overgrowth on the calcitic echinoid plate. These overgrowths only form if the bioclast is composed of a single crystal, and is thus analogous to the syntaxial overgrowths of detrital quartz grains. (x40) (F) Dolomite. This is a coarsely crystalline secondary variety with intercrystalline porosity. Zechstein, Upper Permian, North Sea. It replaces a limestone which was probably a bryozoan "reef". (x20)

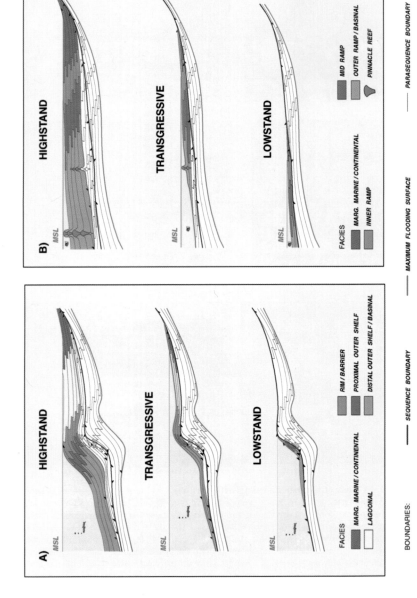

Plate 8 Sequence stratigraphic terminology for carbonate facies and systems tracts. (A) Rimmed carbonate platform. (B) Carbonate ramp. For full explanation see Section 6.4.3. (From Spring and Hansen, 1998, courtesy of the Geological Society of London.)

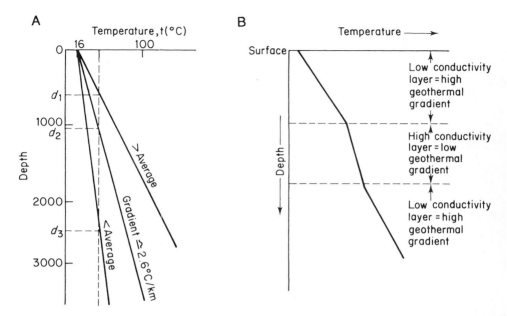

Fig. 7.2. Depth versus temperature graphs illustrating geothermal gradients. (A) Note how a critical temperature t is reached at different depths (d_1, d_2, and d_3) for different geothermal gradients. (B) Note how geothermal gradients may vary vertically through the crust, and are inversely proportional to the thermal conductivity the formations encountered.

Even allowing for regional variations in heat flow, and for the varying thermal conductivities of different rocks, there are still geothermal anomalies in the earth's crust. These are identified by detailed temperature measurements of both surface and subsurface mine and borehole locations. Geothermal contour maps may be compiled from the collated data. The principal causes of geothermal anomalies are igneous intrusions, fluid flow, and the nonplanar geometry of sediments. As a basin undergoes burial, and its sediments are compacted, hot deep water will be expelled. Faults often provide conduits for emigrating fluids, and may thus cause local thermal anomalies along the trend of the fault. Conversely, when basins are uplifted a hydrostatic head may build up in the aquifers of mountainous terrain. Cool meteoric water may thus move deep into the basin along permeable beds to cause anomalously low temperatures

Hot spots may also occur over salt domes and igneous intrusions, and within overpressured clay diapirs, but for different reasons in each case. Salt has a high thermal conductivity. This favors the rapid transfer of heat up the dome toward the surface. This means that over the crests of salt domes porosity is lower and source rocks are more mature than down-flank. Conversely, salt domes generate cool anomalies beneath them, so porosity may be higher and source rocks may be less mature than away from the salt dome (Mello *et al.*, 1995).

Igneous intrusions are naturally very hot when first emplaced. Even long after intrusion they may serve as thermal conduits because of their good conductivity. Overpressured clay has a low thermal conductivity because of the high water content. For the same reason overpressured clays have a lower density than compacted normally pressured ones. Overpressured clay formations thus tend to flow as they are displaced

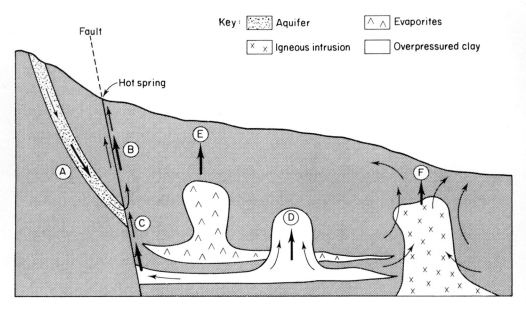

Fig. 7.3. Illustration through a basin showing causes of geothermal anomalies. Thin arrows show directions of water flow. Thick upward pointing arrows indicate positive anomalies ("hot spots"), thick downward pointing arrows indicate negative anomalies. (A) Negative anomaly due to deep meteoric water penetration. (B) Positive anomaly due to upward discharge of hot water along a permeable fault. (C) Positive anomaly due to discharge of superheated water from overpressured clay formation. (D) Positive anomaly within low conductivity overpressured clay diapir. (E) Positive anomaly above highly conductive salt dome. (F) Positive anomaly above igneous intrusion; note peripheral convection currents.

by the overlying denser sediments. In this way overpressured clay may generate diapirs that are analogous to salt domes. Water moves to the crestal position of mud diapirs where it is retained in the closed pressure system. Thus heat energy is retained and domes are positive geothermal anomalies. Where hot water is able to escape from overpressured systems (up faults for example), local positive geothermal anomalies will occur (Fig. 7.3).

An understanding of geothermal gradients and anomalies is useful because the rate of diagenetic reactions is thermally controlled. As will be shown when discussing porosity gradients in sands (see Section 8.5.3.2), porosity tends to be lost faster in areas of high geothermal gradient than in areas with low gradients.

7.2 SUBSURFACE PRESSURES

It is axiomatic that pressure increases with depth beneath the earth's crust. According to Terzaghi's law, the effective overburden pressure is equal to the total vertical pressure exerted by the rock and pore fluid, minus the fluid pressure (Terzaghi, 1936; see also Skempton, 1970). This relationship is also expressed in the Hubbert and Rubey (1959) equation, which states that the total, or overburden pressure, is equal to the sum of the lithostatic and fluid pressures. The lithostatic pressure is the pressure of just the

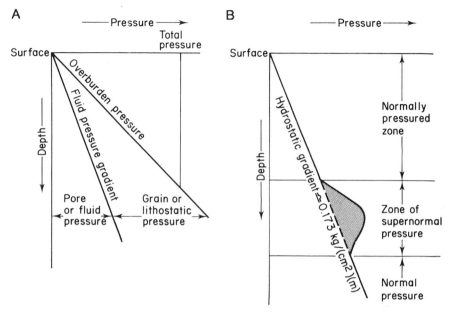

Fig. 7.4. Depth versus pressure graphs illustrating the concept of pressure gradients. (A) Total effective over-burden pressure is the sum of the lithostatic (grain) pressure and the fluid (pore) pressure. (B) Variation in pressure gradient caused by a zone of superpressure, or overpressure, that exceeds the normal hydrostatic pressure gradient.

rock, transmitted through grain contacts. It is thus referred to also as grain pressure. The fluid pressure is that due to the column of fluid within the pores of the sediment (Fig. 7.4). The fluid pressure may be hydrostatic or hydrodynamic. The hydrostatic pressure is that exerted by the vertical column of water at rest. Hydrodynamic pressure is that due to moving fluid. For pure water the hydrostatic gradient is 0.173 kg/(cm)(m). In the subsurface environment of course the connate fluids contain dissolved salts, so the gradient is somewhat higher than this value. Fluid pressure is related to the density of the fluid, and thus varies with salinity and temperature.

Pressures less than hydrostatic, termed "subnormal," are rarely encountered. Pressures above hydrostatic, termed "supernormal," or "overpressure" are quite common. Overpressure can develop in a sedimentary basin in many ways (Plumley, 1980; Osborne and Swarbrick, 1997). One of the most common is where a clastic wedge progrades too fast to allow clays to compact and dewater (Fig. 7.5). Overpressure is of great significance both in sediment diagenesis and petroleum generation and migration.

The potentiometric, or piezometric, surface is defined as the level to which water will rise in a water wall or borehole in the subsurface, and corresponds to the water table at shallow depths. The potentiometric surface may be mapped across a basin using water well and borehole data (Fig. 7.6). The potentiometric surface normally slopes from mountains toward the sea. Hubbert (1953) showed that the piezometric surface could also be expressed as the fluid potential, calculated as follows:

$$\text{Fluid potential} = Gz + \frac{p}{\rho},$$

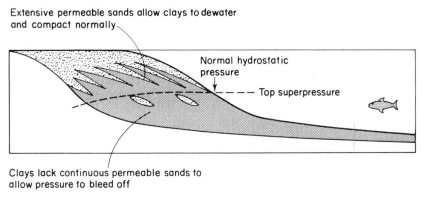

Fig. 7.5. Naivogram through a delta showing how overpressure may develop in prodelta clays because they lack permeable sand layers to permit pore pressure to decline during compaction.

where G is the acceleration due to gravity, z is the datum elevation at the site of pressure measurement, p is the static fluid pressure, and ρ is the density of the fluid. More simply, the fluid pressure is the product of the head of water and the acceleration due to gravity. For a full account of the hydrodynamics of groundwater movement, see Dahlberg (1982).

Normally rock density increases with depth, and porosity decreases with depth. These changes reflect compaction due to the increasing overburden pressure. There are two exceptions to this general statement. Overpressured clays, as already expounded, have anomalously high porosity. This is because pore fluid cannot escape and allow compaction to take place. Any isolated sand beds within a sealed pressure system will also have higher porosities and lower densities than their normally pressured counterparts.

Evaporites are the second exception to the rule that density increases with depth. Halite has density of about 2.03 g/cm^3. Evaporites are deposited with negligible porosity. They do not compact during burial so their density is unchanged. Evaporites are denser than surface sediments. But sands and muds compact as they are buried and, at about 800 m depending on the burial curve, evaporites are less dense than other sediments. They are thus displaced by denser overburden and may flow upward into salt domes. These ideas of density and porosity curves and their anomalies are illustrated in Fig. 7.7.

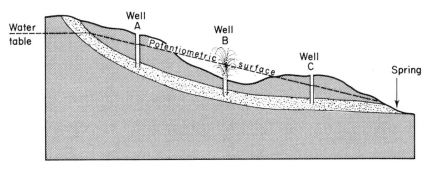

Fig. 7.6. Cross-section illustrating the potentiometric or piezometric surface for a permeable aquifer (stippled)

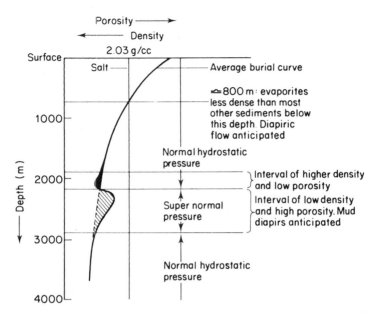

Fig. 7.7. Depth versus density graph showing typical sediment burial curve and porosity relationships. Note how porosity is preserved in overpressured intervals. There is often an overcompacted zone immediately above. Evaporites do not compact significantly, and at about 800 m they are thus less dense than the overburden and, therefore, salt diapirs may be anticipated.

7.3 SUBSURFACE FLUIDS

Diagenesis is controlled both by temperature and pressure, the fundamental geological aspects of which have just been outlined. Now it is approprite to consider the chemistry of the fluids that infill the pores of sediments. These are many and varied, both gaseous and liquid. They are listed and classified in Table 7.1 and are now described.

7.3.1 Nonhydrocarbon Gases

Sediments above the water table will have their pores infilled with atmospheric gases. These normally generate an oxidizing environment at the present day. It was noted earlier, however, when discussing the origin of Pre-Cambrian uraniferous conglomerates (see Section 6.3.2.2.4), that this was not always so. The earth's early atmosphere was enriched in carbon dioxide, and was probably reducing until the rise of the algae whose photosynthetic reactions modified the atmosphere.

   Traces of the inert gases helium, argon, krypton, and radon occur in the subsurface, normally dissolved in connate fluids. They are believed to form from the decay of radioactive minerals in the mantle and to move up into the sedimentary cover along faults and fractures (Sugisaki *et al.,* 1983; Gold, 1979, 1999; Gold and Soter, 1982). Nitrogen, carbon dioxide, hydrogen sulfide, and hydrogen also occur in the subsurface, again normally dissolved in connate waters. They appear to be of mixed origin. These gases have been recorded emanating from volcanoes, and may thus originate from the mantle. They may also form through the diagenesis of organic matter both at shallow and great

Table 7.1
Classification of Subsurface Fluids[a]

	Fluid	Origin
Inert gases	Helium⎫ Argon⎬ Krypton⎭ Radon	Mantle outgassing
Petroleum	Nitrogen⎫ Carbon dioxide⎪ Hydrogen sulfide⎬ Hydrogen⎪ Methane⎭ Condensate Crude oil	Mixed Organic
Water	Meteoric Connate Juvenile	Surface runoff Evolved from runoff or seawater Magmatic

Gases ←———
Liquids ——→

[a] Condensates are varieties of petroleum that are gaseous in the subsurface, but condense to liquids at normal temperatures and pressures. They include ethane, propane, and butane.

depths. The shallow reactions are normally caused by bacterial action, the deeper ones by thermal effect. When these gases occur in volcanic emanations they may in part be of recycled sedimentary origin as well as mantle derived. Nitrogen is essentially inert, but carbon dioxide and hydrogen sulfide play a very important part in diagenesis. Carbon dioxide content affects the pH of the pore fluid, and hence the solubility of many minerals, whereas hydrogen sulfide plays an important part in the formation of metallic sulfide ores (see Section 9.6.5).

7.3.2 Petroleum Fluids

Petroleum is the name given to fluid hydrocarbons. These occur in both gaseous and liquid states, partly controlled by their chemical composition, and partly according to the pressure and temperature of their environment. Methane, often referred to as "dry gas," is of diverse origin. Some methane forms biogenically near the surface of the earth through the bacterial degradation of organic matter. This is often referred to as "marsh gas." As discussed earlier this is a possible cause of "birds-eye" fabric in shallow water carbonate muds (see Section 3.2.2.3.2). Methane also forms from the thermal alteration of buried organic matter, and this is discussed in more detail later. Finally, some methane is derived from the mantle (Gold, 1979, 1999; Gold and Soter, 1982). The diverse origin of methane can be established from its ^{12}C–^{13}C ratio, because this varies according to its source. Particularly high amounts of mantle-derived methane have been encountered emerging from both continental and oceanic rift valleys (MacDonald, 1983).

"Condensate" is the term given to petroleum that is gaseous at high subsurface temperatures, but condenses to a liquid at normal temperatures and pressure at the earth's surface. Chemically, condensate includes ethane, propane, and butane. Liquid petroleum is referred to as crude oil, and is a complex mix of paraffinic, aromatic, naphthenic, and other compounds. The origin of petroleum is beyond the scope of this text. It must be considered briefly, however, because it is a common pore fluid and obviously an important one. The stimulus for most sedimentological research lies in the quest for petroleum. From the days of Mendele'ev down to Gold (1979, 1999), chemists and astronomers have argued for an inorganic abiogenic origin for petroleum. They note that complex hydrocarbons occur in space, and on earth in chondritic meteorites, and escaping from volcanoes. It has been argued that hydrocarbons may form in the mantle in a reaction analogous to that which produces acetylene from calcium carbide and water:

$$CaC_2 + 2H_2O = C_2H_2 + Ca(OH)_2.$$

The evidence for mantle outgassing of methane and radiogenic gases has already been discussed. It is further argued, however, that the range of petroleum compounds encountered in sediments form from mantle-derived methane by polymerization. Space does not permit this theory to be reviewed, but it is a fact that commercial quantities of petroleum normally occur in sedimentary rocks. Where large amounts of petroleum occur in igneous and metamorphic rocks there are sediments nearby. Detailed chemical analysis shows the original biogenic origin of petroleum and can identify the source formation from which it was derived (Tissot and Welte, 1984; Hunt, 1996; Selley, 1998). Petroleum geologists now generally believe that some organic matter is buried in various anoxic environments (see Sections 6.3.2.4.4 and 6.3.2.7.1). On burial, organic matter evolves into **kerogen,** defined as hydrocarbon, insoluble in normal petroleum solvents, that generates liquid petroleum when heated. There are different types of kerogen, some (such as coal) generate only gas, some kerogen gives off both oil and gas, some principally oil. As kerogen is heated it undergoes decarboxylation, generating carbon dioxide in solution as carbonic acid. This may have an important role in generating secondary porosity in sandstones (see Section 8.5.3.4). After decarboxylation, oil is generated between about 60 and 120°C, and gas between about 120 and 220°C. Above these temperatures the kerogen evolves into graphite and is essentially inert. Petroleum emigrates from the kerogen bearing shales that normally serve as source beds. The exact mechanics of migration, whereby petroleum emigrates from an impermeable formation, is still a matter for debate (Baker, 1996). There is now, however, a growing opinion that the primary migration of petroleum from a source bed takes place during the episodic expulsion of fluids from overpressured shales (e.g., Cartwright, 1994; Roberts and Nunn, 1995). Once in permeable strata, however, petroleum is lighter than the ambient connate fluid. It will thus migrate upward in response to buoyancy. Petroleum may ultimately reach the earth's surface and be dissipated as an oil or gas seep.

Occasionally, however, petroleum may be trapped in the subsurface. There are many types of petroleum traps. Some are due to structure, such as folds or faults; some are in domes that are overly salt or overpressured clay diapirs; and some are due to stratigraphy. Stratigraphic traps include depositional pinchouts and truncations, as well as channels, sandbars, and reefs (Fig. 7.8).

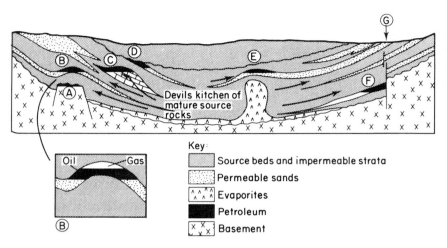

Fig. 7.8. Illustration of petroleum migration and entrapment. Arrows indicate directions of petroleum migration from thermally mature organic-rich source beds in the basin center. Various types of trap are shown as follows: (A) Weathered basement on fault block, (B) anticlinal trap, (C) limestone reef, (D) truncation trap, (E) closure over salt dome, (F) trap due to sealing fault, (G) surface petroleum seep. The detail of trap (B) shows how gas, oil, and water undergo gravity segregation in traps due to their different densities.

7.3.3 Subsurface Waters

Table 7.1 shows that there are three types of water that infill sediment pores: meteoric, connate and juvenile. Meteoric water originates from the precipitation of rain and snow from the atmosphere (Toth, 1999). Connate water has evolved from the original water associated with deposition, and juvenile waters are of magmatic origin.

Rainwater is normally oxidizing and acidic. The causes of the acidity are various. It is due to nitrous acid formed during thunderstorms, to carbonic acid from the atmosphere, and to sulfurous acid from volcanic gases. As rainwater seeps through the soil it may also acquire humic acids from decaying organic matter.

Thus when meteoric water enters sediments it has considerable potential for causing chemical reactions. Within the shallow sedimentary profile there is normally an upper zone, where the pores are open to the atmosphere. This is termed the "vadose zone." It is absent in marshes, but may be many tens of meters deep in deserts. The vadose zone is separated by the water table from the "phreatic zone," in which pores are saturated with liquid. The water table is effectively synonymous with the piezometric surface discussed earlier. Locally the water table may appear as a planar surface, but regionally it slopes toward the surface of seas, lakes, and rivers. Impermeable strata may serve, however, as permeability barriers to groundwater flow. These are termed "aquacludes," and may give rise to perched water tables with their own local piezometric surfaces (Fig. 7.9).

Connate water was once defined as the original (normally marine) water with which sediment was deposited. It is now realized that this is far too simple a definition because of the chemical reactions that take place between a sediment and its pore fluids immediately after deposition. A more appropriate definition is that connate waters are those

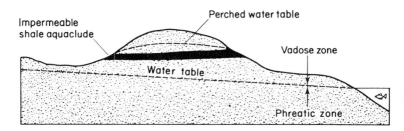

Fig. 7.9. Cross-section illustrating vadose and phreatic zones, and normal and perched water tables. In the vadose zone the pores are open to the atmosphere, but are subjected to infiltrating rain water. In the phreatic zone the pores are permanently saturated.

"which have been buried in a closed hydraulic system and have not formed part of the hydraulic system for a long time" (White, 1957).

A distinction is often made between "compactional water," which is evolved connate water moving through sediments as a result of compaction, and "thermobaric water," which is hot, highly pressured water of deep basinal origin, normally formed by clay dehydration (Galloway, 1984). Connate waters are markedly different from meteoric waters in their temperature, salinity, and chemistry (Drever, 1997). The boundary between meteoric and connate waters is sometimes quite sharp. Fresh meteoric water is less dense than saline connate water. Thus islands of highly permeable limestone may have a biconvex lens of freshwater that floats on more dense connate water. The density contrast is obviously accentuated by the proximity of saline seawater. One hazard of producing water wells in such situations is that, after being used for some time, saline water may cone up to the well bore (Fig. 7.10). In some situations, however, meteoric water, though less dense than connate water, may actually displace it. This occurs when a high hydrostatic head develops in upland areas. Today, for example, meteoric water flows northward through extensive Saharan aquifers to emerge as submarine freshwater springs beneath the surface of the Mediterranean (Fig. 7.11).

The complex chemistry of connate waters has been described by many geologists both from the petroleum and mineral exploration aspects (see Collins, 1975, 1980, and Eugster, 1985, respectively). Important variables to consider are the pH, Eh, salinity, and

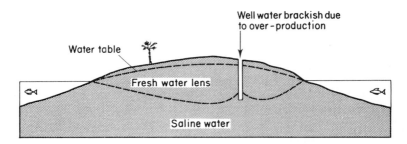

Fig. 7.10. Cross-section through a permeable limestone island, such as a coral atoll, to show how a biconvex freshwater lense may overlie denser saline water. Diagenetic reactions are different in the various zones (see Section 9.2.5).

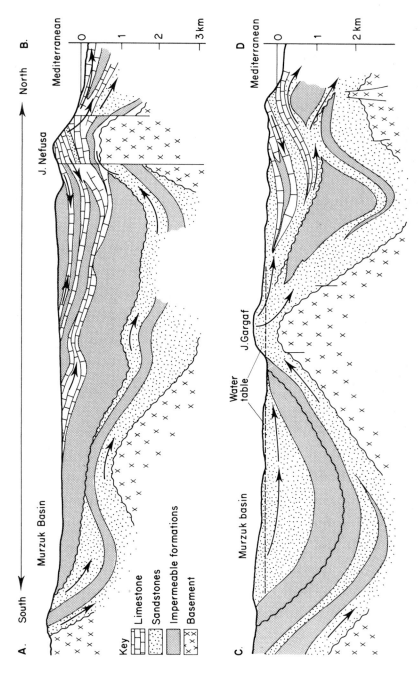

Fig. 7.11. Cross-sections through western Libya to show groundwater flow across the Sahara toward the Mediterranean. Salinity gradually increases with depth and from south to north, except for areas of recharge in the jebels Nefusa and Gargaf. Note the greatly exaggerated scale. (From Pallas, 1980. Courtesy of Academic Press.) For location of sections, see continuation of Fig. 7.11.

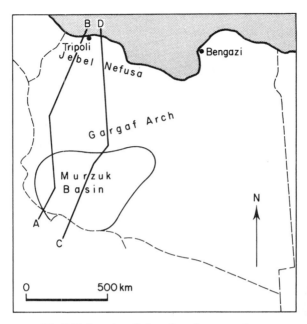

Fig. 7.11 (*continued*) Locality of cross-sections.

composition of connate fluids. Calculation of these parameters from subsurface waters is extremely difficult. Water samples recovered from boreholes have often been contaminated by drilling fluids. Nowadays, however, it is possible to recover microscopic samples from fluid inclusions in authigenic minerals. The retained fluids include gas, connate water, and sometimes even petroleum (Fig. 7.12). These samples can be analyzed to measure the temperature at which the crystal grew and the chemical composition of the fluids from which it was precipitated (Goldstein and Reynolds, 1994).

Fluid inclusion studies show that connate water has a wide range of pH and Eh values. Acidic oxidizing pore fluids occur where there is a large component of meteoric water (Raffensperger and Garvan, 1995a,b). Such fluids are responsible for uranium rollfront ore bodies (see Section 6.3.2.2.4). Acid brines, but of deeper origin, characterize the connate fluids from which Mississippi Valley sulfide ores were formed (see Section 9.6.5). The fluids responsible for exhalative lead–zinc sulfides are normally alkaline and reducing. Oil field brines are also normally alkaline and reducing. The reduction is due to the proximity of hydrocarbons. Oil field waters show a wide range of Eh and pH values, however, and are even oxidizing and acidic in shallow fields with meteoric bottom waters (Fig. 7.13).

Similar variability is seen in the salinity of connate fluids. Salinity, expressed as the total dissolved solids, is measured in parts per million (p.p.m.), but is more conveniently expressed as milligrams per liter (mg/l):

$$mg/l = \frac{p.p.m.}{density}.$$

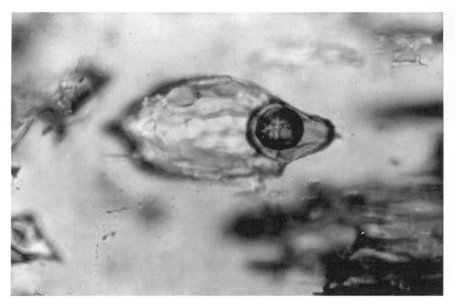

Fig. 7.12. Photomicrograph of an authigenic quartz crystal with a fluid inclusion that contains a globule of oil. (Courtesy of A. Rankin.) It is possible to analyse the inclusion to discover the temperature and chemistry of the fluids in which the crystal grew.

Seawater, which may be taken as a benchmark, normally has a salinity of about 35,000 p.p.m. Subsurface waters range from 0.0 near the surface to more than 600,000 in exceptional cases. Normally salinity increases with depth. Anomalously high salinities are recorded adjacent to evaporite formations. Anomalously low salinities are recorded

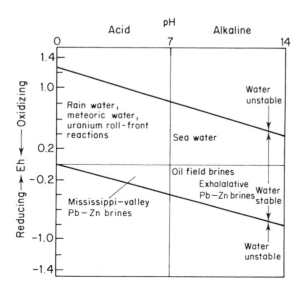

Fig. 7.13. Eh vs pH graph showing the characteristics of rainwater, meteoric water, seawater, and the various types of connate fluid. No attempt has been made to define the boundaries of the different fluids, because of their wide ranges.

beneath unconformities, in overpressured clays, and where there has been extensive infiltration by meteoric water.

The composition of connate water varies considerably from seawater. It is normally enriched in potassium, sodium, and chlorides, but contains fewer sulfates and magnesium. Connate waters also often contain less calcium than seawater, probably due to precipitation of carbonate minerals. These differences demonstrate the evolution that seawater undergoes when it is trapped in the pores of a subsiding sedimentary pile.

The last type of pore water to consider is juvenile or hydrothermal water. This is defined as water that is of deep magmatic origin, and has only just entered the hydrologic cycle. Formerly, almost all epigenetic mineralization was attributed to precipitation from magmatic fluids. Several lines of evidence, including fluid-inclusion analysis, show that many epigenetic mineral deposits are precipitated from hot brines. Stable isotope analysis shows that these brines are not always of magmatic origin, but have evolved from connate waters. This is consistent with the fact that many epigenetic deposits, such as Mississippi Valley ores, occur peripheral to evaporite basins. The source of the metals is often volcaniclastic or from normal sediments or even seawater. The metals have then been concentrated by complex reactions in which organic matter and clay minerals often play catalytic roles. It has long been noted that oil field waters contain significant concentrations of metallic ions. The following reactions are often cited (e.g., Dunsmore, 1973; Eugster, 1985):

$$CaSO_4 + CH_4 \rightarrow CaCO_3 + H_2S + H_2O.$$

The hydrogen sulfide may then react with metallic chlorides:

$$MeCl_2 + H_2S \rightarrow MeS + 2HCl,$$

where Me is a metallic cation.

This brief review of pore fluids shows their diversity and the complexity of their interrelationships. So far they have been considered to be at rest and in isolation. Now it is time to discuss how they move within sedimentary basins.

7.4 FLUID FLOW IN SEDIMENTARY BASINS

There are three ways in which fluids move in sedimentary basins: by hydrostatic head, by sediment compaction, and by convection (Fig. 7.14). These are now considered in turn.

7.4.1 Meteoric Flow

The concept of the piezometric surface, and the idea of meteoric water displacing denser connate water, has already been described and illustrated (Fig. 7.11). This type of fluid movement is naturally most important at shallow depths and in basins that have been uplifted above sea level (Ingebritsen and Sanford, 1998). Hydrodynamic flow is important as a mechanism for generating secondary porosity in weathering zones that may ultimately be preserved beneath unconformities. It sometimes traps petroleum in flexures that lack true vertical closure (termed "hydrodynamic traps") (Dahlberg, 1982), and it is responsible for uranium roll-front mineralization (see Section 6.3.2.2.4).

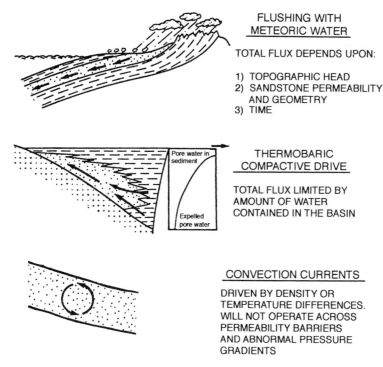

FLUSHING WITH
METEORIC WATER

TOTAL FLUX DEPENDS UPON:

1) TOPOGRAPHIC HEAD
2) SANDSTONE PERMEABILITY
   AND GEOMETRY
3) TIME

THERMOBARIC
COMPACTIVE DRIVE

TOTAL FLUX LIMITED BY
AMOUNT OF WATER
CONTAINED IN THE BASIN

Pore water in
sediment

Expelled
pore water

CONVECTION CURRENTS

DRIVEN BY DENSITY OR
TEMPERATURE DIFFERENCES.
WILL NOT OPERATE ACROSS
PERMEABILITY BARRIERS
AND ABNORMAL PRESSURE
GRADIENTS

Fig. 7.14. Sketches illustrating the three types of fluid flow in sedimentary rocks. (From Burley, 1993, by courtesy of the Geological Society of London.)

7.4.2 Compactional Flow

Once sediment has been deposited it compacts. Compaction rates vary, according to sediment type. They are low in carbonates and sands, and high in clays. The conventional view is that excess pore water is squeezed out of the compacting sediment as the overburden pressure increases during burial. Van Elsberg (1978) points out, however, that this is only a partially correct view. Considered on a basin-wide scale, what really happens is not that the water moves, but that sediment continues to sink through the pore fluids. A series of different layers of connate water has been identified in terrigenous sedimentary basins. These are described in the section on sandstone diagenesis (see Section 8.5.3).

Locally, pore water flows from compacting clays into permeable sands maintaining a hydrostatic pressure regime. In such situations there need be no further fluid movement due to compaction, though other flow mechanisms then may take over. In thick impermeable clay sequences compaction cannot occur and overpressure develops, as previously discussed. It is in these closed systems that superheated, highly pressurized thermobaric connate waters evolve. As a basin matures these waters slowly bleed off when mud diapirs begin to penetrate overlying sand beds and superficial faults that may serve as pressure drains (Berner, 1980; Galloway, 1984).

7.4.3 Convective Flow

Convection currents are an important process in nature. They generate thunderstorms in the atmosphere, and convection cells in the mantle are believed to be responsible for plate tectonics. There is some evidence for convective flow of pore fluids in sedimentary basins. On a small scale convection currents have been shown to exist within oil fields a few kilometers across (Combarnous and Aziz, 1970). There is evidence, however, that convective flow may occur on a large scale in sedimentary basins. If this is true it must have an important effect on diagenesis, and on the emplacement of petroleum and epigenetic ore bodies. Positive geothermal anomalies that generate "hot spots" may well cause fluid to expand, decrease in density, and thus move upward. It was once believed that the mineralization around igneous intrusions was solely due to the magmatic fluids. Stable isotope analysis shows, however, that mineral emplacement was partly due to fluids from the adjacent sedimentary cover. A popular model for intrusive-related minerals invokes a peripheral convection cell of connate fluids (see Jensen and Bateman, 1979, p. 61, for sources). These fluids are drawn from the country rock into the metamorphic aureole and stock, within which mineralization takes place (see Fig. 7.3F).

Wood and Hewitt (1982, 1984) have argued for the existence of basin-wide convection cells. These may cause connate waters to flow for considerable distances, aiding petroleum migration, and leaching minerals into solution in some parts of the basin, to precipitate them as cements elsewhere. Such a large convection cell has been identified in the Tertiary Gulf Coast basin of the USA (Blanchard and Sharp, 1985).

Note that convection cells will only develop in basins with laterally extensive permeable formations. They are unlikely to evolve in impermeable shale or evaporitic basin centers.

To conclude, fluid flow in the subsurface environment may be due to meteoric waters flowing in response to a hydrostatic head, or due to connate waters flowing in response to compaction or convection (Fig. 7.15). These three processes vary in importance

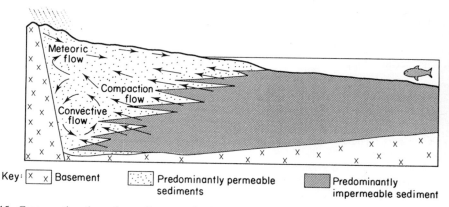

Fig. 7.15. Cross-section through a sedimentary basin illustrating the locations and mutual relationships of the three main types of fluid flow systems: meteoric penetration, connate water flow due to compaction, and that due to convection. Note that the strength of meteoric flow will be related to the hydrostatic head, compactional flow will decline through the life of a basin, and convection cells will best develop where there are laterally extensive permeable beds.

through the history of an evolving sedimentary basin (Cousteau, 1975; Neglia, 1979; Bonham, 1980). Attempts are now being made to model mathematically fluid movement through basins with time, as an aid to predicting the distribution of petroleum and mineral deposits (Bethke, 1985). This aspect will be discussed in Chapter 10. The study of subsurface fluids and their dynamics is evolving rapidly at the present time. It is now unifying geological thought, pointing out the related origin of sediment diagenesis, petroleum migration, and epigenetic mineral emplacement.

SELECTED BIBLIOGRAPHY

For subsurface temperature see:

Deming, D. (1994). Overburden rock, temperature and heat flow. *Mem. — Am. Assoc. Pet. Geol.* **60,** 165–188.

For subsurface pressure see:

Dahlberg, E. C. (1982). "Applied Hydrodynamics in Petroleum Exploration." Springer-Verlag, Berlin. 161pp.
Osborne, M. J., and Swarbrick, R. E. (1997). Mechanisms for generating overpressure in sedimentary basins: A re-evaluation. *AAPG Bull.* **81,** 1023–1041.

For groundwater see:

Ingebritsen, S. E., and Sanford, W. E. (1998). "Groundwater in Geological Processes." Cambridge University Press, Cambridge, UK. 341pp.
Toth, J. (1999). Groundwater as a geologic agent: An overview of the causes, processes and manifestations. *Hydrogeol. J.* **7,** 1–14.

For petroleum see:

Baker, C. (1996). "Thermal Modelling of Petroleum Generation: Theory and Applications." Elsevier, Amsterdam.
Hunt, J. M. (1996). "Petroleum Geochemistry and Geology," 2nd ed. Freeman, New York. 743pp.
Selley, R. C. (1998). "Elements of Petroleum Geology," 2nd ed. Academic Press, San Diego, CA. 470pp.
Tissot, B., and Welte, D. H. (1984). "Petroleum Formation and Occurrence," 2nd ed. Springer-Verlag, Berlin. 530pp.

REFERENCES

Baker, C. (1996). "Thermal Modelling of Petroleum Generation: Theory and Applications" Elsevier, Amsterdam.
Berner, R. A. (1980). "Early Diagenesis." Princeton University Press, Princeton, NJ. 241pp.
Bethke C. M. (1985). A numerical model of compaction-driven groundwater flow and its applications to the palaeohydraulics of intracratonic sedimentary basins. *J. Geophys. Res.* **90,** 6817–6828.
Blanchard, P. E., and Sharp, J. M. (1985). Possible free convection in thick Gulf Coast sandstone sequences. *Trans. S. W. Sect. Am. Assoc. Pet. Geol.,* pp. 6–12.
Bonham, L. C. (1980). Migration of hydrocarbons in compacting basins. *AAPG Bull.* **64,** 549–567.

Burley, S. D. (1993). Models of burial diagenesis for deep exploration plays in Jurassic fault traps of the Central and Northern North Sea. *In* "Petroleum Geology of Northwest Europe" (J. R. Parker, ed.), pp. 1353–1375. Geol. Soc. London, London.

Cartwright, J. C. (1994). Episodic basin-wide hydrofracturing of overpressured Early Cenozoic mudrock sequences in the North Sea basin. *Mar. Pet. Geol.* **11**, 587–607.

Collins, A. G. (1975). "Geochemistry of Oilfield Waters." Elsevier, Amsterdam. 296pp.

Collins, A. G. (1980). Oilfield brines. *In* "Developments in Petroleum Geology" (D. G. Hobson, ed.), pp. 139–188. Applied Science, London.

Combarnous, M., and Aziz, K. (1970). Influence de la convection naturelle dans les reservoirs d'huille ou de gas. *Rev. Ins. Fr. Pet.* **25**, 1335–1354.

Cousteau, H. (1975). Classification hydrodynamiques des Bassins Sedimentaires. *Proc. — World Petrol. Congr.* **9** (2), 105–119.

Dahlberg, E. C. (1982). "Applied Hydrodynamics in Petroleum Exploration." Springer-Verlag, Berlin. 161pp.

Drever, J. I. (1997). "The Geochemistry of Natural Waters," 3rd ed. Prentice-Hall, Hemel Hempstead, England. 474pp.

Dunsmore, H. E. (1973). Diagenetic processes of lead-zinc emplacement in carbonates. *Trans. — Inst. Min. Metall., Sect. B* **82**, B168–B173.

Eugster, H. P. (1985). Oil shales, evaporites and ore deposits. *Geochim. Cosmochim. Acta* **49**, 619–635.

Evans, T. R. (1977). Thermal properties of North Sea rocks. *Log Anal.* **18**, 3–12.

Galloway, W. E. (1984). Hydrologic regimes of sandstone diagenesis. *Mem. — Am. Assoc. Pet. Geol.* **31**, 3–14.

Gold, T. (1979). Terrestrial sources of carbon and earthquake outgassing. *J. Pet. Geol.* **1**, 3–19.

Gold, T. (1999). "The Deep Hot Biosphere." Springer-Verlag, New York. 235pp.

Gold, T., and Soter, S. (1982). Abiogenic methane and the origin of petroleum. *Energy, Explor. Exploit.* **1**, 89–104.

Goldstein, R. H., and Reynolds, T. J. (1994). Systematics of fluid inclusions in diagenetic minerals. *SEPM Short Course* **31**, 1–199.

Hubbert, M. K. (1953). Entrapment of petroleum under hydrodynamic conditions. *Bull. Am. Assoc. Pet. Geol.* **37**, 1954–2026.

Hubbert, M. K., and Rubey, W. W. (1959). Role of fluid pressure in mechanics of overthrust faulting. *Bull. Am. Assoc. Pet. Geol.* **70**, 115–166.

Hunt, J. M. (1996). "Petroleum Geochemistry and Geology," 2nd ed. Freeman, New York. 743pp.

Ingebritsen, S. E., and Sanford, W. E. (1998). "Groundwater in Geological Processes." Cambridge University Press, Cambridge, UK. 341pp.

Jensen, M. L., and Bateman, A. M. (1979). "Economic Mineral Deposits," 3rd ed. Wiley, Chichester. 593pp.

Lillie, R. J. (1999). "Whole Earth Geophysics." Prentice Hall, London. 361pp.

MacDonald, G. J. (1983). The many origins of natural gas. *J. Pet. Geol.* **5**, 341–362.

Mello, U. T., Karner, G. D., and Anderson, R. N. (1995). Role of salt in restraining the maturation of subsalt cource rocks. *Mar. Pet. Geol.* **12**, 697–716.

Neglia, S. (1979). Migration of fluids in sedimentary basins. *AAPG Bull.* **63**, 573–597.

North, F. K. (1985). "Petroleum Geology." Allen and Unwin, London. 607pp.

Osborne, M. J., and Swarbrick, R. E. (1997). Mechanisms for generating overpressure in sedimentary basins: A re-evaluation. *AAPG Bull.* **81**, 1023–1041.

Pallas, P. (1980). Water resources of the Socialist People's Libyan Arab Jamahiriya. *In* "The Geology of Libya" (M. J. Salem and M. T. Busrewil, eds.), pp. 539–554. Academic Press, London.

Plumley, W. J. (1980). Abnormally high fluid pressure: survey of some basic principles. *AAPG Bull.* **64**, 414–430.

Press, F., and Siever, R. (1982). "Earth," 3rd ed. Freeman, New York. 613pp.

Raffensperger, J. P., and Garvan, G. (1995a). The formation of unconformity-type uranium ore deposits. 1. Coupled groundwater flow and heat transport modelling. *Am. J. Sci.* **295**, 581–636.

Raffensperger, J. P., and Garvan, G. (1995b). The formation of unconformity-type uranium ore deposits. 1. Coupled hydrochemical modelling. *Am. J. Sci.* **295**, 639–696.

Roberts, S. J., and Nunn, J. A. (1995). Episodic fluid expulsion from geopressured sediments. *Mar. Pet. Geol.* **12,** 195–204.

Selley, R. C. (1998). "Elements of Petroleum Geology," 2nd ed. Academic Press, San Diego, CA. 470pp.

Skempton, A. W. (1970). The consolidation of clays by gravitational compaction. *Q. J. Geol. Soc. London* **25,** 375–412.

Sugisaki, R., Id, M. T., Takeda, H., and Isobe, Y. (1983). Origin of hydrogen and carbon dioxide in fault gases and its relation to fault activity. *J. Geol.* **91,** 239–258.

Terzaghi, K. (1936). The shearing resistance of natural soils. *Proc. Int. Conf. Soil Mech., 1st,* Harvard, pp. 54–56

Tissot, B., and Welte, D. H. (1984). "Petroleum Formation and Occurrence," 2nd ed. Springer-Verlag, Berlin. 530pp.

Toth, J. (1999). Groundwater as a geologic agent: An overview of the causes, processes and manifestations. *Hydrogeol. J.* **7,** 1–14.

Van Elsberg, J. N. (1978). A new approach to sediment diagenesis. *Bull. Can. Pet. Geol.* **26,** 57–86.

White, D. E. (1957). Magmatic and metamorphic waters. *Geol. Soc. Am. Bull.* **68,** 1659–1682.

Wood, J. R., and Hewitt, T. A. (1982). Convection and mass transfer in porous sandstones a theoretical model. *Geochim. Cosmochim. Acta* **46,** 1707–1713.

Wood, J. R., and Hewitt, T. A. (1984). Reservoir diagenesis and convective fluid flow. *Mem.— Am. Assoc. Pet. Geol.* 99–110.

8 Allochthonous Sediments

8.1 INTRODUCTION: THE CLASSIFICATION OF SEDIMENTARY ROCKS

Natural processes tend to separate the various products of weathering and, after erosion and transportation, deposit sand, mud, and carbonates in many different environments. Thus sediments are grouped and segregated spontaneously on the earth's surface. The next two chapters describe how sediments are turned into rock, first the allochthonous sediments (those that are transported into a sedimentary basin), and then the autochthonous sediments (those that form within a basin).

To begin with, however, it is necessary to review how the various sedimentary rocks are named and classified. There are two main reasons for such exercises. First, effective communication requires a uniformity of nomenclature. Secondly, in a particular study, it is often necessary to differentiate, compare, and contrast rock types. The following section describes first the problems of sediment classification in general, and then the classification of the allochthonous rocks in particular. Sediment is "what settles at the bottom of a liquid; dregs; a deposit" (Chambers Dictionary, 1972 edition). This definition is itself unacceptable to most geologists since it would exclude, for example, eolian deposits and biogenic reefs from the realm of sedimentology. Similar dilemmas will be encountered with most other descriptive terms applied to sedimentary petrography.

Essentially, five main genetic classes of sediment can be recognized: chemical, organic, residual, terrigenous, and pyroclastic (Hatch *et al.,* 1971). The chemical sediments are those that form by direct precipitation in a subaqueous environment. Examples include evaporites such as gypsum and rock salt, as well tufa and some lime muds.

The organic sediments are those composed of organic matter of both animal and vegetal origin. Examples include skeletal limestones and coal. The residual sediments are those left in place after weathering, and examples include the laterites and bauxites described in Chapter 2.

The terrigenous sediments are those whose particles were originally derived from the earth, and include the mudrocks, siliciclastic (as opposed to carbonate) sands, and conglomerates. Pyroclastic sediments are the product of volcanic activity. Examples include ashes, tuffs, volcaniclastic sands, and agglomerates. Additional terms to introduce at this point are clastic, detrital, and fragmental. These all tend to be used in the same way — describing a rock as formed from the lithifaction of discrete particles, a sediment no less.

Table 8.1
Classification of Sedimentary Rocks

Group	Class
I. Autochthonous sediments	(a) *Chemical precipitates*— the evaporites: gypsum, rock salt, etc.
	(b) *Organic deposits*— coal, limestones, etc.
	(c) *Residual deposits*— laterites, bauxites, etc.
II. Allochthonous sediments	(d) *Terrigenous deposits*— clays, siliciclastic sands, and conglomerates
	(e) *Pyroclastic deposits*— ashes, tuffs, volcaniclastic sands, and agglomerates.

Like most classifications of geologic data this one has several inconsistencies. Note particularly that many limestones, though organic in origin, are detrital in texture. Cannel, drift, or boghead coals are not truly autochthonous. Many evaporites are diagenetic in origin.

The five genetic classes previously defined are untenable when closely scrutinized. For example, are gypsum sand dunes chemical or terrigenous deposits? Are phosphatized bone beds chemical, organic, or residual deposits, and so on? These five main genetic classes of sedimentary rocks can be divided into two separate types: the allochthonous and the autochthonous deposits. The allochthonous sediments are those that are transported into the environment in which they are deposited. They include the terrigenous and pyroclastic classes, together with rare reworked (allodapic) carbonates. The autochthonous sediments are those that form within the environment in which they are deposited. They include the chemical, organic, and residual classes. Table 8.1 shows the relationship between these various terms. The end-member concept is one of the most useful ways of classifying rocks. Sediments containing three constituents can be classified in a triangular diagram in which each apex represents 100% of one of the three constituents (Fig. 8.1). Four component sediment systems can be plotted in three dimensions within the faces of a tetrahedron (Fig. 8.2). Because sedimentary rocks contain many components, neither of these systems is entirely satisfactory. Enthusiasts for classificatory schemes will have to indulge in statistical aerobics using factor analysis, etc. (e.g., Harbaugh and Merriam, 1968, p. 157).

8.2 ALLOCHTHONOUS SEDIMENTS CLASSIFIED

The allochthonous sediments, as previously defined, consist of the terrigenous and pyroclastic classes. The quartzose terrigenous sands are sometimes termed "siliciclastic" to differentiate them from the bioclastic and pyroclastic detrital sediments. The pyroclastic sediments, those derived from volcanic activity, are included for the sake of completeness. They are not a volumetrically significant part of the present sedimentary cover of the earth and are not described in detail. The allochthonous sediments may conveniently be classified using the end-member triangle shown in Fig. 8.3. The scheme attempts to reconcile the inconsistency that the nomenclature of sedimentary rocks is founded on two irreconcilable parameters: grain size and composition. It is interesting to see the different emphasis placed on sediment nomenclature in modern and ancient

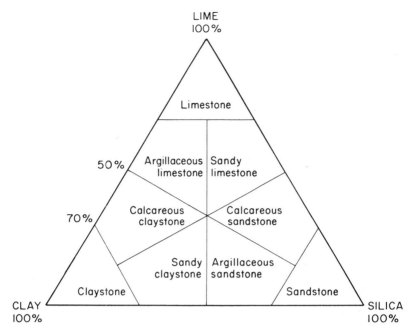

Fig. 8.1. An example of the end-member classification style. This triangle differentiates sands, claystones, and limestones.

deposits. Studies of modern sediments are primarily concerned with the hydrodynamics of transport and deposition. Sediment description predominantly emphasizes grain size and texture rather than mineral composition. Hence a scheme similar to that in Fig. 8.4 is generally used. In ancient sediments, however, emphasis is placed much more

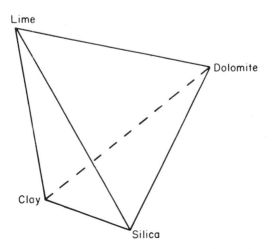

Fig. 8.2. An example of the end-member classification style. This tetrahedron differentiates four-component systems, in this case limestones, dolomites, sandstones, and clays.

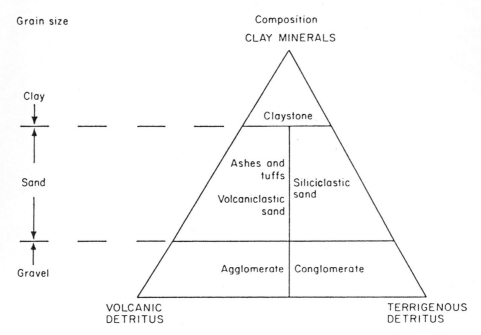

Fig. 8.3. A classification of the allochthonous sediments based on grain size and composition.

on composition and this is reflected by the abundance of rock names that stress composition rather than texture (e.g., arkose, tuff).

The four main types of allochthonous sediments are now described. These may be grouped conveniently, if illogically, into the mudrocks (composed dominantly of clay minerals), the pyroclastics, the sandstones, and the rudaceous rocks.

8.3 MUDROCKS

The term "mud" is ill defined and loosely used. In Recent deposits, sediments referred to as mud are wet clays with a certain amount of silt and sand. Its lithified equivalent is termed "mudstone." More rigidly defined, however, are clay and silt in the Wentworth grade scale. Clays are sediments with particles smaller than 0.0039 mm. Silts have a grain size between 0.0039 and 0.0625 mm. Their lithified equivalents are claystone and siltstone, respectively.

Shale is another term applied to fine-grained sediment. It is ill defined and does not differentiate silt from clay-grade sediment. Shaley parting, on the other hand, is a valid description for fissility in a fine sediment. Generally this is due to traces of mica aligned on laminae. The term "shale," however, could perhaps usefully be abandoned by geologists, except when communicating to engineers or management.

As discussed later, clays are deposited with a primary water-saturated porosity of up to 80%. Most of this is quickly lost, first by dewatering (synaeresis, see Section 5.3.5.4) and later by compaction. Other major constituents of mudrocks include clay minerals,

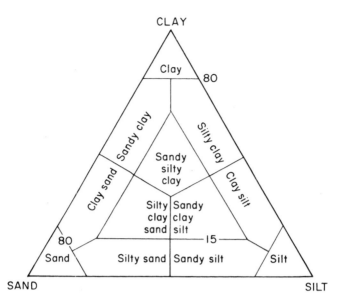

Fig. 8.4. An end-member triangle for classifying unconsolidated sediment on the basis of grain size. (From Shepard, 1954. Courtesy of the Society for Sedimentary Geology.)

detrital grains, organic matter, and carbonates. The clay minerals are described in some detail later. The detrital grains consist of fine angular particles of quartz, micas, and heavy minerals such as zircon and apatite. Organic matter in mudrocks is chemically very complex and the nomenclature organic mudrocks is not rigidly defined.

Essentially mudrocks can be named with reference to an end-member triangle whose apices represent pure organic matter, pure lime, and pure clay minerals (Fig. 8.5). Mudrocks composed largely of admixtures of various clay minerals can be termed "claystones," or more pedantically, "orthoclaystones," to indicate their purity. With increasing lime content, claystones grade into marls and on into micrites, which are pure lime mudrocks.

8.3.1 Sapropelites, Oil Shales, and Oil Source Rocks

Organic matter is present in at least small quantities in all sediments. It is, however, most abundant in the mudrocks and has been studied intensively, both because organic-rich shales are commonly thought to be the source of crude oil, and because certain mudrocks, termed "oil shales," release crude oil on heating (Yen and Chilingarian, 1976). **Sapropelite** is a name given to organic-rich mudrocks.

Essentially the organic matter in sediments is of four types: kerogen, asphalt, crude oil, and natural gas. These are organic compounds of great complexity. Their nomenclature is ill defined and these four groups span a continuous spectrum of hydrocarbons. Kerogen is a dark gray-black amorphous solid that is present in varying quantities in mudrocks; when pure it is termed "coal." It contains between 70 and 80% carbon, 7 and 11% hydrogen, 10 and 15% oxygen, and traces of nitrogen and sulfur. The exact molecular structure of kerogen is not well known, but, by definition, kerogen includes

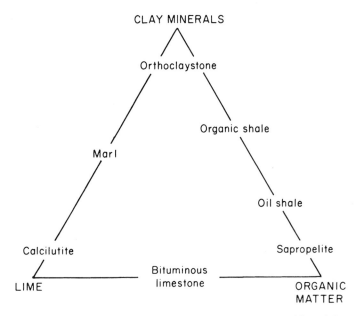

Fig. 8.5. Triangular diagram illustrating the nomenclature and composition of the mudrocks.

those hydrocarbon compounds which are not soluble in normal petroleum solvents, such as ether, acetone, benzene, or chloroform (Dott and Reynolds, 1969, p. 79). Kerogen is the major constituent of the organic-rich mudrocks (see also Tissot and Welte, 1984; Hunt, 1996).

Asphalt, or bitumen, is similar to kerogen, but it is soluble in normal petroleum solvents. It contains 80–85% carbon, 9–10% hydrogen, and 2–8% sulfur, with traces of nitrogen and negligible oxygen. Asphalt occurs both in infilling sediment pores and in fractures.

Hydrocarbons that are liquid at normal temperatures and pressure are termed "crude oil." Oil contains 82–87% carbon, 12–15% hydrogen, and traces of sulfur, nitrogen, and oxygen. Oils are extremely variable in molecular composition but generally consist of varying proportions of four main groups: paraffins, aromatics, naphthenes, and asphalts. Crude oil occurs in pore spaces of many rocks in favorable circumstances.

Table 8.2 summarizes the properties and composition of these various hydrocarbons, including natural gas for the sake of completeness. Mudrocks with substantial traces of organic matter are simply referred to as organic claystones. They are generally dark colored. Oil shales are mudrocks that are sufficiently rich in organic matter to yield free oil on heating (Duncan, 1967). They form in restricted anaerobic low-energy environments, both marine and nonmarine. A notable oil shale occurs in the Green River Formation of Utah, Colorado, and Wyoming. This was deposited in a series of Eocene lakes (Donell *et al.,* 1967). The presence of desiccation cracks and other shallow water features together with juxtaposition to trona (sodium carbonate) suggest a continental sabkha depositional environment (Surdam and Wolfbauer, 1973).

Another notable oil shale occurs in the Carboniferous deltaic facies of the Midland

Table 8.2
Properties and Composition of the Main Groups of Organic Hydrocarbons

Organic matter	Properties	Average composition (% weight)		
		C	H$_2$	S + N + O$_2$, etc.
Kerogen	Solid at surface temperatures and pressures. Insoluble in normal petroleum solvents.	75	10	15
Asphalt	Solid or plastic at surface temperatures and pressures. Soluble in normal petroleum solvents.	83	10	7
Crude oil	Liquid at surface temperatures and pressures.	85	13	2
Natural gas	Gaseous at surface temperatures and pressures.	70	20	10

Valley of Scotland (Greensmith, 1968) and was one of the first sources of mineral oil extracted in the world. This oil shale is also lacustrine in origin. Torbanite is a particularly pure variety of these Scottish oil shales, consisting almost entirely of the alga *Botryococcus*. This rock is sometimes termed "boghead coal" because it is closely allied to the sapropelic coals, also termed "cannel coals" (see Section 9.3.2).

Thus with increasing organic content, the claystones grade from organic claystone into oil shale (or, strictly, oil claystone) and thence into the dominantly carbonaceous sapropelites, torbanites, and coals. The organic-rich mudrocks have been the subject of intensive research because they are popularly thought to be the source rocks from which petroleum is generated.

It is now realized that a mudrock needs to contain over 1.5% organic carbon to be a significant source rock. The type of kerogen determines the type of hydrocarbon which may be generated. Terrestrial humic kerogen tends to be gas prone. Algal kerogen tends to be oil prone. Mixed kerogen can generate both oil and gas. Temperature is also important, however, critical thresholds need to be crossed for oil and gas generation. Oil generation normally takes place between 60 and 120°C, gas generation occurs between about 120 and 220°C, above which the kerogen has been reduced to inert carbon (Fig. 8.6). The subject of oil source rocks is covered in far greater depth in the textbooks of Tissot and Welte (1984) and Hunt (1996).

8.3.2 Orthoclaystones and Clay Minerals

The orthoclaystones are clay-grade rocks composed mainly of the clay group of minerals (Moore and Reynolds, 1997; Schieber *et al.*, 1998a,b; Srodon, 1999). The clays are an extensive and complex mineral suite, which, as discussed in Chapter 2, form largely by chemical degradation of preexisting minerals during weathering (Wilson, 1999). The

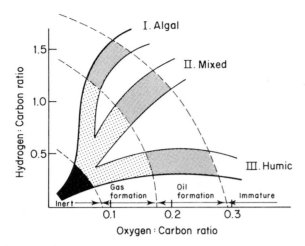

Fig. 8.6. Graph of hydrogen:carbon ratio plotted against oxygen:carbon ratio showing the diagenetic pathways of the different types of kerogen. Note that the generation of oil and gas is contingent both on their level of maturation and the original composition of the kerogen. Type I kerogen generates principally oil, type II, both oil and gas, and type III, principally gas.

mineralogy and origin of the main clay minerals is described before we discuss the claystone rocks.

The three principal groups of clay minerals are the illites, the montmorillonites or smectites, and kaolinite. To these three may also be added the chlorites and glauconite. These last two differ somewhat from the other clay minerals in mode of formation, but show similarities in composition and atomic structure. All five of these mineral groups are hydrous aluminosilicates. Crystallinity is unusual in all but the chlorites and kaolinites. Where detectable, it is of the monoclinic system with the possible exception of glauconite, which occurs amorphously almost without exception (Grim, 1968; Chamley, 1989). Essentially the clays are arranged in layers of tetrahedra of oxygen and hydroxil radicals. Within the spaces of the tetrahedra lie silica and alumina radicals. In the kaolinite clays no other elements are present within this two-layer lattice framework. The illites and smectites are based on a three-layer lattice framework of tetrahedra with substitution of alumina radicals by potassium, calcium, and magnesium. The chlorites and glauconites are more complex still, with a structure of mixed two- and three-layer lattices to accommodate their more complex radicals of iron and magnesium.

8.3.2.1 Kaolinite Group

The kaolinite group of clay minerals includes kaolinite, dickite, nacrite, halloysite, and allophane. The differences between these minerals are both chemical and structural. Some have identical composition, but different crystal structures. Kaolinite is the simplest clay mineral in structure and the purest in composition. It is formed from feldspars both by hydrothermal alteration and by superficial weathering. Kaolinite is a common detrital clay mineral in sediments derived from granitic and gneissose sources. Kaolinite often undergoes considerable crystallization during diagenesis to form characteristic "books." Aggregates of long accordion or worm-like crystals are also found in some sediments. Kaolinite is an important authigenic cement in some sandstones (see Section 8.5.3.3.3).

In certain circumstances kaolinite sedimentation is sufficiently abundant to form a pure claystone termed "kaolin," also known as pipe clay, ball clay, china clay, and fire clay. The Oligocene Bovey Tracey lake beds of Devon are composed largely of pure kaolinite derived from the adjacent Dartmoor granite (Edwards, 1976).

Tonsteins are another distinctive rock type, composed largely of kaolinite. These occur intimately associated with coals in the Carboniferous strata of Europe. Individual beds are thin yet laterally very extensive. While it is possible that tonsteins formed from the clays of weathered granite, their regional persistence favors an origin due to alteration of volcanic ash falls (Price and Duff, 1969; Bohor and Triplehorn, 1993). Tonsteins are tough indurated rocks but the younger kaolinite claystones are generally white and plastic.

Kaolinite is extensively used in the ceramic, paper-making, and pharmaceutical industries. Pure kaolinite rocks are nonmarine in origin because kaolinite quickly transforms to more complex clays in the presence of seawater.

8.3.2.2 Illite Group

The illite clays, sometimes termed the hydromicas, are three-layer aluminosilicates with up to 8% K_2O. This potassium may either be present due to the incomplete degradation of potash feldspars to kaolinite, or to diagenesis of kaolinite within a marine environment. Illite is the most abundant clay mineral in sediments but it is less obvious than kaolinite because it is seldom present in crystals that can be seen with an optical microscope. Even under an electron microscope, illite crystals are smaller and less well developed than those of kaolinite (see Section 8.5.3.3.3).

8.3.2.3 Smectite Group

The third group of clay minerals is the smectites, of which montmorillonite is the chief example. These are three-layer lattice types which have the unusual property of expanding and contracting to adsorb or lose water. Montmorillonite can contain up 20% water as well as calcium and magnesium. Mudrocks composed largely of smectite clays are termed bentonites. Bentonite deposits may be recognized at outcrop by their powdery cauliflower-like surface. Placed in water, montmorillonite lumps expand visibly and fall apart. Bentonites are formed by the alteration of volcanic ash in place. This may occur in both marine and continental environments. Fragments of detrital glass, frequently devitrified, are generally present, together with microscopic grains of quartz, micas, feldspars, and heavy minerals.

The type montmorillonite comes from Montmorillon in France. Montmorillonite is also the chief constituent of the Jurassic and Cretaceous Fuller's Earth in southern England (Robertson, 1986). Bentonites occur in the Upper Cretaceous rocks of Arkansas, Oklahoma, Texas, and Wyoming, notably the Mowry Shale of the Big Horn basin (Slaughter and Earley, 1965).

Because of their peculiar properties, bentonites are a major constituent of circulating mud systems used in rotary drilling. Conversely, montmorillonite is an unfortunate clay mineral to have as a matrix in an oil reservoir. During production, water entering the reservoir may cause the clay matrix to expand and thus destroy permeability.

8.3.2.4 Chlorite

The chlorites are similar to the clay minerals just described in many ways, but also share affinities with the mica mineral group. The chlorites are mixed-layer lattice clays with up to 9% FeO and 30% MgO.

Chlorites occur as an alteration product of primary micas, and are a common accessory detrital mineral in immature sands and mudrocks. Conversely chlorite replaces illite and other clay minerals at the point where diagenesis merges into metamorphism; it is a characteristic constituent of the microcrystalline matrix of greywackes.

8.3.2.5 Glauconite

The fifth clay mineral to consider is glauconite. This term is used in two ways. It is applied to pretty, rounded green grains commonly seen in marine sediments and to a particular mineral (Plate 3A). Analyses of the former show them to contain a mixture of clays whose lattices are in various stages of ordering. Glauconite proper is a three-layer clay mineral containing magnesium, iron, and potassium. Unlike the other clay minerals, glauconite does not form from the hydrothermal or terrestrial weathering of preexisting minerals. Glauconite occurs in dark green amorphous grains seldom larger than fine-sand grade. It is found both in mudrocks and sandstones. Where especially abundant in sands, the rock is commonly named "greensand." Glauconite formation accompanied by greensand sedimentation occurred at particular times on a worldwide basis, notably in the Cambrian and through the late Cretaceous and early Tertiary (Pettijohn *et al.*, 1972, p. 229).

Glauconite grains appear to have formed by the replacement of fecal pellets and by infilling foraminiferal tests which have subsequently been destroyed. Glauconite also occurs infilling larger shells and replacing detrital micas. It is a matter of observation that glauconite occurs in ancient sediments of marine origin. It is easily weathered and, with one or two exceptions, does not occur in any abundance as a reworked mineral.

There has been considerable debate both as to the nature of the chemical reactions responsible for glauconite formation and of the parameters that control the reaction. Glauconite has generally been considered to form by the transformation of degraded smectitic or illitic-type sheets into more ordered mixed-layer lattices. Several studies, however, have argued for neoformation of glauconite within voids (Odin, 1972, 1988; Bjerkli and Ostmo-Saeter, 1973; Chamley, 1989). A polygenetic origin for glauconite formation seems to be clearly proven.

Geochemical evidence suggests that glauconite formation, by whatever process, occurs in seawater at low temperatures and in an environment that is neither strongly oxidizing nor reducing. Optimum depth for glauconite genesis appears to be between 50 and 1000 m. However, these parameters are hard to define because, once formed, glauconite is stable in seawater and can survive transportation in a marine environment. Thus glauconite is found disseminated in marine mudrocks, in clean well-sorted cross-

Table 8.3
Summary of the Salient Features of the Main Groups of Clay Minerals and Their Associates

Mineral	Composition (additional to hydrated alumino-silicate) Ca	Mg	Fe	K	Atomic lattice structure	Source	Rock name
Montmorillonite					Three-layer	Volcanics	Bentonite
Chlorite					Four-layer	Mafic minerals	
Glauconite					Three-layer	Submarine diagenesis	
Illite					Three-layer	Feldspars	
Kaolin					Two-layer	Feldspars	China clay / Fire clay / Tonstein

bedded sand shoals, and as a minor constituent of basinal turbidites. Table 8.3 summarizes the salient features of the main clay minerals.

8.3.2.6 Compaction of Clays

Figure 8.7 shows that clays are deposited with porosities ranging between 50 and 80%. Water is lost from most clays immediately after deposition by consolidation, the process whereby a clay is changed to claystone. This includes both cementation and dehydration, as well as compaction due to overburden pressure. It is important to note that dewatering of clays at shallow depths is not due solely to overburden pressure. Dewatering also occurs in many clays as a spontaneous process termed "synaeresis" (White, 1961), forming small desiccation cracks at the mud:water interface (see also Section 5.3.5.4).

The compaction of clay at shallow depths has been intensively studied by engineering geologists. This is because it is critical to know the physical properties of a clay if it is to be a foundation for civil engineering projects, such as the building of high-rise blocks, motorways, and dam sites. Furthermore, it is important not only to know the physical properties of the clay at the preliminary stage of site investigation, but also to be able to predict the degree of compaction to be expected if the site is drained.

Clay compaction curves have been studied by many geologists. Reviews and data have been given by Magara (1980), Dzevanshir et al. (1986), and Addis and Jones (1986). These data are summarized in Fig. 8.7. Note that porosity diminishes from some 80% down to 20% within the first 2 km of burial. Thereafter, porosity loss is at a much slower rate. This is very different from sandstone burial curves in which the porosity loss is linear with depth (see Fig. 8.17 later).

As the rate of water expulsion and porosity loss in clays decelerates with increasing depth of burial, the problems of clay compaction move from the field of engineering geology to petroleum exploration. The expulsion of oil from compacting muds has been

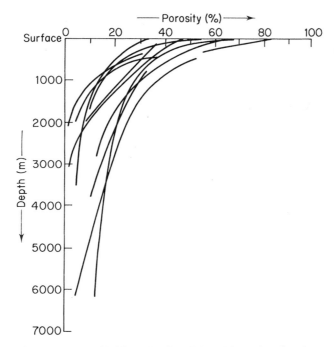

Fig. 8.7. Clay compaction curves, compiled from data in Addis and Jones (1986) and Dzevanshir *et al.* (1986). Note that porosity is lost very rapidly in the first 2 km of burial. Thereafter, the rate is much slower and approximately linear. Contrast with the sandstone burial curves in Fig. 8.17.

implicit in most theories of petroleum generation in the Western world. As clays compact and bury they also undergo a number of geochemical changes. These occur in three main stages: diagenesis, catagenesis, and metagenesis. (Note that organic geochemists appear to restrict the term "diagenesis" to much shallower depths than most geologists; compare, for example, Tissot and Welte, 1984, with Pettijohn, 1957.)

For the purposes of this discussion diagenesis refers to processes operating in the first 2 km of burial. As already seen this is the zone in which porosity is lost rapidly by compaction. Chemical reactions take place at relatively low temperatures and pressures. They are intimately related to bacterial reactions of organic matter. Curtis (1978) and Gautier and Claypool (1984) have shown that these take place in a regular sequence as burial progresses (Fig. 8.8). The shallowest diagenetic zone is one of oxygenation. This is very thin or completely absent in subaqueously deposited muds. In continental environments, such as playa lakes, however, subaqueous mud deposition may be followed by long periods of dehydration while the clay remains above the water table. In such situations organic matter may be oxidized and removed. Iron may develop as limonite, which may then dehydrate to red ferric oxides. Where continental muds are deposited close to the water table in arid environments groundwater may move to the surface by capillarity. As this water evaporates it will precipitate carbonate and evaporite minerals. Such muds may thus become lithified without undergoing significant compaction.

In subaqueous environments, however, an oxygenated zone may be absent, and

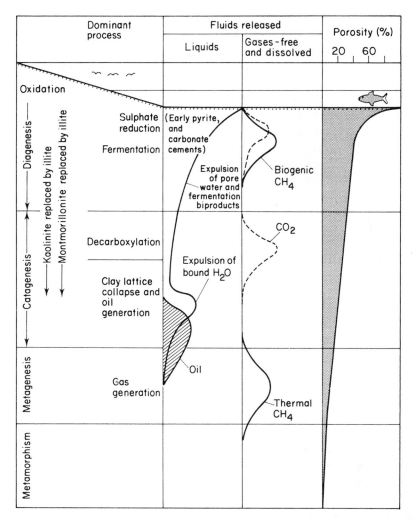

Fig. 8.8. Burial depth diagram showing the phases of clay diagenesis and the maturation of contained organic matter. No attempt has been made to show a vertical scale on this figure because many of the reactions extend over a wide range of temperatures. Thus, as geothermal gradients vary, so may the depths of the different diagenetic phases.

sulfate-reducing bacteria active. In this situation sulfur may react with ferrous hydroxide to form pyrite:

$$Fe(OH)_2 + 2S \rightarrow FeS_2 + H_2O).$$

Sulfate ions may react with organic matter to form hydrogen sulfide:

$$(SO_4^+ + 2CH_2O) \rightarrow 2HCO_3^- + H_2S,$$

where $2CH_2O$ is used to represent organic matter.

Bacterial fermentation of organic matter also takes place during shallow burial. These reactions generate water, carbon dioxide, and biogenic methane. This increases the pH of the pore fluids, permitting carbonate precipitation. Early carbonate cements include both calcite and siderite. These cements are often patchily developed as concretions that may occur intermittently along beds. The early age of their formation can often be proved because the concretions contain whole fossil shells, while in adjacent laminae the little shells have been flattened and shattered. Sometimes, the nodules have undergone subsequent dehydration, with carbonate cement growth in the shrinkage cracks (Fig. 8.9). Fluid inclusion and stable isotope analyses can give much information on the physical and chemical environment in which this cementation took place.

As temperature and pressure increase with burial depth, the organic fraction of the mud evolves into kerogen, and bacterial processes diminish in importance. Diagenesis merges into catagenesis toward about 2 km as the temperature exceeds 50°C. In this zone important clay mineral reactions take place. Recall that montmorillonitic clays contain structured water within their atomic lattice. At temperatures of about 100–110°C the montmorillonite lattice collapses, to expel a large volume of pore water, and hence increase the pore pressure. Montmorillonite and kaolinite change into illite. These reactions take place at the depths and temperatures in which oil generation is known to occur.

Space does not permit the problem of primary oil migration to be discussed. This is one of the last great geological mysteries: How does petroleum emigrate from an impermeable clay? At these depths shale porosity is down to some 10–20%, and simple squeezing cannot explain the mechanism of oil emigration. For reviews of the many theories, see North (1985, pp. 225–240), Hunt (1996, pp. 238–267), and Selley (1996, pp. 214–225). There must obviously be a close and complex relationship between clay mineral diagenesis, kerogen maturation, and petroleum emigration. As kerogen matures it undergoes decarboxylation, giving off carbon dioxide. This goes into solution in the water as carbonic acid. These acid solutions are thus expelled from the clays ahead of petroleum. This is a process of great importance to sandstone diagenesis (see Section 8.5.3.4.2). With increasing temperature kerogen generates oil at temperatures of 60–120°C (Tissot and Welte, 1984; Hunt, 1996). As burial continues and temperatures approach 200°C catagenesis phases out into the third phase, termed "metagenesis." During this phase oil generation from kerogen ceases, to be replaced by dry gas formation. Finally, metagenesis merges into metamorphism. Kerogen has now given up all its petroleum, and is nothing but carbon (graphite). Clay minerals begin to recrystallize into recognizable crystals of mica and chlorite.

8.3.2.7 Clays and Clay Diagenesis: Summary

The compaction and consolidation of clays at shallow depths is of significance to the engineering geologist. The changes that occur in clays as they change to claystones at greater depths is of interest to the petroleum geologist because this may be relevant to theories of petroleum genesis and migration. Similarly, mining geologists have considered that low-temperature ore bodies may have been derived from the residual fluids of compacting clays aided by brines acting as transporting media (e.g., Davidson, 1965; Amstutz and Bubinicek, 1967).

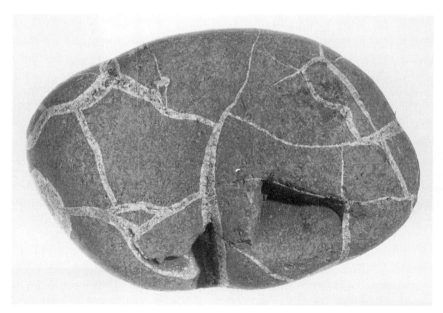

Fig. 8.9. (Upper) Carbonate concretion in Kimmeridge Clay (Upper Jurassic) Burning Cliffs, Dorset. Note the way in which clay laminae converge away from the concretion, thus demonstrating that it formed as a result of localized cementation during shallow burial, prior to compaction. (Lower) Septarian nodule from the Lias (Lower Jurassic), Charmouth, Dorset. This concretion developed shrinkage cracks that were then infilled by calcite. In both cases the concretions contain complete fossils, while flattened shells occur in adjacent shales. This is further evidence for early precompaction cementation.

In concluding this review of clay minerals, the following points should be noted. The clay mineral suite that is found in a particular rock at a particular time is a result of four main variables. The nature of the source rock controls the input of clay minerals. There is a higher probability of kaolinite and illite forming from a granite source than from a volcanic hinterland. Similarly, the more intense the weathering the higher the probability of kaolinite occurring at the expense of illite. Size-sorting during transportation may also segregate the various clay mineral species. Kaolinite crystals tend to be larger than illites and illite to be larger than montmorillonite. This fact is complicated, however, by the tendency of clay particles to form floccules when they are carried from acidic freshwater to neutral seawater. Simultaneously, various transformations of clay minerals take place within seawater both during transportation and early burial (Keller, 1970).

Thus while most studies of modern marine coasts show some regular zonation of clay minerals, it is hard to demonstrate whether this results from differential transportation or incipient diagenesis (e.g., Porrenga, 1966). More exhaustive accounts of clays and clay minerals will be found in Muller (1967), Grim (1968), Millot (1970), Mortland and Farmer (1978), Potter *et al.* (1980), Shaw (1980), and Chamley (1989).

8.4 PYROCLASTIC SEDIMENTS

Figure 8.3 showed how the allochthonous sediments are divisible into the claystones (defined by their grain size and clay mineral composition), the siliciclastic sands of terrigenous silica minerals, and the volcaniclastic sediments. The volcaniclastic sediments are relatively rare by volume in the earth's crust, but deserve mention for the sake of completeness. Many minerals of volcanic rock are unstable at surface temperatures and pressures. For this reason, therefore, detritus derived from volcanic activity is commonly only preserved interbedded with lava flows and can seldom survive detrital transportation far from the volcanic center from which it came (Orton, 1996). The volcaniclastic sediments can be classified into three groups according to their particle size.

Agglomerates are the counterpart to conglomerates. They are formed both by explosive eruptions and by scree movement of volcanic detritus both within a caldera and on the flanks of volcanoes. Sand-grade volcaniclastic sediment is of two types. Erosional volcaniclastic sands are produced by normal subaerial or subaqueous processes acting on eruptive rocks. The pyroclastic sediments in contrast are ejected into the atmosphere during volcanic eruptions. Pyroclasts include, therefore, "bombs" which fall close to the vent, sands which fall around the vent for a distance of kilometers, and dust which may be carried into the upper atmosphere and transported around the world. In many cases it is impossible to distinguish whether an ancient volcaniclastic sediment was produced by normal erosion of lavas or by pyroclastic action. Large clasts cannot normally be transported far from their source. Pumice provides an exception to this because it floats. As illustrated by the frontispiece of this book, pumice can be transported across whole oceans.

Volcaniclastic sands are generally referred to as tuffs or ashes. They may be subaerial or subaqueous. Volcaniclastic sands are composed essentially of crystals, glass, and rock fragments (Plate 3B). The crystals are of minerals associated with the eruption, such as olivine and quartz. Glass occurs both as globules and angular irregularly shaped shards.

Rock fragments are composite grains of volcanic minerals and glass. Volcaniclastic sands are generally poorly sorted because if extensively transported and reworked they are rapidly destroyed. Eolian volcaniclastic sands are an exception to this rule. Dunes of basalt sand occur, for example, in Iceland, and around the volcanic crater of Waw en Namus in the Libyan Sahara.

Fine-grained volcanic ash tends to undergo intensive postdepositional alteration. This may give rise to the montmorillonitic bentonites and kaolinitic tonsteins discussed in the previous section. Palagonite is a mineraloid produced by the alteration of basalt glass. It is sometimes sufficiently abundant in the Pacific to make a sediment termed palago-nite mud, and also occurs lithified in Pacific volcanic islands. Traces of volcanic ash are a common constituent of modern pelagic sediments.

Volcaniclastic sands may have good primary porosity and permeability, especially if deposited on beaches or dunes. They lose porosity on burial very quickly, however, be-cause of their unstable mineralogy. During shallow burial they undergo hydration and carbonation leading to the formation of authigenic carbonate, clay, laumontite, and other zeolites. Further burial results in dehydration of the clays, albitization of feld-spars, and the development of additional zeolites (Galloway, 1974). Thus volcaniclastic rocks are generally poor-quality aquifers or petroleum reservoirs.

Volcanic and volcaniclastic rocks are of interest in mineral exploration, however, be-cause they appear to be the primary source of some metalliferous deposits; the metals having been originally disseminated in the volcanic rocks, subsequently mobilized dur-ing weathering, and concentrated and reprecipitated during shallow meteoric diagene-sis. Uranium roll-fronts and some red-bed stratiform copper deposits are associated with volcanic centers (see Section 6.3.2.2.4). More detailed accounts of the genesis and petrology of volcaniclastic sediments are found in Fisher and Schmincke (1984), Suthren (1985), and Orton (1996).

8.5 SANDSTONES

This section describes the important group of sedimentary rocks, the sandstones. Spe-cifically, it is concerned with the quartzose or siliciclastic sands as distinct from the vol-caniclastic and carbonate sands. About 30% of the land's sedimentary cover is made of terrigenous sand and sandstone. This group of rocks is a fruitful topic for study both for itself and to aid the exploitation of economic materials within sandstones. Because they are often highly porous, sandstones are frequently major aquifers and petroleum reser-voirs. They are more uniform in stratigraphy and petrophysical character than carbon-ates and it is, therefore, easier to predict their geometry and reservoir performance.

The following account discusses the problems of sandstone nomenclature and classi-fication, summarizes the features of the more common sandstone types, and concludes by discussing the effect of diagenesis on sand porosity and permeability.

8.5.1 Nomenclature and Classification of Sandstones

The nomenclature and classification of sandstones has always been a popular academic pursuit. Indeed Klein (1963) and Pettijohn *et al.* (1972, pp. 149–174) have written re-views of this topic that include classifications of sandstone classifications. It is important that sandstones be named with a terminology that is tolerably familiar to, and agreed on by, practicing geologists. Any nomenclatural system has to have arbitrary bounding parameters that separate one rock type from another. These bounding parameters are most useful when based on some underlying concepts of sand genesis. The basic problem of classifying sands is that they can be grouped according to their physical composition (i.e., grain size and matrix content) or according to their chemical composition (i.e., mineralogy). There are more textural and mineralogical components deemed to be significant in sandstone nomenclature than can be conveniently represented in an end-member triangle or tetrahedron.

Higher, statistically based classificatory schemes, such as factor analysis, may be more logical. On the other hand, they lack the simplicity and visual appeal of end-member classifications. Thus the majority of sand classifications are based on end-member triangles, the three components generally being chosen from quartz, clay, feldspar, or lithic content.

One of the most fruitful concepts on which sandstone nomenclature is based is the idea of maturity. The maturation of a sand takes place in two ways. It matures chemically and it matures physically. Sediments form from the weathering of mineralogically complex source rocks. Throughout weathering and transportation relatively unstable minerals are destroyed and chemically stable minerals thus increase proportionally. Quartz is the most abundant stable mineral and feldspar is a common example of an unstable mineral. An index of the chemical maturity of a rock might, therefore, be the ratio of quartz to feldspar. As sediments are reworked, perhaps through two or more cycles of sediment, they thus tend to mature to pure quartz sands.

Physical maturity, on the other hand, describes the textural changes that a sediment undergoes from the time it is weathered until it is deposited. These changes involve both an increase in the degree of sorting and a decrease in matrix content. Thus an index of the degree of physical maturation might be the ratio of grains to matrix. Total clay content is a useful index of textural maturation, given certain reservations to be discussed shortly.

Both physical and chemical maturation occur during the history of a sand population, but they are not closely related (Johnsson and Basu, 1993). Thus a chemically mature sand may be physically immature and vice versa. This is because chemical composition is essentially a result of provenance, while textural composition is a result of process. From the preceding analysis, it would appear that the most appropriate triangular classification to adopt would have stable grains at one apex, matrix at a second and unstable grains at the third (Fig. 8.10). As a sand population increases in textural maturity it would move away from the matrix apex. As it improves in mineralogical maturity it would move away from the unstable grain apex. Since both types of maturation occur simultaneously, albeit at different rates, the net tendency is for a sediment to move to the stable grain apex. This may not be achieved in a single sedimentary cycle, but it is the ultimate destination of any sand.

These concepts of maturity can be used as a basis for sandstone nomenclature (Fig. 8.11). The total amount of clay material in a sand is obviously the best indicator of matrix content. Feldspar is a common and often volumetrically abundant unstable mineral which may be used as an index of chemical immaturity. Quartz is the obvious choice as the index mineral for the apex of chemical stability.

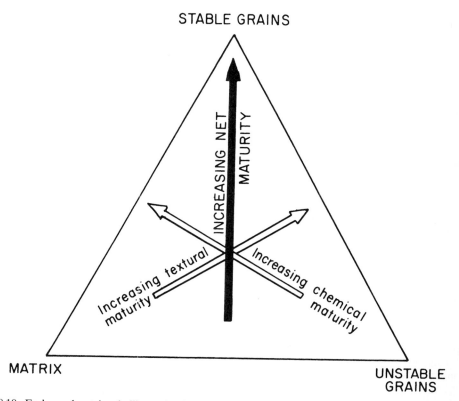

STABLE GRAINS

INCREASING NET MATURITY

Increasing textural maturity

Increasing chemical maturity

MATRIX

UNSTABLE GRAINS

Fig. 8.10. End-member triangle illustrating how the composition of a sand is a function both of its textural maturity, expressed as matrix content, and of chemical maturity, expressed as unstable grain content.

Using quartz, feldspar, and clay as end-members, a sand classification scheme can be drawn up as shown in Fig. 8.12. This shows that sands may be divided into two broad textural types, the arenites and the wackes. Arenites, with a matrix content of less than 15%, are texturally mature. Wackes, with a matrix content of between 15 and 75%, are immature. Rocks with more than 75% matrix are not sandstones but mudrocks.

Similarly, sands can be divided into chemically mature arenites and chemically mature wackes. These both have less than 25% of feldspar. These two sandstones are designated protoquartzite and quartz-wacke. Logically the mature arenites should be (and have been) called quartz-arenites, but protoquartzite is a long-established name for sands of this general composition. Space is found in the quartz apex for orthoquartzite sands, the purest and most mature of all. Sands with more than 25% feldspar are chemically immature and are divided into the arkoses and greywackes for the arenites and wackes, respectively. Logical synonyms for these two names would be feldspar-arenite and feldspar-wacke, but the terms arkose and greywacke are too well established to be omitted.

Essentially, therefore, this scheme arbitrarily divides sands into four main groups depending on the degree of physical and chemical maturation. This system is similar to many others which juggle the choice of end-members, rock names, and percentage

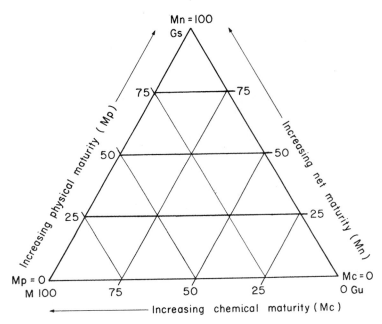

Fig. 8.11. End-member triangle in which a sand may be plotted to illustrate its maturity indices expressed as the percentage composition of matrix, stable and unstable grains. For full explanation see text. Gs, chemically stable grains; Gu, chemically unstable grains; M, matrix content; Mp, physical maturity index, Mc, chemical maturity index; Mn, net maturity index.

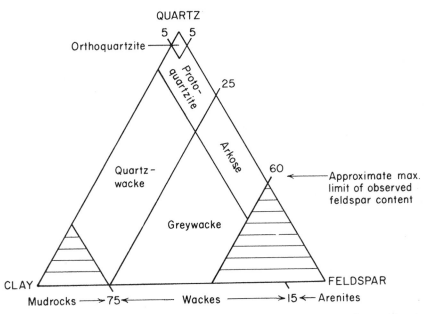

Fig. 8.12. A classification of sandstones based on the use of clay as an indicator of textural maturity, and feldspar as an index of chemical maturity.

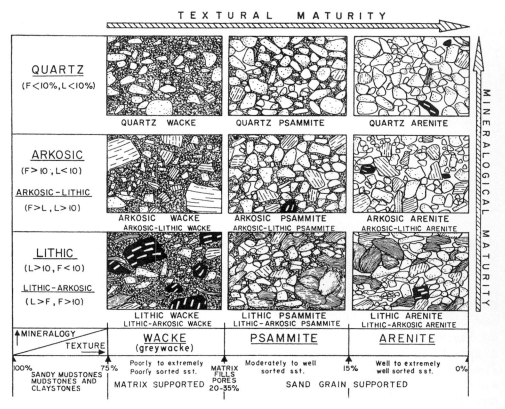

Fig. 8.13. A classification of sandstones proposed by Nagtegaal (1978). This scheme follows generally accepted lithological subdivisions, but classes arkosic and lithic sandstones at 10% of feldspars and lithics, respectively, because of the significance of these components in diagenesis. Texturally, "psammites" are recognized as an intermediate class with regard to matrix content. (From Nagtegaal, 1978, courtesy of the Geological Society of London.)

cutoffs within triangles. Most notable of these are schemes proposed by Dott (1964) and Nagtegaal (1978). Both of these authors propose schemes that allow four end-members. They add lithic grain content to quartz, feldspar, and clay, as proposed above (Fig. 8.13). Dickinson *et al.* (1983) have used the end-member classification of sands, with quartz, feldspar, and lithic grains, as a way of relating sand type to plate tectonic setting. Broadly orthoquartzites characterize interior cratonic basins, arkoses characterize areas of basement uplift, and lithic sands characterize undissected arcs.

The four main sandstone clans are now described. First though, a few cautionary remarks are needed concerning clay matrix, rock fragments (known also as lithic grains), and the heavy mineral assemblage of sandstones.

Clay content has been proposed as an index of the degree of textural maturity of a sediment. This is based on the general observation that in modern sediments clay tends to be winnowed out during sand transportation. Thus clay is largely absent in high-energy environments such as shallow marine shelves. Furthermore clay content often increases with decreasing grain size (Fig. 8.14). This results in the curious situation seen

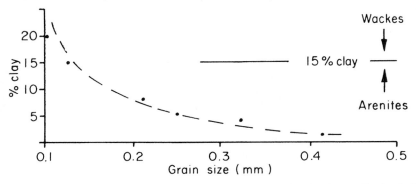

Fig. 8.14. Relationship between grain size and clay content in Torridonian Sandstone, Scotland. (From Selley, 1966. Courtesy of the Geologists' Association.)

in some graded turbidites where, within a few centimeters, an arkose may fine up into a greywacke.

A further important point to note is that the clay content of a lithified sand is not necessarily all syndepositional in origin. Some matrix probably infiltrates pore spaces shortly after deposition. Some clay is transported as silt and sand-grade particles. On compaction they are squashed to form a matrix between more resistant grains. During diagenesis unstable detrital grains break down and form a microcrystalline matrix composed largely of clay minerals. This is a commonly observed feature of greywackes. The question has been asked to what degree do greywackes truly indicate an origin as a muddy sand and to what degree is the matrix due to the decay of labile grains (Cummins, 1962).

An analogous problem is often seen in arkoses. Feldspar grains commonly show varying degrees of alteration to kaolinite, and individual grains of kaolinite can be seen in varying stages of compaction between quartz grains. It is hard to measure feldspar content and clay-matrix content accurately in such specimens This discussion shows, therefore, that clay matrix must only be regarded as a rough guide of textural maturity. In lithified sands the clay content is probably rather higher than the original depositional matrix content.

Another important constituent of sandstones that deserves special mention is rock fragments. Many sands contain grains that are not monomineralic, but which are composite grains. These are called lithic grains or rock fragments. Lithic grains are a popular choice for an end-member of many sandstone classifications, and rock names such as lithic greywacke and litharenite have appeared, as noted earlier. There are two important points to note about lithic grains. The first is that the lithic grain content of a sand is likely to be related to the particle size of the source rocks. Lithic grains are unlikely to be common in an arkose derived from a coarsely crystalline granite. Conversely, lithic grains may be abundant in sands derived from microcrystalline volcanics, metamorphics, or well-indurated mudrocks. The second point to note about lithic grains is that their abundance is dependent on grain size. Figure 8.15 illustrates this very predictable relationship. Obviously the larger a sand grain is then the larger is the probability of it containing more than a single mineral crystal.

A final important constituent of sandstones to consider is the heavy mineral suite.

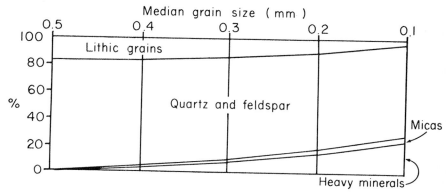

Fig. 8.15. Relationship between grain size and composition in Torridonian Sandstone, Scotland. Lithic grains increase with grain size, conversely mica and heavy minerals decrease. (From Selley, 1966. Courtesy of the Geologists' Association.)

Most sands contain a small (1–2%) but often diverse suite of heavy minerals. These often have an importance out of all proportion to their volume. Heavy minerals are important for two reasons: They may be commercially valuable, and they may indicate the provenance of the sand (e.g., Dill, 1998). A sand may include valuable detrital grains of gold, zircon, ilmentite, and other heavy minerals in sufficient quantities to form placer ores, as discussed elsewhere (see Sections 6.3.2.2.4 and 6.3.2.7.3). Heavy minerals often indicate the provenance of a sand, with characteristic assemblages indicating various igneous and metamorphic sources. The content and composition of a heavy mineral assemblage is controlled, however, not just by provenance, but also by grain size and burial history.

As Fig. 8.15 shows, the heavy mineral content of a sand varies with the grain size, normally increasing with decreasing particle diameter. This is partly due to the original size of the mineral grains, and partly due to the hydraulics of sediment transport and deposition. When a sand is buried heavy minerals leach out due to what is termed "interstratal solution" (Pettijohn, 1957). There tends to be a depth/time-related sequence of heavy mineral destruction (Fig. 8.16). Thus the use of the heavy mineral assemblage of a sand as a provenance indicator must be used with care, noting that it will also reflect the grain size and burial history.

Having pointed to some of the hazards and limitations of sandstone nomenclature, the main groups are now described. For this purpose it is convenient to group them into the wackes (including both quartz-wacke and greywacke), the arkoses, and the quartzites (including both the protoquartzites and the orthoquartzites).

Beyond the world of the lecture theatre and the learned journals, the problems of sand classification and nomenclature seem to vanish like a hangover. In industry the general practice is either to describe a sand as a sand, and no more, or alternatively, as on a well log, a rock may be described more fully using a format similar in content and sequence to this:

LITHOLOGY, Color, hardness, grain size, grain shape, sorting, mineralogy, fossils (if any), porosity, hydrocarbon shows, i.e., stain, fluorescence, or cut.

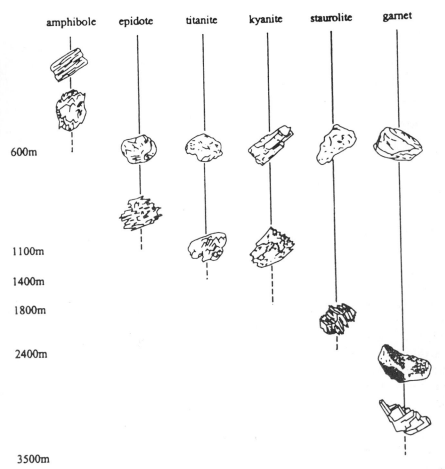

Fig. 8.16. Burial depth distribution of heavy minerals in Upper Paleocene sandstones in the central North Sea, showing the decrease in heavy mineral diversity with increasing burial caused by dissolution of unstable minerals. (Reprinted from Morton, A. C., and Hallsworth, C. R. 1999. Processes controlling the composition of heavy mineral assemblages in sandstones. *Sediment. Geol.* **124**, 3–29. Copyright 1999, with permission from Elsevier Science.)

8.5.2 Sandstones Described

The quartzites, arkoses, and wackes are described in the following subsections.

8.5.2.1 *Quartzites*

Sands that are mature in texture and mineralogy are broadly referable to as the quartzites or quartz-arenites. Various authors have proposed the minimum percentage of quartz necessary for a sand to qualify. The term **quartzite** has been used in metamorphic geology for any sand, regardless of composition, which breaks across grains rather than around grain boundaries. Among sedimentary petrographers, however, a quartzite generally refers to a quartz-rich sand irrespective of its degree of lithifaction. As defined in the scheme in Fig. 8.12 the quartzites have less than 25% feldspar and less than 15% matrix. The quartzites are generally arbitrarily divisible into the protoquartzites, which contain some feldspars and matrix, and the orthoquartzites, which are almost pure silica. The quartzites are, therefore, by definition, sands that are mature in texture and mineralogy. Typical quartzites are white, pale gray, or pink in color. They range from unconsolidated to splintery in their degree of lithifaction. Grain size is variable, but sorting is generally good and individual grains are normally well rounded.

The main detrital grains are quartz, derived from igneous and metamorphic rocks, and cherts (made of quartz, chalcedony, or chrystobalite) reworked from sediments. Rare feldspar and mica grains may be present. Heavy minerals are generally the stable residue such as zircon, tourmaline, apatite, and garnet. Intraformational autochthonous detrital grains are often quite common in quartzites. Glauconite, phosphate pellets, and skeletal debris are typical examples. Probably the majority of glauconitic sands are referrable to the quartzite group (were it not for the presence of that mineral).

Because of their uniform grain size, rounded grain shape, and low clay content, quartzites possess high porosities and permeabilities at the time of their deposition. Cementation is generally by calcite or secondary silica, but where it is absent, quartzites make the best aquifers and hydrocarbon reservoirs of all the sandstone types. It is probable that most, if not all, quartzites are polycyclic in origin. That is to say that they have been through more than one cycle of weathering, erosion, transportation, and deposition to achieve the necessary maturity to qualify. It is a matter of observation that the majority of quartzites were deposited in marine sand-shoal environments. Extensive quartz sand sedimentation occurred in the Lower Paleozoic shelf seas around the border of the Canadian Shield and the northern margin of the Saharan Shield (e.g., Bennacef et al., 1971). Quartzose sands are also found in eolian deposits such as the Permian Rotliegende of the North Sea basin (Glennie, 1972) and the early Mesozoic dune sand formations of the Colorado Plateau (Baars, 1961). Eolian and marine shoal environments provide the optimum conditions for quartzites to be deposited by selective winnowing. Nevertheless, they do also occur in many other environments, including modern deep-sea sands (Hubert, 1964) and ancient turbidites (Sturt, 1961). Plate 3C illustrates a thin section of the Simpson Group (Silurian) orthoquartzite of Oklahoma.

8.5.2.2 Arkoses

The term **arkose** was first proposed by Brogniart (1826) for a coarse sand composed of quartz and substantial quantities of feldspar from the Auvergne in France. This name is still used for rocks of that general description. Essentially the arkoses are sands which are relatively mature in texture (i.e., low in clay matrix) and are immature in mineralogy as shown by the abundance of feldspar. The clay content must generally be less than 15% and there must be more than about 25% of feldspar. Few arkoses have more than 60% feldspar because this would presuppose a source in which feldspar was more abundant than quartz. The converse is generally true and the ratio of quartz to feldspar begins to rise as soon as the source rock is submitted to weathering.

Arkoses, therefore, are the product of the incomplete degradation of acid igneous and metamorphic rocks such as granites and gneisses. It was once argued that arkoses were indicators of cold and/or arid climates in which physical weathering processes dominated chemical ones. Krynine (1935) showed, however, that modern alluvial arkoses occur in tropical rain forest climates. These form because rapid runoff and erosion cut gullies through the weathered zone and eroded unweathered granite. Nevertheless, some arkoses, such as the Torridonian (Pre-Cambrian) example of Scotland, contain well-rounded feldspar grains (Plate 3D). It is unlikely that these grains are polycyclic. The implication is that rounding occurred within the first and only cycle of sedimentation, and it is known that eolian action is far more effective at this than running water (see Section 3.1.2). There is a strong presumption that the Torridonian arkoses were wind-rounded eolian sands that were deposited fluvially in an arid climate.

The typical arkose is a pink or red colored rock; less commonly it is gray. Pink coloration is due to the feldspars, but the red colored examples owe this feature to absorption of red ferric oxide into the clay matrix. Arkoses show a wide range of grain sizes and are often poorly sorted. Arkose often forms in place on granites forming a transitional weathering zone, termed "granite wash," in which it is hard to distinguish sediment from igneous rock. This situation is a well-site geologist's nightmare. It is very hard to distinguish arkose from granite on the basis of drill cuttings, yet the contact can easily be picked subsequently from geophysical logs. The grains of arkoses are typically angular to subrounded and there is generally a significant amount of clay matrix. For this reason arkoses seldom stay unconsolidated for long, like the quartzites. Most ancient arkoses show some degree of lithifaction due to clay bonding. In extreme cases porosity can be completely obliterated by a silica or carbonate cement. The feldspars in arkose are of various types, depending on the nature of the source rock, but microcline and albite tend to be more abundant than the less stable calcic feldspars. Alteration of the feldspars to kaolinite and sericite is a common feature, but it is hard to tell whether this occurred by hydrothermal alteration of the source rock, during weathering, or through subsequent burial diagenesis.

Micas are a common accessory mineral and arkoses often contain a diverse suite of heavy minerals. These may give some indication of the source terrain, specifically whether it was igneous or metamorphic. In addition to the ubiquitous stable suite of zircon, apatite, garnet, and tourmaline, it is not uncommon to find opaque iron ores as well as many other heavy minerals.

Huckenholz (1963) described the petrography of the type arkoses of the Auvergne. Most arkoses seem to occur in alluvium adjacent to granitoid basement. They are, therefore, the characteristic sediment type of fault-bounded intracratonic basins. Because such sediments are deposited in an oxidizing alluvial fan environment, many arkoses are of what is termed the "red bed" facies assemblage.

Typical examples of arkoses formed in this setting occur on the Pre-Cambrian shields all over the world. Specific examples include the Torridonian (Pre-Cambrian) of Scotland (Selley, 1966) and the adjacent Jotnian, Dala, and Sparagmite Pre-Cambrian series of the Scandinavian Shield. The Triassic Newark group of the Connecticut trough, USA, is another classic example of an arkose. Though probably the bulk of arkoses occur in fluvial environments, they are also found elsewhere. Arkoses mixed with coralgal debris occur on modern beaches of the fault-bounded Red Sea. Ancient analogs occur on the oil-soaked, reef-crowned granite highs of the Sirte basin, Libya.

8.5.2.3 Wackes

The wackes are, by definition, texturally immature sands with more than 15% matrix. Chemically mature wackes, with less than 25% feldspar, are termed "quartz-wackes." Chemically immature wackes are the classic greywackes. The name "greywacke" comes from the German "grauewacke," which was applied to the Paleozoic sandstones of the Harz Mountain. Petrographic descriptions of examples from the type area have been given by Helmbold (1952) and Mattiatt (1960).

The **greywackes** are characteristically hard, dark gray-green rocks that break with a hackly fracture. Under the hand lens, greywackes are very poorly sorted with particles ranging from very coarse sand grains down into clay-grade matrix. They have been aptly described as microconglomerates. Grains are commonly angular and of poor sphericity. Quartz is overshadowed by an abundance of other detrital minerals. Feldspar is present, but so also are mafic grains such as horneblende and, occasionally, pyroxenes. Some of the larger grains are lithic rock fragments, and, depending on the source, these may have been derived from volcanics or older metasediments such as quartzite or slate. Micas are abundant and include both muscovite and biotite, as well as microcrystalline diagenetic chlorite and sericite. A diverse suite of unstable heavy minerals is also typical. Plate 3E illustrates a Jurassic greywacke from the northern North Sea.

All of these detrital grains are set in the abundant matrix. This is a microcrystalline paste of clay minerals, chlorite, sericite, quartz, carbonate (often siderite), pyrite, and occasionally carbonaceous matter. Corrosion of detrital grain boundaries is sometimes seen in the form of a characteristic *chevaux de frise* of micas. This rims not only the unstable grains but sometimes even quartz.

The quartz-wackes differ from the typical greywacke just described in that they lack the diverse suite of unstable detrital minerals. Their absence is coupled with an increase in the number of quartz and sedimentary lithic grains, though the clay paste is still present, and in hand specimen quartz-wackes and greywackes are hard to distinguish. The quartz-wackes correspond in part to the subgreywackes and lithic wackes of some texts (Plate 3F).

Particular attention has been given to the origin of the matrix in wackes (e.g., Dott, 1964). The poor sorting of these sandstones presupposes that the matrix may be largely syndepositional. On the other hand, the presence of authigenic minerals, and the corrosion of detrital grains, shows that some of the matrix is diagenetic in origin (Cummins, 1962; Brenchley, 1969). In this context it is important to note that the typical wackes are largely Pre-Cambrian and Paleozoic in age. This suggests that time, deep burial, and/or high geothermal gradients are needed to generate greywackes. These are thus perhaps metamorphic rocks. This argument is of particular importance when considering deepwater sands. Most ancient greywackes occur in flysch facies, which are commonly interpreted as turbidites. Many modern deep-sea sands are clean and well sorted. These may have been transported into deep water by turbidity flows, but then undergone reworking by geostrophic contour currents (Hubert, 1964). This reworking may have removed any clay matrix that may have originally been present. The problem is threefold: Is clay necessary for turbidity flows or not? If it is necessary, then modern deep-sea sands may be due to normal bottom traction currents.

The greywackes, as stated in the previous paragraphs, are commonly found in pre-Mesozoic flysch facies. These typically occur in subductive troughs and it is apparent, both from their petrology and regional setting, that greywackes are often derived from the rising island arcs of volcanic origin. Hence the unstable suite of mafic minerals and the relatively high percentage of iron and magnesia in greywackes.

The Steinmann Trinity is a picturesque name given to a commonly observed association of greywackes, cherts, and volcanics (Steinmann, 1926). This is often present in the early phase of sea floor spreading (see Section 10.1.3). Crustal thinning is held responsible for submarine volcanics and the generation of ultrabasic ophiolites and pillow lavas. It has been argued that the cherts form from blooms of radiolaria caused by an abundance of volcanically generated silica, whereas the greywackes are formed from sediments shed off rising volcanic island arcs.

Quartz-wackes, on the other hand, are more usually derived from preexisting sediments. Sands contribute the quartz, shales produce the clay, and the lithic fraction comes from the indurated equivalents of both. Thus quartz-wackes are typically found, not so much in flysch settings, but in proximal continental deposits. Quartz-wackes occur in fanglomerate and alluvial environments. One example is provided by the continental Mesozoic sandstones of the Sirte basin, Libya, which were derived from Paleozoic sediments. Another example is the Cretaceous sands of offshore South Africa, which were drived from the Paleozoic sediments of the Cape Fold Belt.

Continental quartz-wackes are often red-brown in color due to impregnation of the clay matrix by red ferric oxide. With increasing transportation, the quartz-wackes lose some of their clay content and assume a rock type often termed **subgreywacke.** This is found in both fluvial and deltaic deposits. Examples of subgreywackes occur in the fluvial Devonian and deltaic Pennsylvanian (Upper Carboniferous) sandstones on both sides of the North Atlantic.

This brief review of sandstone petrography shows that the composition of a newly deposited sand is a product of provenance and process. The chemical maturity of a recently eroded sediment will depend on the source rock and the extent of weathering. Chemically mature sediments are generally polycyclic in origin and owe their maturity to derivation from preexisting sedimentary formations. Chemically immature sands are generally first-cycle material derived from igneous and high-grade metamorphic rocks. Recently eroded sediment is commonly poorly sorted and rich in argillaceous matrix. Eolian and aqueous processes increase the textural maturity of a sand, glacial processes may reverse it.

The next section of this chapter shows how postdepositional changes affect sandstone composition, and attempts to relate these to the evolution of porosity and permeability in the terrigenous sands. After deposition a sand may be buried and turned into sandstone by lithifaction. This consists of physical compaction and chemical diagenesis. These are now discussed in turn.

8.5.3 Diagenesis and Porosity Evolution of Sandstones

8.5.3.1 Introduction

The term **diagenesis** has been applied in varying ways to the postdepositional, yet premetamorphic processes that affect a sediment. Dunoyer de Segonzac (1968) has reviewed the history and semantics of this term. For the purposes of the following account the definition of Pettijohn (1957, p. 648) is used:

> Diagenesis refers primarily to the reactions which take place within a sediment between one mineral and another, or between one or several minerals and the interstitial or supernatant fluids.

This definition limits diagenesis to essentially chemical processes distinct from physical processes, such as compaction. Note that this definition is broader than that used by organic geochemists. The diagenesis of sandstones has been described in great detail by Larsen and Chilingar (1962), Folk (1968), Pettijohn *et al.* (1972), Marshall (1987), Meshri and Ortoleva (1990), Morse (1994), Giles (1997), and Montanez *et al.* (1997).

The following account principally concerns those aspects of sandstone diagenesis that affect porosity and permeability. Before proceeding to the details of sandstone cementation, however, it is necessary to consider the chemistry of the fluids that move through the pores, and their effect on cementation and solution. The following account begins by considering sandstone diagenesis from the broad aspect of porosity gradients. It then proceeds to consider the details of how porosity is diminished by cementation and enhanced by solution.

8.5.3.2 Porosity Gradients in Sandstones

As a general statement it is true to say that the porosity of sandstones decreases with depth. The porosity of a sandstone at a given depth may be expressed thus (Selley, 1978):

$$\phi^d = \phi^p - G \cdot D,$$

where ϕ^d is the porosity at depth D, ϕ^p is the original porosity at the surface, G is the porosity gradient, and D is the depth below surface. The original porosity at the surface will depend on depositional process and provenance (as discussed in Section 3.2.3). The porosity gradient will be a function of sandstone composition, pore fluid composition and history, temperature, and time (Selley, 1978). There are now a number of studies of porosity gradients in sandstone basins, notably by Fuchtbauer (1967), Wolf and Chilingarian (1976), Nagtegaal (1978), Magara (1980), and Schmoker and Gautier (1988), who have produced an improved formula for calculating porosity as an exponential function of depth:

$$\phi = a \cdot e^{bz},$$

where ϕ is the porosity, z is the depth, and a and b are constants representing particular sediment properties. Figure 8.17 shows some examples of sandstone porosity gradients that have been recorded. It is interesting to note that porosity loss is largely linear with depth. The rapid loss of porosity with shallow burial shown by clays is largely absent. This is because compaction is of less importance in clean sands (Lundegard, 1992). Intuitively one would expect wackes to lose porosity faster than arenites. This is because they are not as well sorted as arenites and therefore have lower primary porosities. Furthermore, because of their clay content, the wackes may undergo more compaction than clean arenites. Of the wacke sands the chemically stable quartz-wackes may lose porosity at a lower rate than the chemically unstable greywackes; similarly, one might expect arkoses to lose porosities faster than quartz-arenites. As Fig. 8.18 illustrates, there is some truth to these generalities, but the story is often more complicated than this, as will be revealed shortly.

Geothermal gradient is another factor that may be expected to affect sandstone porosity loss, because cementation will proceed faster in hot basins than in cool ones.

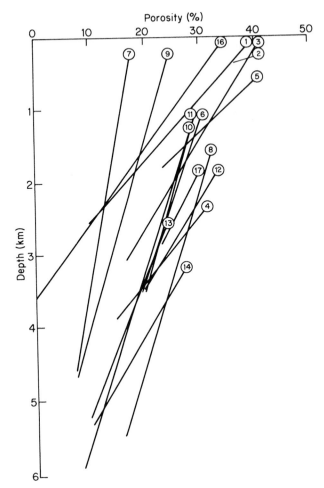

Fig. 8.17. Sandstone porosity gradients from around the world. Note that sandstone porosity gradients are linear with depth, unlike the burial curves of clays (see Fig. 8.7). Sources of data: 1–3 (Galloway, 1974); 4 (Selley, 1978); 5 (Mayuga, 1970); 6–9 (Maxwell, 1964); 10–15 (Loucks *et al.*, 1984); 16 (Koithara *et al.*, 1980); and 17 (Maher, 1980).

Figure 8.19 shows a weak correlation between geothermal gradient and porosity gradient, but there are obviously other factors, such as mineralogy, that are also important. Overpressure can preserve porosity to greater depths than anticipated and the early invasion of hydrocarbons inhibits porosity loss by excluding connate fluids from the pores of a sand (e.g., Philipp *et al.*, 1963; Fuchtbauer, 1967). From Maxwell (1964) to Scherer (1987) and Giles (1997), geologists have produced mathematical formulas that attempt to relate the various factors that control porosity gradients. General statements can be made to the effect that porosity will be preserved to greater depths in cool basins of quartzose sands than in hot volcaniclastic ones. Accurate gradients for a particular case

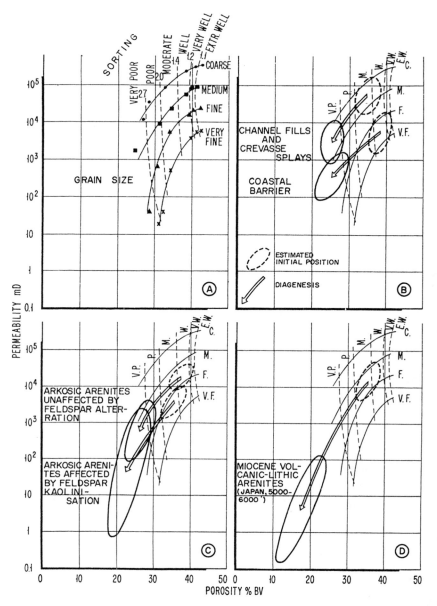

Fig. 8.18. Illustrations of the changes in porosity and permeability for sandstones of different mineralogical composition and for different burial depth. (From Nagtegaal, 1978, courtesy of the Geological Society of London.) (A) Reference grid for porosity and permeability values of various grain size and sorting mixes of unconsolidated clay-free sands. (B) Porosity/permeability plots for Miocene quartz arenites at 11,800–12,000 ft (3596–3657 m) burial, (12 samples) Niger delta. (C) Porosity/permeability plots for arkosic arenites with and without feldspar kaolinisation, at 9,000–10,000 ft (2743–3047 m) (20 samples) and 10,000–11,000 ft (3048–3352 m) (34 samples) burial, respectively. Jurassic sands, North Sea. (D) Porosity/permeability plots for Miocene volcanic lithic arenites (greywackes, according to many classifications) at 5000–6000 ft (1524–1828 m) burial (18 samples), Japan.

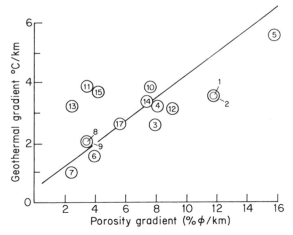

Fig. 8.19. Graph of porosity gradients plotted against geothermal gradient. Numbers refer to the porosity gradients in Fig. 8.17. A general correlation between geothermal and porosity gradients is apparent, but not perfect. This is because porosity gradients are also related to mineralogy, pressure, and the presence of petroleum. In particular, chemically unstable sands will often lose porosity faster than quartzose ones.

can only be established from well data in which all the variables (mineralogy, facies, geothermal and pressure gradients) are known. Thus it is extremely difficult to use geological parameters to predict the porosity gradient in a given situation. Geophysics is now able to identify porous sands, and even to identify the presence of petroleum, using the seismic method, but this lies beyond the scope of this book.

8.5.3.3 Porosity Loss by Cementation

The previous account of porosity gradients showed how sandstones lose porosity with increasing burial depth. In sands, unlike clays, the effect of compaction is subordinate to cementation. The ways in which porosity is lost by cementation are described next.

Recall from Chapter 7 that pore fluids can be placed in three groups according to their origin. Meteoric water is that of the rivers and rain. It is commonly present in the pores within the top 100 m or so of the earth's surface, but may also be preserved fossilized below unconformities. Meteoric water contains relatively low concentrations of dissolved salts. Meteoric groundwater tends to have a positive Eh, due to dissolved oxygen, and a low pH, due to carbonic and humic acids.

Connate water was originally defined as residual seawater within the pores of marine sediment. Seawater is oxidizing and neutral. It is now realized that seawater trapped in pores undergoes considerable modification as chemicals are precipitated or dissolved. The term "connate water" is retained, however, for such deeply buried fluids. The Eh and pH of connate fluids vary widely, extending from acidic and oxidizing, where they mingle with meteoric waters, to alkaline and reducing, where they are associated with petroleum accumulations. The salinity of pore fluids gradually increases with depth, going from the meteoric to the connate zone.

Juvenile waters are those of hydrothermal origin, characterized by high temperature and bizarre chemistry. Figure 8.20 shows the Eh:pH ranges of these fluids. The dia-

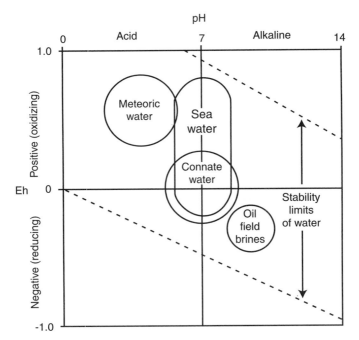

Fig. 8.20. Diagram showing the Eh and pH of the various types of water that move through pores and thus affect diagenesis.

genetic history of a sandstone will obviously be controlled by the chemistry of the fluids that have moved through its pore system. The factors that determine mineral precipitation or solution include the chemistry of the sediment and the composition, concentration, Eh, and pH of the pore fluids. Many chemical reactions occur during sandstone diagenesis. Only about half a dozen of these are of major significance in sandstone cementation and porosity evolution: those which control the precipitation and solution of silica, calcite, and the clay minerals. These are considered in turn.

The solubility of calcite and silica are unaffected by Eh, but are strongly affected in opposing ways by pH. Silica solubility increases with pH, whereas calcite solubility decreases. Thus in acid pore fluids, such as meteoric waters, calcite tends to dissolve and quartz overgrowths to form, whereas in alkaline waters calcite cements develop and may even replace quartz. For mildly alkaline fluids (pH between 7 and 10) quartz and calcite cements may both develop (Fig. 8.21). Clay minerals are similarly sensitive. Kaolinite tends to form in acid pore fluids, while illite develops in alkaline conditions. This is a very important point which significantly controls the permeability of a sand. Siderite, glauconite, and pyrite are all stable in reducing conditions. Figure 8.22 summarizes the stability fields of these various minerals. A cement is a crystalline substance which is precipitated with the pores of a sediment after it is deposited. It should be distinguished from the matrix, which is microgranular material that occurs in pores and is of syndepositional origin.

The three most common cements of sandstones are carbonates, silica, and clays. Many other authigenic minerals occur in sandstones but they are seldom sufficiently abundant

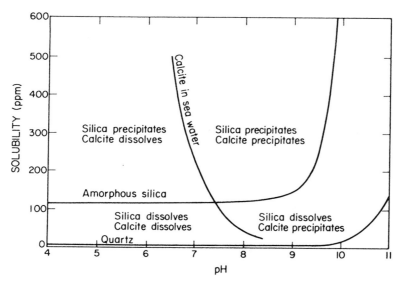

Fig. 8.21. Diagram showing the way in which the solubility of calcite and quartz are related to pH. (From Figure 9-7. "Origin of Sedimentary Rocks, Second Edition." Blatt/Middleton/Murray © 1980. Reprinted by permission of Prentice-Hall, Inc., Upper Saddle River, NJ.)

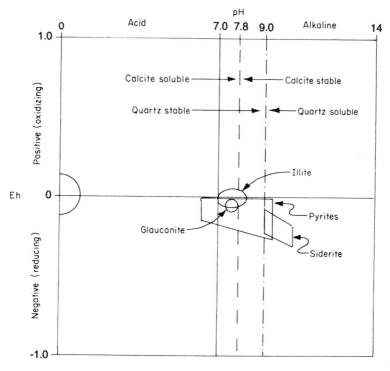

Fig. 8.22. Eh versus pH plot showing the approximate stability fields of important sedimentary minerals. Compare with Fig. 8.20.

to form the dominant cementing mineral. These minerals include barytes, celestite, anhydrite (gypsum at outcrop), halite, hematite, and feldspar.

The net effect of all these mineral cements is to diminish or completely destroy the primary intergranular porosity and the permeability of the sandstone. Because of their abundance and significance, the origin of carbonate, silica, and clay cements is described next.

8.5.3.3.1 Carbonate cements

Carbonate cements in sandstones consist of calcite, dolomite, and occasionally siderite. Modern sediments have been found with cements of both aragonite and calcite (e.g., Allen *et al.,* 1969; Garrison *et al.,* 1969, respectively). These reports indicate that carbonate cements can form at surface temperatures and pressures. Their genesis does not require high temperatures or pressures. These cements are precipitated from solutions which gained their calcium carbonate both from connate water expelled by compaction and from the dissolution of shells.

In ancient sandstones aragonite is largely unknown as a cement due to reversion to calcite, the stable form of calcium carbonate. Carbonate cements range from fringes of small crystals rimming detrital grains through sparite-filled pores, to single crystals, centimeters across, which completely envelop the sand fabric. This latter type of texture is termed "poikilitic" or "poikiloblastic" (Plate 4A). It is easily identified in hand specimens because the sandstone tends to break along cleavage fractures which twinkle in the sunlight. This is known as "lustre mottling." Calcite cements can, therefore, be present in a sandstone in sufficient quantities to infill all primary intergranular porosity.

Dolomite is the second common type of carbonate cement found in sandstones. It occurs typically in rhomb-shaped crystals which, by themselves, seldom completely destroy porosity. In argillaceous sandstones microcrystalline calcite, dolomite, and siderite are often present within the clay matrix. As previously discussed, the presence of a carbonate cement indicates that the sand has been bathed in alkaline pore fluids.

Carbonate cements will form in alkaline pore fluids, irrespective of Eh, as shown in Figs. 8.21 and 8.22, respectively. These conditions are commonplace in the subsurface, but are especially found over breached petroleum accumulations. It is believed that bacterial oxidation of petroleum liberates CO_2, causing the precipitation of carbonate mineral cements. Experienced drillers commonly report a slowdown in rate of penetration of the bit when drilling through the cemented shale cap rock of a petroleum accumulation. This type of carbonate cemented capping, commonly referred to as an HRDZ (hydrocarbon-related diagenetic zone) can sometimes be imaged seismically (O'Brian and Woods, 1995). When the escaping fluids reach the sea floor, carbonate mud mounds, termed "cold seeps," form from methanogenic bacteria and higher carbonate-secreting life forms. These have been particularly well documented from the North Sea and the Gulf of Mexico (see Hovland *et al.,* 1987, and Sassen *et al.,* 1993, respectively).

The spatial distribution of carbonate cements is much more varied than that of silica (Morad, 1998). Cement distribution may be uniform, layered (Fig. 8.23A), concretionary (colloquially referred to as "doggers") (Fig. 8.23B), or occuring as envelopes at sand:shale contacts (Fig. 8.23C). Carbonate-cemented envelopes at sand:shale

Fig. 8.23. Illustrations of the spatial distribution of carbonate cements in sandstones. (A) Carbonate-cemented and cement-free layers in the Bridport Sand (Lower Jurassic), Burton Bradstock, Dorset. For discussion of mode of formation see Bryant *et al.* (1988). (B) Isolated carbonate concretion (colloquially termed "dogger") in Bencliff Grit (Middle Jurassic), Osmington Mills, Dorset. (C) Carbonate-cemented envelopes at sand:shale contacts, Campos basin, offshore Brazil. (Reprinted from Carvalho, M. V. H., De Ros, L. F., and Gomez, N. S. Carbonate cementation patterns and diagenetic reservoir facies in the Campos basin Cretaceous turbidites eastern offshore Brazil. *Mar. Pet. Geol.* **12,** 741–758. Copyright 1995, with kind permission from Elsevier Science Ltd.)

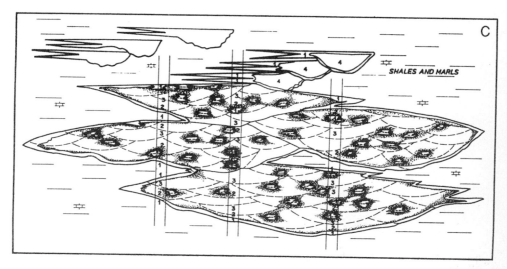

SHALES AND MARLS

C

Fig 8.23 — *Continued*

boundaries have been described by many geologists (e.g., Fothergill, 1955; Fuchtbauer, 1967; Moncure *et al.,*1984; Sullivan and McBride, 1991; Carvalho *et al.,* 1995). Selley (1992a) described the recognition of cemented envelopes on geophysical well logs and suggested two modes of origin (Fig. 8.24). They might either result from the precipitation from fluids expelled from adjacent compacting shales, or remain after the leaching

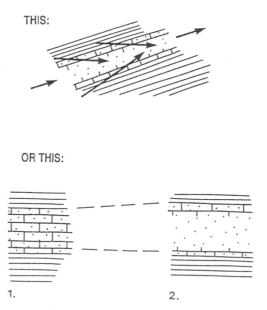

THIS:

OR THIS:

1.   2.

Fig. 8.24. Illustrations two possible mechanisms for the formation of cemented envelopes at sand:shale contacts. They may either develop from the precipitation of minerals where fluids enter a sand from an adjacent compacting shale, or as a residue from a migrating diagenetic front. (From Selley, 1992a. Courtesy of the Geological Society of London.)

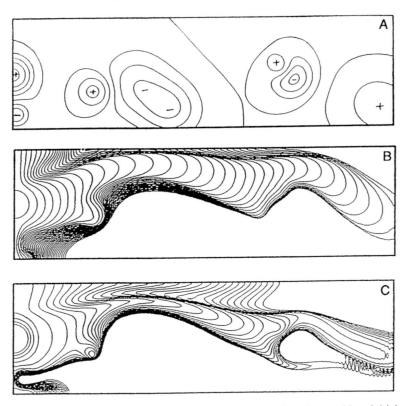

Fig. 8.25. Temporal development of a reaction front in calcite-cemented sandstone with an initial nonuniform texture. The size of the system is 10 cm × 3 cm; the grid is 201 × 61. (A) Porosity contour map at the initial time $t_0 = 0$. The contour interval is 2%. The plus (minus) sign means that the porosity there is higher (lower) than the average porosity of the system (chosen to be 10%). (B) Temporal development of the reaction front under a driven flow $v = 1 \times 10^{-3}$ cm/s. $D_t = 5 \times 107$ s. (C) Temporal development of the reaction front under driven flow $v = 1 = 10^{-2}$ cm/s, 10 times faster than that used for (B). $D_t = 2 \times 106$ s. (Reprinted from Chen, W., and Ortoleva, P. 1990. Reaction front fingering in carbonate-cemented sandstones. *Earth Sci .Rev.* **29**, 183–198. Copyright 1990, with permission from Elsevier Science.)

of an earlier cement. In the latter case they are the carbonate analogs of uranium roll-front ores described earlier (see Section 6.3.2.2.4). This mechanism has been endorsed by Chen and Ortoleva (1990), who worked out the physical processes and chemical reactions that occur as a diagenetic front moves through a permeable sand (Fig. 8.25).

8.5.3.3.2 Silica cements

Sandstones are commonly cemented to varying degrees by silica. Rarely this is in the form of amorphous colloidal hydrated silica, opal. This occurs in younger rocks at low pressures but sometimes at high temperatures, as in some hot springs. Opal dehydrates with age to microcrystalline quartz, termed "chalcedony." This is quite a common cement in sandstones of various ages. By far the most common type of silica cement, however, is quartz overgrown in optical continuity on detrital quartz grains. These authigenic overgrowths develop in a variety of styles. During early cementation euhedral overgrowths form with well-defined crystal facies (Plate 4B). Sorby (1880) first drew attention to this phenomenon in the Permian Penrith sandstone of England (see also Waugh, 1970a,b).

As cementation continues, however, pore spaces are completely infilled with quartz (Plate 4C). It is not always apparent whether the resultant fabric has been produced only by continuous development of the initial quartz overgrowths. Alternatively, solution may have taken place where grains are in contact, and the dissolved silica reprecipitated in the adjacent pore space. This process is termed **pressure solution.**

The genesis of secondary silica cement has been extensively studied because this is the most common type of porosity destroyer in sandstones (McBride, 1990; Worden, 2000). Particular attention has been paid to finding the depth below which effective porosity is absent in a particular sedimentary basin. This may be used to predict the "economic basement" below which it would be futile to search for aquifers or hydrocarbon reservoirs.

Attention has been directed toward the source of silica, the physicochemical conditions that govern its precipitation, and the relationship between silica cementation and pressure solution. There is no doubt that silica cements may have been precipitated from solutions which derived their silica from organic debris such as radiolaria, diatom tests, and siliceous sponge spicules. Likewise, some silica-rich solutions must have been expelled from compacting clays. A number of successful attempts have been made to grow silica overgrowths artificially. These have been achieved at high temperatures and pressures (Heald and Renton, 1966; Paraguassu, 1972), and also at normal temperatures and pressures too (Mackenzie and Gees, 1971). Study of the relationship between secondary silica, porosity, and depth of burial is inextricably linked with the phenomenon of pressure solution, or pressure welding. Rittenhouse (1971) has given a quantitative analysis of porosity loss that integrates pressure solution with grain shape and packing. Taylor (1950) showed how, with increasing depth of burial, the number of grain contacts per grain increased from about one or two near the surface up to five or more at great depth. Simultaneously, Taylor showed how the nature of the grain contacts changed with increasing depth of burial. At shallow depths tangential or point contacts are typical. These grade down into long contacts where grain margins lie snugly side by side. At greater depths still, concavo-convex and sutured grain boundaries prevail where there has been extensive pressure solution. These changes in the number and nature of grain contacts are accompanied by a gradual decrease in porosity (Fig. 8.26). Great care must be taken in studying the relationship of secondary silica to pressure solution. Sippel (1968) and Sibley and Blatt (1976) have shown that cathodoluminescence examination of sands reveals far greater amounts of secondary quartz than revealed by examination with a polarizing microscope. Many sands that appear to have lost porosity by extensive pressure solution have, in fact, lost it by extensive secondary quartz cementation. This calls into question the accuracy of all modal analysis of sandstone composition and most studies of pressure solution and silica cementation.

8.5.3.3.3 Clay cements

Minor amounts of authigenic clay in a sand inhibit the precipitation of carbonate and silica cements. But significant amounts of clay may not only occupy pores, but may also

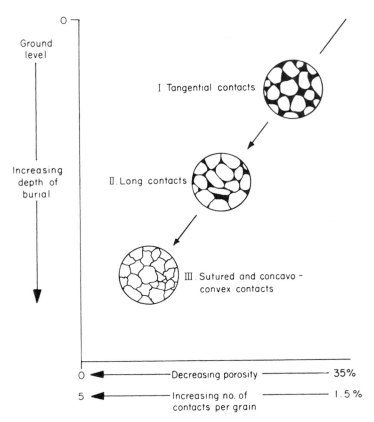

Fig. 8.26. The destruction of porosity in sandstones with increasing depth. Previously interpreted as largely due to pressure solution, cathodoluminescence shows that diagenetic silica cementation is the main process.

significantly diminish permeability (Wilson and Pittman, 1977; Guvan *et al.*, 1980). There are four main types of clay to consider: kaolinite, illite, montmorillonite, and chlorite.

Kaolinite grows as discrete crystals often with a shape like an accordian (Fig. 8.27A). These crystals are sometimes so extended that they look like a worm (described as "vermicular" by the learned). Kaolinite cement obviously destroys porosity, but it does not do as much damage to permeability as other authigenic clays, which, though they may diminish porosity by the same amount, volume for volume, are far more destructive of permeability.

Illite tends to form fibrous crystals that grow radially from the edges of sand grains (Fig. 8.27B). These furry illite jackets extend across the throat passages, thus diminishing permeability.

Montmorillonite has the worst effect on permeabilitity because its lattice structure can absorb water, expand, and destroy permeability (Fig. 8.27C). Thus, for example, a petroleum reservoir with a montmorillonitic clay may be quite permeable, except where it has been damaged by mud filtrate around a well bore.

The productivity of a petroleum reservoir may be may be controlled by its clay mineralogy. In the Lower Permian (Rotliegende) gas fields of the North Sea, illitic cemented

Fig. 8.27. Scanning electron micrographs illustrating different authigenic minerals in sandstone. (A) Authigenic quartz crystals encrusted by later kaolinite "accordian" crystals. (B) Fibrous crystals of illite, a major permeability diminisher. (C) Montmorillonite (smectite), the dreaded swelling clay whose lattice can absorb water. (D) Authigenic dolomite rhombs, with platey chlorite clay cement. (Courtesy of J. Huggett and Petroclays.)

sands tend to be subcommercial, and kaolinite-cemented ones economic (Stadler, 1973). Thus it is important to be able to predict the distribution of authigenic kaolinite and illite. Because of its stability in low pH waters, kaolinite tends to be deposited in continental environments; it may be converted to illite during diagenesis in the presence of alkaline connate waters.

Illite, by contrast, is more commonly the detrital clay found in marine deposits. Recrystallization of illite from detrital flakes to furry coatings may take place in alkaline connate waters. Conversely, kaolinite may form authigenically either from the breakdown of detrital grains, such as feldspar, or by the recrystallization of illite fibers to kaolinite crystals. These changes occur in low pH conditions in several ways. Meteoric water may flush recently deposited sands during a marine regression or lithified rocks undergoing weathering. The permeable kaolinitic zone may then be preserved beneath an unconformity. Kaolinization may also be caused by carbonic acid-rich waters produced by the devolatilization of coal beds.

Chlorite is a ferro-magnesian rich clay mineral that is produced by the weathering of basic igneous rocks, such as gabbro and basalt. It is stable in reducing alkaline fluids and is susceptible to oxidation. Chlorite is only found at the earth's surface today at high polar latitudes where chemical weathering is negligible. Authigenic chlorite cement in sandstones has a characteristic platey habit (Fig. 8.27D). Because of its high iron and magnesium content it is commonly found in volcaniclastic sandstones.

8.5.3.4 Porosity Enhancement by Solution

There is considerable evidence for solution porosity development in sandstones. The petrographic criteria have been outlined by Schmidt *et al.* (1977). The evidence includes cement or grains in various stages of leaching. Poikilotopic calcite cements sometimes leach out to form sands with excellent porosity. This is because when the calcite cement originally grew it corroded the quartz and thus diminished grain size. The postsolution fabric thus has a catenary pore system with excellent continuity. When grains leach out they may leave large isolated pores, thus increasing the total porosity, but not necessarily improving permeability significantly (Plate 4D). Finally, of course, secondary porosity may be generated in sandstones by fractures. Figure 8.28 illustrates the way in which the different types of secondary pore systems affect porosity and permeability.

Note that the amount of secondary porosity due to the leaching of grains is related to the percentage of chemically unstable grains. There is a gradual increase in porosity

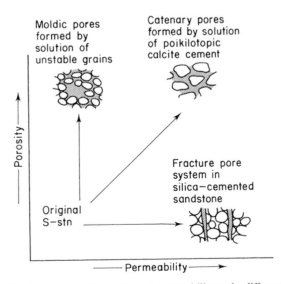

Fig. 8.28. Graph illustrating the response of porosity and permeability to the different types of secondary pore systems in sandstones.

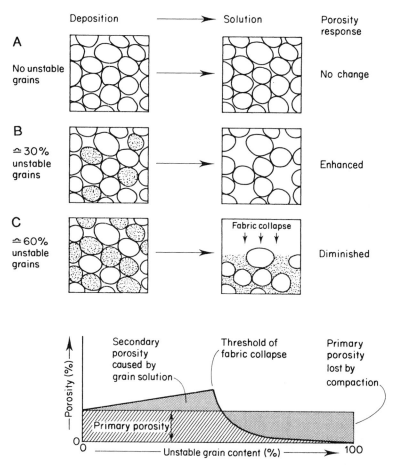

Fig. 8.29. Sketch illustrating how secondary porosity due to grain solution is related to the percentage of unstable grains. Note that there is a gradual increase in porosity with increasing unstable grain content until a critical threshold is reached. Above this point the original depositional fabric of the rock will collapse. For illustrative purposes cement and the effect of cement solution are omitted.

with unstable grain content until a critical threshold is reached, above which the fabric of the rock will collapse. There is an optimum percentage of unstable grains in a sand to generate maximum solution porosity (Fig. 8.29). In practical terms this explains why volcaniclastic sands are such poor reservoirs, because their unstable grain content commonly exceeds the fabric collapse threshold. Solution porosity is absent in quartzites (apart from that due to cement leaching). Solution porosity is best developed in arkoses and quartz sands with a minor component of carbonate grains.

The cause of solution porosity has long been a matter for debate. The grains that leach out include feldspar, glauconite, and carbonate clasts. It is commonly carbonate cement that leaches out. These observations suggest that the leaching solutions are acidic (see Fig. 8.22). Acid solutions can generate secondary porosity in sandstones in two principal ways: epidiagenesis and kerogen decarboxylation. These are considered in turn.

8.5.3.4.1 Secondary porosity by epidiagenesis

The term **epidiagenesis** is applied to the reactions that take place in a previously lithifed sediment (Fairbridge, 1967). Weathering was described in some detail in Chapter 2. This process is well understood and it is known to affect sediments down to over a 100 m below the surface. Geologists are familiar with the development of solution porosity in carbonates, and its preservation in the subsurface beneath unconformities (Fig. 6.65, V). Geophysicists are familiar with the static correction that has to be made in seismic processing to allow for low velocities in the superficial weathered zone.

The epidiagenetic processes are now considered. First, physical weathering, the release of overburden pressure and mass movement on sloping ground, can generate fracture porosity. At the same time diverse chemical changes can generate secondary porosity by leaching. In carbonate-cemented sandstones, groundwater rich in humic and other acids will leach out the cement and carry the carbonate away in solution. This may leave an unconsolidated sand with a porosity approaching that which it had when it was first deposited. In one case a sand buried to some 3 km has 10% porosity, but this is locally as high as 30% beneath an unconformity (Fig. 8.30).

Weathering profiles in modern deserts provide excellent examples of secondary porosity development in sandstones. Unfortunately, due to the great age and, often, long

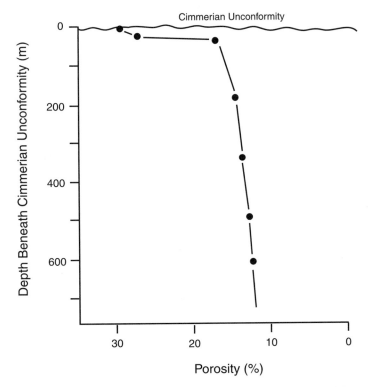

Fig. 8.30. Graph of porosity versus depth beneath the Cimmerian unconformity for 10 Brent (Middle Jurassic) sand samples in the Statfjord field, North Sea. Topmost data point subsumes 4 samples. (From data in Shanmugam, 1990)

exposure of the rocks, it is not possible to prove the time or climatic conditions that produced the weathering profile seen at the present time. On the intensely weathered bare rock surfaces of the Saharan hamadas, sandstones typically show a dark brown ferruginous layer a few millimeters thick. Beneath this crust two zones can be distinguished. The upper zone is one of increased porosity; the lower zone is one of decreased porosity. In the upper zone iron and carbonate are removed in solution; micas, feldspars, and illitic clays are altered to kaolinite and the total clay content is reduced by leaching. Detrital silica grains can be corroded in the more porous sand, though silica cementation may occur in the less permeable sands. This upper zone of leaching, and hence increased secondary porosity, varies from decimeters to hundreds of meters in thickness. The second lower zone is one where porosity is decreased by precipitation of the minerals percolating down from the zone above. Silica is the dominant cement. This is often opaline hydrous silica in modern weathering profiles, though this ages to chalcedony in ancient examples. Iron may also be precipitated in this zone, particularly in the form of ferruginous crusts at the contacts between permeable sands and impermeable shales. The overall effect of these reactions is to decrease the porosity and permeability of this zone.

The previous account, based largely on a detailed petrographic study by Hea (1971), is summarized in Fig. 8.31. An understanding of the process of epidiagenesis, briefly reviewed above, is of some significance in the search for porous sand bodies such as aquifers and hydrocarbon reservoirs. Every modern weathering profile is essentially a potential unconformity and, because of epidiagenesis, petrographic studies of rocks based on outcrop samples are unrepresentative of the formation as a whole. Epidiagenetically

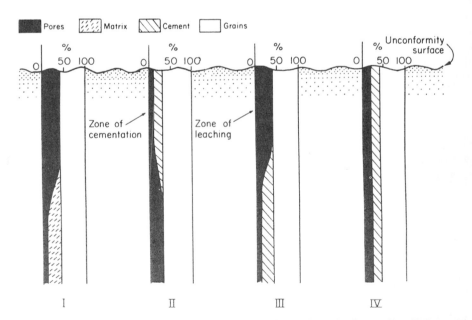

Fig. 8.31. Varieties of epidiagenesis and porosity development beneath weathering profiles. (I) Extensive porosity may form from the weathering of an argillaceous sand. (II) Porosity may be destroyed by cementation in a clean friable sand. (III) Porosity may form by the solution of calcite cement. (IV) Silica-cemented sand may undergo little modification when subjected to weathering.

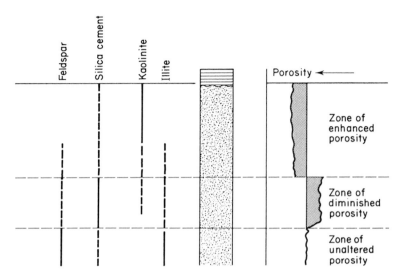

Fig. 8.32. Mineralogy and porosity variation in the Sarir Sandstone (?Lower Cretaceous) beneath the regional Cretaceous unconformity in the Sirte Basin, Libya. This is the reservoir of the Sarir, Messla, and other major oil fields. The thickness of the zone of enhanced porosity varies according to the extent of erosion down to the cemented zone. (Based on data in Hea, 1971 and Al-Shaeib *et al.*, 1981.)

induced porosity is generally destroyed after burial by compaction and cementation (Al-Gailani, 1981). It can be preserved, however, by the early invasion of petroleum and/or overpressure. Epidiagenesis is thus one of several factors that makes unconformity zones favored sites for hydrocarbon accumulation (Shanmugam, 1998). The Prudhoe Bay field of Alaska (Shanmugam and Higgins, 1988), the Sarir and similar fields in Libya, and many of the Jurassic oil fields of the northern North Sea have all undergone epidiagenetic porosity enhancement (Figs. 8.32 and 8.33). The zone of enhanced porosity, and

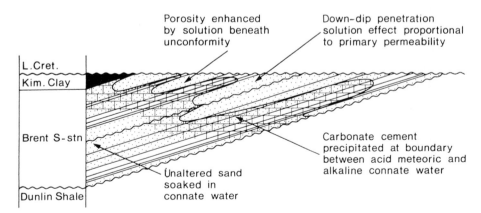

Fig. 8.33. Illustration of epidiagenetic porosity enhancement of the Middle Jurassic Brent Group sandstones beneath the Cimmerian unconformity of the northern North Sea. Kaolinitic sandstones with secondary porosity, immediately beneath the unconformity, pass down through carbonate-cemented sandstones, into illitic sands with unaltered feldspar and no secondary porosity. Shale interbeds and reservoir heterogeneity complicate the pattern, with the meteoric invasion deepest in the thickest, coarsest and most permeable sand beds. (From Selley, 1984.)

its underlying cemented zone, may be noted on geophysical well logs (Selley, 1992a). This boundary may also have sufficient velocity contrast to be imaged as a reflecting horizon on seismic sections, cross-cutting stratigraphic reflectors, as in the Murchison field of the North Sea (Ashcroft and Ridgway, 1996).

8.5.3.4.2 Kerogen decarboxylation

An alternative process to epidiagenesis for the development of secondary porosity was suggested by Rowsell and De Swardt (1974, 1976). They postulated that the acidic fluid necessary for the leaching may have been produced by the **decarboxylation** of kerogen dispersed in shales. The earlier account of clay diagenesis showed that maturing kerogen emits carbon dioxide prior to petroleum generation. The carbon dioxide goes into solution in the connate water as carbonic acid (see Fig. 8.8). This water is expelled from the shale into permeable formations, leaching out unstable grains and cement, enhancing porosity, and preparing reservoirs to receive the petroleum. Unequivocal evidence that this process works has come from the Gulf Coast basin of the USA. Here solution porosity occurs in deeply buried sediments. Because this basin has subsided steadily from the end of the Cretaceous Period, there is no likelihood that the secondary porosity is due to epidiagenesis (Loucks *et al.,* 1984).

8.5.3.5 Sandstone Diagenesis and Porosity Evolution: Summary

The preceding pages reviewed the factors that control the petrophysical characteristics of sandstones. The evolution of porosity in sandstones is much simpler than in carbonates because of the greater chemical stability of silica. The porosity of a sand is a reflection of its texture, mode of deposition, and extent of diagenesis. The grain size, grain shape, sorting, and packing of a sediment play an important role in determining primary intergranular porosity (see Section 3.2.3). Pryor (1973) has shown how these vary for different environments and has documented the spatial variation of porosity and the vectorial variation of permeability in different types of sand body.

Studies of the Mackenzie delta of Arctic Canada and of the U.S. Gulf Coast basins have led to attempts to explain diagenesis in a series of regular stages as sandstones are progressively buried (e.g., Overton, 1973; Van Elsberg, 1978; Surdam *et al.,* 1984 Surdam and Crossey, 1987). The following account attempts to synthesize these ideas. When a sand has been deposited it first goes through the essentially physical processes of compaction and dewatering. The early chemical diagenetic changes are dominated by reduction or oxidation, hence this is termed the "redoxomorphic" phase. These reactions primarily concern oxygen, naturally, iron, sulfur, and organic matter. They are largely the consequence of bacterial action. Essentially, a sand that has high permeability and is deposited above the water table will be subjected to oxidizing reactions. This is because the pore system will be subjected both to free air and oxygenated groundwater. Organic matter is oxidized, and sulfur compounds are oxidized and carried off as soluble sulfate ions. Iron tends to be preserved as ferric oxide. This is red in color and impregnates the clay pellicles of detrital grains. This is why the majority of (but not all) red sandstones are of continental origin, both eolian and fluvial.

By contrast, low-permeability argillaceous sands, and those deposited below the water table, undergo early diagenesis in a reducing alkaline pore fluid. This is because of the relative shortage of free oxygen. This is largely because of bacterial action. Organic matter may be preserved, and iron and sulfur combine to form pyrites. This combination of substances, together with the lack of red ferric oxide imparts an overall drab gray-green color to the sediment. Early pore-filling cements of siderite, calcite, and quartz may develop. A considerable amount of work has been done on the mineralogy of these oxidizing and reducing iron reactions (Van Houten, 1973; Folk, 1976; Walker, 1976; Allen, 1986). The consensus of opinion is that the iron soon achieves equilibrium with the diagenetic environment, irrespective of whether it was deposited in the ferric or ferrous state. Field relationships and petrographic studies show that these reactions are completed at an early stage of diagenesis. A clear example of this occurs in the Pre-Cambrian Torridon Group of northwest Scotland. Here there is a sequence of three deposits: scree, lacustrine, and braided alluvium. These deposits are banked against an irregular basement topography with marginal fanglomerate development. The three facies are red, gray-green, and red in vertical sequence. These colors extend into the marginal fanglomerate, indicating that the iron coloration is a primary feature. The reducing water-logged conditions of the lake allowed gray-green chloritic cement to develop in both it and the adjacent fanglomerate. The oxidizing environment of the earlier scree and later braided deposits caused a red ferric oxide to develop in them and their lateral conglomeratic equivalents (Fig. 8.34). Other examples are described by Friend (1966), Walker (1967), and Van Houten (1968). Early red ferric oxide cements can change to gray-green ferrous iron during later burial. This is very characteristic around oil fields in red sandstone reservoirs. The strongly reducing oil field brines often form a gray-green halo in the sandstones of the water zone around the field (e.g., the Lyons Sand of the Denver Basin, Levandowski *et al.*, 1973).

Throughout the redoxomorphic phase of diagenesis porosity of a sand is lost slowly, but this is due primarily to the effects of compaction and dehydration rather than to the

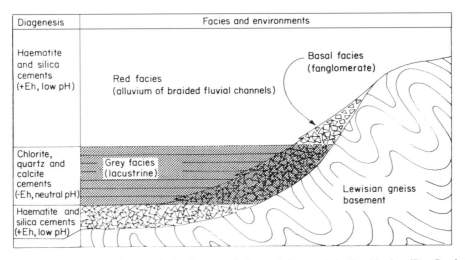

Fig. 8.34. Diagram showing the correlation between facies and diagenesis in Torridonian (Pre-Cambrian) sediments of northwest Scotland. This suggests that cementation was early and directly related to the chemistry of syndepositional pore fluids. (From Selley, 1966. Courtesy of the Geologists' Association.)

chemical effects of diagenesis. This is in contrast to the second phase of sandstone diagenesis. With increasing burial the newly cemented sandstone moves down from the shallow zone of reducing and alkali conditions into the catagenic zone. Here the sands are flushed by connate waters expelled from compacting clays. Where the clays are rich in organic matter the emitted fluids may contain carbonic acid. These may leach out early carbonate cements and unstable grains, thus generating secondary porosity. The minor importance of secondary solution porosity by decarboxylation during deep burial is clearly shown by the straight lines of the sandstone porosity gradients shown earlier in Fig. 8.17.

Once the carbonic acids have flushed through the sands, the pore fluid returns to an alkaline state, and a second phase of carbonate cementation may take place. If burial continues further the sandstone may enter the metagenic zone. Carbonate cements are leached out again as the sandstone becomes a tightly cemented quartzite, clays recrystallize to mica, and the metamorphic zone has been reached. Evidence suggests that major sandstone diagenesis takes place in a relatively narrow thermal zone between 80 and 120°C. For an average geothermal gradient of 2.6°C/km, this is between 2500 and 4200 m.

Two events may interrupt this regular sequence. Petroleum invasion of pore spaces expels connate water and preserves the rock from further cementation or solution (e.g., De Souza and De Assis, Silva, 1998). Thus porosity is often higher within petroleum reservoirs than in adjacent water bearing zones (Fig. 8.35). Geophysical well logging is beyond the scope of this text, but it should be noted that geophysical well logs can identify cemented envelopes, unconformity motifs, and the change in porosity at a petroleum:water contact (Selley, 1992a) Diagenetic phenomena may also be imaged seismically. The seismic imaging of the zone of enhanced porosity beneath the Cimmerian unconformity of the North Sea has been described by Ashcroft and Ridgway (1996). It is also possible to identify petroleum:water contacts on seismic data. These are horizontal reflectors that cross-cut stratigraphic reflecting horizons. They are variously termed "flat spots," "bright spots," and "DHIs" (**direct hydrocarbon indicators**). Not all seismically defined flat spots actually occur beneath petroleum accumulations. It is believed that these "phantom" flat spots mark paleo-petroleum:water contacts, where cementation occured beneath a petroleum reservoir. The trap leaked, allowing the petroleum to escape, leaving the residual cemented zone to trap unwary petroleum explorationists. Francis *et al.* (1997) describe an excellent example of a phantom flat spot from the Irish Sea (Fig. 8.36). Figure 8.37 summarizes the porosity and seismic responses of some of these diagenetic phenomena.

With increasing burial temperature and pressure increase, petroleum degrades to carbon. Clays recrystallize, quartz cement becomes pervasive and porosity is almost completely destroyed. At any stage during burial, however, the sandstone may be uplifted, experience epidiagenetic porosity enhancement, undergo renewed burial, and be recycled through the above diagenetic sequence (Fig. 8.38).

8.6 RUDACEOUS ROCKS

The rudaceous rocks are sediments at least a quarter of whose volume is made of particles larger than 2 mm in diameter. They grade down through the granulestones into

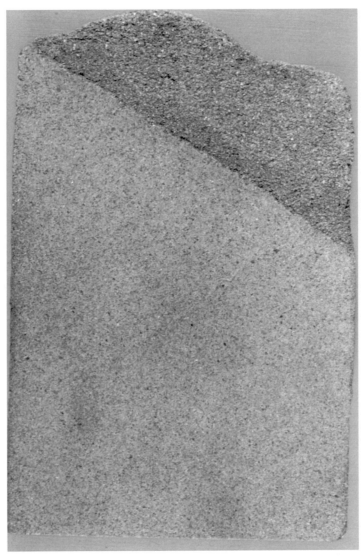

Fig. 8.35. Core cut across a petroleum:water contact. (The well was deviated, the contact is horizontal, truly.) Friable uncemented upper zone has gradually been lost during transportation and subsequent student erosion. Lower cemented zone below the petroleum:water contact has a higher preservation potential.

Fig. 8.36. Real (upper) and modeled (lower) seismic section from the Irish Sea, showing real and phantom direct hydrocarbon indicators (DHIs). A DHI occurs beneath a gas accumulation on the crest of the anticline in the left-hand fault block. In the right-hand fault block, however, the DHI is a phantom. No petroleum is present. The DHI marks a paleo-petroleum:water contact where porosity was preserved by petroleum in a trap, and diminished by cementation beneath the petroleum:water contact. Subsequently, the petroleum leaked out of the trap, but a phantom DHI marks the boundary between lower tight and higher more porous sand. (From Francis *et al.*, 1997, courtesy of the Geological Society of London.)

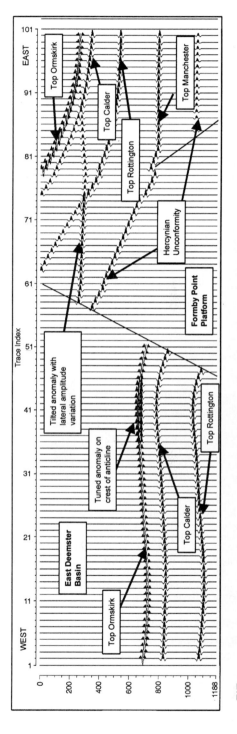

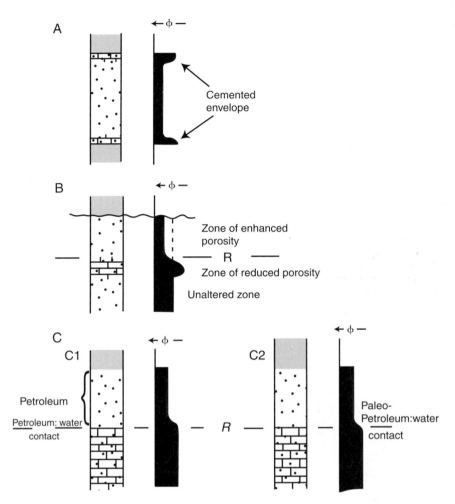

Fig. 8.37. Illustrations of sandstone diagenetic motifs and their porosity responses. Porosity (ϕ) increases to the left. All sand:shale boundaries may be seismic reflectors, but R = seismic reflecting horizons due to diagenetic fronts within sandstone. (A) Envelope effect, showing carbonate cement at sand:shale contacts. (B) Unconformity motif, showing zones of enhanced and diminished porosity analogous to "gossan" ore bodies (see Section 2.3.4.2). (C) Petroleum:water contact diagenetic effects. C1 illustrates a valid trap, still containing petroleum, which inhibits cementation above the petroleum:water contact, whereas C2 illustrates a trap which once contained petroleum that inhibited cementation, but which has since leaked, leaving a phantom flat spot at the diagenetic front of the paleo-petroleum:water contact. Examples of all the above motifs are cited in the text.

the very coarse sandstones. Traditionally the rudaceous deposits are divided into breccias, whose particles are angular, and conglomerates, whose particles are rounded. Breccias are rare rocks that occur principally in faults (tectonic breccias) and in some screes. The majority of rudaceous rocks are conglomerates (Koster and Steel, 1984).

8.6.1 Conglomerates

The conglomerates can be divided into three groups by their composition. The volcaniclastic conglomerates, termed agglomerates, have already been described (see Sec-

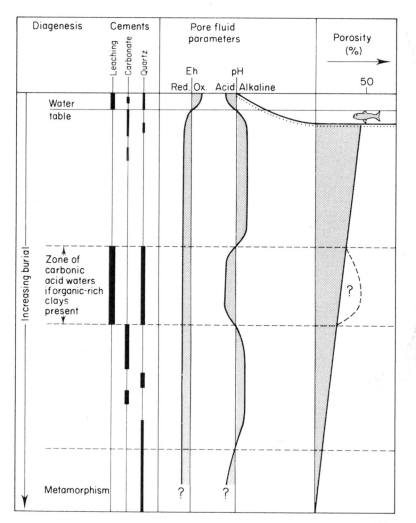

Fig. 8.38. An attempt to display the changes in pore fluid chemistry and diagenetic response of sandstones during diagenesis. For a full explanation and documentation see text. Any attempt to produce a simple scheme for such complex processes is fraught with problems. Note particularly that the sequence may be interrupted by uplift, meteoric flushing, and epidiagenesis, or by the expulsion of connate water from pores by petroleum invasion. For details of clay mineral reactions, see Fig. 8.8.

tion 8.4). The other two groups are the carbonate conglomerates (known sometimes as the calcirudites) and the terrigenous conglomerates (silicirudites).

Continental carbonate conglomerates are rare because of their solubility in acid groundwaters. Marine carbonate conglomerates are more common. The best known examples are probably the "coral rock" boulder beds that form submarine screes around reef fronts. Thin layers of carbonate clasts, often phosphatized, overlie chalk "hardground" horizons (see Section 9.2.5.1).

The following account is concerned primarily with the terrigenous conglomerates. Texturally, conglomerates may be conveniently divided into two types: grain-supported

Fig. 8.39. Photographs illustrating different conglomerate textures. (A) Matrix-supported conglomerate from the Hassaouna Sandstone Group (Cambrian), Jebel Mourizidie, southern Libya. (B) Clast-supported conglomerate with very well-rounded cobbles. The Budleigh Salterton Pebble Bed (Triassic), Devon.

and mud-supported conglomerates (Fig. 8.39). The mud-supported rudaceous rocks are properly termed the "diamictites." In these the pebbles are seldom in contact with one another, but are dispersed through a finer grained matrix like currants in a bun. The origin of these rocks was discussed earlier in Section 4.4. It is apparent that the diamictites, or pebbly mudstones, originate from various processes that are not well understood. Some are due to mud flows that occur both in subaerial environments and underwater, where they are sometimes termed "fluxoturbidites." Other diamictites are glacial in origin. Notable examples are the Pleistocene boulder clays.

The second class of conglomerates, defined by texture, are grain supported. In these the individual pebbles touch one another, but the intervening spaces are generally infilled by a matrix of poorly sorted sand and clay. In fluvial conglomerates it is quite likely that this was a primary depositional feature. In marine conglomerates, on the other hand, such as those beach gravels that mark marine transgressions, it is more likely that there was no depositional matrix. The matrix that is commonly present is probably due to infiltration from overlying sediment. Conglomerates are deposited with high porosities but, because of their large throat passages, they have excellent permeability. This high permeability enables porosity to be quickly destroyed by matrix infiltration even before any cementation can occur.

Two further classes of conglomerate are defined according to the compositions of their pebbles. Polymictic conglomerates are composed of pebbles of more than one type. Oligomictic conglomerates have pebbles of only one rock type. Whereas the polymictic conglomerates are of diverse composition, the oligomictic conglomerates are generally quartzose. This is because of the chemical stability of silica. Thus the conglomerate of polycyclic sediments is commonly made of pebbles of vein quartz, quartzite, and chert. Polymictic conglomerates are generally the product of aggradation where tectonically active source areas shed wedges of fanglomerates. Oligomictic conglomerates, by contrast, are generally the product of degradation where tectonic stability allows extensive reworking to produce the laterally extensive basal conglomerates at unconformities.

A third bipartite division of conglomerates is according to the source of the pebbles. Extraformational or exotic conglomerates are composed of pebbles that originated outside the depositional basin. Intraformational conglomerates contain pebbles of sediment that originated within the depositional basin. The majority of limestone conglomerates are intraformational in origin for the reason already stated. Intraformational sand conglomerates are rare because unconsolidated sand lacks cohesion and disaggregates on erosion. Occasionally, however, intraformational conglomerates of sand occur adjacent to petroleum seeps, where the oil acts as a binding agent, allowing sand cobbles to be transported for some distance (Fig. 8.40).

Mud intraformational conglomerates are commonly called "shale flake" or "shale pellet" conglomerates. These are volumetrically insignificant, occurring in beds generally only one or two clasts thick. Sedimentologically they are significant, however, because they indicate penecontemporaneous erosion close to the site of deposition. Intraformational shale pellet conglomerates are often present at the base of turbidite units and of channels (Fig. 8.41).

This brief review of conglomerates shows that there is no consistent scheme of nomenclature and classification. Table 8.4 shows how descriptive terminology is based on divisions of texture, composition, and source.

8.6.2 Sedimentary Breccias

Breccias, defined as rocks composed of angular cobbles, are formed from meteoritic impact and tectonic and sedimentary processes. It is not always easy to differentiate their origin (Reimold, 1997). Only the sedimentary breccias are dealt with here. Sedimentary

Fig. 8.40. Intraformational conglomerate of oil-cemented sand cobbles on a Lower Cretaceous fluvial channel floor, adjacent to the Mupe Bay paleo-seep, Dorset. (For further details see Selley, 1992b).

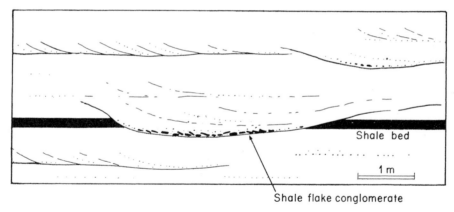

Shale bed

1 m

Shale flake conglomerate

Fig. 8.41. Intraformational conglomerate of claystone pellets. The channel cuts through a clay unit, demonstrating the penecontemporaneous origin of the conglomerate.

Table 8.4
Nomenclature of Conglomerates

I.	Texture	*Orthoconglomerate*—grain-supported
		Paraconglomerate (syn. diamictite)—mud-supported
II.	Composition	*Polymictic*—pebbles composed of several rock types
		Oligomictic—pebbles composed of one rock type
III.	Source	*Intraformational*—pebbles originate within the basin
		Extraformational (syn. exotic)—pebbles extrabasinal in origin

There is no satisfactory classification of conglomerates. Nomenclature and description are based on the criteria of texture, composition, and source.

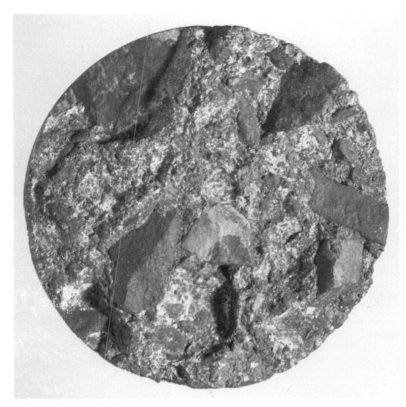

Fig. 8.42. A 10-cm diameter core of Rotliegende (Lower Permian) basal breccia immediately above the unconformity with Coal Measures (Upper Carboniferous), Southern North Sea.

breccias originate in two main ways. They either form in very proximal environments, where newly eroded detritus has not yet undergone attrition, abrasion, and rounding, or as evaporite collapse deposits. Most sedimentary breccias are of the first type. They are deposited immediately adjacent to unconformities in terrestrial and subaqueous screes and in more extensive fanglomerate environments. Figure 8.42 illustrates a typical example. Such sedimentary breccias normally pass rapidly downcurrent into conglomerates, as clasts are rounded during transportation.

Sedimentary breccias due to evaporite solution are rarer, but more interesting. When evaporites are leached by meteoric water they may give rise to curious residual breccias of insoluble rocks. There is often a brecciated cap on the crest of many shallow salt domes. The clasts are often polygenetic, reflecting the diverse formations through which the salt dome may have penetrated. Failure to recognize such formations for what they are can cause both geological and paleontological confusion when drill cuttings of such breccias are examined.

Some evaporites are interbedded with other sediments, normally carbonates. If the evaporites leach out then the intervening strata will break up to form a solution collapse breccia. A notable example of such a formation is provided by the so-called "Broken Beds" that occur in the Purbeck Beds (Upper Jurassic) on the Dorset coast (Fig. 8.43). The Purbeck Beds consist of interbedded shales and fecal pellet limestones. Locally

Fig. 8.43. The "Broken Beds" within the Purbeck Beds (Upper Jurassic), Dorset, southern England. This is an evaporite solution collapse breccia of lagoonal fecal pellet lime mudstones.

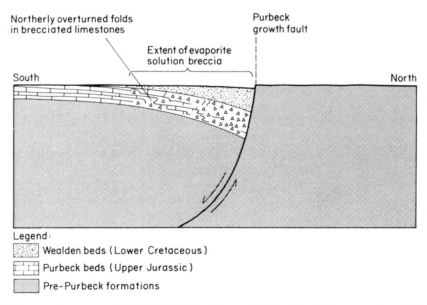

Fig. 8.44. Cross-section across the Purbeck fault line, Dorset, England. Both the extent of evaporite deposition, and the resultant solution collapse breccia of the Purbeck Beds, were controlled by syndepositional movement along a growth fault. Datum pre-Albian unconformity. (For further details see Selley and Stoneley, 1987.)

they contain nodular anhydrite (replaced by gypsum near the surface) and salt pseudo-morphs (see Fig. 5.34). The Purbeck Beds are interpreted as sabkha deposits laid down in hypersaline and brackish lagoons. Locally the Purbeck Beds contain a spectacular evaporite collapse breccia, termed the "Broken Beds." These only occur in the rollover zone of a major growth fault. They are locally contorted into folds that are overturned toward the fault plane (Fig. 8.44). It appears that, not only may the extent of the sabkha have been restricted by the rollover hollow, but also that the northerly paleoslope may have controlled the movement of this solution collapse breccia (Selley and Stoneley, 1987).

SELECTED BIBLIOGRAPHY

General:

Giles, M. R. (1997). "Diagenesis: A Quantitative Perspective." Kluwer Academic Press, Dordrecht, The Netherlands. 526pp.

For clays see:

Schieber, J., Zimmerle, W., and Selhi, P. S., eds. (1998). "Shales and Mudstones I: Basin Studies, Sedimentology and Palaeontology." E. Schweizerrbart'sche Verlagsbuchhandlung, Stuttgart. 875pp.
Schieber, J., Zimmerle, W., and Selhi, P. S., eds. (1998). "Shales and Mudstones II: Basin Studies, Petrography, Petrophysics, Geochemistry, and Economic Geology." E. Schweizerrbart'sche Verlagsbuchhandlung, Stuttgart. 753pp.

For volcaniclastic sediments see:

Orton, G. J. (1996). Volcanic environments. *In* "Sedimentary Environments: Processes, Facies and Stratigraphy" (H. G. Reading, ed.), 3rd ed., pp. 486–567. Blackwell, Oxford.

For sandstones see:

Morse, D. G. (1994). Siliciclastic reservoir Rocks. *Mem.—Am. Assoc. Pet. Geol.* **60,** 121–139.

For conglomerates and breccias see:

Koster, E. H., and Steel, R. J., eds. (1984). "Sedimentology of Gravels and Conglomerates," Mem. No. 10. Can. Soc. Pet. Geol. Calgary.
Reimold, W. V. (1997). Exogenic and endogenic breccias: A discussion of major problematics. *Earth Sci. Rev.* **43,** 25–47.

REFERENCES

Addis, M. A., and Jones, M. E. (1986). Volume changes during diagenesis. *Mar. Pet. Geol.* **2,** 241–246.
Al-Gailani, M. B. (1981). Authigenic mineralization at unconformities: Implications for reservoir characteristics. *Sediment. Geol.* **29,** 89–115.
Allen, J. R. L. (1986). Time scales of colour change in late Flandrian intertidal muddy sediments of the Severn Estuary. *Proc. Geol. Assoc.* **97,** 23–28.

Allen, R. C., Gavish, E., Friedman, G. M., and Sanders, J. E. (1969). Aragonite-cemented sandstone from outer continental shelf off Delaware Bay. *J. Sediment. Petrol.* **39,** 136–149.

Al-Shaeib, Z., Ward, W. C., and Shelton, J. W. (1981). Diagenesis and secondary porosity evolution of Sarir Sandstone, Southeastern Sirte Basin, Libya. *AAPG Bull.* **65,** 889–890.

Amstutz, G. C., and Bubinicek, L. (1967). Diagenesis in sedimentary mineral deposits. *In* "Diagenesis in Sediments" (G. Larsen and G. V. Chilingar, eds.), pp. 417–475. Elsevier, Amsterdam.

Ashcroft, W. A., and Ridgway, M. S. (1996). Early discordant diagenesis in the Brent Group, Murchison Field, UK North Sea, detected in high values of seismic-derived acoustic impedance. *Pet. Geosci.* **2,** 75–82.

Baars, D. L. (1961). Permian blanket sandstones of Colorado Plateau. *In* "Geometry of Sandstone Bodies" (J. A. Peterson and J. C. Osmond, eds.), pp. 79–207. Am. Assoc. Pet. Geol., Tulsa, OK.

Bennacef, A., Beuf, S., Biju-Duval, B., de Charpal, O., Gariel, O., and Rognon, P. (1971). Example of cratonic sedimentation: Lower Paleozoic of Algerian Sahara. *Bull. Am. Assoc. Pet. Geol.* **55,** 25–245.

Bjerkli, K., and Ostmo-Saeter, J. S. (1973). Formation of glauconite in foraminiferal shells on the continental shelf off Norway. *Mar. Geol.* **14,** 169–178.

Blatt, H., Middleton, G., and Murray, R. (1980). "Origin of Sedimentary Rocks," 2nd ed. Prentice-Hall, Englewood Cliffs, NJ. 782pp.

Bohor, B. F., and Triplehorn, D. M., eds. (1993). "Tonsteins: Altered Volcanic Ash Layers in Coal-bearing Sequences," Spec. Publ. No. 285. Geol. Soc. Am. Boulder, CO. 48p.

Brenchley, P. J. (1969). Origin of matrix in Ordovician greywackes, Berwyn Hills, North Wales. *J. Sediment. Petrol.* **39,** 1297–1301.

Brogniart, A. (1826). L'arkose, caractères minéralogiques et histoire géognostique de cette roche. *Ann. Sci. Nat. (Paris),* **8,** 113–163.

Bryant, I. D., Kantorowitz, J. D., and Love, C. F. (1988). The origin and recognition of laterally continuous carbonate-cemented horizons in the Upper Lias of southern England. *Mar. Pet. Geol.* **5,** 108–133.

Carvalho, M. V. H., De Ros, L. F., and Gomez, N. S. (1995). Carbonate cementation patterns and diagenetic reservoir facies in the Campos basin Cretaceous turbidites eastern offshore Brazil. *Mar. Pet. Geol.* **12,** 741–758.

Chamley, H. (1989). "Clay Sedimentology." Springer-Verlag, Berlin. 623pp.

Chen, W., and Ortoleva, P. (1990). Reaction front fingering in carbonate-cemented sandstones. *Earth Sci. Rev.* **29,** 183–198.

Cummins, W. A. (1962). The greywacke problem. *Liverpool Manchester Geol. J.* **3,** 51–72.

Curtis, C. D. (1978). Possible links between sandstone diagenesis and depth-related geochemical reactions occurring in enclosing mudstones. *Q. J. Geol. Soc. London* **135,** 107–117.

Davidson, C. F. (1965). A possible mode of origin of strata-bound copper ores. *Econ. Geol.* **60,** 942–954.

De Souza, R. S., and De Assis Silva, C. M. (1998). Origin and timing of carbonate cementation of the Namorado Sandstone (Cretaceous), Albacora Field, Brazil: Implications for oil recovery. *In* "Carbonate Cementation in Sandstones: Distribution Patterns and Geochemical Evolution" (S. Morad, ed.), pp. 309–325. Blackwell Science, Oxford.

Dickinson, W. R., Beard, L. S., Brakenbridge, G. R., Erjavek, J. L., Ferguson, R. C., Inman, K. F., Klepp, R. A., Lindberg, F. A., and Ryberg, P. T. (1983). Provenance of North American Phanerozoic sandstones in relation to tectonic setting. *Geol. Soc. Am. Bull.* **94,** 222–235.

Dill, H. D. (1998). A review of heavy minerals in clastic sediments with case studies from the alluvial fan through to the nearshore marine environments. *Earth Sci. Rev.* **45,** 103–132.

Donell, J. R., Culbertson, W. C., and Cashion, W. B. (1967). Oil shale in the Green River formation. *Proc.—World Pet. Congr.* **7** (3), 699–702.

Dott, R. H. (1964). Wacke, greywacke and matrix-what approach to immature sandstone classification. *J. Sediment. Petrol.* **34,** 625.

Dott, R. H., and Reynolds, M. J. (1969). "Sourcebook for Petroleum Geology," Mem. No. 5. Am. Assoc. Pet. Geol., Tulsa, OK. 471pp.

Duncan, D. C. (1967). Geologic setting of oil shale deposits and world prospects. *Proc.—World Pet. Congr.* **7** (3), 659–667.

Dunoyer de Segonzac, G. (1968). The birth and development of the concept of diagenesis (1866–1966). *Earth Sci. Rev.* **4,** 153–201.

Dzevanshir, R. D., Buryakovskiy, L. A., and Chilingar, G. V. (1986). Simple quantitative evaluation of porosity of argillaceous sediments at various depths of burial. *Sediment. Geol.* **46,** 169–176.

Edwards, R. A. (1976). Tertiary sediments and structure of the Bovey Tracey basin, South Devon. *Proc. Geol. Assoc.* **87,** 1–26.

Fairbridge, R. W. (1967). Phases of diagenesis and authigenesis. *In* "Diagenesis in Sediments" (G. Larsen and G. V. Chilingar, eds.), pp. 19–28. Elsevier, Amsterdam.

Fisher, R. V., and Schmincke, H. U. (1984). "Pyroclastic Rocks." Springer-Verlag, Berlin. 472pp.

Folk, R. L. (1968). "Petrology of Sedimentary Rocks." Hemphill's Book Store, Austin, TX. 170pp.

Folk, R. L. (1976). Reddening of desert sands. *J. Sediment. Petrol.* **46,** 604–613.

Fothergill, C. A. (1955). The cementation of oil reservoir sands. *Proc.—World Pet. Congr.* **4** (Sect. 1), 300–312.

Francis, A., Millwood Hargrave, M., Mullholland, P., and Williams D. (1997). Real and relict direct hydrocarbon indicators in the East Irish Sea Basin. *In* "Petroleum Geology of the Irish Sea and Adjacent Areas." (N. S. Meadows, S. P. Trueblood, M. Hardman, and G. Cowan, eds.) *Geol. Soc. London. Sp. Pub. 124,* 185–194.

Friend, P. F. (1966). Clay fractions and colours of some Devonian red beds in the Catskill Mountains, U.S.A. *Q. J. Geol. Soc. London* **122,** 273–292.

Fuchtbauer, H. (1967). Influence of different types of diagenesis on sandstone porosity. *Proc.—World Pet. Congr.* **7** (3), 353–369.

Galloway, W. E. (1974). Deposition and diagenetic alteration of sandstone in northeast Pacific arc-related basins: implications for greywacke genesis. *Geol. Soc. Am. Bull.* **85,** 379–90.

Garrison, R. E., Luternauer, J. L., Grill, E. V., MacDonald, R. D., and Murray, J. W. (1969). Early diagenetic cementation of Recent sands, Fraser River delta, British Columbia. *Sedimentology* **12,** 27–46.

Gautier, D. L., and Claypool, G. E. (1984). Interpretation of methanic diagenesis in ancient sediments by analogy with processes in modern diagenetic environments. *Mem.—Am. Assoc. Pet. Geol.* **37,** 111–123.

Giles, M. R. (1997). "Diagenesis: A Quantitative Perspective." Kluwer Academic Press, Dordrecht, The Netherlands. 526pp.

Glennie, K. W. (1972). Permian Rotliegendes of Northwest Europe interpreted in light of modern desert sedimentation studies. *Bull. Am. Assoc. Pet. Geol.* **56,** 1048–1071.

Greensmith, J. T. (1968). Palaeogeography and rhythmic deposition in the Scottish oil-shale group. *Proc. U.N. Symp. Dev. Util. Oil Shale Res.* Tallin, pp. 1–16.

Grim, R. E. (1968). "Clay Mineralogy," 2nd ed. McGraw-Hill, New York. 596pp.

Guvan, N., Hower, W. F., and Davies, D. K. (1980). Nature of authigenic illites in Sandstone Reservoirs. *J. Sediment. Petrol.* **50,** 761–766.

Harbaugh, J. W., and Merriam, D. F. (1968). "Computer Applications in Stratigraphic Analysis." Wiley, Chichester. 282pp.

Hatch, F. H., Rastall, R. H., and Greensmith, J. T. (1971). "Petrology of the Sedimentary Rocks," 5th rev. ed. Thomas Murby, London. 502pp.

Hea, J. P. (1971). Petrography of the Paleozoic-Mesozoic sandstones of the southern Sirte basin, Libya. *In* "The Geology of Libya" (C. Grey, ed.), pp. 107–125. University of Libya, Tripoli.

Heald, M. T., and Renton, J. J. (1966). Experimental study of sandstone cementation. *J. Sediment. Petrol.* **36,** 977–991.

Helmbold, R. (1952). Beitrag zur Petrographie der Tanner Grauwacken. *Heidelb. Beitr. Mineral. Petrogr.* **3,** 253–280.

Hovland, M., Talbot, M. R. M., Qvale, H., Olaussen, S., and Assberg, L. (1987). Methane-related carbonate cements in pock-marks of the North Sea. *J. Sediment. Petrol.* **57,** 81–92.

Hubert, J. F. (1964). Textural evidence for deposition of many western North Atlantic deep-sea sands by ocean-bottom currents rather than turbidity currents. *J. Geol.* **72,** 757–785.

Huckenholz, H. G. (1963). Mineral composition and textures in greywackes from the Harz Mountains (Germany) and in arkoses from the Auvergne (France). *J. Sediment. Petrol.* **33,** 914–918.

Hunt, J. M. (1996). "Petroleum Geochemistry and Geology," Second Edition. Freeman, San Francisco. 743pp.

Johnsson, M. J., and Basu, A., eds. (1993). "Processes Controlling the Composition of Clastic Sediments," Spec. Publ. No. 284. Geol. Soc. Am. Boulder, CO. 352p.

Keller, W. D. (1970). Environmental aspects of clay minerals. *J. Sediment. Petrol.* **40**, 788–813.

Klein, G. de Vries (1963). Analysis and review of sandstone classifications in the North American geological literature. *Geol. Soc. Am. Bull.* **74**, 555–576.

Koithara, J., Bisht, J. S., and Raj, H. (1980). Statistical analysis of density and porosity of subsurface rock samples from Cauvery Basin. *Bull. Indian Oil & Gas Comm. Dehra Dun.* **17**, 103–107.

Koster, E. H., and Steel, R. J., eds. (1984). "Sedimentology of Gravels and Conglomerates," Mem. No. 10. Can. Soc. Pet. Geol. 441pp. Calgary.

Krynine, P. D. (1935). Arkose deposits in the humid tropics, a study of sedimentation in southern Mexico. *Am. J. Sci.* **29**, 353–363.

Larsen, G., and Chilingar, G. V., eds. (1962). "Diagenesis in Sediments." Elsevier, Amsterdam. 808pp.

Levandowski, D., Kaley, M. E., Silverman, S. R., and Smalley, R. G. (1973). Cementation in Lyons sandstone and its role in oil accumulation, Denver Basin, Colorado. *Am. Assoc. Pet. Geol. Bull.* **57**, 2217–2244.

Loucks, R. G., Dodge, M. M., and Galloway, W. E. (1984). Regional controls on diagenesis and reservoir quality in Lower Tertiary sandstones along the Texas Gulf Coast. *Mem. — Am. Assoc. Pet. Geol.* **37**, 15–46.

Lundegard, P. D. (1992). Sandstone porosity loss — a "big picture" view of the importance of compaction. *J. Sediment. Petrol.* **39**, 12–17.

Macdonald, A. M. (1972). Chambers Twentieth Century Dictionary. W. R. Chambers. Edinburgh. 1639pp.

Mackenzie, F. T., and Gees, R. (1971). Quartz: synthesis at earth-surface conditions. *Science* **3996**, 533–535.

Magara, K. (1980). Comparison of porosity-depth relationships of shale and sandstone. *J. Pet. Geol.* **3**, 175–185.

Maher, C. E. (1980). Piper oil field. *Mem. — Am. Assoc. Pet. Geol.* **30**, 131–172.

Marshall, J. D., ed. (1987). "Sediment Diagenesis," Spec. Publ. 36. Geol. Soc. London, London. 360pp.

Mattiatt, B. (1960). Beitrag zur Petrographie der Oberharzer Kulmgrauwacke. *Heidelb. Beitr. Mineral. Petrogr.* **7**, 242–280.

Maxwell, J. C. (1964). Influence of depth, temperature and geologic age on porosity of quartzose sandstone. *Bull. Am. Assoc. Pet. Geol.* **48**, 697–709.

Mayuga, M. N. (1970). Geology of California's giant Wilmington field. *Mem. — Am. Assoc. Pet. Geol.* **14**, 148–184.

McBride, E. (1990). Quartz cement in sandstones: A review. *Earth Sci. Rev.* **26**, 26–112.

Meshri, I. D., and Ortoleva, P., eds. (1990). "Prediction of Reservoir Quality Through Chemical Modelling," Mem. No. 49. Am. Assoc. Pet. Geol., Tulsa, OK. 175pp.

Millot, G. (1970). "Geology of Clays." Springer-Verlag, Berlin. 429pp.

Moncure, G. K., Lahann, R. W., and Seibert, R. N. (1984). Origin of secondary porosity and cement distribution in a sand/shale sequence from the Frio Formation (Oligocene). *Mem. — Am. Assoc. Pet. Geol.* **37**, 151–162.

Montanez, I. P., Gregg, J. M., and Shelton, K. L. (1997). Basin-wide diagenetic patterns: Integrated petrologic, geochemical and hydrologic considerations. *Spec. Publ. — Soc. Econ. Paleontol. Mineral.* **57**, 1–302.

Moore, D. M., and Reynolds, R. C. (1997). "X-ray Diffraction and the Identification and Analysis of Clay Minerals." Oxford University Press, New York. 503pp.

Morad, S., ed. (1998). "Carbonate Cementation in Sandstones: Distribution Patterns and Geochemical Evolution." Blackwell Science, Oxford. 576pp.

Morse, D. G. (1994). Siliciclastic reservoir rocks. *Mem. — Am. Assoc. Pet. Geol.* **60**, 121–139.

Mortland, M. M., and Farmer, V. C. (1978). "International Clay Conference," Dev. Sedimentol., Vol. 27. Elsevier, Amsterdam. 662pp.

Morton, A. C., and Hallsworth, C. R. (1999). Processes controlling the composition of heavy mineral assemblages in sandstones. *Sediment. Geol.* **124,** 3–29.

Muller, G. (1967). Diagenesis in argillaceous sediments. *In* "Diagenesis in Sediments" (G. Larsen and G. V. Chilingar, eds.), pp. 127–178. Elsevier, Amsterdam.

Nagtegaal, P. J. C. (1978). Sandstone-framework instability as a function of burial diagenesis. *J. Geol. Soc., London* **135,** 101–106.

North, F. K. (1985). "Petroleum Geology." Allen & Unwin, London. 607pp.

O'Brian, G. W., and Woods, E. P. (1995). Hydrocarbon-related diagenetic zones (HRDZs) in the Vulcan sub-basin, Timor Sea: Recognition and exploration implications. *APEA J.,* pp. 220–252.

Odin, G. S. (1972). Observations on the structure of glauconite vermicular pellets: A description of the genesis of these granules by neoformation. *Sedimentology* **19,** 285–294.

Odin, G. S. (1988). "Green Marine Clays." Elsevier, Amsterdam. 446pp.

Orton, G. J. (1996). Volcanic environments. *In* "Sedimentary Environments: Processes, Facies and Stratigraphy" (H. G. Reading, ed.), 3rd ed., pp. 485–567. Blackwell, Oxford.

Overton, H. L. (1973). Water chemistry analysis in sedimentary basins. *14th Annu. Logging Symp., Soc. Prof. Well Log Anal. Assoc., Pap.* L, pp. 1–15.

Paraguassu, A. B. (1972). Experimental silicification of sandstone. *Geol. Soc. Am. Bull.* **83,** 2853–2858.

Pettijohn, F. J. (1957). "Sedimentary Rocks." Harper Bros, New York. 718pp.

Pettijohn, F. J., Potter, P. E., and Siever, R. (1972). "Sand and Sandstone." Springer-Verlag, Berlin. 618pp.

Philipp, W., Drong, H. J., Fuchtbauer, H., Haddenhorst, H. G., and Jankowsky, W. (1963). The history of migration in the Gifhorn trough (N.W. Germany). *Proc.—World Pet. Congr.* **6** (Sect. 1, Pap. 19), 457–481.

Porrenga, D. H. (1966). Clay minerals in Recent sediments of the Niger Delta. *Clays Clay Miner., Proc. Conf.* **14,** 221–233.

Potter, P. E., Maynard, B., and Pryor, W. A. (1980). "Sedimentology of Shales." Springer-Verlag, New York. 306pp.

Price, N. B., and Duff, P. McL. D. (1969). Mineralogy and chemistry of tonsteins from Carboniferous sequences in Great Britain. *Sedimentology* **13,** 45–69.

Pryor, W. A. (1973). Permeability-porosity patterns and variations in some Holocene sand bodies. *Am. Assoc. Pet. Geol. Bull.* **57,** 162–189.

Reimold, W. V. (1997). Exogenic and endogenic breccias: A discussion of major problematics. *Earth Sci. Rev.* **43,** 25–47.

Rittenhouse, G. (1971). Pore space reduction by solution and cementation. *Bull. Am. Assoc. Pet. Geol.* **55,** 80–91.

Robertson, R. H. S. (1986). "Fuller's Earth: A History of Calcium Montmorillonite," Miner. Soc. Occac. Publ. Volturna Press, Hythe, England. 421pp.

Rowsell, D. M., and De Swardt, A. M. J. (1974). Secondary leaching porosity in Middle Ecca-Sandstones. *Trans. Geol. Soc. S. Afr.* **77,** 131–140.

Rowsell, D. M., and De Swardt, A. M. J. (1976). Diagenesis in Cape and Karoo sediments, South Africa, and its bearing on their hydrocarbon potential. *Trans. Geol. Soc. S. Afr.* **79,** 81–153.

Sassen, R., Roberts, H. H., Aharon, P., Larkin, J., Chinn, E. W., and Carney, R. (1993). Chemosynthetic bacterial mats at cold hydrocarbon seeps, Gulf of Mexico continental slope. *Org. Geochem.* **20,** 77–89.

Scherer, M. (1987). Parameters influencing porosity in sandstones: A model for sandstone porosity prediction. *AAPG Bull.* **71,** 485–491.

Schieber, J., Zimmerle, W., and Selhi, P. S., eds. (1998a). "Shales and Mudstones I: Basin Studies, Sedimentology and Palaeontology." E. Schweizerrbart'sche Verlagsbuchhandlung, Stuttgart. 875pp.

Schieber, J., Zimmerle, W., and Selhi, P. S., eds. (1998b). "Shales and Mudstones II: Basin Studies, Petrography, Petrophysics, Geochemistry, and Economic Geology." E. Schweizerrbart'sche Verlagsbuchhandlung, Stuttgart. 753pp.

Schmidt, V., McDonald, D. A., and Platt, R. L. (1977). Pore geometry and reservoir aspects of secondary porosity in sandstones. *Bull. Can. Geol.* **25,** 271–290.

Schmoker, J. W., and Gautier, D. L. (1988). Sandstone porosity as a function of thermal maturity. *Geology* **16**, 1007–1010.

Selley, R. C. (1966). Petrography of the Torridonian rocks of Raasay and Scalpay, Invernessshire. *Proc. Geol. Assoc.* **77**, 293–314.

Selley, R. C. (1978). Porosity gradients in North Sea oil-bearing sandstones. *J. Geol. Soc., London* **135**, 119–131.

Selley, R. C. (1984). "Porosity Evolution of Truncation Traps: Diagenetic Models and Log Responses," Proc. Norw. Offshore North Sea Conf., Stavanger, Pap. G3, pp. 1–17. Norw. Pet. Soc., Oslo.

Selley, R. C. (1992a). The third age of wireline log analysis: Application to reservoir diagenesis. *Spec. Publ.— Geol. Soc. Am.* **65**, 377–388

Selley, R. C. (1992b). Petroleum seepages and impregnations in Great Britain. *Mar. Pet. Geol.* **9**, 226–244.

Selley, R. C. (1996). "Elements of Petroleum Geology," 2nd ed. Academic Press, San Diego, CA. 470pp.

Selley, R. C., and Stoneley, R. (1987). Petroleum habitat in south Dorset. *In* "Petroleum Geology of North West Europe" (J. Brooks and K. Glennie, eds., pp. 139–148. Graham & Trotman, London.

Shanmugam, G. (1990). Porosity prediction in sandstones using erosional unconformities. *Mem.— Am. Assoc. Pet. Geol.* **49**, 1–23.

Shanmugam, G. (1998). Origin, recognition, and importance of erosional unconformities in sedimentary basins. *In* "New Perspectives in Basin Analysis" (K. L. Kleinspehn and C Paola, eds.), pp. 83–108. Springer-Verlag, Berlin.

Shanmugam, G., and Higgins, J. B. (1988). Porosity enhancement from chert dissolution beneath Neocomian unconformity; Ivishak Formation, North Slope, Alaska. *AAPG Bull.* **72**, 523–535.

Shaw, H. F. (1980). Clay minerals in sediments and sedimentary rocks. *In* "Developments in Petroleum Geology" (G. D. Hobson, ed.), Vol. 2, pp. 53–85. Applied Science, Barking, England.

Shepard, F. P. (1954). Nomenclature based on sand-silt-clay ratios. *J. Sediment. Petrol.* **24**, 151–158.

Sibley, D. F., and Blatt, H. (1976). Intergranular pressure solution and cementation of the Tuscarora Quartzite. *J. Sediment. Petrol.* **46**, 881–896.

Sippel, R. F. (1968). Sandstone petrology, evidence from luminescence petrography. *J. Sediment. Petrol.* **28**, 530–554.

Slaughter, M., and Earley, J. W. (1965). Mineralogy and geological significance of the Mowry Bentonite, Wyoming. *Spec. Pap.— Geol. Soc. Am.* **83**, 1–116.

Sorby, H. C. (1880). On the structure and origin of non-calcareous stratified rocks. *Proc. Geol. Assoc., London* **36**, 46–92.

Srodon, J. (1999). Use of clay minerals in reconstructing geological processes: Recent advances and some perspectives. *Clay Miner.* **34**, 27–37.

Stadler, P. J. (1973). Influence of crystallographic habit and aggregate structure of authigenic clay minerals on sandstone permeability. *Geol. Mijnbouw* **52**, 217–220.

Steinmann, G. (1926). Die ophiolitischen Zoren in den Mediterranen Kettengebirgen. *C.R. Int. Geol. Congr. 14th,* Madrid.

Sturt, B. A. (1961). Discussion in: Some aspects of sedimentation in orogenic belts. *Proc. Geol. Assoc.* **1587**, 78.

Sullivan, K. B., and McBride, E. F. (1991). Diagenesis of sandstones at shale contacts and diagenetic heterogeneity, Frio Formation, Texas. *AAPG Bull.* **75**, 121–138.

Surdam, R. C., and Crossey, L. J. (1987). Integrated diagenetic modelling. *Annu. Rev. Earth Planet. Soc.* **15**, 141–170.

Surdam, R. C., and Wolfbauer, C. A. (1973). Depositional environment of oil shale in the Green River formation, Wyoming. *Am. Assoc. Pet. Geol. Bull.* **57**, 808 (abstr.).

Surdam, R. C., Boase, S. W., and Crossey, L. J. (1984). The chemistry of secondary porosity. *Mem.— Am. Assoc. Pet. Geol.* **37**, 127–150.

Suthren, R. J. (1985). Facies analysis of volcanic sediments: A review. *In* "Sedimentology: Recent Advances and Applied Aspects" (P. J. Brenchley and B. P. J. Williams, eds.), pp. 123–146. Blackwell, Oxford.

Taylor, J. M. (1950). Pore space reduction in sandstones. *Bull. Am. Assoc. Pet. Geol.* **34,** 701–716.

Tissot, B., and Welte, D. H. (1984). "Petroleum Formation and Occurrence," 2nd ed. Springer-Verlag, Berlin. 530pp.

Van Elsberg, J. N. (1978). A new approach to sediment diagenesis. *Bull. Can. Soc. Pet. Geol.* **26,** 57–86.

Van Houten, F. B. (1968). Iron oxides in red beds. *Geol. Soc. Am. Bull.* **79,** 399–416.

Van Houten, F. B. (1973). Origin of red beds, a review — 1961–1972. *Annu. Rev. Earth Planet. Sci.* **1,** 39–61.

Walker, T. R. (1967). Formation of red beds in modern and ancient deserts. *Geol. Soc. Am. Bull.* **78,** 353–368.

Walker, T. R. (1976). Diagenetic origin of Continental red beds. *In* "The Continental Permian in Central, West and Southern Europe" (H. Falke, ed.), pp. 240–282. Reidel, Dordrecht, The Netherlands.

Waugh, B. (1970a). Petrology, provenance and silica diagenesis of the Penrith Sandstone (Lower Permian) of northwest England. *J. Sediment. Petrol.* **40,** 1226–1240.

Waugh, B. (1970b). Formation of quartz overgrowths revealed by scanning electron microscopy. *Sedimentology* **14,** 309–320.

White, G. (1961). Colloid phenomena in sedimentation of argillaceous rocks. *J. Sediment. Petrol.* **31,** 560–565.

Wilson, M. D., and Pittman, E. D. (1977). Authigenic clays in sandstones: Recognition and influence on reservoir properties and paleoenvironmental analysis. *J. Sediment. Petrol.* **47,** 3–31.

Wilson, M. J. (1999). The origin and formation of clay minerals in soils: Past, present and future perspectives. *Clay Miner.* **34,** 1–27.

Wolf, K. H., and Chilingarian, G. V. (1976). "Compaction of Coarse-grained Sediments," Vol. II, Elsevier, Amsterdam. 808pp.

Worden, R., ed. (2000). "Quartz Cementation in Sandstones." Blackwell Science, Oxford. 396pp.

Yen, T. F., and Chilingarian, G. V., eds. (1976). "Oil Shale." Elsevier, Amsterdam. 292pp.

9 Autochthonous Sediments

9.1 INTRODUCTION

The second great group of sedimentary rocks is variously referred to as the chemical or autochthonous sediments group. This group consists of rocks that form within a depositional basin, as opposed to the terrigenous sands and muds that originate outside the basin. The chemical rocks are sometimes divided into organic and inorganic groups. Carbonate skeletal sands are a good example of the first type; evaporites of the second. Biochemical research shows, however, that there is no clearly defined boundary between these two groups. For example, apparently spontaneous precipitates of lime mud occur in response to chemical changes in seawater due to the activity of plankton and bacteria (see Section 9.2.3.2). It is a matter of semantics whether such sediments are organic or inorganic in origin. Though the chemical rocks are directly precipitated within a sedimentary basin they can be subjected to minor reworking. There are, therefore, examples of detrital (i.e., clastic) chemical sediments, but these must be carefully distinguished from the detrital terrigenous sediments of extra-basinal origin.

Table 9.1 shows the main chemical sedimentary rocks. Of these the carbonates are volumetrically the most important. The carbonate rocks include the limestones made of calcium carbonate ($CaCO_3$), and the dolomites, composed of the mineral dolomite ($CaMg(CO_3)^2$). Some pedants prefer to restrict the term "dolomite" to the mineral and refer to the rock as "dolostone." The carbonate rocks form by organic processes, by direct inorganic precipitation, and by diagenesis. Carbonates are important aquifers and hydrocarbon reservoirs because of the high porosity which they sometimes contain. Porosity distribution is complex, however, and has merited considerable research. For these reasons carbonate rocks will subsequently be described and discussed in some detail.

A second important group of the chemical sediments are the evaporites. These form both by inorganic crystallization and by diagenesis. The most common evaporite mineral is anhydrite, calcium sulfate ($CaSO_4$), and its hydrated product gypsum ($CaSO_4 \cdot 2H_2O$). Less common evaporites are rocksalt or halite (NaCl) and a whole host of potassium and other salts. The sedimentary ironstones are a much less abundant chemical sedimentary rock. They also form both as precipitates and by diagenesis. The common ferruginous sedimentary minerals are pyrite (FeS_2) and siderite ($FeCO_3$). The sedimentary iron ores, however, include the oxides goethite and hematite, and chamosite, a complex ferrugi-

394

Table 9.1
Major Types of Chemical Rocks

Carbonates	{ Dolomites
	{ Limestones
Evaporites	{ Anhydrite/gypsum
	{ Halite/rock salt
	{ Potash salts, etc.
Siliceous rocks	− Chert, radiolarite, novaculite
Carbonaceous rocks	{ Humic group — coal series
	{ Sapropelitic group — oil shales and cannel coals
Sedimentary ironstones	
Phosphates	

The chemical rocks are those which form within the depositional environment. They include direct chemical precipitates, such as some evaporites, and formation by organic processes, such as coal and shell limestones. Not all chemical sediments are syndepositional. Diagenetic processes are important in the genesis of some evaporites, dolomites, cherts, ironstones, and phosphates.

nous aluminohydrosilicate. These rocks are discussed in Section 1.3.2.2 on synsedimentary ores. Phosphorite is a sedimentary rock composed dominantly of phosphates. It forms largely during early diagenesis in sediment, immediately beneath the sediment/water interface, aided by reworking and concentration of incipient pellets and concretions. The phosphate minerals, like the ironstones, are chemically very complex. Examples include collophane, dahllite, francolite, and fluorapatite. These are calcium phosphates combined with various other radicals. Phosphorites are discussed in Section 9.5. Coal is a sedimentary rock formed entirely by biochemical processes. It originates from the accumulation under anaerobic conditions of vegetable detritus as peat, in swamps, marshes, meres, and pools. Extensive coal beds are especially characteristic of ancient deltaic deposits (see Section 9.3). The last of the chemical sediments to be considered are the siliceous rocks, termed "chert." These are composed largely of microcrystalline quartz and chalcedony, a variety of silica with a spherulitic habit. Hydrous silica, opal, occurs in Tertiary rocks, but appears to dehydrate on burial to silica (Ernst and Calvert, 1969). The various chemical sediments are now described.

9.2 CARBONATES

9.2.1 Introduction

The carbonate rocks are extremely complex in their genesis, diagenesis, and petrophysics. There are a number of reasons for this (Ham and Pray, 1962). Carbonate rocks are intrabasinal in origin. Unlike terrigenous sediments they are easily weathered and their weathering products are transported as solutes. Carbonate rocks are, therefore, deposited at or close to their point of origin. Most carbonate rocks are organic in origin. They contain a wide spectrum of particle sizes, ranging from whole shells to lime mud of diverse origin. These sediments are deposited with a high primary porosity. The carbonate minerals are chemically unstable, however. This combination of high primary porosity and permeability, coupled with chemical instability, is responsible for the complicated diagenesis of carbonate rocks, and hence for the problems of locating aquifers and hydrocarbon reservoirs within them.

The following brief account of carbonates first defines their mineralogy, then describes their petrography and classification and concludes by showing the relationship between diagenesis and porosity development.

The carbonate rocks have generated a vast literature. Key works include Chilingar *et al.* (1967a,b), Milliman (1974), Bathurst (1975), Reijers and Hsu (1986), Scoffin (1987), and Tucker and Wright (1990).

9.2.2 Carbonate Minerals

It is necessary to be familiar with the common carbonate minerals to understand the complex diagenetic changes of carbonate rocks. Calcium carbonate $(CaCO_3)$ is the dominant constituent of modern carbonates and ancient limestones. It occurs as two minerals, aragonite and calcite. **Aragonite** crystallizes in the orthorhombic crystal system, while calcite is rhombohedral. **Calcite** forms an isomorphous series with magnesite $(MgCO_3)$. A distinction is made between high- and low-magnesium calcite, with the boundary being arbitrarily set at 10 mol%.

Ancient limestones are composed largely of low magnesium calcite, while modern carbonate sediments are made mainly of aragonite and high-magnesian calcite. Aragonite is found in many algae, lamellibranchs, and bryozoa. High magnesian calcite occurs in echinoids, crinoids, many foraminifera, and some algae, lamellibranchs, and gastropods (Table 9.2). An isomorphous series exists between calcite and magnesite $(MgCO_3)$. Skeletal aragonite and calcite also contain minor amounts of strontium, iron, and other trace elements. The relationship between carbonate secreting organisms, the mineralogy of their shells, and their contained trace elements has been studied in detail (e.g., Lowenstam, 1963; Milliman, 1974). These factors are important because their variation and distribution play a controlling part in the early diagenesis of skeletal sands. **Dolomite** is another important carbonate mineral, giving its name also to the rock. Dolomite is calcium magnesium carbonate $(CaMg(CO_3)^2)$. Isomorphous substitution of some magnesium for iron is found in the mineral termed ferroan dolomite or **ankerite** $Ca(MgFe)(CO_3)^2$. Unlike calcite and aragonite, dolomite does not originate as skeletal material. Dolomite is generally found either crystalline, as an obvious secondary replacement of other carbonates, or as a primary or penecontemporaneous replacement mineral in cryptocrystalline form. The problem of dolomite genesis is elaborated on later.

Siderite, iron carbonate $(FeCO^3)$, is one of the rarer carbonate minerals. It occurs, apparently as a primary precipitate, in ooliths. These "spherosiderites," as they are termed, are found in rare restricted marine and freshwater environments. Spherosiderite is often associated with the hydrated ferrous aluminosilicate, chamosite, in sedimentary iron ores (see Section 9.4). Siderite also occurs as thin bands and horizons of concretions in argillaceous deposits, especially in deltaic deposits. It is not uncommon to find siderite bands contorted and fractured by slumping. Siderite clasts are also found in intraformational conglomerates. These facts suggest that siderite forms diagenetically during early burial while the host sediment is still uncompacted. Its formation is favored by al-

Table 9.2
Mineralogy of the Main Carbonate-Secreting Organisms[a]

Taxon	Aragonite	Calcite
Algae		
Red		C
Green	A	
Coccoliths		C
Foraminifera		
Benthonic	(A)	C
Planktonic		C
Sponges	(A)	C
Coelenterates		
Stromatoporoids	A	C?
Corals		
Rugosa		C
Tabulata		C
Scleractinia	(A)	C
Alcyonaria	(A)	C
Bryozoa	(A)	C
Brachiopods		C
Mollusks		
Lamellibranchs	A	C
Gastropods	A	C
Pteropods	A	
Cephalopods (Ammonite opercula)	A	(C)
Belemnoids		C
Annelids	A	C
Arthropods		C
Echinoderms		C

[a]Brackets denote minor occurrence. Note that some groups secrete both aragonite and calcite. High and low magnesian calcite varieties not distinguished. The original composition of the stromatoporoids is questionable because they are now extinct. Based on Scholle (1978), with some simplifications and embellishments.

kaline reducing conditions (Fig. 8.22). Table 9.3 summarizes the salient features of the main carbonate minerals.

To conclude this brief summary of a large topic, the following points are important. The principal carbonate minerals are the calcium carbonates, calcite and its unstable polymorph aragonite; and dolomite, calcium magnesium carbonate. Modern carbonate sediments are composed of both aragonite and calcite. Only calcite, the more stable variety, occurs in lithified limestones. Dolomite does not occur as a biogenic skeletal mineral. It forms as a secondary replacement mineral or, rarely, by primary precipitation or penecontemporaneous replacement of other carbonates.

9.2.3 Physical Components of Carbonate Rocks

Carbonate rocks, like sandstones, have four main components: grains, matrix, cement, and pores. Unlike sandstones, however, the grains of carbonate rocks, though commonly

Table 9.3
Summary of the Common Carbonate Minerals

Mineral	Formula	Crystal system	Occurrence
Aragonite	} $CaCO_3$ {	{ Orthorhombic	Present in certain carbonate skeletons
			Unstable and reverts to stable polymorph calcite
Calcite		{ Hexagonal	Present in certain carbonate skeletons, as mud (micrite) and cement (sparite)
Magnesite	$MgCO_3$	Hexagonal	Present in minor amounts within the lattices of skeletal aragonites and calcites
Dolomite	$CaMg(CO_3)_2$	Hexagonal	Largely as a crystalline diagenetic rock, also penecontemporaneously associated with evaporites
Ankerite (ferroan dolomite)	$Ca(MgFe)(CO_3)_2$	Hexagonal	A minor variety of dolomite
Siderite	$FeCO_3$	Hexagonal	Found as concretions and ooliths (spherosiderites)

monominerallic, are texturally diverse and polygenetic. The various grain types, matrix, and cement are described next. They are tabulated in Table 9.4.

9.2.3.1 Grains

Grains are the particles that support the framework of a sediment. They are thus generally of sand grade or larger. As Table 9.4 and Fig. 9.1 show, carbonate grains are of

Table 9.4
Main Components of Carbonate Rock

I. Grains	{ (a) Detrital grains	{ Lithoclasts Intraclasts
	(b) Skeletal grains	
	(c) Peloids (including fecal pellets)	
	(d) Lumps	{ Composite grains Algal lumps

	(e) Coated grains	{ Ooliths
		  Pisoliths
		  Algally encrusted grains

II. Matrix	{ Micrite Clay	
III. Cement	Sparite	
IV. Pores		

After Leighton and Pendexter (1962).

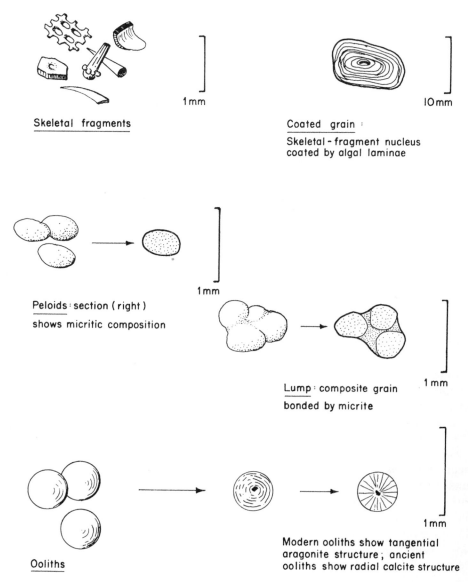

Fig. 9.1. Sketches of some of the principal types of carbonate grain. For full descriptions see text. Photomicrographs of recent and ancient examples are shown in plates Plates 6 and 7, respectively.

many types, as briefly described here. First are the detrital grains, which are of two types. They include rock fragments, or **lithoclasts.** These are grains of noncarbonate material which originated outside the depositional basin. Quartz grains are a typical example of lithoclasts and, as they increase in abundance, limestones grade into sandy limestones and thence to calcareous sandstones. The second type of detrital grains are **intraclasts.** These are fragments of reworked carbonate rock which originated within the depositional basin (Plate 6B). Early cementation followed by penecontemporaneous erosion is a common feature of carbonate rocks, and is responsible for the generation of intraclasts. Intraclasts may range from sand size, via intraformational conglomerates of penecontemporaneously cemented **beach rock** to, arguably, fore-reef slump blocks.

The most important of all grain types is skeletal detritus, individual grains of which are termed **bioclasts** (Plate 6C). As pointed out in the previous section, this is composed of aragonite or calcite with varying amounts of trace elements. The actual crystal habit of skeletal matter is varied too, ranging from the acicular aragonite crystals of lamellibranch shells to the single calcite crystals of echinoid plates. The size of skeletal particles is naturally very variable, ranging down from the largest shell to individual disaggregated microscopic crystals. Continued abrasion of skeletal debris by wave and current action and by biological processes, such as boring, is responsible for the very poor textural sorting characteristic of carbonate sediments.

Peloids are a third major grain type (Plate 6B). These were first defined as structureless cryptocrystalline carbonate grains of some 20–60 μm in diameter (McKee and Gutschick, 1969). Peloids form in many different ways. **Pellets** are peloids of fecal origin, excreted by marine invertebrates. Pelletoids are peloids formed by the micritization of skeletal grains through the action of endolithic algae. These colonize carbonate grain surfaces, bore into them, and change their original fabric into structureless micrite (Taylor and Illing, 1969). Several other processes have been proposed for peloid formation (MacIntyre, 1985; Chafetz, 1986). The genesis of peloid formation is important because this grain type is sometimes a major constituent of limestone formations. Peloidal deposits are especially characteristic of lagoons and other sheltered shallow inner-shelf environments.

Lumps are another important carbonate grain type. These are botryoidal grains which are composed of several peloids held together. They are sometimes termed "composite grains" or **grapestone.** Grains such as these are probably formed by the reworking of peloidal sediment that has already undergone some lithifaction. They are thus nascent intraclasts (Illing, 1954).

Last of the grain types to consider are the **coated grains.** These are grains that show a concentric or radial arrangement of crystals about a nucleus. The most common coated grains are **ooids.** These are rounded grains of medium to fine grain size which generally occur gregariously in sediments termed oolites, devoid of other grain types or matrix. Ooids contain a nucleus. This is normally a quartz grain or a shell fragment. Modern ooids are generally composed of concentric layers of tangentially arranged aragonite. Modern oolites occur in high-energy environments, such as sand banks and tidal deltas. Like their ancient analogs they are generally well sorted, matrix free, and cross-bedded. These data all suggest that ooids form by the bonding of aragonite crystals around nuclei, such as quartz or skeletal grains, in a high-energy environment (Plate 6D). The physicochemical processes that cause ooid formation are unclear, but it is noted that they generally tend to form where cool dilute seawater mixes with warm concentrated waters of lagoons and restricted shelves (e.g., the modern Bahamas platform). Recent ooids are covered with a mucilaginous jacket of blue-green algae which serves as a site for aragonite precipitation.

Pisoids are coated grains several millimeters in diameter (Fig. 9.2A). They form in caverns (cave pearls). A rock composed of pisoids is termed a "pisolith." Vadose pisoliths form in caliche crusts beneath the weathered zones of soils ancient and modern (Dunham, 1969).

Oncoids are a third type of coated grain (Fig. 9.2B). These are several centimeters in diameter and irregularly shaped. The laminae are discontinuous around the grain. Oncoids are formed by primitive blue-green algae growing on a grain and attracting carbonate mud to their sticky surface. Intermittent rolling of the grain allows the formation of discontinuous lime-mud films. In contrast to ooids therefore, pisoids and oncoids are indicators of low-energy environments. A rock composed of oncoids is termed an "oncolite."

9.2.3.2 Micrite

Carbonate mud is termed **micrite.** The upper size limit of micrite is variously taken as 0.03–0.04 mm in diameter. Micrite may be present in small quantities as a matrix within a grain-supported carbonate sand, or may be so abundant that it forms a carbonate mud-rock, termed "micrite" or "calcilutite" (Plate 6A). Most modern lime muds are composed of aragonite; their lithified fossil analogs are made of calcite.

Several processes appear to generate lime mud. The action of wind, waves, and tides will smash shell debris and ultimately may abrade them into their constituent crystals. Fecal pellets may be similarly disaggregated. Biological action is also efficient at breaking up carbonate particles to form crystal muds. This includes the parrot fish which eats coral, shell-munching benthonic and burrowing invertebrates, and especially some blue-green algae. These form pits within skeletal grains and lead to the micritization of the grain surface. Grains so softened will tend to break up and release the micrite. The calcareous algae, such as *Halimeda,* also secrete aragonite needles within their mucilagenous tissues. On death the mucilage rots to release the aragonite needles.

Some evidence suggests that direct inorganic precipitation of aragonite muds may sometimes take place. In the Bahamas Banks and the large gulf between Iran and Arabia, "whitings" have been described by Milliman *et al.* (1993) and Wells and Illing (1964), respectively. These are temporary cloudy patches of lime mud disseminated in seawater. They are attributed to spontaneous abiogenic precipitation of aragonite from seawater. Some geochemical data are inconsistent with this interpretation, however, and there are other explanations for these phenomena.

Micrite can also form as a cryptocrystalline cement in certain circumstances. Because of this it must be used with care as an index of depositional energy. In conclusion, it would appear that lime mud, micrite, can form by a variety of processes.

9.2.3.3 Cement

The third component of carbonate rocks is cement. By definition this applies to crystalline material that grows within the sediment fabric during diagenesis. The most common cement in limestones is calcite, termed "spar" or **sparite** (Plate 7D). Other cements in carbonate rocks include dolomite, anhydrite, and silica.

It is the custom to restrict the term "cement" to the growth of crystals within a pore space. This has also been termed "drusy crystallization." This type of spar is genetically distinguishable from neomorphic spar which grows by replacement of preexisting carbonate (Folk, 1965). Neomorphism is itself divisible into two varieties. Polymorphic transformations are those which involve a mineral change, as in the reversion of

Fig. 9.2. Photographs illustrating some of the larger varieties of carbonate particles. (A) Pisolitic wackestone, Late Pre-Cambrian. Ella Island, East Greenland. The pisoids show concentric laminae and have diameters of 0.5–1.2 mm. (B) Algal oncoids, Inferior Oolite (Middle Jurassic), Dorset. These particles, up to 15 cm in diameter, have a bioclast or intraclast as a nucleus, and a rind of algal laminae. Algal oncoids typically form in

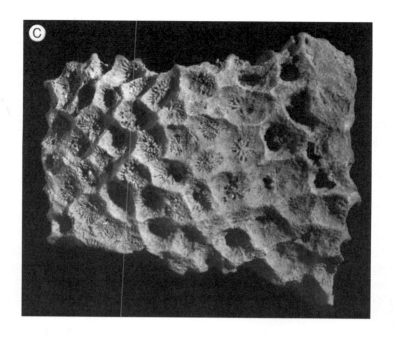

shallow lagoons. (C) A 10-cm-wide colonial coral, Dam Formation (Miocene), Dhahran dome, Saudi Arabia. (D) A 15-cm-wide vertical section of laminated algal stromatolite, Carboniferous Limestone (Lower Carboniferous), Mendip Hills, England.

aragonite to calcite. Recrystallization is the development of calcite spar by the enlargement of preexisting calcite crystals. Additional discussion of cement is contained in Section 9.2.5 on diagenesis.

9.2.4 Nomenclature and Classification

The nomenclature and classification of carbonate rocks was once a topic of great confusion and debate. There were several good reasons for this, including the fact that so many parameters can be used to define carbonate rock types. These include chemical composition (e.g., limestone, dolomite), grain size, particle type, type and amount of porosity, degree of crystallinity, quantity of mud, and so on. The concepts on which present-day carbonate nomenclature are based are contained in a volume of papers edited by Ham (1962). Two of these articles deserve special mention because they proposed a series of terms and groupings which are widely used today.

Folk (1965) divided limestones into four main classes (Table 9.5). The first two classes include rocks composed largely of grains (allochems); these he jointly termed the allochemical limestones. One class is dominated by sparite cement, the other by micrite matrix. The third class is for rocks lacking grains, termed the "orthochemical" limestones. This group includes micrite lime mud carbonates. The fourth group is for rocks made of in-place skeletal fabrics. This group, the autochthonous reef rocks, includes coral biolithite and algal stromatolites (Figs. 9.2C and D). Folk proposed a bipartite nomenclature for the allochemical rocks. The prefix defined the grain type and the suffix denoted whether sparite or micrite predominated. Thus were born words like "oosparite" and "pelmicrite." Where more than one allochem type was present, two should be used, with the major one first, as in "bioоosparite." When both sparite and micrite were present, they could both be compounded with the dominant constituent first: as in "pelbiomicsparite." Thus was born the rock name "biooointrapelmicsparite." Joking apart, this is a logical and flexible scheme for classifying and naming carbonate rocks.

Dunham's (1969) approach to the problem was quite different, but equally instructive. Like Folk, he placed the in-place reef rocks in a class of their own — the **boundstones.** A second class was erected for the crystalline carbonates whose primary depositional fabric could not be determined. Dunham divided the rest of the carbonates into four groups according to whether their fabric was grain supported or mud supported (Table 9.6).

Grainstones are grain-supported sands with no micrite matrix. **Packstones** are grain-supported sands with minor amounts of matrix. **Wackestones** are mud-supported rocks

Table 9.5
Major Grouping of the Carbonate Rocks

Allochemical rocks	Composed largely of detrital grains
	I Sparite cement dominant
	II Micrite matrix dominant
Orthochemical rocks	III Composed dominantly of micrite
Autochthonous reef rock	IV Biolithite

After Folk (1962).

Table 9.6
Classification of Carbonate Rocks

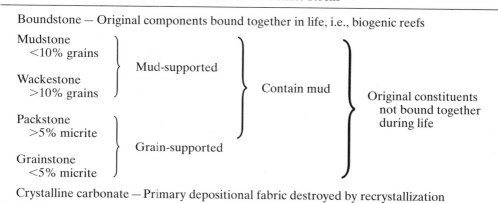

Boundstone — Original components bound together in life, i.e., biogenic reefs

Mudstone <10% grains ⎫
Wackestone >10% grains ⎬ Mud-supported

Packstone >5% micrite ⎫
Grainstone <5% micrite ⎬ Grain-supported

⎫ Contain mud ⎬ Original constituents not bound together during life

Crystalline carbonate — Primary depositional fabric destroyed by recrystallization

After Dunham (1962).

with a significant but dispersed number of grains, and mudstones are carbonate muds (Plates 6A–E). This scheme too has much to commend it. The nomenclature is simple and identification can be made with only a hand lens. Furthermore, this system, by drawing attention to fabric and matrix content, gives an index of the depositional energy. Thus the mud-supported limestones may be taken to indicate deposition in a low-energy environment. By contrast, the matrix-free grain-supported rocks suggest deposition in high-energy environment in which no mud could come to rest.

The concept of textural maturity was discussed earlier when considering siliciclastic sediments. It must, however, may be applied to carbonates with certain reservations. The polygenetic origin of micrite has been mentioned already. It is possible for a clean carbonate sand to be deposited in a high-energy environment. Subsequently, however, micrite may develop by bioturbation, algal micritization, cementation, and by infiltration due to high permeability.

Falls and Textoris (1972) analyzed the granulometry of some Recent carbonate beach sands using the same statistical techniques as those applied to siliciclastic sediment, yet they reported difficulties in analyzing the fine fraction due to micritization. Similarly, the large initial size of carbonate skeletal material makes it dangerous to use grain size as an energy index in the same way as it can be used in terrigenous deposits. Consider, for example, the oyster reefs of modern lagoons. In terms of particle size these are essentially low-energy conglomerates (Fig. 9.3). Nevertheless, some studies do show that skeletal debris can be comminuted, size graded, and winnowed to produce granulometric characteristics which mimic those of terrigenous sands (e.g., Hoskins, 1971).

Subject to the above limitations the textural maturity of carbonate sediments may be considered in a way that is analogous to the maturity of terrigenous sands. A mature carbonate sediment is one that is composed exclusively of one grain type. The different carbonate grains result from different processes operating in different environments. Thus a carbonate composed of a single grain type indicates genesis from a single process, one with two grain types reflects those with two processes, and so forth (Smosna, 1987).

Fig. 9.3. Photograph of an oyster biostrome, Marada Formation (Miocene), Jebel Zelten, Libya. This demonstrates a major difference between siliciclastic and carbonate rocks. Because some of the oysters are up to 15 cm long, this biostrome is technically a conglomerate, in terms of grain size, but it formed in place without transportation. Carbonate grain size (and texture) must therefore be used as indicators of energy level and maturity with care.

Thus grain size, sorting, and matrix content can only be used with reservations as indicators of hydrodynamic environment in carbonate rocks. Nevertheless, the classifications and nomenclature of carbonates proposed by Folk and Dunham are extremely useful and, used in conjunction, encompass most varieties of limestones with flexibility and finesse (Fig. 9.4).

Sadly neither of the schemes outlined above embrace what, for many geologists, is the most important aspect of carbonates, namely, their reservoir characteristics, in terms of porosity and permeability. Archie (1952) classified carbonate reservoirs based on the concept that pore-size distribution controls permeability and petroleum saturation, and that pore-size distribution is related to the fabric of the rock. This scheme was developed by Lucia (1999). Lucia classifies carbonates into three petrophysical classes according to their pore-type: (1) interparticle, (2) separated vugs, and (3) communicating vugs. These three petrophysical classes may then be subdivided acording to Dunham's terminology of grainstone, packstone, wackestone, and mudstone (Fig. 9.5).

9.2.5 Diagenesis and Porosity Evolution of Limestones

9.2.5.1 Diagenesis and Petrophysics

It has already been pointed out that the diagenesis of carbonate rocks is very complex. This is basically because of their unstable mineralogy and because their high initial per-

	GRAIN TYPES			
MUDSTONE < 10% grains	Lime mud, micrite, calcilutite, chalk			
	PELLETS	SHELL DEBRIS	OOLITHS	INTRACLASTS
WACKESTONE > 10% grains, mud supported	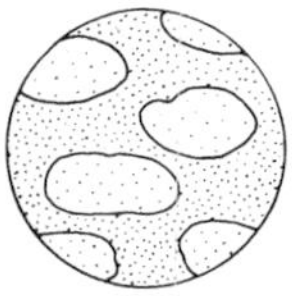 Pelmicrite	Biomicrite	Oomicrite	Intramicrite
PACKSTONE > 5% mud, grain supported	Pelmicsparite	 Biomicsparite	Oomicsparite	Intramicsparite
GRAINSTONE < 5% mud	Pelsparite	Biosparite	 Oosparite	Intrasparite
BOUNDSTONE original components bound together	Reef rock, biolithite			

Fig. 9.4. Nomenclature and classification of limestones according to the mudstone, wackestone, packstone, grainstone and boundstone scheme of Dunham (1962), and Folk (1962) the rest. For modern examples of grain types see Plate 6; for ancient examples of limestones, see Plate 7.

meability makes them susceptible to percolating reactive fluids. Choquette and Pray (1970) have pointed out the salient petrophysical characteristics of carbonate rocks: Primary porosity is generally higher than in terrigenous sands, running at between 40 and 70%. It consists of both interparticle and intraparticle porosity. The ultimate porosity in a carbonate reservoir, however, may be only 5–15%. Little of this is primary porosity. Most of it is of the diverse secondary porosity varieties described in Chapter 3; namely, moldic, vuggy and intercrystalline. The size and shape of individual pores is extremely variable within any one rock and, unlike sandstones, there is little correlation between pore volume, pore geometry and grain size, shape, and sorting. Because of the

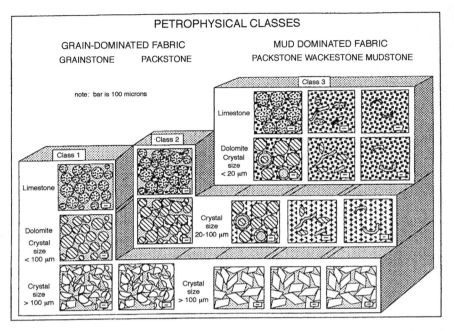

Fig. 9.5. Petrophysical/rock fabric classification of carbonates proposed by Lucia. This classifies carbonates according to whether porosity is intergranular (class 1), vuggy isolated (class 2) and vuggy connected (Class 3). Each class is subdivided according to Dunham's (1962) classification scheme of mudstone, wackestone, packstone, and grainstone. (From Lucia, 1999, Fig. 21, p. 48, of *Carbonate Reservoir Characterization* © 1999. Courtesy of Springer-Verlag.)

erratic petrophysical variations within a small volume of rock, it is necessary to measure porosity and permeability from whole cores rather than from small plugs. Extensive coring of hydrocarbon reservoirs is necessary for accurate calculations of reserves and for effective production. For detailed accounts of the petrophysics of carbonate oil reservoirs see Chilingar *et al.* (1972), Langres *et al.* (1972), Reeckmann and Friedman (1982), Burchette and Britten (1985), Schneidermann and Harris (1985), and Schroder and Purser (1986).

The following account of the relationship between porosity and diagenesis of carbonate rocks is drawn freely from the writings of Murray (1960), Larsen and Chilingar (1967), Choquette and Pray (1970), Bathurst (1975), Tucker and Wright (1990), and Lucia (1999). An account of the main diagenetic processes and their petrophysical effects is followed by examination of the diagenetic characteristics of some of the major carbonate rock types.

The earlier brief account of cement drew attention to the different modes of formation of sparite-crystalline calcite. Crystallization, used in its restricted sense, means the infilling of primary inter- and intraparticle porosity by the drusy growth of sparite out from the pore walls. This naturally results in a decrease in porosity. "Neomorphism" is the term applied to describe the recrystallization or replacement of a mineral (Folk, 1965). Neomorphism can lead to both increasing, or unaltered porosities. Recrystallization, defined as neomorphism in which the mineralogy is unchanged, does not significantly alter the amount or type of porosity. However, polymorphic transformations, in which one mineral replaces another, can have large effects on rock porosity. As already mentioned, one of the earliest diagenetic changes is the transformation of aragonite to calcite. This results in an increase in total rock volume of 8%.

Another important diagenetic process is the replacement of one mineral by another, as for example calcite by dolomite. Dolomitization can cause an overall contraction of the rock by as much as 13% (Chilingar and Terry, 1964). It is the intercrystalline porosity caused by this replacement that makes secondary dolomites such attractive reservoir rocks. Conversely, a decrease in porosity is caused by the transformation of dolomite to calcite; dedolomitization or calcitization as it is called (Shearman *et al.*, 1961).

Leaching is one of the most important processes giving rise to secondary porosity. Solution porosity may be due to the selective solution of matrix, cement, or specific grain types. Vuggy porosity results from the solution of pores whose boundaries cross-cut the fabric. Moldic porosity describes the selective solution of one particular grain type, for example, oomoldic porosity or biomoldic porosity. Lastly, silicification is another characteristic diagenetic process in carbonate rocks. The silicification of lime muds is described elsewhere (see Section 9.7) and is not of petrophysical significance. In calcarenites and reef rocks, however, silicification can be an important destroyer of primary porosity when it develops as a chalcedonic cement. Silica also occurs either as a wholesale replacement of the rock or selectively to produce silicified fossils to delight paleontologists. This change is generally a one-way process. Table 9.7 and Fig. 9.6 summarize the major diagenetic processes in carbonate rocks and illustrate their various effects of porosity.

The various diagenetic processes just described take place in response to changes in pore fluid chemistry. These reactions are still imperfectly understood, but this section will attempt to outline the major ones. The various pore fluids, their pH, and Eh have already been described in the section dealing with sandstone diagenesis (see Section 8.5.3.3). Reference back to Figs. 8.20 and 8.21 shows that calcite is soluble in acid meteoric waters but stable, and may be precipitated, in alkaline connate waters. It is also necessary, however, to consider the stability fields of aragonite and dolomite. Their solubility is also controlled by pH, but salinity and Ca:Mg ratio are significant parameters.

Table 9.7
Summary of the Main Diagenetic Processes in Carbonate Rocks
and Their Effects on the Amount and Type of Porosity

Diagenetic process			Porosity response
1. Drusy crystallization			Decrease of primary porosity
2. Neomorphism	Recrystallization	Calcite–Calcite	No change
	Polymorphism	Aragonite–Calcite	8% decrease
	Replacement	Calcite–Dolomite	13% increase
		Dolomite–Calcite	13% decrease
3. Leaching			Increase in porosity of moldic and vuggy types
4. Silicification	Chalcedonic pore filling		Decrease in primary porosity
	Replacement		No change

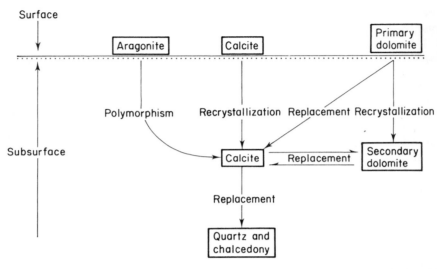

Fig. 9.6. Flow chart illustrating the main diagenetic processes in carbonates and their terminology. Note that the arrows indicate which reactions are reversible and which are not.

Calcite:aragonite:dolomite stability fields are shown in Fig. 9.7. This shows that aragonite is the stable polymorph of calcium carbonate for saline waters with relatively high Mg:Ca ratios. Dolomite can form in a wide range of Mg:Ca ratios and salinities. The diagenesis and petrophysical evolution of calcarenites, calcilutites, and dolomites are now described and discussed in turn.

9.2.5.2 Diagenesis of Calcarenites

For the purpose of this discussion calcarenites will be deemed to include not only lime sands, but also reefs, that is, those carbonate sediments which originally contain both high primary porosity and permeability. From the preceding discussion, remember that these sediments are predominantly aragonitic, and that this reverts to calcite during diagenesis, while at the same time the original pore space will tend to be infilled with a calcite cement. The exact way in which these changes take place depends on the fluids with which the pores are bathed.

The following account attempts to synthesize a diffuse literature and is based particularly on papers by Friedman (1964), Purser (1978), and Longman (1980). If one considers a cross-section through a carbonate island there are five major pore fluid environments (Fig. 9.8). If the topography is sufficient there may be raised ground where carbonate sediment, or previously lithified limestone, occurs in the vadose zone above the water table. The pores of the vadose zone will be full of air and thus chemically inert. When rain falls, however, the pores will be flushed with acidic meteoric water. This will tend to corrode the carbonate minerals, generating moldic and vuggy porosity and enlarging preexisting fractures in lithified rock. This is, of course, epidiagenesis, analogous to that discussed earlier for sandstones (see Section 8.5.3.4). If this process continues uninterrupted then cavernous porosity may develop, leading ultimately to karstic topography, such as the "cockpit" country of parts of the Caribbean.

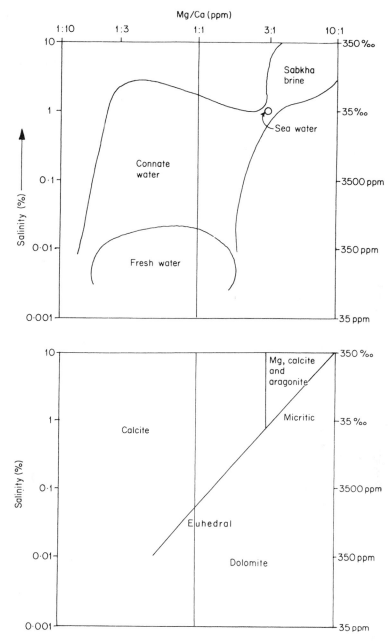

Fig. 9.7. Diagrams showing (upper) the salinity and Mg:Ca ratios encountered in various environments and (lower) the stability fields of carbonate minerals. (From Folk and Land, 1974. American Association of Petroleum Geologists Bulletin, AAPG © 1974, reprinted by permission of the American Association of Petroleum Geologists whose permission is required for further use.)

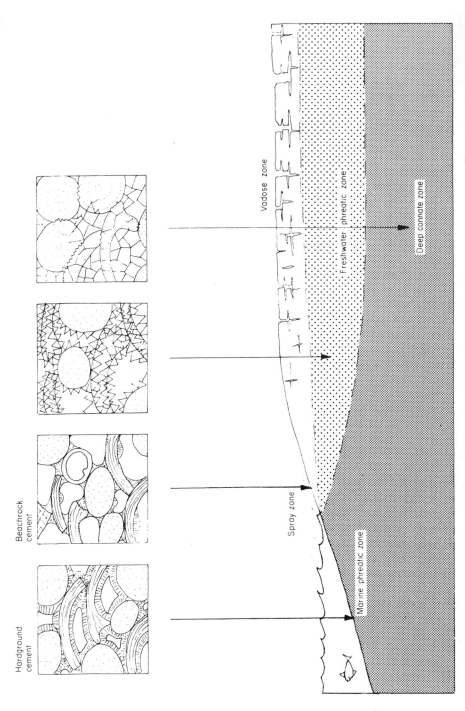

Fig. 9.8. Diagrammatic cross-section through a carbonate shoreline showing the different pore fluid environments and resultant diagenetic fabrics. For explanation, see text.

A second type of vadose diagenesis occurs near the shore in the intertidal and supra-tidal spray zone, where sediment pores are intermittently flushed, not by freshwater, but by seawater. As this evaporates in the pores the salinity increases and generates a characteristic type of cement. Because the pore fluids are only filled intermittently the cement tends to be irregular. Uniform isopachous rim coats are unusual. Cement is often restricted to throat passages (meniscus fill), or depends from the upper surface of pores like stalactites. Cement fabrics are generally fibrous or micritic and are of aragonitic or high magnesium calcite composition. This is referred to as **beachrock** cementation (Plate 7A).

Below the water table, in the phreatic zone, the pores are soaked in water. For a carbonate bank or island a biconvex lens of meteoric freshwater overlies denser seawater (near the coast) which merges inland to connate water of modified marine and meteoric origin (Fig. 9.8). In the freshwater phreatic zone percolating water becomes carbonate saturated, so an even cement of dog-tooth calcite may grow out into the pores. At the same time, however, aragonite becomes unstable, and thus goes into solution. Because some shells are aragonitic, and some calcitic, a selective biomoldic porosity develops. These biomolds also develop a drusy cement along their internal surfaces. This can end up as a curious fabric in which drusy calcite crystals are aligned back to back along the rim of a biomold. They grow out from an organic pellicle or from an algally formed micrite rim.

The overall effect of diagenesis at this stage has thus been to increase the total porosity of the rock. Because of the isolated nature of the biomoldic pores, little of this new porosity is effective. In fact, the drusy calcite has destroyed some of the intergranular porosity, and by bridging pore throats, has also diminished permeability (Plate 7B).

Dedolomitization may also occur in the freshwater phreatic zone, leaving dolomoldic pores which may later be infilled by calcite pseudomorphs (Evamy, 1967). Immediately beneath the sea floor, where the pores are full of seawater, a second phreatic zone occurs with its own distinctive diagenetic fabric. Here an aragonitic rim cement develops around the grains. The aragonite develops either as micrite (i.e., cryptocrystalline aragonite, not a depositional matrix) or as a radial acicular fabric. In either case the cement is isopachous, rimming the grains evenly (Fig. 9.8). This **hardground** type of cement generally develops as a layer only a meter or so deep below the sea floor. In modern examples the carbonate sand beneath is unconsolidated.

Below the sea-floor "hardground" and beneath the freshwater phreatic zone is the deep connate environment. Here sediment that has escaped the various surface diagenetic processes will still be unconsolidated and highly porous to begin with. Because these sands have not been lithified already they will be very susceptible to porosity loss by compaction, whereas those sediments which have already lost some porosity during early shallow diagenesis will not undergo compaction to the same degree. The effects of compaction are manifest by signs of pressure solution at the point of contact of grains, and by stylolites due to wholesale solution of rock (Park and Schot, 1968). Simultaneously the pores become infilled with a coarse sparite mosaic. Unlike the spar fill of freshwater cementation, this does not grow inward from dog-tooth crystals, but develops crystals of uniform size, though these may sometimes be seen cross-cutting "ghosts" of grains and earlier shallow diagenetic fabrics (Plate 7C).

Particular attention has been directed toward finding the source of the calcium carbonate for this deep connate type of cement. Two major ones have been proposed: internal and external. The internal source is provided by pressure solution, the external by migrating connate fluids. These are discussed in turn. There is evidence for considerable intrastratal solution in deeply buried carbonates. The evidence is both microscopic and macroscopic. Pressure solution can often be seen to have taken place in some deeply buried skeletal limestones. Stylolites provide evidence of solution on a much larger scale (Carozzi and von Bergen, 1987). A **stylolite** is a sutured boundary between two rocks (Fig. 9.9A). A microstylolite is one between two grains. Stylolites are commonly seen in limestones and orthoquartzites. Microstylolites are seen at quartz:quartz and calcite:calcite grain contacts. Stylolites are usually parallel or subparallel to bedding. Microstylolites occur at burial depths as shallow as 90 m (Shlanger, 1964), stylolites as shallow as 60 m (Dunnington, 1967). It is axiomatic that stylolites are caused by solution. Estimates of the amount of solution can be made from both the maximum thickness of a stylolite or from the offset of planar surfaces, such as fractures or clay laminae (Fig. 9.9B). Stylolite surfaces are commonly marked by clay and/or organic matter left behind by the solution process. Stylolites normally occur in nonporous impermeable formations, so there are problems in establishing not only how the material is dissolved, but also how it is transported. Studies of the Dukhan field of Qatar, and the Murban field of Abu Dhabi, showed that as much as 30–40% of the original rock volume was lost by stylolitization of limestone (Dunnington, 1967). Stylolites are absent in the porous oil-bearing part of the reservoirs, but are abundant in the water-bearing intervals. Porosity drops off markedly below the oil:water contacts of the fields (these are two of many cases that demonstrate how petroleum preserves porosity).

Stylolitization seems to be restricted to monomineralic rocks with less than 10% clay. When this figure is exceeded there is still evidence of extensive solution. This is seen by the gradual development of wispy argillaceous laminae that pass gradationally into nodular limestones. These look superficially like conglomerates of limestone clasts in clay matrix (Robin, 1978). This process, together with stylolitization, provides evidence for extensive solution of carbonate during deep burial. This must contribute to the large amount of carbonate cement seen in many limestones.

The second source for the calcium carbonate needed for deep calcite cement could be connate fluids. The ions may have been derived from the solution of shallower limestones, as previously discussed, or from fluids squeezed from compacting clays. There is a problem with invoking an external source for the cement. It is possible to calculate how much calcium carbonate is required, and the necessary concentrations of calcium and carbonate ions in pore fluids. These calculations show that whole oceans of water must pass through a carbonate sediment before it is completely cemented (Weyl, 1958).

This discussion of the source of calcite cement is important. Because a modern reef may have up to 80% porosity and a skeletal sand some 60%, it follows that over half of a completely nonporous limestone is made of cement. It appears most likely that in deeply buried lime sands porosity is lost very largely by intrastratal solution and reprecipitation. As burial develops permeability diminishes, and the role of migrating fluids as agents of cementation declines.

The foregoing account of diagenesis and porosity evolution applies only to carbonate

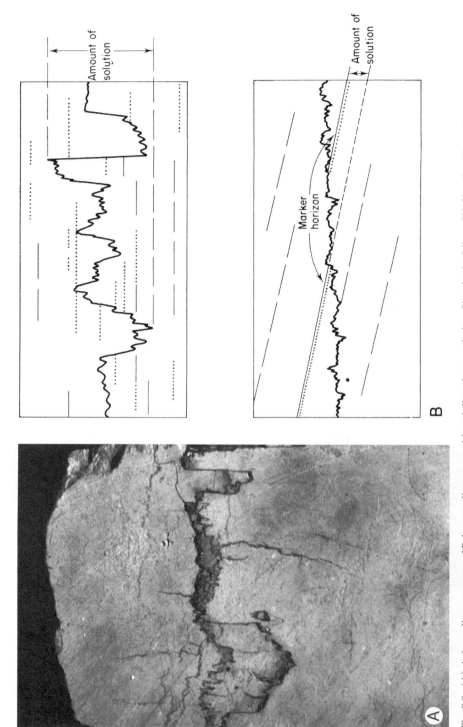

Fig. 9.9. (A) A 6-cm-diameter core of Paleocene limestone with stylolites due to solution. Sirte basin, Libya. (B) Sketches illustrating how to measure the amount of vertical displacement due to stylolitic solution.

sands and reefs, which are highly permeable when first deposited. Lime muds, which though initially porous are of low permeability, follow different diagenetic pathways.

9.2.5.3 Diagenesis of Lime Muds

The diagenesis and petrophysics of lime muds are far simpler than for calcarenites. The main reason for this is that though lime muds are often as porous as calcarenites, they are far less permeable due to the small size of the throat passages. They are thus not nearly as susceptible to flushing by fluids of diverse chemistry. An important distinction must be made at the outset between lime muds made of aragonite and those made of calcite.

From the Pre-Cambrian until the Cretaceous, lime muds were made almost entirely of aragonite. When buried the aragonite reverts to calcite. As already seen this reaction results in an 8% increase in volume, and a corresponding decrease in porosity (Note, this is not an 8% decrease in porosity, but an increase of 8% by volume of rock, i.e., an aragonite mud with 40% porosity and 60% grains would lose 48% porosity by the aragonite:calcite reversion.) The rearrangement of crystals, coupled with compaction, however, generally leads to a total loss of porosity and permeability. Thus it is a matter of observation that most lime mudstones are hard, tight splintery rocks of negligible porosity and permeability. They can only become reservoirs if secondary porosity has been generated by fracturing, dolomitization, or solution.

An exception to this general rule is provided by a particular type of lime mud known as **chalk.** Chalk is a fine-grained limestone composed largely of coccoliths, the calcitic plates of coccospheres (Fig. 9.10). These are skeletal remains of a group of nannoplanktonic golden-brown algae (Black, 1953). Coccolithic limestones first became important toward the end of the Jurassic Period. By the middle of the Cretaceous Period vast quantities of coccolith muds began to be deposited across the continental shelves of the ocean basins of the world. These gave rise to the Chalk Group of northwest Europe (Hakansson et al., 1974; Hancock, 1998) and its worldwide equivalents, such as the Austin Chalk of Texas (Ager, 1993). Chalk is deposited with porosities and permeabilities similar to those of aragonitic lime muds. Because chalks are of calcitic composition, however, they do not undergo the early diagenetic recrystallization of aragonite to calcite. Thus chalks generally remain as "chalky" friable rock, retaining porosities on the order of 20–30%. They have considerable storage capacity and may serve as aquifers or petroleum reservoirs (Scholle, 1977; Scholle et al., 1983). To actually yield their contained fluids, chalk must have permeability. This may occur due to fracturing. Alternatively, chalks can still retain some original intergranular permeability. This is best preserved when compaction has been inhibited by abnormally high pore pressures (relieving the stress at grain contacts) or petroleum invasion, as was noted earlier when discussing sand reservoirs. These conditions are responsible for the productivity of the chalk reservoirs of the Ekofisk group of fields in the Norwegian sector of the North Sea (Byrd, 1975; Heur, 1980). Here the Cretaceous chalk has been fractured over salt domes. Simultaneously, rapid burial and high heat flow favored overpressuring and hydrocarbon generation. Early diagenesis of chalks is minimal, but worthy of comment. Horizons which show evidence of early cementation are widespread stratigraphically, though not volumetrically important. These penecontemporaneously cemented layers are termed "hardgrounds."

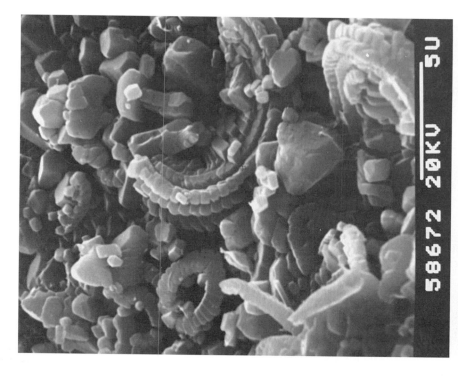

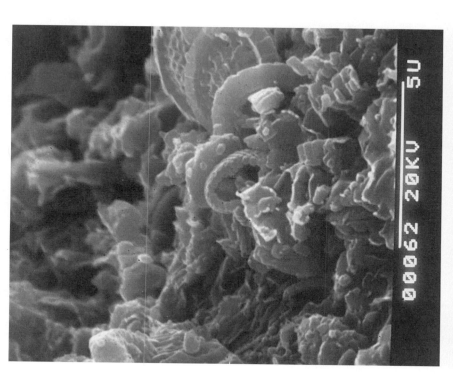

Fig. 9.10. Scanning electron micrographs of calcitic chalk showing coccolithic rings and skeletal fragments of algal nannoplankton. Unlike aragonitic lime mud, chalk does not undergo neomorphism, so porosity may be preserved, though permeability is low, due to fine particle size. (Left) Lower Campanian chalk, Fannin County, Texas. (Right) Coniacian chalk, Kiplingcotes Station quarry, Yorkshire. (Courtesy of J. M. Hancock.)

Evidence for early cementation is provided by the observation that the upper part of the beds is often extensively affected by the activities of boring organisms, or colonized by organisms such as oysters, which required a rigid substrate on which to attach themselves. These bored surfaces are frequently immediately overlain by thin intraformational conglomerates of chalk, often phosphatic and glauconitic, in a marly matrix (Fig. 9.11). Examples of hardground horizons of great lateral continuity, have been

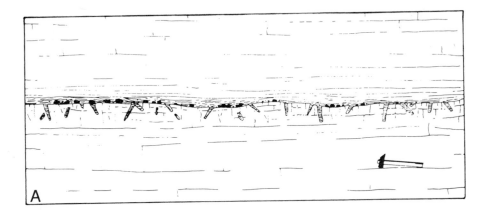

Fig. 9.11. Examples of "hardground" surfaces in limestone sequences indicating penecontemporaneous cementation, erosion, and renewed deposition. (A) Field sketch of hardground in Eocene chalk, Jordan. A pure micrite is abruptly overlain by an argillaceous micrite. The contact is bored and immediately overlain by an intraformational conglomerate of phosphatized clasts with some glauconite. (B) Photograph of "hardground" in Inferior Oolite (Middle Jurassic) limestone, Dorset. Note the borings penetrating the limestone beneath the erosion surface (marked by the base of the hip flask). Note the limestone pebbles and large disarticulated shell fragments immediately above the contact.

described from the Upper Cretaceus chalk of northwest Europe (Jefferies, 1963; Bromley, 1967).

These chalk hardgrounds act as permeability barriers that seal off the underlying sediments, rendering them an essentially closed system, inhibiting the migration of connate fluids and hence cement precipitation. Diagenesis may be restricted to pressure solution at coccolith contacts (Scholle *et al.*, 1998).

Thus hardgrounds may play a role in preserving porosity by restricting the extent of early cementation in the underlying sediment. The presence of hardgrounds within a limestone formation that is otherwise porous and permeable may cause vertical permeability barriers within the succession. As such they may influence the migration and entrapment of oil and affect the reservoir engineering characteristics of the formation as a whole.

Figure 9.12 summarizes the diagenetic pathways of lime muds, showing the important control exerted on porosity and permeability by the original mineralogy and also the beneficial effect of fracturing.

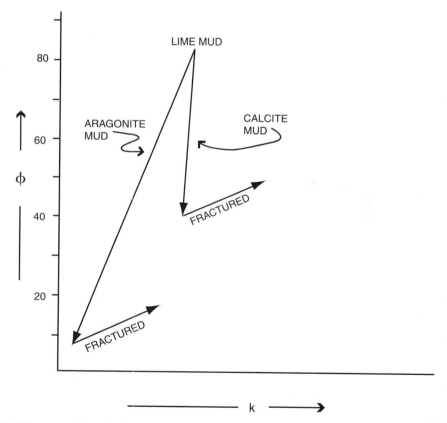

Fig. 9.12. Diagrammatic graph of permeability versus porosity showing the diagenetic pathways for aragonitic and calcitic lime muds. Both start with the same high porosities, but aragonitic muds lose porosity rapidly during burial as they change to calcite, with an 8% volumetric expansion. Calcite mud loses some porosity due to burial compaction. Neither has reservoir potential in the subsurface unless fractured, then calcitic chalk, with its retained porosity, may serve well.

9.2.6 Dolomite

9.2.6.1 Introduction: Chemical Constraints on Dolomite Formation

The term **dolomite** is applied both to the mineral $Ca \cdot Mg(CO_3)^2$ and to the rock of this mineralogical composition. The term **dolostone** is sometimes used for the latter. The exact genesis of dolomite is still a fruitful field for research despite many years of field observation and laboratory research (Morrow, 1982a,b; Wells, 1986; Purser *et al.*, 1994). Figure 9.7 showed that a high ratio of magnesium to calcium is not necessarily a prerequisite for dolomite precipitation. Normal seawater has an Mg:Ca ratio of about 3:1. Dolomite forms in supersaline environment where Mg:Ca ratios exceed this value. It is noteworthy, however, that dolomite may form at the expense of calcite for Mg:Ca ratios less than 1:1 if the salinity is very low (Folk and Land, 1974).

Conditions necessary for dolomite formation appear to include initial permeability within the host sediment, coupled with sufficient pressure differential to permit pore fluid movement, an adequate and continuous supply of magnesium ions, and a fluid which is undersaturated with respect to calcium ions. The last two of these conditions seem to be fulfilled in the so-called Dorag model of Badiozamani (1973). Seawater and freshwater may both be saturated with respect to dolomite and calcite, but mixtures with between 5 and 50% seawater are undersaturated with respect to calcite and supersaturated with dolomite (Fig. 9.13). This suggests that dolomitization may be expected where marine and fresh waters mix.

Having now examined the theoretical constraints for dolomite formation it is pertinent to examine the observational evidence for dolomite formation. It has long been known that there are two main types of dolomite, primary or syngenetic, and secondary or diagenetic. These are considered in turn.

9.2.6.2 Primary Dolomites

Primary dolomites are defined as those which formed at the time of deposition. There is discussion as to whether genuine direct precipitation of dolomite occurs, or as to whether it is in fact a replacement of previously formed minerals, that is, penecontemporaneous rather than strictly primary. Recent dolomite deposits have been described from many arid hypersaline coasts (termed "sabkha" from the Arabic for salt marsh), and from warm humid coasts too (Budd, 1997). In some examples it is believed that the dolomite is a direct precipitate within the pore spaces of aragonite mud. Friedman (1979) has described such a case from marginal pools of the Red Sea, and von der Borch (1976) has cited another from the Coorong Lagoon of Australia. Figure 9.14 illustrates a third. In other instances, however, the modern dolomite has been interpreted as a replacement of preexisting aragonite or calcite; see, for example, the accounts by Butler (1969) and McKenzie (1981) of the sabkha dolomite of Abu Dhabi.

In all of these cases there is general agreement that the dolomites are primary, or penecontemporaneous. These modern examples are all micritic and cryptocrystalline with a grain size of less than 1–20 μm. Petrophysically they are like chalk, porous, but of low permeability. Analogs of these Recent primary dolomites occur in ancient carbonate sequences. They too are characterized by a cryptocrystalline texture and low permeability. Evidence for their primary origin is provided by their bedded concordant nature, as opposed to the irregular discordant occurrence of secondary dolomites. They

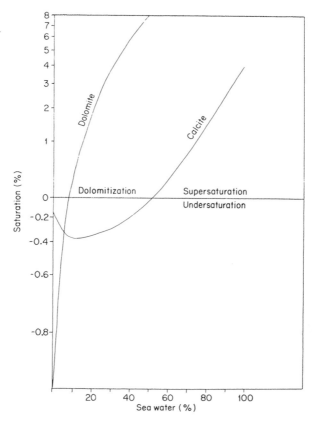

Fig. 9.13. Graph of saturation curves for calcite and dolomite for various mixes of seawater and freshwater. Dolomitization occurs where the percentage of seawater varies between 5 and 50%. (After Badiozamani, 1973.)

Fig. 9.14. Air photo of Ambergris Cay, Belize. Modern dolomite is forming in supratidal flat muds behind the barrier reef. (Courtesy of E. Purdy.)

occur in sabkha facies, interbedded with fecal pellet muds, stromatolitic algal limestones, and evaporites. Associated sedimentary structures include desiccation cracks and "tepee" structures (the wigwam-like buckling of bedding due to penecontemporaneous hydration and expansion of anhydrite; see, for example, Kendall, 1969).

9.2.6.3 Secondary Dolomites

Secondary dolomites are defined as those that are obviously of postdepositional origin. This is clearly shown by the way in which such dolomites have an irregular distribution, discordant to bedding and cross-cutting sedimentary structures. Unlike primary dolomite this type has crystals of more than 20 μm in diameter, which are occasionally euhedral or idiomorphic and cross-cut relic microfabrics of the original limestone (Plate 1C). These secondary dolomites have a characteristic sugary texture (sometimes referred to as sucrosic or saccharoidal). This has resulted in part from the bulk volume shrinkage (as calcite is replaced by dolomite) and in part from the dissolution of residual calcite during the final stages of dolomitization. Thus secondary dolomites are frequently porous with intercrystalline pores connected to one another by planar throat passages. Unlike primary dolomites these secondary dolomites can act as excellent hydrocarbon reservoirs.

Several models for secondary dolomitization have been proposed, and there are supporters and critics of each. Most of the models are based on the idea that dolomitization takes place when brines of high Mg:Ca ratio flow through permeable limestone. The reaction is a straightforward replacement according to the formula:

$$2CaCO_3 + Mg^{2+} \rightleftharpoons CaMg(CO_3) + Ca^{2+}.$$

Note that this is a reversible reaction — dedolomitization (or calcitization) is also known to occur (Shearman et al., 1961). One of the most popular models for secondary dolomitization by magnesium enriched brine is the "seepage reflux" or "leaky dam" mechanism. Consider a lagoon or restricted embayment in an arid climate. Seawater flows into the lagoon. Calcium ions are removed from the lagoonal water, both by the secretion of lime by organisms and, as salinity increases, by the precipitation of gypsum. Thus the Mg:Ca ratio increases. Continued evaporation may increase the density of the water until it begins to flow downward and seaward through the permeable reef or carbonate sand barrier. The magnesium-enriched brines dolomitize the limestones as they pass through them (Fig. 9.15). The Bonaire Lagoon in the Antilles has been cited as a modern example of this process (Deffayes et al., 1965), and the Stettler Formation of the Williston basin, Canada, cited later as an ancient analog (see Section 9.6.3). Nonetheless, as with every proposed model of dolomite formation, the "leaky dam" scenario has its critics (e.g., Wells, 1986; Budd, 1997).

The one fact that should not be lost sight of in all the foregoing is that primary dolomites tend to be microcrystalline and porous, but impermeable. Secondary dolomites, on the other hand, tend to be coarsely crystalline, with good intercrystalline porosity. Because of their coarse crystals they have large pores and pore throats, so they may be highly permeable. Primary dolomites do not make good petroleum reservoirs, without leaching and/or fracturing, whereas secondary dolomites do make excellent reservoirs.

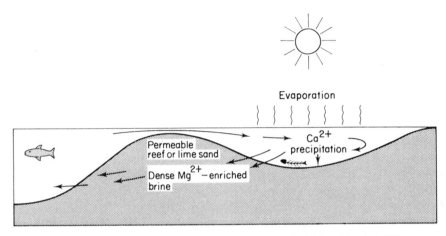

Fig. 9.15. Illustration of the "leaky dam" or "seepage reflux" model of dolomitization. This argues that, in an arid climate, the Mg:Ca ratio increases in lagoonal waters as calcium is depleted by organic secretion of lime and the inorganic precipitation of gypsum. As salinity increases, the dense magnesium-enriched brine seeps downward and seaward through the permeable barrier limestones, replacing them with dolomite. Like every model of dolomitization, the "leaky dam" theory has its critics.

9.2.7 Diagenesis and Porosity Evolution of Carbonates: Summary

Finally, it is relevant to enquire how all this work on limestones and dolomites aids in the prediction of porosity distribution in the subsurface. Figure 9.16 shows some carbonate burial curves. It is interesting to compare these with those for clays and sandstones previously presented (Figs. 8.7 and 8.17). Carbonate burial curves are of little help, because these rocks characteristically show rapid vertical and lateral variations in reservoir quality (Reeckmann and Friedman, 1982). Feazel and Schatzinger (1985) have discussed the factors that govern carbonate porosity. These include minimal burial, reduced burial stress (by overpressure), rigid framework preventing compaction (as in a reef, or a carbonate bank that undergoes early but minor cementation), stable mineralogy (such as a high ratio of calcite to aragonite), permeability barriers to inhibit fluid movement (such as hardgrounds), the solution of temporarily filled pores (as for example by halite), and finally the presence of petroleum.

The foregoing list of complex variables explains why some limestones are porous and others are not. But it does little to help predict porosity distribution. In some carbonates porosity is still largely primary. In such cases, therefore, it may help to interpret depositional environments so as to predict facies trends. With increasing diagenesis, however, limestones become, in turn, cemented, and then leached, with secondary porosity that is often unrelated to facies (Fig. 9.17).

From Levorsen (1934) onward, unconformities have been recognized as an important factor in controlling porosity development. Secondary solution porosity in subunconformity sands was considered in some detail in the earlier section on sandstone diagenesis. In carbonates, with their greater solubility than quartz, subunconformity secondary porosity is even more extensively developed (Fig. 9.18). Carbonate sedimentologists have thus embraced sequence stratigraphic concepts with the fervor of neophytes (Loucks and Sarg, 1993). Sea level low stands allow the ingress of meteoric water into

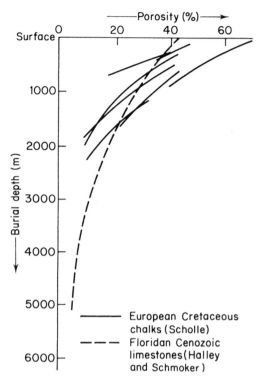

Fig. 9.16. Carbonate burial curves. Compiled from data in Scholle *et al.,* 1983, and Halley and Schmoker (1984).

Relationship between facies, diagenesis and φ in carbonate reservoirs

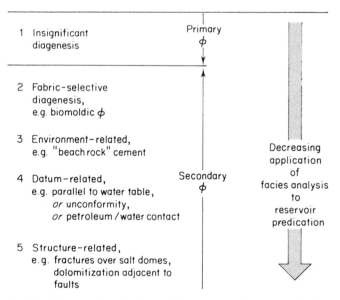

Fig. 9.17. Diagram showing that, as carbonate diagenesis increases, environmental interpretation diminishes in relevance as a tool for predicting porosity.

Fig. 9.18. Photograph of sequence boundary between Portland Limestone skeletal oolite and overlying Purbeck algal laminite (both late Jurassic), Isle of Portland, Dorset, England. Note biomoldic porosity beneath the contact. Adjacent to this locality fossil trees occur in growth position on the Portland oolite, testifying to the penecontemporaneous emergence and meteoric flushing of the oolite shoals.

emergent carbonate shelf sediments. This generates extensive secondary porosity and, in appropriate situations, causes dolomitization in the exposed rocks. High stands allow the porous strata to be buried by a transgressive systems tract, hopefully permitting the burial and preservation of porosity beneath the unconformity (Plate 8).

Complexities occur where the unconformity cross-cuts different carbonate facies, which may have undergone distinctive early burial diagenesis. For instance, secondary solution porosity occurs beneath three major Cretaceous unconformities that truncate carbonates on the Arabian shield. These unconformities are a major factor in the migration and entrapment of petroleum in this region (Harris *et al.,* 1984).

Studies of modern limestone caves reveal additional complexities in pore system development. In modern limestones groundwater percolates down toward the water table enlarging vertical joints by solution. Below the water table, however, water flows sub-horizontally, and thus tends to enlarge pore systems parallel to the bedding planes. Thus a petroleum reservoir may have an upper zone in which vertical permeability exceeds horizontal permeability, and a lower zone in which the situation is reversed. These two zones may cross-cut stratigraphy and the petroleum:water contact (Fig. 9.19). This is a challenging situation for a reservoir engineer to grasp. As limestone dissolution continues, extensive cave development may give rise to karst topography. Collapse of the caves forms breccias which may serve as petroleum reservoirs, as in the Auk field of the North Sea, and the Casablanca field of offshore Spain (Brennand and Van Veen, 1975, and Watson, 1982, respectively).

The tool for locating sequence boundaries and their associated porosity is provided

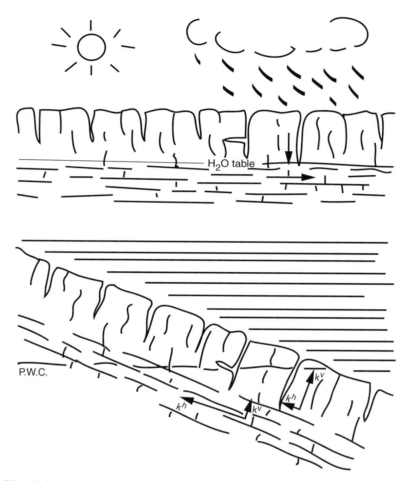

Fig. 9.19. (Upper) Sketch of modern limestone weathering profile showing vertical enlargement of pores above the water table, and horizontal enlargement beneath. With increasing dissolution cavernous porosity may form, and progress to the development of karstic topography, followed by the formation of collapse breccias. (Lower) sketch of unconformity-truncated tilted carbonate petroleum reservoir. Note that there is an upper zone within which the vertical permeability (K_v) is greater than the horizontal permeability (K_h). In the lower zone the situation is reversed ($K_h > K_v$). The boundary between these two zones, the paleo-water table, cross cuts stratigraphy and the petroleum:water contact. This situation produces quite a challenge for petroleum reservoir engineers.

by seismic geophysical surveys, a topic beyond the scope of this text. Suffice for now to note that acoustic velocity varies with porosity. Thus in a basin infilled with 100% pure limestone, seismically measured variations in velocity indicate variations in porosity. In the real world, however, such a situation is rare. Minor impurities in the limestone, such as clay, sand, dolomite, or evaporites will influence velocity and complicate the velocity/porosity conversion. Once velocity is calibrated with lithology data from a well, however, it may be possible to map porosity variations across the basin.

9.3 COAL

9.3.1 Introduction

There is no doubt that coal is of vegetable origin because coals not only contain recognizable plant remains, but transitions can be found between obvious accumulations of vegetable matter, for example, peat, through lignites or brown coals, into true coals and on into anthracite. This series, from peat, through lignite, into the humic coals and finally anthracite is called the "coal series." The position of a coal in the series is termed its "rank." Thus lignite is a very low-rank coal, while at the other extreme, anthracite is a very high-rank coal. Physical appearance, physical properties, and chemical composition of coals change with rank as do their utilization characteristics, such as calorific value, coking properties, and gas generation potential (Gayer and Harris, 1996).

Thus knowledge of the rank of a coal can be a guide to its utilization. The changes that vegetable matter undergo in the course of alteration to coal are termed "maturation" or "coalification." Maturation takes place in two stages: the peat stage and the burial stage. In the peat stage the plant material suffers a measure of biochemical degradation, and when it is buried, progressive increase in both overburden load and temperature bring about dynamothermal maturation that slowly turns the peat into coal. The peat stage is an essential prerequisite for the formation of coal. Under normal circumstances when plants die, they are exposed to air and are broken down primarily by oxidation and also by various organisms, particularly the fungi and aerobic bacteria. Where plant remains accumulate in swamp or bog environments, however, they become water saturated. Aerobic decay soon depletes the water of oxygen, the aerobic organisms die off and anaerobic bacteria take over. The anaerobic bacteria operate without oxygen but they are equally as capable of breaking down organic matter as the aerobic forms. Because of the stagnant nature of swamps and bogs, however, the waste products of the bacteria are not flushed away, but build up in interstitial waters and ultimately render the environment sterile. Bacterial activity is thus curtailed and the partially decomposed plant material remains in a state of arrested decay. In this state the material is peat. If the peat is drained the toxic materials are flushed out, decomposition sets in again, and the peat may ultimately be destroyed. If the peat is not drained, however, but is buried under relatively impermeable sediments, its geological preservation becomes possible.

9.3.2 Coal Petrography

Coal is a mature Type III variety of kerogen (see Section 7.3.2). Chemically, coals comprise the three elements carbon, hydrogen, and oxygen with minor proportions of sulfur and nitrogen and mineral impurities. The latter remain as ash after the coal has been burned and, clearly, high ash content is undesirable. Inherent sulfur is normally present in a very small amount and was probably derived from sulfur proteins in the original plants. Some coals have a high sulfur content due to the presence of disseminated pyrites: this is deleterious because it generates sulfurous fumes on combustion of the coal. Although the nitrogen content of coal is small, it was economically important in the production of ammonia as a by-product of the coal-gas industry.

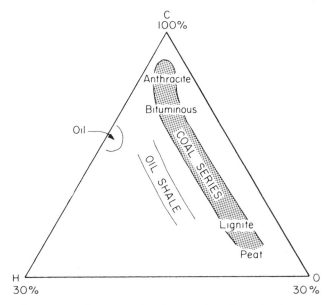

Fig. 9.20. Triangular diagram showing the coal series and the relationship between coal, oil shale, and crude oil.

Dynamothermal diagenesis of coal on burial is expressed by the changes in carbon, hydrogen and oxygen contents that accompany maturation. If the elemental composition of coal is plotted in terms of these three elements (Fig. 9.20) it will be seen that the coals lie within a narrow zone, called the coal belt. Carbon content increases progressively with increase in rank, but the proportion of hydrogen remains fairly constant, at between 5 and 7% in the humic coals, before it falls rapidly in the semianthracites and anthracites. Oxygen decreases with increase in rank.

The chemical changes that accompany maturation or coalification are not well understood but they no doubt involve production of gaseous carbon dioxide and methane. Some occurrences of natural gas, for example, that in the Permian Rotliegendes of Holland and the southern parts of the North Sea, appear to have been derived from devolatilization of underlying Westphalian coals (Patijn, 1964).

In general terms the rank of coals tends to increase with age, in the sense that most lignites or brown coals are Tertiary or Mesozoic in age, whereas the Upper Paleozoic occurrences are true coals. Age is only coincidental, however, and the prime control appears to be thermal, which is generally related to depth of burial. In any normal vertical succession of humic coals the carbon content, that is, the rank, increases with depth. This relationship is called Hilt's law (Hilt, 1873), and it is likely that increase in temperature with burial is an important factor in maturation. There is debate about the causes of the change from coal into anthracite. In some instances it may be the natural continuation of thermal maturation, but in many occurrences coals change laterally into anthracites as they approach zones of tectonic deformation. The latter relationship is found in the South Wales coal field where the coals pass westward into anthracite as they enter the zone of shearing associated with the Ammonford compression. Similarly in North America the Pennsylvanian coals pass laterally into anthracites as they enter the Appalachian fold belt.

Coals may be devolatilized by thermal metamorphism where they are cut by igneous intrusions. Examples of this are seen where Tertiary dykes cut Westphalian coal measures in northeast England. Although the physical appearance of the coals is only affected for a few meters on either side of the intrusions, devolatilization as expressed by increase in carbon content spreads out regionally.

Most brown coals or lignites are obviously woody, lusterless, and not well jointed; humic coals on the other hand do not normally show a woody appearance in hand specimen. They are well bedded and display various degrees of luster from bright to dull. On the basis of luster and physical appearance, the humic coals are divided into four "rock" types: vitrain, clarain, durain, and fusain (Stopes, 1919). Vitrains are the bright coals which have a vitreous luster. Clarains are less bright with a silky appearance, whereas the durains are dull. Fusain is sooty black and very friable so that, unlike vitrain, clarain, and durain, it readily soils the hands like charcoal.

It is important to distinguish between coal type and rank. Coal type is the expression of the composition of the original plant materials, whereas rank is a measure of the extent of maturation or diagenesis that the vegetable matter has undergone. Some coal seams consist mainly of one rock type, but many are composite and comprise alternations of layers of vitrain, clarain, durain, and fusain of varying thicknesses. Coal normally breaks in three directions into roughly prismatic blocks. Two of the surfaces are joints and are termed "cleat" and "end," respectively. The cleat is the more strongly developed joint, and in the days of manual extraction its direction often determined the layout of the underground workings. The third surface is parallel to bedding and usually coincides with a fusain layer. It is these bedding surfaces of fusain on a block of coal that soil the fingers.

As with most other rocks, the "rock" types of coal are each composed of a series of discrete components, termed "macerals" (the prefix "mac-" implying that they are macerated plant remains, and the termination "-erals" indicating that they are analogous to minerals). Macerals are named with the suffix "-inite" to distinguish them from the rock type of coal, termination "-ain." The macerals were all plant tissues or plant degradation products that were modified chemically by diagenetic processes, that is, by maturation. The common ones include vitrinite, the main constituent of the coal type vitrain, of which there are two varieties — tellinite and collinite. Under the microscope, tellinite shows a compressed cellular structure of what was formerly woody tissue or xylem, whereas collinite which has similar optical properties is structureless. Suberinite, cutinite, and exinite are macerals formed by coalification of bark, leaf cuticles, and spore jackets, respectively. Fusinite stands apart from other macerals in that it is nearly pure carbon. It often shows cellular structure under the microscope. The cell walls are usually preserved as a mass of broken fragments, "bogenstrukture." This contrasts markedly with cellular tissues that are preserved as tellinite, where the cell walls have been deformed but not broken by compaction. It appears therefore that where cellular tissues are preserved as fusinite, the alteration must have taken place very early in the maturation history, certainly before significant compaction due to overburden load. It has been suggested that fusinite may have formed during the peat stage by a process analogous to "dry rot" or by forest fires. "Micrinite" is the general term applied to aggregates of very fine-grained macerated plant materials most of which are too small to be identified. The coal type vitrain is composed mainly of the maceral vitrinite and with an increase in the proportion of other macerals, vitrain grades into clarains and durains. The durains are characterized by micrinite and/or exinite, and some of the grey durains are very rich in exinite.

The degree of maturation of these macerals, vitrinite and exinite, can be assessed optically under the microscope by change in their color under ordinary transmitted light. In the humic coals, vitrinite changes from translucent yellow, through orange and red to deep red with increase in rank of coal. Exinite retains shades of yellow throughout much of the range of humic coals and only darken to orange and red with increase in rank of coal. Exinite retains shades of yellow throughout much of the range of humic coals and only darkens to orange and red in the higher ranks, that is, semi-anthracite. The color transmission of the macerals is affected by thickness of the thin section and determinations have to be made on very thin sections, with strict control of thickness. Vitrinite can also be studied by reflected light and, using the techniques of ore microscopy, its reflectance can be measured. The reflectance of vitrinite changes with increase in rank, and can be used as an index of maturation. Spores and scraps of vitrinite are present in minor traces in many sediments other than coal, and where they occur determination of their degree of maturation can provide a guide to the thermal history of the rocks that contain them. This knowledge is important in the subsurface exploration of potential oil-bearing formation because maturation of liquid hydrocarbon runs parallel to that of coal.

9.3.3 Environments of Coal Deposition

Most coal seams appear to have formed in place. That is to say, the coal-forming peat accumulated where the plants lived and died. Some coal, however, appears to be of "drift" origin, and the plant remains were transported, such as log rafts, to the site of deposition. An in-place origin is demonstrated by the presence of roots and rootlets that lead down from the coal into an underlying fossil soil or "seat earth" (refer back to Fig. 2.6). Seat earths are of interest in their own right, by virtue of their refractory properties. Clayey seat earths, termed "fire clays," consist mainly of kaolin minerals (disordered kaolinite) and are of value for manufacture of refractory bricks. Sandy seat earths, "ganisters," are almost pure quartz rocks and are used for producing silica fire bricks. There can be little doubt that the seat earths were deposited as normal clays and sands, but that they were leached of alkali metals and reconstituted by humic acids from the overlying peats (see also Section 2.3.1.2).

Certain coals, such as some in the Karro Group (Permo-Carboniferous) of southern Africa, lack any form of seat earth, abruptly overlying shale. It has been argued that such coals may have formed from floating islands of vegetation, analogous to those seen today in the Sud of Sudan, and the Tigris-Euphrates delta at the head of the large gulf between Iran and Arabia.

Great attention has been paid to the environments of deposition of coals and their precursors (Dapples and Hopkins, 1969; Wanless *et al.*, 1969; Ethridge and Flores, 1981; Galloway and Hobday, 1983; Rahman and Flores, 1984; Fielding, 1985; Lyons and Rice, 1986; Merritt and McGee, 1986; Scott, 1987). The physicochemical conditions that fa-

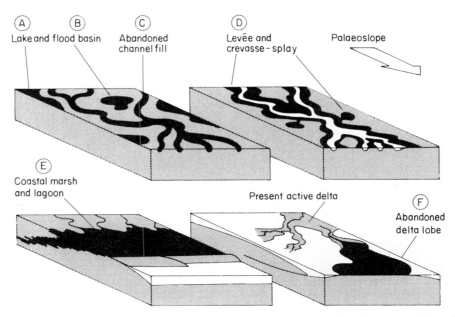

Fig. 9.21. Block diagrams illustrating the diverse depositional environments of coal formation. Examples of A–F are cited in the text. Note that to predict the geometry of an individual coal bed it is important to diagnose the subenvironment, rather than the overall depositional environment of the associated sediments. Channel abandonment coals, for example, occur in fluvial, deltaic, and intertidal settings. Note also that coal-bed geometry is controlled, not only by depositional environment, but also by subsequent erosion, especially channeling.

vor peat formation have already been outlined. These occur in a wide range of environments that include fluvial, lacustrine, deltaic, and barrier beach coasts. It is probably more important to identify the subenvironment of a particular coal rather than that of the overall facies. This is because it is the smaller scale local environment that controls the geometry of an individual coal bed. These geometries are many and varied (Fig. 9.21). To begin with, some coals form in lacustrine environments (Fig. 9.21A). These obviously have an irregular sheet geometry that corresponds to the outline of the lake. The nature of the deposit may vary from an algal sapropelite in the lake center to floating peats and then rooted peats toward the lake shoreline (see Section 6.3.2.4.2). With diminishing size lakes grade into flood basins in low-lying parts of alluvial plains. These waterlogged environments are also suitable for peat formation (Fig. 9.21B). Peat formation is associated with channels in several ways. Some channels are abandoned and then become sites for peat formation. Such deposits have irregular shapes and, though they may meander, will essentially be elongated down the paleoslope (Fig. 9.21C). Tributary systems will occur on the alluvial plain, and distributary system on the delta plain. In the Upper Carboniferous (Pennsylvanian) coal basin of Illinois channel-fill deposits can be traced for distances of 200 km (Trask and Palmer, 1986). An alternative to this situation is found where sand-filled channels meander down the middle of ribbons of levee coals that thin away from the channel margin (Fig. 9.21D). These are where peat formation took place on raised ground between flood basins and channels.

Some of the largest lignite deposits in the world occur in the Cretaceous basins of the

Fig. 9.22. A modern example of coal formation in setting E in Fig. 9.21. Bed of peat forming in coastal salt marsh, drained by tidal channels, Poole Harbour, Dorset, England.

Rocky Mountains that extend from Alaska down to Colorado (Kent, 1986; Merritt and McGee, 1986). These have a broadly similar mode of formation. They developed in lagoons and marshes sheltered from the sea by barrier island sands (Fig. 9.21E). As the shoreline regressed these peat deposits prograded over the coastal barrier sand. This resulted in the deposition of lignites tens of meters thick. In the Powder River and Wind River basins of Wyoming and Utah individual formations are some 15 km wide and can be traced for some 40 km along the paleoshoreline. Modern analogs of this habitat occur today along the Nigerian coast to the east of Lagos. A more modest example is illustrated in Fig. 9.22.

Finally, blanket peats may form over whole abandoned delta lobes (Fig. 9.21F). Elliott (1975) has documented such delta abandonment coals in the Yoredale (Carboniferous) deltaic deposits of northern England. Individual coals may be traced for lateral distances of 10 km. The coals have subordinate rootlet horizons that discordantly overlie the deposits of many different on-delta sedimentary environments. The geometry and extent of delta abandonment coals is only limited by the geometry and dimensions of the delta. Note, however, that delta abandonment coals, like all others, are susceptible to localized erosion by subsequent channel downcutting.

This brief review of the environments of coal formation shows how the analysis of their environments of deposition may be used to predict their scale and geometry.

9.4 SEDIMENTARY IRON ORES

The sedimentary iron ores are an economically important group of sedimentary rocks that have been studied in great detail (Clemmey, 1985; Cox and Singer, 1986; Young and

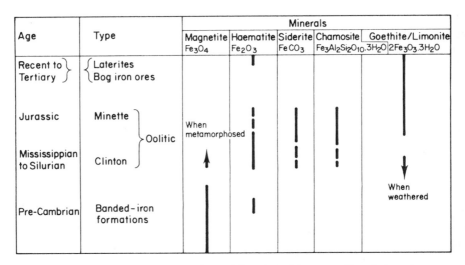

Fig. 9.23. Table illustrating the mineralogy and age of the principal iron ores. Note that any iron ore may be metamorphosed to magnetite or weathered to limonite.

Taylor, 1989). There are four major groups of sedimentary iron ores. They appear to have developed in a consistent time series, as shown both by their mineralogy and by their environments of formation (Fig. 9.23).

Hydrated iron oxides characterize the lateritic deposits that are known in modern and Cenozoic weathering profiles. These need not be discussed here because they were dealt with in Chapter 2. The bog iron ores are also relatively young and consist of limonite. They form in peat bogs and lacustrine environments by the oxidation of ferrous bicarbonate, $Fe(HCO_3)^2$. Siderite ores, both bedded and concretionary, are characteristic of deltaic mudstone environments. Though no longer economically important, the Wealden ores of southeastern England and the Carboniferous iron ores of south Wales are of this type.

The two other major sedimentary iron deposits are the oolitic and banded iron ore formations. These are described next in some detail.

9.4.1 Oolitic Iron Ores

Two main periods of oolitic iron formation occurred. One period ranged from the Silurian to the Early Carboniferous (Mississippian). These ores are loosely referred to as of Clinton type, and include the iron deposits of Birmingham (Alabama), Wabana (Newfoundland), Wadi Shatti (Libya), and Gara Djebilet (Algeria).

The second main phase of oolitic iron ore formation was in the Jurassic Period. These include the **minette** ores of Chile, Lorraine (France), and Northampton (England). The Clinton and "minette" oolitic ores share a number of common features. The iron minerals include the iron carbonate siderite, the iron silicate chamosite, and the oxides goethite and hematite. The deposits contain a marine fauna, and range from sparite-cemented hematite grainstones to chamosite-siderite wackestones (Plate 5B).

Sedimentological studies suggest environments of deposition that range from freshwater lagoon to marine embayment and restricted inner shelf. It has been noted that the two main periods of oolitic iron formation are episodes when the world's oceans were largely anoxic (Van Houten and Arthur, 1989). There is general agreement that the iron was brought into the marine realm by rivers that drained a lateritized hinterland. There is, however, considerable debate as to the exact mode of ore formation. The transformist/metasomatic school argues that primary aragonitic ooids were later replaced by the iron minerals (Sorby, 1856; Kimberley, 1979). The syngenetic school argues for primary deposition of iron minerals (Taylor, 1949; Hallam, 1963; Bubinicek, 1971).

In support of the metasomatic school it should be noted that many hematic ores occur beneath unconformities, such as the Forest of Dean deposits (England), and the Shatti Valley ores of Libya (Goudarzi, 1971; Turk et al., 1980). This could reflect epidiagenetic replacement of limestones by ferruginous groundwaters. Furthermore, in both the Lorraine and Gara Djebilet deposits there are Nubecularid foraminifera, whose tests are normally cryptocrystalline calcite, but now are replaced by iron, and whose chambers are infilled with hematite and apatite (Champetier et al., 1985).

In support of the primary precipitate theory it should be noted that there are considerable differences between the ooids of iron ores and those of limestones (Section 9.2.3.1). Ferruginous ooids are often not spherical, but flattened ("flax seed" ore of the Clinton deposit). Many ore bodies are not high-energy, cross-bedded grainstone shoal deposits, like most oolitic limestones. They tend to be poorly sorted wackestones in which both grains and matrix are mineralized. Evidence of primary deposition includes the occurrence of iron ore intraclasts and ooids in adjacent and overlying normal sediments, the occurrence of unreplaced calcite shells within ferruginous material, and the fact that the marine fauna are occasionally stunted, suggesting a hostile environment (Fig. 9.24). There seems no simple answer to the origin of oolitic iron ores. For a thoughtful review of this problem see Hallam (1981, pp. 107–110).

9.4.2 Pre-Cambrian Banded Ironstone Formations

Oolitic ores are rare in the Pre-Cambrian, but are replaced in economic importance by the banded ironstone formations, often referred to colloquially as BIFs. Pre-Cambrian banded ironstones are known on all the continents (Fig. 10.8). They contain the largest

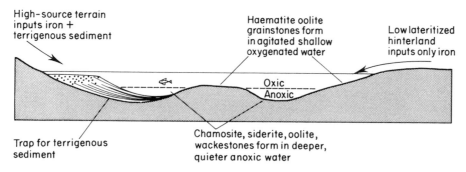

Fig. 9.24. Illustration of the environments of deposition of oolitic iron ores. Note that terrigenous sediment may be absent from environments of iron formation either because of low-lying hinterland (right) or because sand and mud may be rapidly deposited close to the coast (left). For sources of the ideas on which this figure is based, see Hallam (1981).

and richest iron deposits in the world. Notable examples include the Animikie Series of the Lake Superior region, North America, the Hammersley basin of Western Australia, and the Transvaal basin of South Africa. There is naturally a huge literature on these deposits. The following account draws on publications by Trendall (1968), Gross (1972), Goodwin (1973), UNESCO (1973), and Mel'nik (1982).

The Pre-Cambrian banded ironstones are remarkably similar wherever they occur in the world. They all formed between 1900 and 2500 million years ago. The deposits are found in laterally extensive sequences with the following succession from base to top: dolomite, quartzite, red shale, black ferruginous shale, banded iron formation, and black shale. The ore bodies themselves are sometimes several hundred meters thick and consist of rhythmically interbedded chert and iron ore (Fig. 9.25). The ore is a complex

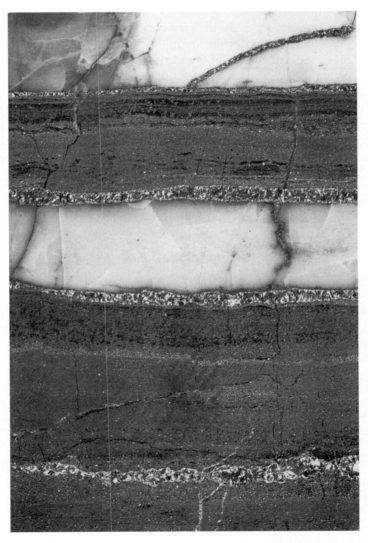

Fig. 9.25. A 5-cm-wide polished slab of Pre-Cambrian banded iron formation (BIF), consisting of interlaminated chert and taconite (the iron ore) from Mount Goldworthy, Western Australia.

mineral, termed "taconite," in which the iron occurs in the reduced ferrous state, though recent weathering may have oxidized it to ferric iron (Guilbert and Park, 1986).Two types of BIF are recognized. The Algoma type is of Archean age and is always related to submarine volcanism. The Superior type is of early Proterozoic age and is sometimes, but not invariably, associated with volcanism.

The origin of these deposits has aroused much speculation. The age and reduced state of the iron minerals is consistent with formation when the earth's atmosphere lacked oxygen (Nunn, 1998), and is shortly before the evolution of algae changed the atmosphere of the earth from reducing to oxic conditions (Holland, 1972).

There are two main theories to explain the formation of banded ironstones: the clastic sedimentary and the volcanogenic sedimentary. The clastic sedimentary theory advocates that the iron was transported into the basins from a deeply weathered lateritic hinterland. Deposition was then by normal sedimentation in a shallow aqueous environment. The rhythmic nature of the deposits may be seasonal or in response to longer alternations of oxidizing and reducing conditions. The volcanogenic sedimentary theory believes that the source for the iron and the silica was the volcanic rocks that are commonly associated with the deposits.

The first theory seems able to explain the Lake Superior type of banded iron ore, which is developed across extensive continental shelves. The volcanogenic theory seems more appropriate for the Algoma type of banded iron ore, which occurs in subductive zones with extensive igneous activity and has generally undergone subsequent metamorphism.

9.5 PHOSPHATES

9.5.1 Mode of Occurrence of Phosphates

The element phosphorus is an essential constituent of all living matter, both plant and animal. Phosphate minerals are, therefore, extensively used as agricultural fertilizers. They are a valuable natural resource that must be located by geologists (Nriagu and Moore, 1985; Cook, 1986; Cook and Shergold, 1986; Slansky, 1986; Notholt *et al.*, 1989; Burnett and Riggs, 1990; Ilyin, 1998). Table 9.8 summarizes the mode of occurrence of

Table 9.8
Main Modes of Occurrence of Economic Phosphates

Type		Mineral	% Total world production
Primary igneous		Apatite	24
Sedimentary	{ Bedded / Placer }	Collophane and apatite	74
Guano		Complex phosphates and nitrates	2
			100

Production percentages from McKelvey (1967).

Fig. 9.26. Phosphatized coprolite (12-mm calibre) from the Chalk (Upper Cretaceous), Beachy Head, south-ern England.

economic phosphates. This summary shows that, though some phosphates occur in ig-neous rocks as apatite, the majority of economic phosphates are in sedimentary rocks.

The mineralogy of phosphates is complex and obscure because of their tendency to occur as microcrystalline aggregates. Typical phosphate minerals are admixtures of the phosphate radical (PO_4) with calcium, water, and traces of fluoride and uranium. Phos-phates occur in sedimentary rocks as matrix and as nodules, ooliths, pellets and phos-phatized shells, bones, teeth, and coprolites (Fig. 9.26). They also occur as a replacement of limestones (Bentor, 1980).

The bulk of the world's bedded phosphates occur in the famous phosphate belt, which stretches from Syria through the Levant, Sinai, Egypt, Morocco, and into Mauritania.

Phosphates occur interbedded with Upper Cretaceous and Eocene chalks and cherts. There can be little doubt that these phosphates resulted from the upwelling of oceanic currents from Tethys onto the broad continental shelves along its southern shore. Another feature responsible for the accumulation of phosphates at this time is believed to be a warm humid climate. It has been argued that this increased the rate of continental weathering and raised the amount of phosphate transported into the sea by rivers (Follini, 1996).

The Phosphoria Formation (Permian), which contains some of the main phosphate rocks of the USA, occurs in a similar setting. This formation extends across about 260,000 km^2 of Idaho, Wyoming, and Utah. It was deposited on a marine shelf bounded by the Cordilleran trough to the west. The phosphorites occur interbedded with dolomite and chert. Chert and mudstone increase westward into the Cordilleran foredeep. The dolomites grade eastward into red beds and evaporites (Campbell, 1962; McKelvey, 1967).

Pedley and Bennett (1985) have presented a particularly elegant account of phosphatization of the Miocene limestones of Malta in the Mediterranean. They differentiated three zones: a zone of in-place phosphatization passed downcurrent into a zone of phosphatized hardgrounds, with phosphate conglomerates. This zone in turn passed downcurrent into a distal zone of limestones with scattered phosphate nodules (Fig. 9.27). In addition to these bedded-shelf phosphates, it is important to remember the other modes of occurrence of phosphate.

Phosphates have been recorded from the modern sea bed, notably off the western coasts of America and Africa. Detailed studies show the wide range of occurrence of these deposits. Some phosphates have been recorded forming diagenetically within organic-rich diatomaceous oozes of the South West African shelf (Baturin, 1970). On the South African shelf phosphatized Miocene limestones are eroded to form phosphate gravel placers (Parker and Siesser, 1972). Further north, off the West African

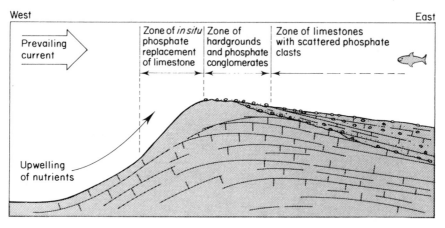

Fig. 9.27. Illustration of the mode of formation of phosphates by penecontemporaneous replacement of Maltese Miocene platform limestones. (Simplified from Pedley, H. M., and Bennett, S. M. 1985. Phosphorites, hardgrounds, and syndepositional solution subsidence. *Sediment. Geol.* **45**, 1–34. Copyright 1985, with permission from Elsevier Science.)

shelf, phosphate placers occur that derived from Eocene Moroccan and Pliocene Saharan phosphorites (Tooms et al., 1971).

Detrital phosphate gravels not only occur on marine shelves, but also as alluvial gravels. The "river pebble" deposits of South Carolina and Florida are a case in point.

Guano is a deposit rich in phosphates and nitrates formed from the excreta of sea birds and bats (Plate 5C). Guano has been a significant source of these minerals notably from the Chilean coasts and on Pacific islands such as Nauru, where solutions rich in phosphates have percolated down from guano to replace reefal limestones.

These examples illustrate three characteristic features of phosphate minerals: their genesis by replacement of carbonates, their ability to be eroded and reworked, and their occurrence on continental shelves.

9.5.2 Mode of Formation of Phosphates

Seawater is generally nearly saturated with phosphate ions, ranging from 0 to 3 ppm PO_4 in deep cold water to about 0.01 ppm in warm surface water. The solubility of phosphate decreases with increasing temperature and increasing pH. These changes occur, and phosphates thus tend to be precipitated, where deep cold oceanic water wells up into shallower warmer waters. These conditions are fulfilled in several situations (Fleming, 1957). The most significant locus at the present time appears to be along the western coasts of South America and Africa, where the cold currents of the Humbolt and Benguela move northward. Rich in nutrients, including phosphates, these waters generate blooms of phytoplankton, which in turn support shoals of fish and flocks of sea birds. Phosphate removed from the seawater by organisms returns again when they die, and settles on the sea bed within miscellaneous organic matter. Phosphates become concentrated during early compaction of the mud. Constant agitation winnows out the lighter material to leave denser incipient phosphate mud pellets. These continue to become enriched with phosphate, as do bones, teeth, and shell debris. In this manner, bedded phosphate rock — phosphorite — is formed (Fig. 9.28).

Eight factors favor phosphate formation: a broad shallow shelf, an adjacent major ocean, a low latitude (<40°), high organic productivity, shallow marine sedimentation, minimal terrigenous input, a marine transgression, and a suitable environmental trap, such as a bay, estuary or carbonate bank (Brown, 1994).

Cognizant of these factors it is now possible to model episodes of phosphate formation throughout the earth's history, by integrating paleoclimate, paleooceanography, and plate movement. This is done in conjunction with studies of total organic carbon productivity and preservation, in an attempt to locate petroleum source beds (Parrish, 1995).

9.6 EVAPORITES

9.6.1 Introduction

The evaporites include the mineral salts such as anhydrite and halite. As their name implies it was once widely assumed that these rocks form by the evaporation of salt-rich

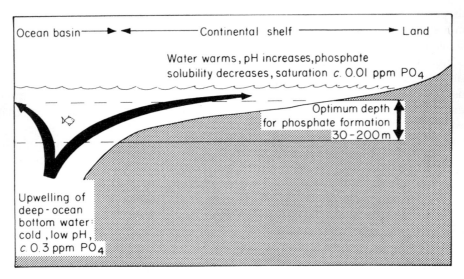

Fig. 9.28. Diagram illustrating a general model for phosphate formation on a marine shelf. (Figures for optimum water depth from Buschinski, 1964.)

fluids. Table 9.9 lists some of the main evaporite minerals. This is a partial documentation, however, because the total number of evaporite minerals is vast.

For many years it was widely accepted that evaporites formed largely by the precipitation or crystallization of salts at the sediment:water interface (e.g., Borchert and Muir, 1964). The replacement textures shown by the microscopic fabrics of evaporites, however, point to extensive diagenetic changes. These are to be expected in view of the chemical instability of the evaporite minerals.

It is now largely accepted, however, that the genesis of evaporites by diagenesis is the rule rather than the exception (Kirkland and Evans, 1973; Kendall, 1984; Sonnenfeld, 1984).

The following account of the evaporite rocks first documents their gross geologic characteristics, then reviews their genesis and concludes with discussion of their economic importance.

Table 9.9
Some of the More Common Evaporite Minerals

Name	Composition	
Anhydrite	$CaSO_4$	
Gypsum	$CaSO_4 \cdot 2H_2O$	Sulfates
Polyhalite	$CaSO_4 \cdot MgSO_4 \cdot K_2SO_4 \cdot 2H_2O$	
Epsomite	$MgSO_4 \cdot nH_2O$	
Halite	$NaCl$	
Sylvite	KCl	Chlorides
Carnallite	$KMgCl_3 \cdot 6H_2O$	
Bischofite	$MgCl_2 \cdot 6H_2O$	

Table 9.10
Major Constituents of Seawater as Weight Percentages of Dissolved Materials

Cations			Anions		
Sodium	Na^+	30·61	Chloride	Cl^-	55·04
Magnesium	Mg^{2+}	3·69	Sulphate	SO_4^{2+}	7·68
Calcium	Ca^{2+}	1·16	Bicarbonate	HCO_3^-	0·41
Potassium	K^+	1·10	Bromine	Br^-	0·19
Strontium	Si^{2+}	0·03			

9.6.2 Gross Geologic Characteristics

It appears most probable that evaporites form from saline-rich fluids, that is, brines. Brines may be generated by concentration of seawater, by evaporation or freezing, or as residual connate fluids in the subsurface. Secondary brines can form where meteoric groundwater passes through and dissolves previously formed evaporites. Normal ocean water contains 3.45% by weight of dissolved substances; 99.9% of the dissolved material comprises the nine ions shown in Table 9.10.

Some of the earliest work on the genesis of evaporites was to study salts formed from the evaporation of seawater (Usiglio, 1849; Van't Hoff and Weigert, 1901). Particular attention was paid to volume and composition of the minerals which formed at particular temperatures and phases of evaporation. These studies demonstrated two main facts; that inconceivable quantities of seawater were necessary to form observed volumes of evaporites in a closed system, and that the observed percentages of salts in an evaporite assemblage differ somewhat from those produced by the evaporation of seawater.

To amplify the first of these points: a column of seawater 1000 m high would evaporate out to form 14.85 m of salts. Many evaporite basins, however, are thousands of meters thick and thus simplistically require improbably large volumes of seawater to beget them.

Figure 9.29 shows the observed percentages of salts in normal seawater compared with those found in the Permian Zechstein basin of the North Sea. An attractive explanation for these two points is that evaporite formation occurs in a silled basin. It is a matter of observation that evaporite formations characteristically occur in basins that had restricted access to the sea. Examples include the Zechstein of the North Sea (Fig. 9.30), the Michigan basin, the Paradox salt basin, and the Canadian Devonian evaporites.

In a restricted basin it is easy to see how seawater from the open ocean may flow into the basin. Here excessive evaporation concentrates the seawater. The incipient brine sinks to the basin floor because of its higher density. The sill prevents drainage of the brine out to the open sea. Continuous recycling of the brine increases concentration to the point at which evaporites begin to crystallize on the basin floor (Fig. 9.31). This process would be aided by the fluctuating sea level, which allows repeated influxes of water over the sill, followed by a drop in water level so as to completely restrict the body of brine. This is the classic "evaporating dish" mechanism for evaporite genesis (Sloss, 1969).

Supporting evidence for this mechanism includes the fact that evaporites tend to be

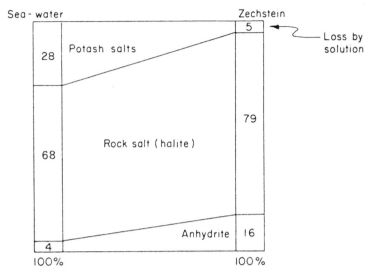

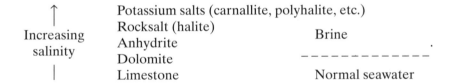

Fig. 9.29. Comparative sections of the percentages of evaporite minerals produced by the evaporation of average seawater, and the average observed percentages of minerals in the Zechstein (Lower Permian) evaporites of the North Sea basin.

zonally arranged within a basin, with salts requiring higher salinity for their formation occurring toward the depocenter. Similarly, evaporite minerals tend to be cyclically arranged in the same motif, that is,

↑
Increasing
salinity
|

Potassium salts (carnallite, polyhalite, etc.)
Rocksalt (halite)   Brine
Anhydrite
Dolomite   _ _ _ _ _ _ _ _ _ _ _ _ _
Limestone   Normal seawater

This cyclicity is classically demonstrated in the Zechstein evaporites of the North Sea basin (Fig. 9.30), but is also found in most other examples. These cycles are sometimes hundreds of meters thick when fully developed. Cyclicity is also present, however, on a much smaller scale. The monotonous repetition of interlaminated couplets of dolomite with anhydrite and of halite with potash salts is examined more closely in the next section.

Returning again to the gross geology of evaporites, it is noticeable that they occur in two particular tectonic settings. The first of these are the intracratonic basins (defined in Section 10.2.2), which lie within stable cratonic shields. These basins are characterized by gradual downwarping over a prolonged period of time, accompanied by infilling with diverse continental and shallow marine deposits including evaporites. Within a single basin these often range over a considerable span of time. Thus in the Michigan basin of North America, salt formations range in age from Silurian to Early Carboniferous (Mississippian). Similarly, in the Williston basin athwart the Canadian/USA border, evaporites formed intermittently from the Devonian through to the Permian.

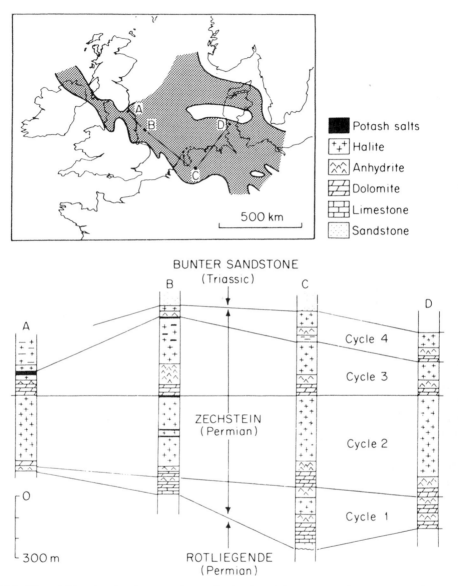

Fig. 9.30. (Upper) Map showing the approximate distribution of the Zechstein (Lower Permian) evaporites of the North Sea basin. (Lower) Cross-section demonstrating the lateral continuity of four major evaporite cycles.

An additional characteristic feature of these basins is that the evaporites are closely associated with reefal limestones. Sometimes a distinct reef belt occurs around an evaporite infilled depocenter, as in the Silurian Michigan basin, and in the Permian Delaware basin of West Texas. In other instances, such as the Middle Devonian Elk Point basin, pinnacle reefs occur within the main salt depocenter (Fig. 9.32). Barrier reefs may also contribute to the sill on the seaward side of an intracratonic basin.

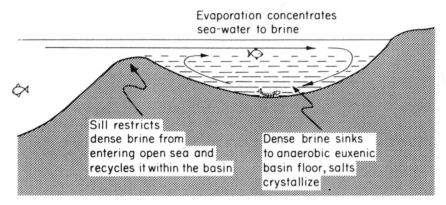

Fig. 9.31. Illustration of the classic barred basin model for evaporite genesis by direct crystallization from brine at the sediment/fluid interface.

The second common tectonic setting for evaporites is in the ocean margin coastal basins and their rift valley precursors. Chapter 10 explains how tension in the earth's crust generates rift valley systems. Where these occur in cratons, rift basins are infilled by continental clastics and volcanics. As tension develops, accompanied by crustal thinning, the rift floor sinks to sea level. At this time conditions are favorable for extensive intermittent marine incursions and hence for evaporite formation. Ultimately the rift is split in two, and an incipient ocean forms along the axis. The fault-bounded coastal basins of the ocean margin show a vertical sequence of continental clastics, evaporites, and normal marine deposits. Evaporites are found in this setting on both sides of the

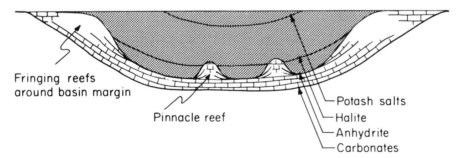

Fig. 9.32. Diagrammatic cross-section of an intracratonic evaporite basin and associated carbonates. This highlights the problem of whether evaporite crystallization occurred from euxenic brines on the basin floor synchronous with carbonate reef growth in surface waters of normal salinity. Alternatively, the evaporites may have formed by the replacement of sabkha supratidal carbonates when sea level dropped to the level of the basin floor.

Atlantic from the Grand Banks of Newfoundland, the Gulf of Mexico (the Louann salt), to Brazil on the west and also along the coast of west and southwest Africa (see Section 10.2.4.2).

A further particular feature of the evaporites is their high degree of plasticity. When buried beneath an overburden of younger sediment, salt can act as a lubricant to permit the cover to deform into eccentric structural styles. The salt acts as a zone of "decollement" along which the mobile cover becomes detached from the rigid basement. Notable examples of this phenomenon occur in Iran and in the Jura Mountains of southern France.

As discussed earlier, evaporites are less dense than other buried sediments, and thus may flow upward to form discrete pillows, walls, and domes (Alsop *et al.,* 1996). Salt domes or diapirs may be only a few kilometers across, but they can extend vertically up through a sedimentary cover thousands of meters thick. The crest of the dome often contains a cap rock of limestone, dolomite, anhydrite, gypsum, native sulfur, and diverse sulfide minerals (Kyle and Posey, 1991). The tops of diapirs sometimes overhang sediments, in the shape of a mushroom. In some cases the diapir becomes detached from the mother bed, and in still further extreme instances a whole new layer of salt can be intruded and then detached from a deeper mother bed, with the conduits by which it was emplaced barely identifiable.

The adjacent sediments are generally extensively faulted over the crest of the dome and a characteristic "rim syncline" may encircle the pillar (Fig. 9.33). Salt domes such as these are extensively recorded from the large gulf between Iran and Arabia (Jackson, 1991), from north Germany and from the African coastal basins previously discussed. The examples from the Gulf Coast of Louisiana and Texas are particularly well documented (e.g., Halbouty, 1967; Jackson *et al.,* 1996).

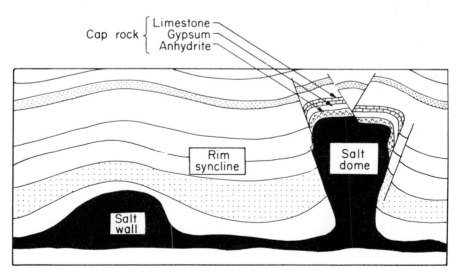

Fig. 9.33. Diagrammatic cross-section showing the morphology of salt-deformation structures. Salt domes (diapirs), though sometimes only a few kilometers in diameter, can penetrate over 1000 m of sedimentary cover.

9.6.3 Carbonate-Anhydrite Cycles

The general relationships of carbonate-anhydrite cycles are well exemplified in the Stettler Formation (Upper Devonian) of western Canada (Fuller and Porter, 1969). Followed in the subsurface the Upper Devonian sediments pass laterally from red beds under Saskatchewan, through anhydrite and carbonate-anhydrite rocks into open marine sediments under western Alberta (Fig. 9.34). In one borehole core, 13 rhythmic alternations of carbonate and anhydrite rocks were counted in a 15.24-m interval from the upper part of the formation. The carbonate members are in part limestone and in part dolomite. Near Calgary, a wedge of marine limestone extends eastward into the

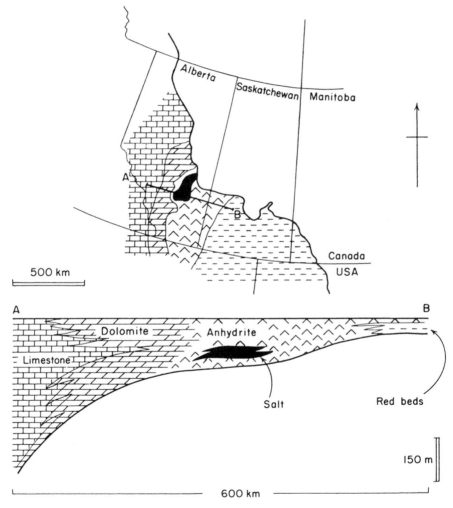

Fig. 9.34. Map and cross-section showing the distribution of evaporites and associated carbonates of the Stettler Formation (Middle Devonian), northwestern Williston basin. See how the evaporite depocenter occurs in a restricted embayment separated by dolomite from open marine limestones to the west. (After Fuller and Porter, 1969. American Association of Petroleum Geologists Bulletin, AAPG © 1969, reprinted by permission of the American Association of Petroleum Geologists whose permission is required for further use.)

evaporites, and this limestone, the Crossfield member, is the reservoir rock of the Olds gas field of Alberta. Anhydrite rocks are for the most part "tight" with respect to oil and gas, and in consequence make excellent seals to hydrocarbon reservoirs.

Another example of carbonate-anhydrite cycles is found in the upper part of the Madison Limestone Formation of southeast Saskatchewan (Fuller, 1956). The anhydrite units die out westward toward the open marine facies, while some of the limestones tend to die out eastward in the direction of the inferred shoreline. The carbonate-anhydrite cycles of the Mississippian are much thicker than those of the Middle Devonian Stettler Formation: Some cycles are over 30 m thick. Oolites and carbonate mud-pellet rocks are abundant in the limestones, suggesting that they were deposited in shallow-water environments. Some of the limestones have undergone only limited cementation, and are oil reservoirs, for example, the Midale cycle (Fig. 9.35). Locally the overlying anhydrite rocks form the caps to the reservoirs, but the situation is complicated by the fact that the Madison Limestone Formation underlies a major unconformity. This is not deleterious, however, because the rocks above the unconformity also act as oil and gas seals.

Another well-documented example of carbonate-anhydrite cycles is that of the Arab-Darb Formation (Jurassic) of the offshore oil field of Umm Shaif in the large Gulf between Iran and Arabia (Wood and Wolfe, 1969). Seven carbonate-anhydrite cycles were encountered within a 15-m interval. In each cycle the limestones shallow upward and the top of each limestone shows an assemblage of features that characterize shallow subtidal and intertidal zone sedimentation (Fig. 9.36). The limestones pass up into anhydrite rocks, without a significant break, but at the top of many of the anhydrite units there is a clearly marked erosion surface. Each cycle is, therefore, a carbonate-anhydrite cycle and not an anhydrite-carbonate cycle. Because each carbonate unit is a "shallowing up" entity, with the upper part becoming intertidal, this poses the problem of the environment of formation of the anhydrite.

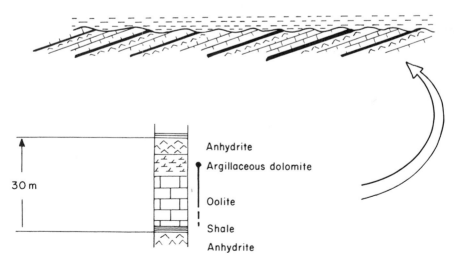

Fig. 9.35. Midale evaporite cycle of the Madison Formation, Mississippian (Lower Carboniferous), Saskatchewan. This shows the hydrocarbon trapping mechanism provided by the sub-Mesozoic unconformity. (After Fuller, 1956.)

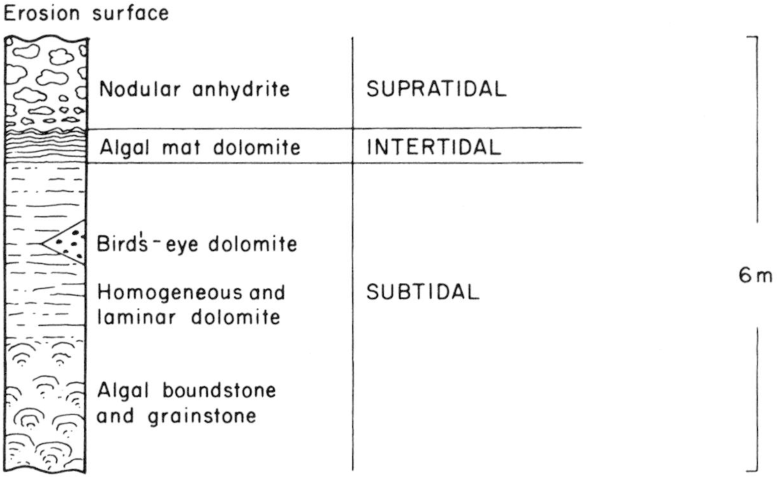

Erosion surface

	Nodular anhydrite	SUPRATIDAL
	Algal mat dolomite	INTERTIDAL
	Bird's‐eye dolomite	
	Homogeneous and laminar dolomite	SUBTIDAL
	Algal boundstone and grainstone	

6 m

Fig. 9.36. Regressive sabkha cycle in Arab-Darb Formation (Jurassic), United Arab Emirates. (From Wood, G. V., and Wolfe, M. J. 1969. Sabkha cycles in Arab/Darb formation of the Trucial Coast of Arabia. *Sedimentology* **12,** 165–191. Courtesy of Blackwell Science.)

It is important at this stage to refer to the essential features of the anhydrite. Although the anhydrite forms units that are interbedded with the limestones, the anhydrite is characteristically not bedded but is usually nodular. At one extreme the nodules may occur scattered in a background of carbonate, while at the other the nodules are tightly packed and separated only by thin films of carbonate mud and organic matter (Fig. 9.37). In some instances the nodules occur in layers that simulate bedding and locally the layers show remarkable contortions, termed "enterolithic structure," because of the resemblance to the coils of an intestine (Shearman, 1966). The nodules comprise masses of tiny plate-like crystals, the arrangement of which often changes from place to place within any one nodule — in one part the arrangement may be subparallel, whereas elsewhere it may be decussate. The nodular form of the anhydrite argues against the anhydrite having been precipitated from a standing body of brine, and this led some earlier

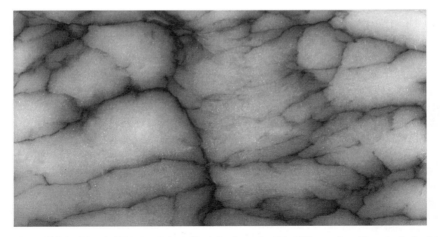

Fig. 9.37. Polished 9-cm slab of nodular anhydrite from the Khuff Formation (Permian), Arabia.

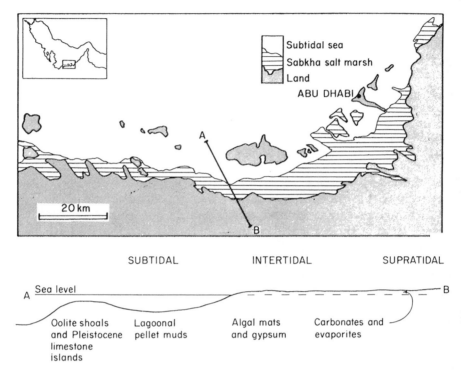

Fig. 9.38. Map and cross-section showing the distribution of the broad sabkha salt marsh along the coast of the United Arab Emirates, and its associated deposits.

workers to argue that the nodules grew displacively within the sediment that now contains them.

Although marine evaporites are apparently not forming to any great extent at the present day, they are locally developed in the coastal areas of some desert regions. It was one of these, the United Arab Emirates of the Arabian Gulf, that provides the answers to some of the problems posed by the carbonate-anhydrite cycles of the ancient rocks (Fig. 9.38). The UAE is a shoal-water complex of islands and lagoons some 200 m in lateral extent in which the whole spectrum of shallow-water carbonate sediments is being formed (Evans *et al.,* 1969). The sediments include aragonite muds, pelleted muds, oolite sands, skeletal sands, and a small coral patch reef. By virtue of the shallow water and onshore wind and waves, a wide supratidal flat has developed along much of the coast. For the greater part the flat stands only 0.5–1.0 m above normal high-tide level and in places extends inland for 20 km or so. It is a completely flat barren desert, and the arabic word "sabkha" can be used to describe it (Fig. 9.39). In terms of geomorphology, however, this coastal sabkha is simply the desert zone analog of the salt marshes of temperate regions (refer back to Fig. 6.51). Trenches dug in the sabkha expose an upward succession from earlier subtidal carbonate sediments, through intertidal into supratidal sediments. This is a simple regressive sequence of shoreline sedimentation. In places the present intertidal zone is marked by a wide belt of sediment-trapping algal mats, and similar algal mats occur buried under large areas of the sabkha flat. It is evident that the intertidal and supratidal zones have been prograding seaward, so that the sediments of these facies change in age laterally, that is, they are diachronous (Fig. 9.40).

Fig. 9.39. The sabkha of Abu Dhabi, United Arab Emirates. Anhydrite, gypsum, dolomite, magnesite, and intermittently halite form in this environment, both replacing and displacing lagoonal carbonate mud, as the sabkha advances seaward across the shelf. Sabkhas such as this have existed along the edge of the Arabian platform since the deposition of the Khuff Group in the Permian Period. Their deposits are characterized by cyclicity and great lateral continuity (Al-Jallal, 1994).

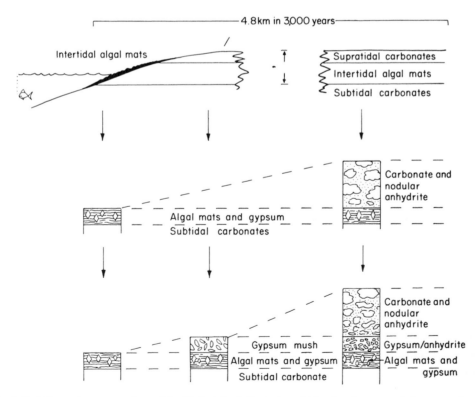

Fig. 9.40. Schematic cross-section of the United Arab Emirates sabkha showing present-day regressive sequence with superimposed early diagenetic sabkha evaporites.

Within the lagoonal waters calcium ions are removed by the organic precipitation of aragonite skeletal debris. Thus the water becomes progressively enriched in ions of magnesium, potassium, and sulfates. Simultaneously salinity increases due to the high rate of evaporation. Thus, although the background sediment is basically carbonate, evaporite minerals occur in abundance in the sediments of the intertidal and supratidal facies. Gypsum and dolomite are present in the sediments of the present intertidal and low supratidal zones, and they are also found in earlier buried intertidal sediments, while nodules of anhydrite are locally abundant in the sediments of the supratidal facies. In all essential respects this present-day nodular anhydrite is strictly comparable with that found in anhydrite units of the ancient carbonate-anhydrite cycles. It is evident in the field that the nodules grew within the host carbonate sediment by displacement, and that features such as enterolithic folds are growth structures that resulted from the demand for space as more and more anhydrite crystals grew within layers of coalesced nodules. The groundwaters of the coastal sabkha are marine derived, and due to heat and aridity they have become concentrated by capillarity and evaporation to the point where they are now highly concentrated brines. It is from these interstitial brines that the evaporite minerals are formed.

Prior to the discovery of this presently forming anhydrite of the UAE, it had been thought that the nodular anhydrite of the ancient carbonate-anhydrite cycles may have been formed in shallow, highly saline coastal lagoons. A supratidal origin now appears more likely; the carbonate-anhydrite cycles are desert zone carbonate-shoreline regressive cycles of sedimentation in which evaporites were emplaced in the intertidal and supratidal sediments by penecontemporaneous diagenesis.

Vertical repetition of the cycles, and the total thickness of the succession, is the expression of the diastropic background of relative subsidence. It is convenient to refer to these cycles as "sabkha cycles." In a sense, sabkha cycles can be thought of as being desert zone analogs of coal measures. Along parts of the UAE coast the sabkha passes into a complex of continental dune sands and inland sabkhas, which in turn pass back into alluvial fans at the front of the Oman Mountains (Fig. 9.41). The lateral transition from continental desert zone sediments, through coastal sabkhas with evaporites, out into open marine sedimentation, is essentially that displayed by the Upper Devonian Stettler Formation of western Canada. In some ancient sabkha cycles, such as those in the

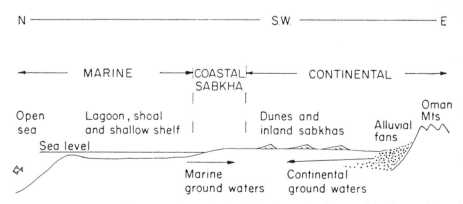

Fig. 9.41. Schematic cross-section across the United Arab Emirates showing transition from arid continental to open marine environments.

Windsorian (Carboniferous) of Nova Scotia, the nodular anhydrite units are overlain by red beds. In these cases it appears that the regressions were followed by the establishment of continental conditions.

Carbonate-anhydrite sabkha cycles are present in the Purbeck evaporites of southern England (West, 1964; Shearman, 1966), and thick developments have been encountered in borehole cores in the Lower Carboniferous of parts of the east Midlands of England (Llewellyn et al., 1969).

It is appropriate to refer briefly to the chemical process that appears to operate in the formation of sabkha-type evaporites. The essential feature is that the evaporite minerals grow interstitially within earlier formed sediment. Dolomite and gypsum develop in the sediments of the present-day high intertidal and low supratidal environment, and they are also present in the older buried intertidal sediments under the sabkha plain. Anhydrite characterizes the supratidal concentration of the interstitial brines inward through the intertidal into the supratidal zone. It appears that concentration is an important factor in determining which of the two calcium sulfate minerals will be formed. Gypsum forms during the early stages of concentration but anhydrite is not generated until high concentrations are achieved. Although much of the anhydrite appears to have formed directly as anhydrite (Plate 5D), there are occurrences where the anhydrite pseudomorphs gypsum. In the latter cases it would appear that gypsum, formed during the early stages of concentration of the groundwaters, became made over into anhydrite as the concentration increased. The dehydration of gypsum to anhydrite is a reversible reaction:

$$CaSO_4 \cdot 2H_2O \rightleftharpoons CaSO_4 + 2H_2O$$
$$\underline{\hphantom{xxxxxxx}}\ \text{Increasing temperature and salinity}\ \longrightarrow.$$

Dehydration of gypsum occurs with increasing salinity and temperature; hydration of anhydrite occurs with decreasing salinity and temperature (MacDonald, 1953). Thus anhydrite can only form at the surface of the earth in arid hypersaline environments. Gypsum may form in cooler, less saline environments, but dehydrates to anhydrite on burial. Murray (1964) showed that, for an average salinity and geothermal gradient, the dehydration of gypsum takes place at about 1000 m below the surface of the earth. Anhydrite hydrates to gypsum at a similar depth.

The dolomite in the sabkha forms by reaction between the brines and the host carbonate sediment, and the reaction releases calcium ions. Normal seawater carries more sulfate ions than are required to satisfy the calcium in seawater, so that as seawater is concentrated by evaporation, the calcium will be precipitated as calcium sulfate, but the excess sulfate ions remain in the brine. This excess sulfate is available to combine with the calcium ions released by dolomitization of the carbonate sediments, and a further crop of calcium sulfate is generated. Thus by virtue of dolomitization, the sabkha mechanism of evaporite genesis may lead to production of almost twice the amount of calcium sulfate minerals as would be formed by simple evaporation of the same volume of seawater.

Halite deposits are uncommon in carbonate-anhydrite evaporite sequences. They do occur occasionally, however, as for example in the Middle Devonian Stettler Formation of western Canada (Fig. 9.34). Such a large lens of halite could be accounted for if local subsidence developed at the back of an otherwise emergent sabkha plain. The groundwaters, already concentrated sodium chloride brines, would break surface and form a brine pool from which halite would be precipitated.

9.6.4 Halite-Potash Evaporite Successions

The halite-potash evaporite successions differ from the cyclic carbonate-anhydrite sequences in a number of important respects. Although many of them contain significant proportions of carbonate and anhydrite rocks, they are characterized by a substantial thickness of halite. Generally they comprise extensive basin-shaped accumulations that appear to have formed in large, partially enclosed embayments that had only restricted access to the open sea. They are exemplified by the Permian Zechstein evaporites of northwest Europe, by the Middle Devonian evaporite complex of the Elk Point basin of western Canada, the Silurian evaporites of the Michigan basin, the Pennsylvanian evaporites of the Paradox basin, Utah, and the Cambrian evaporites of Siberia.

In a general way the succession of mineral salts tends to be cyclic, in the sense that they pass up from carbonate-anhydrite rocks into thick piles of halite and in some instances terminate with potassium salts. Four such cycles are developed in the Zechstein evaporites of Germany (refer back to Fig. 9.30). It is the thicknesses of halite that are the remarkable feature of these deposits. The Prairie Halite of the Elk Point basin in Saskatchewan is approximately 200 m thick, and some 500 m of halite are present in the Zechstein of Germany. Vast quantities of seawater had to be processed to form these thick accumulations and space had to be provided to accommodate them. The latter consideration has led to long controversy as to whether the halite was formed in deep brine-filled basins or in shallow brine pools against a background of subsidence.

A common rock type in the thick halite successions is the so-called "layered halite rock." This consists of repeated alternations of layers of halite, 2–10 cm in thickness, separated by 1-mm-thick laminae of anhydrite, or anhydrite and dolomite sometimes with organic matter. Some geologists have interpreted these cyclic alternations of anhydrite and halite as recording annual evaporation cycles. Such cycles are termed **Jahresringe** by German geologists. On this basis it has been argued that the 200 m thickness of salt of the Prairie Halite of Saskatchewan was deposited in 4000 years, and that the 500 m of halite in the Zechstein accumulated in 10,000 years. The Jahresringe concept argues against a shallow-water origin, because to accommodate the observed thicknesses would require a background of subsidence of approximately 5 cm a year. Such a rate of subsidence is greatly in excess of what could reasonably be expected in any tectonic or diastropic setting, and it becomes necessary, therefore, to postulate an initially deep basin. Keep in mind, however, that the deep brine hypothesis depends largely on the validity of the Jahresringe concept. The petrology of the Prairie Halite has been described by Wardlaw and Schwerdtner (1966). Each halite layer is an admixture of two types of halite crystals. Some of the crystals carry abundant tiny brine inclusions; these are arrayed in planes parallel to the cube faces, and give the crystals a zoned appearance. In thin sections, under the microscope, the zoned crystals are elongated upward and the zones appear as chevrons with their apices directed upward. This type of fabric is indicative of competitive growth of crystals upward from a substrate, and the abundant brine inclusions suggest intermittent episodes of rapid growth. The other type of halite crystal is clear and free of inclusions, and the mutual relationships of the two types of crystals suggest that the clear halite replaced halite with inclusions.

Layered halite rocks comprising layers of halite 2–10 cm in thickness separated by thin laminae of gypsum are forming at the present day, or formed in the Recent geological past, in brine pools on coastal salt flats at the head of the Gulf of California, Mexico. The deposit is rarely more than 25 cm thick. The halite layers are built of zoned crystals that have the same arrangement as those in the layers of the Prairie Halite, but the rock is riddled with small dissolution hollows. The salt flats dry out frequently, and when they are periodically flooded by the sea, this incoming water pipes its way down into the halite rock. The way in which the dissolution pipes corrode the crystals of zoned halite is very similar to the manner in which the clear halite appears to replace the zoned halite in the layered Prairie Halite. Indeed, if the dissolution hollows in the present-day occurrence could be filled with clear halite, the two rocks would be identical. If the apparent replacement of zoned halite by clear halite in the layers of the Prairie Halite is the record of piping of the rock by dissolution and subsequent filling by a later generation of clear halite, then this would argue against a deep-water origin for the salt, because it is difficult to conceive how piping could take place beneath a deep standing body of brine. Another feature evident in the Recent occurrence in Mexico is that the layers are not annual but may have taken long periods of time to form.

Thus interpretations of the environment of deposition of salt deposits, like the 2-m thickness of the Prairie Halite, range from the extremes of, on the one hand, a brine-filled basin approximately 200 m deep with rapid deposition, to slow accumulation in very shallow water against a background of gentle subsidence on the other.

Where potash salts are present in the major halite evaporite sequences, as for example in the Middle Devonian of the Elk Point basin of Canada or the Permian Zechstein of northwest Europe, the potash salts usually occur in the upper part of the succession or, in the case of the Zechstein, in the upper part of the major evaporite cycles. It would seem at first sight that they were precipitated residual lakes of brine left after the deposition of halite. However, the deposits pose problems. One of these is that their overall chemical composition is not that that which would be expected by simple evaporation of seawater. In the theoretical direct evaporation of seawater, precipitation of halite should be followed by, first, precipitation of $MgSO_4 \cdot nH_2O$ (epsomite), then by KCl (sylvite), and finally by $MgCl_2 \cdot 6H_2O$ (bischofite). In most occurrences, the epsomite, or its mineralogical equivalent, is only weakly developed or is absent, and the bischofite is also characteristically absent. Although it is reasonable to argue that bischofite is so soluble that it is unlikely to survive even if its precipitation was achieved, the depletion with respect to magnesium and sulfate demands explanation. The Middle Devonian potash deposits of Saskatchewan provide an interesting example, because they consist solely of sylvite (KCl) and carnallite ($KMgCl_3 \cdot nH_2O$). Not only is sulfate absent, but the proportion of magnesium is much lower than would be predicted. Evidently the brines had been conditioned and their chemistry modified earlier in their history. The absence of sulfate ions could be accounted for by the activities of sulfate-reducing bacteria, but such a process cannot explain the combined deficiency in both sulfate and magnesium.

It is of interest at this stage to refer back to the reactions that are taking place in the formation of the Recent evaporites of the UAE sabkhas. Recall that dolomitization of the carbonate sediments releases calcium ions, and these promote precipitation of more calcium sulfate than would occur with simple evaporation of seawater. The resultant sabkha brines are, in consequence, stripped of their sulfate and depleted with respect to magnesium. Brines of this composition could generate the restricted mineral assemblage found in the potash zones of the Devonian of Saskatchewan. The Middle Devonian evaporites of the Elk Point basin are separated from the open marine Mackenzie limestone and shale basin that lay to the north by a limestone dolomite complex, the Presqu'ile Formation. It is a matter of opinion whether the Presqu'ile Formation was present as a physical barrier (i.e., a leaky dam with a deep-water evaporite basin behind) or whether the carbonate complex built up as shoal banks coevally with the accumulation of shallow-water evaporites behind. Whichever of the alternatives applied, the seawater that entered the Elk Point evaporite basin had to do so through the Presqu'ile carbonate barrier. The Presqu'ile Formation is largely dolomite with a complex of dolomites and anhydrite rocks behind it, and these in turn pass back mainly into halite. The possibility has to be considered that the incoming seawater was conditioned chemically by dolomitization and associated precipitation of sulfate as it passed through the barrier. In consequence the brines that passed on into the distal parts of the basin would only have been capable of generating halite and the observed potash assemblage of silvite and carnallite.

Some potash deposits evidently accumulated in highly saline lakes of residual brine that remained after the virtual drying out of the evaporite "basin." In other instances the lakes may have been formed after complete drying out, by the bleeding of interstitial brines into tectonic depressions. However, in some potash deposits the crystal fabrics are not those of precipitates, but of diagenetic replacements, for example, the Permian potash of Texas and New Mexico, and the Zechstein potash of northwest England. The evidence suggests that the potash minerals were emplaced by reaction between interstitial potassium and magnesium chloride brines and earlier formed minerals.

9.6.5 Economic Significance of Evaporites

Evaporite minerals are of great economic importance for three reasons. They are an economic material in their own right, they are closely related to the genesis and entrapment of hydrocarbons, and there is a strong presumption that evaporite associated brines play an important role in the genesis of certain metallic ores (Melvin, 1991). These three aspects are now examined.

Evaporites are a natural resource of great importance. They supply a large proportion of the world's requirements for the rare earth elements, notably sodium and potassium, for the halogens, principally chlorine and bromine, and for sulfur. Chemical industrial complexes thus tend to be situated adjacent to economic evaporite bodies. The crests of salt domes develop a diagenetic cap rock of limestone, dolomite, anhydrite, gypsum native sulfur, and diverse sulfide minerals (Kyle and Posey, 1991). Sometimes the sulfur is in commercial quantities (Fig. 9.42).

Evaporites are of importance in the search for oil and natural gas for three reasons: source, structure and seal (Buzzalini et al., 1969). The conditions that favor evaporite genesis are unfavorable to biological decay. In a basin with brine in its lower depths, organic matter may be preserved on the basin floor interbedded with evaporites because the conditions are hostile to bacteria. Similarly, in the sabkha environment, algal laminae are preserved interbedded with evaporites and carbonates. Organic laminae are thus a common constituent of evaporites (be they basinal or sabkha in origin) and there is a strong presumption that evaporites are often potential hydrocarbon-source rocks

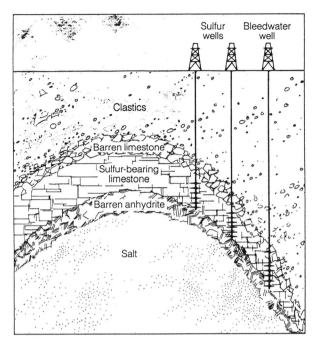

Fig. 9.42. Cross-section through the crest of the Main Pass 299 salt dome, Louisiana, USA, showing location of commercial sulfur deposit, some 50 m thick, at about 540-m depth. (From Christensen *et al.*, 1991, courtesy of Schlumberger Oilfield Review.)

(Sonnenfeld, 1985). There are, however, problems in explaining the migration of petroleum from evaporites, since these rocks are so impermeable.

Secondly, as previously pointed out, the plastic behavior of evaporites enables them to generate structures, even in areas devoid of tectonic activity (Jackson *et al.*, 1996). Salt domes host a series of potential hydrocarbon traps, both domal anticlines above the cap rock and faulted flank traps (refer to Fig. 7.8).

Finally, evaporites are significant because they provide an ideal reservoir seal, combining a maximum of plasticity with a minimum of permeability. Thus evaporites seal reefal reservoirs, such as those of the Williston basin, and provide the seal for the gas in the Rotliegende sandstone reservoirs of the North Sea.

It has already been pointed out that a characteristic evaporite basin is rimmed by reef limestones and that those often contain oil, probably generated from and certainly sealed by, the evaporites. In many instances the reefs also contain sour gas (H_2S) and a particular type of mineral deposit. These are the telethermal (low-temperature) suites of sulfide ores of lead (galena PbS) and of zinc (sphalerite ZnS). Associated minerals include florite (CaF_2), baryte ($BaSO_4$), dolomite, and crystalline calcite. The classic example of this type of ore is in the Mississippi Valley and they are sometimes referred to collectively as "Mississippi Valley type" ores (see Section 6.3.2.8.3).

Many lines of evidence suggest that the sulfur of the sour gas and of the sulfide metals was provided by the reaction between anhydrite and hydrocarbons. In its simplest terms the reaction can be written:

$$CaSO_4 + \text{hydrocarbons} \rightarrow H_2S + Ca^{2+} + CO_3^{2+} + H_2O \text{ etc.}$$

$$\searrow \swarrow$$

$$\downarrow$$

$$CaCO_3$$

This reaction is known to be promoted by sulfate-reducing bacteria, but there is reason to believe that it can also take place in a sterile environment, in the absence of bacteria (Dunsmore, 1973; Eugster, 1985, 1986; Sawlowicz, 1986). In the latter case this would be an exothermic reaction which could generate the high-temperature crystalline fabric seen in the ores and their gangue; hence, the superficial resemblance of these bodies to hydrothermal veins. The implications of this possibility are discussed in Section 6.3.2.8.3.

In conclusion, the evaporites are a very important group of rocks. Their origin is still a matter for debate, centering on the degree to which they form by crystal growth on the sediment/water interface of basins, and the extent to which they form by replacement of carbonate during the diagenesis of brine-soaked supratidal sabkhas.

Regardless of their precise genesis, evaporites are economically important minerals, both in their own right and because of their close association with petroleum and sulfide ores.

9.7 CHERTS

Chert is a nonclastic sediment composed of microcrystalline quartz. It occurs in both beds and nodules in limestones and sandstones (McBride, 1979). Cherts pose three main problems: (1) What is the source of the silica? (2) What is the environment of deposition of bedded cherts? (3) How do nodular cherts replace other sediments?

Four minerals are associated with chert. Opal or opaline silica is amorphous and isotropic. Its formula is variously presented as $SiO_2 \cdot nH_2O$ or $SiO_2 \cdot nSi(OH)_4$. Christobalite is isotropic or felted and has the formula $SiO_2 + SiO_4$ tetrahedra. These minerals only occur in post-Cretaceous cherts. **Chalcedony** is microcrystalline quartz that often shows fibrous or radial habits. Finally, microquartz, with anhedral crystals $10–50\ \mu m$ in diameter, characterizes the pre-Eocene cherts. The results of the Deep Sea Drilling Program have provided much information on the formation of chert (Davies and Supko, 1973; Adachi et al., 1986). Cherts are found below depths of some 60 m beneath modern ocean floors. They occur in calcareous and terrigenous clays, in volcanic sediments, and in siliceous oozes. They are most common in the latter. The source for the silica is partially biogenic, largely from the solution of radiolaria. There appear to be four phases of chert formation. It begins with the infilling of foraminiferal molds by chalcedony. This is followed by the replacement of carbonate matrix by christobalite, and then by the replacement of foraminifera by chalcedony. Finally, all remaining porosity is infilled by quartz, and the christobalite reverts to quartz also.

One of the best known ancient chert deposits is the Paleozoic Caballos Novaculite of the Marathon basin, Texas. While there is general agreement that this formed as a primary chert there is debate as to its environment of deposition. McBride and Thomson (in Folk and McBride, 1976) argue for a bathyal depth, while Folk (1973; and in Folk and McBride, 1976) argues for an intertidal origin.

Bedded cherts are also known from lacustrine deposits. For example, silicified limestones with the freshwater gastropod *Hydrobia* occur in the Continental Mesozoic beds of the Kufra basin, Libya.

Bedded cherts of uncertain origin occur interbedded with Pre-Cambrian ironstones, as previously illustrated in Fig. 9.25, and whose environment of deposition was discussed in Section 9.4.2.

It has often been remarked that cherts tend to be associated with contemporaneous volcanic activity. It has been argued that vulcanism releases large amounts of silica into the environment. This encourages "blooms" of silica-secreting organisms, such as radiolaria, diatoms, and sponges, and thus generates chert formation (e.g., Wenk, 1949; Khvorova, 1968).

Another distinctive type of silica rock is nodular chert (Fig. 9.43). This is especially characteristic of fine-grained limestone, but also occurs in sandstones. Nodular chert beds are commonly found in the late Cretaceous and Tertiary chalks of the Middle East and Europe. This type of chert, often termed **flint,** occurs in irregular rounded nodules of eccentric shape and low sphericity. They are generally concentrated along beds and sometimes plug burrows or replace fossils. Less commonly they infill joints and fractures (Plates 5E and 5F).

The genesis of chert, both bedded and nodular, has attracted considerable speculation (Sieveking and Hart, 1986). The main point at issue is whether chert originates as a colloidal silica gel on the sediment/water interface or whether it occurs by replacement. Undoubtedly both processes can take place in different situations. Primary precipitation of **opal** has been described in ephemeral Australian lakes (Peterson and von

Fig. 9.43. Nodular chert in Cretaceous chalk, Annis's Knob, Beer, Devon. This variety of chert, known to the natives as "flint," appears to have often infilled burrows.

der Borch, 1965). Some ancient cherts are also undoubtedly penecontemporaneous with sedimentation, as, for example, the synsedimentary chert breccias and dykes described by Harris (1958), Steinitz (1970), and Carozzi and Gerber (1978). It has been argued that flint formed some 5–10 m below the sediment/water interface where porosity was still in the region of 75–80%. Anaerobic bacterial sulfate reduction released H_2S, reduced pH, and caused the dissolution of chalk and the concomitant precipitation of silica (Clayton, 1986). In other instances, however, the presence of chert in fractures and of chert-replaced carbonate skeletal debris provide conclusive evidence of a secondary origin. Clearly cherts are polygenetic.

SELECTED BIBLIOGRAPHY

For carbonates in general see:

Jordan, C. F., and Wilson, J. L. (1994). Carbonate reservoir Rocks. *Mem. — Am. Assoc. Pet. Geol.* **60,** 141–158.
Lucia, F. J. (1999). "Carbonate Reservoir Characterization." Springer-Verlag, Berlin. 226pp.
Tucker, M. E., and Wright, V. P. (1990). "Carbonate Sedimentology." Blackwell Science, Oxford. 496pp.

For dolomites see:

Purser, B. H., Tucker, M. E., and Zenger, D. H., eds. (1994). "Dolomites." Blackwell, Oxford. 464pp.

For sedimentary ironstones see:

Young, T. P., and Taylor, W. E. G., eds. (1989). "Phanerozoic Ironstones," Spec. Publ. No. 46. Geol. Soc. London, London. 251pp.

For phosphates see:

Follini, K. B. (1996). The phosphorous cycle, phosphogenesis and marine phosphate-rich deposits. *Earth Sci. Rev.* **40,** 55–124.

For coal see:

Gayer, R., and Harris, I., eds. (1996). "Coalbed Methane and Coal Geology," Spec. Publ. No. 107. Geol. Soc. London, London. 344pp.

For evaporites see:

Melvin, J. L., ed. (1991). "Evaporites, Petroleum and Mineral Resources." Elsevier, Amsterdam. 556pp.

For cherts see:

Sieveking, G. de G., and Hart, M. B., eds. (1986). "The Scientific Study of Flint and Chert." Cambridge University Press, Cambridge, UK. 290pp.

REFERENCES

Adachi, M., Yamamoto, K., and Sugisaki, R. (1986). Hydrothermal chert and associated siliceous rocks from the northern Pacific. *Sediment. Geol.* **47,** 125–148.
Ager, D. V. (1993). "The Nature of the Stratigraphical Record," 3rd ed. Wiley, Chichester. 151pp.

Al-Jallal, I. A. (1994). The Khuff Formation: Its Regional Reservoir Potential in Saudi Arabia and other Gulf countries; depositional and stratigraphic approach. *In* GEO '94. The Middle East Petroleum Geosciences, Vol. 1. pp. 103–119. Gulf Petrolink, Manama, Bahrain.

Alsop, G. I., Blundell, D. J., and Davidson, I., eds. (1996). "Salt Tectonics," Spec. Publ. No. 104. Geol. Soc. London, London. 384pp.

Anonymous (1971). Massive Danian limestone key to Ekofisk success. *World Oil,* May, pp. 51–52.

Archie, G. E. (1952). Classification of carbonate reservoir rocks and petrophysical considerations. *Bull. Am. Assoc. Pet. Geol.* **36,** 278–298.

Badiozamani, K. (1973). The dorag dolomitization model — application to the Middle Ordovician of Wisconsin. *J. Sediment. Petrol.* **43,** 965–984.

Bathurst, R. G. C. (1975). "Carbonate Sediments and Their Diagenesis," 2nd ed. Elsevier, Amsterdam. 658pp.

Baturin, G. N. (1970). Recent authigenic phosphorite formation on the south-west African shelf. *Rep. — Inst. Geol. Sci. (U.K.)* **70/13,** 90–97.

Bentor, Y. K. (1980). Marine phosphorites. *Spec. Publ. — Soc. Econ. Paleontol. Mineral.* **29,** 1–264.

Black, M. (1953). The constitution of the chalk. *Proc. Geol. Soc.* **1491,** LXXXI–LXXXVI.

Borchert, H., and Muir, R. O. (1964). "Salt Deposit: The Origin, Metamorphism and Deformation of Evaporites." Van Nostrand-Reinhold, London. 338pp.

Brennand, T. P., and Van Veen, F. R. (1975). The Auk Field. *In* "Petroleum and the Continental Shelf of Northwest Europe" (A. W. Woodland, ed.), Vol. 1, pp. 275–284. Applied Science, London.

Bromley, R. G. (1967). Some observations on burrows of Thalassinodean Crustacea in chalk hardgrounds. *Q. J. Geol. Soc. London* **123,** 159–182.

Brown, M. (1994). "Discovering Mineral Resources. Nature & Resources," Vol. 30, pp. 9–20. UNESCO, Paris.

Bubinicek, L. (1971). Geologie due Gisement de Fer de Lorraine. *Bull. Cent. Rech. Pau* **5,** 223–320.

Budd, D. A. (1997). Cenozoic dolomites of carbonate islands: Their attributes and origin. *Earth Sci. Rev.* **42,** 1–47.

Burchette, T. P., and Britten, S. R. (1985). Carbonate facies analysis in the exploration for hydrocarbons: A case study from the Cretaceous of the Middle East. *In* "Sedimentology Recent Developments and Applied Aspects" (P. J. Brenchley and B. P. J. Williams, eds.), pp. 311–338. Blackwell, Oxford.

Burnett, W. C., and Riggs, S. R., eds. (1990). "Phosphate Deposits of the World," Vol. 3. Cambridge University Press, Cambridge, UK. 639pp.

Buschinski, G. I. (1964). Shallow-water origin of phosphorite evaporites. *In* "Deltaic and Shallow Marine Sediments" (L. M. J. U. Van Straaten, ed.), pp. 62–70. Elsevier, Amsterdam.

Butler, G. P. (1969). Modern evaporite deposition and geochemistry of coexisting brines, the Sabkha, Trucial Coast, Arabian Gulf. *J. Sediment. Petrol.* **39,** 70–90.

Buzzalini, A. D., Adler, F. J., and Jodry, R. L., eds. (1969). Evaporites and petroleum. *Bull. Am. Assoc. Pet. Geol.* **53,** 775–1011.

Byrd, W. D. (1975). Geology of the Ekofisk Field, offshore Norway. *In* "Petroleum and the Continental Shelf of Northwest Europe" (A. Woodland, ed.), Vol. 1, pp. 439–445. Applied Science, London.

Campbell, C. V. (1962). Depositional environments of Phosphoria Fomation (Permian) in southeastern Bighorn basin, Wyoming. *Bull. Am. Assoc. Pet. Geol.* **46,** 478–503.

Carozzi, A. V., and Gerber, M. S. (1978). Synsedimentary chert breccia: a Mississippian tempestite. *J. Sediment. Petrol.* **48,** 705–708.

Carozzi, A. V., and von Bergen, D. (1987). Stylolitic porosity in carbonates: A critical factor for deep hydrocarbon production. *JPT, J. Pet. Geol.* **10,** 263–282.

Chafetz, H. S. (1986). Marine peloids: A product of bacterially induced precipitation of calcite. *J. Sediment. Petrol.* **56,** 812–817.

Champetier, Y., Hamdadou, E., and Hamdadou, M. (1985). Examples of biogenic support of mineralization in two oolitic iron ores — Lorraine (France) and Gara Djebilet (Algeria). *Sediment. Geol.* **51,** 249–255.

Chilingar, G. V., and Terry, R. D. (1964). Relationship between porosity and chemical composition of carbonate rocks. *Pet. Eng. B* **54,** 341–342.

Chilingar, G. V., Bissell, H. J., and Fairbridge, R. W. (1967a). "Carbonate Rocks," 2 vols. Elsevier, Amsterdam. 471 and 413pp.

Chilingar, G. V., Bissell, H. J., and Wolf, K. H. (1967b). Diagenesis of carbonate rocks. *In* "Diagenesis in Sediments" (G. Larsen and G. V. Chilingar, eds.), pp. 197–322. Elsevier, Amsterdam.

Chilingar, G. V., Mannon, R. W., and Rieke, H. (1972). "Oil and Gas Production from Carbonate Rocks." Elsevier, Amsterdam. 408pp.

Choquette, P. W., and Pray, L. C. (1970). Geological nomenclature and classification of porosity in sedimentary carbonates. *Bull. Am. Assoc. Pet. Geol.* **54,** 207–250.

Christensen, R. J., Davidson, D., Hallford, D., Johnston, D., Scholes, P., Huff, R., Kreis, H., Raihl, A., La Sala C., and Swan, S. (1991). *In situ* mining with oilfield technology. *Schlumberger Oilfield Rev.* **3,** 8–17.

Clayton, C. J. (1986). The chemical environment of flint formation in Upper Cretaceous Chalks. *In* "The Scientific Study of Flint and Chert" (G. de G. Sieveking and M. B. Hart, eds.), pp. 43–54. Cambridge University Press, Cambridge, UK.

Clemmey, H. (1985). Sedimentary ore deposits. *In* "Sedimentology: Recent Developments and Applied Aspects" (P. J. Brenchley and B. P. J. Williams, eds.), pp. 229–248. Blackwell, Oxford.

Cook, P. J. (1986). Genesis of sedimentary phosphate deposits. *In* "Geology in the Real World: The Kingsley Dunham Volume" (R. W. Nesbitt and I. Nichol, eds.), pp. 51–4. Inst. Min. Metall., London.

Cook, P. J., and Shergold, J. H. (1986). "Phosphate Deposits of the World," Vol. 1. Cambridge University Press, Cambridge, UK. 560pp.

Cox, D. P., and Singer, D. A., eds. (1986). "Mineral Deposit Models," Bull. No. 1693. U.S. Geol. Surv. Washington, DC. 379pp.

Dapples, E. C., and Hopkins, M. E., eds. (1969). "Environments of Coal Deposition," Spec. Pap. No. 114. Geol. Soc. Am., Boulder, CO. 204pp.

Davies, T. A., and Supko, P. R. (1973). Oceanic sediments and their diagenesis: Some examples from deep-sea drilling. *J. Sediment. Petrol.* **43,** 381–390.

Deffayes, K. S., Lucia, F. J., and Weyl, P. K. (1965). Dolomitization of Recent and Plio-Pleistocene sediments by marine evaporite waters of south Bonaire, Netherlands Antilles. *Spec. Publ.— Soc. Econ. Paleontol. Mineral.* **13,** 71–88.

Dunham, R. J. (1962). Classification of carbonate rocks according to depositional texture. *In* "Classification of Carbonate Rocks: A Symposium" (W. E. Ham, ed.), pp. 108–121. Am. Assoc. Pet. Geol., Tulsa, OK.

Dunham, R. J. (1969). Vadose pisolite in the Capitan Reef (Permian), New Mexico and Texas. *Spec. Publ.— Soc. Econ. Paleontol. Mineral.* **14,** 182–191.

Dunnington, H. V. (1967). Aspects of diagenesis and shape change in stylolitic limestone reservoirs. *Proc.—World Pet. Congr.* **7,** 339–352.

Dunsmore, H. E. (1973). Diagenetic processes of lead-zinc emplacement in carbonates. *Trans.— Inst. Min. Metall. Sect. B* **82,** B168–B173.

Elliott, T. (1975). The sedimentary history of a delta lobe from a Yoredale (Carboniferous) cyclothem. *Proc. Yorks. Geol. Soc.* **40,** 505–536.

Ernst, W. G., and Calvert, S. E. (1969). An experimental study of the recrystallization of porcellanite and its bearing on the origin of some bedded cherts. *Am. J. Sci.* **267A,** 114–133.

Ethridge, F. G., and Flores, R. M. (1981). Recent and ancient non-marine depositional environments: Models for exploration. *Spec. Publ.— Soc. Econ. Paleontol. Mineral.* **31,** 1–349.

Eugster, H. P. (1985). Oil shales, evaporites and ore deposits. *Geochim. Cosmochim. Acta* **49,** 619–635.

Eugster, H. P. (1986). Reply to the discussion by Zbigniew Salowicz. *Geochim. Cosmochim. Acta* **50,** 1831–1832.

Evamy, B. D. (1967). Dedolomitization and the development of rhombohedral pores in limestones. *J. Sediment. Petrol.* **37,** 1204–1215.

Evans, G. E., Schmidt, V., Bush, P., and Nelson, H. (1969). Stratigraphy and geologic history of the sabkha, Abu Dhabi, Persian Gulf. *Sedimentology* **12,** 145–159.

Falls, D. L., and Textoris, D. A. (1972). Size, grain type and mineralogical relationships in Recent marine calcareous beach sands. *Sediment. Geol.* **7,** 89–102.

Feazel, C. T., and Schatzinger, R. A. (1985). Prevention of carbonate cementation in petroleum reservoirs. *Spec. Publ. — Soc. Econ. Paleontol. Mineral.* **36,** 97–106.

Fielding, C. D. (1985). Coal depositional models and the distinction between alluvial and delta plain environments. *Sediment. Geol.* **42,** 41–48.

Fleming, R. H. (1957). General features of the oceans. *Mem. — Geol. Soc. Am.* **67,** 87–108.

Folk, R. D. (1962). Spectral subdivision of limestone types. *In* "Classification of Carbonate Rocks: A Symposium" (W. E. Ham, ed.), pp. 62–84. Am. Assoc. Pet. Geol., Tulsa, OK.

Folk, R. L. (1965). Some aspects of recrystallization in ancient limestones. *Spec. Publ. — Soc. Econ. Paleontol. Mineral.* **13,** 14–48.

Folk, R. L. (1973). Evidence for peritidal deposition of the Devonian Caballos novaculite, Marathon, Texas. *Am. Assoc. Pet. Geol. Bull.* **57,** 702–725.

Folk, R. L., and Land, L. S. (1974). Mg/Ca ratio and salinity: two controls over crystallization of dolomite. *AAPG Bull.* **59,** 60–68.

Folk, R. L., and McBride, E. F. (1976). The Caballos novaculite revisited. Part I: Origin of novaculite members. *J. Sediment. Petrol.* **46,** 659–669.

Follini, K.B. (1996). The phosphorous cycle, phosphogenesis and marine phosphate-rich deposits. *Earth Sci. Rev.* **40,** 55–124.

Friedman, G. M. (1964). Early diagenesis and lithifaction in carbonate sediments. *J. Sediment. Petrol.* **34,** 777–813.

Friedman, G. M. (1979). Dolomite is evaporite mineral — Evidence from rock record and from sea-marginal pools of Red Sea. *AAPG Bull.* **63,** 453.

Fuller, J. G. C. M. (1956). Mississippian rocks and oil fields in southeastern Saskatchewan. *Rep. — Sask., Dep. Miner. Resour.* **19,** 1–72.

Fuller, J. G. C. M., and Porter, J. W. (1969). Evaporite formations with petroleum reservoirs in Devonian and Mississippian of Alberta, Saskatchewan and North Dakota. *Am. Assoc. Pet. Geol. Bull.* **53,** 909–926.

Galloway, W. E., and Hobday, D. K. (1983). "Terrigenous Clastic Depositional Systems." Springer-Verlag, Berlin. 423pp.

Gayer, R., and Harris, I., eds. (1996). Coalbed Methane and Coal Geology," Spec. Publ. No. 107. Geol. Soc. London, London. 344pp.

Goodwin, A. M. (1973). Plate tectonics and evolution of Pre-Cambrian crust. In "Implications of Continental Drift to the Earth Sciences" (D. H. Tarling and S. K. Runcorn, eds.), Vol. 2, pp. 1047–1069. Academic Press, London and New York.

Goudarzi, G. H. (1971). Geology of the Shatti Valley area iron deposit. *In* "The Geology of Libya" (C. Grey, ed.), pp. 491–500. University of Libya, Tripoli.

Gross, G. A. (1972). Primary features in cherty iron-formations. *Sediment. Geol.* **7,** 241–262.

Guilbert, J. M., and Park, C. F. (1986). "The Geology of Ore Deposits." Freeman, New York. 985pp.

Hakansson, E., Bromley, R. G., and Perch-Neilson, K. (1974). Maastrichtian chalk of north-east Europe — a pelagic shelf sediment. *Spec. Publ. — Int. Assoc. Sedimentol.* **1,** 211–233.

Halbouty, M. T. (1967). "Saltdomes — Gulf Region, United States and Mexico." Gulf Publishing, Houston, TX. 425pp.

Hallam, A. (1963). Observations on the palaeoecology and ammonite sequences of the Frodingham Ironstone (Lower Jurassic). *Palaeontology* **6,** 554–574.

Hallam, A. (1981). "Facies Interpretation and the Stratigraphic Record." Freeman, New York. 290pp.

Halley, R. B., and Schmoker, J. W. (1984). High-Porosity Cenozoic carbonate rocks of south Florida: progressive loss of porosity with depth. *AAPG Bull.* **67,** 191–200.

Ham. W. E., ed. (1962). "Classification of Carbonate Rocks: A Symposium." Am. Assoc. Pet. Geol., Tulsa, OK. 279pp.

Ham, W. E., and Pray, L. C. (1962). Modern concepts and classifications of carbonate rocks. *Mem. — Am. Assoc. Pet. Geol.* **1,** 1–279.

Hancock, J. M. (1998). The Cretaceous. *In* "Introduction to the Petroleum Geology of the North Sea" (E. K. Glennie, ed.), 4th ed., pp. 255–272. Blackwell, Oxford.

Harris, L. D. (1958). Syngenetic chert in the Middle Ordovician Hardy Creek limestone of south-western Virginia. *J. Sediment. Petrol.* **28,** 205–208.

Harris, P. M., Frost, S. H., Seigle, G. A., and Schneiderman, N. (1984). Regional unconformities and depositional cycles, Cretaceous of the Arabian Peninsula. *Mem. — Am. Assoc. Pet. Geol.* **36,** 67–80.

Heur, M. D. (1980). Chalk reservoir of the West Ekofisk field. *In* "Sedimentation of North Sea Reservoir Rocks" (R. Hardman, ed.), pp. 1–19. Norw. Pet. Soc., Oslo.

Hilt, C. (1873). Die Beziehung zwischen der Zusammensetzung und der technischen Eigenschaften de Steinkohlen. *VDIZ.* **17,** 194–202.

Holland, H. D. (1972). The geologic history of seawater — an attempt to solve the problem. *Geochim. Cosmochim. Acta* **36,** 637–652.

Hoskin, S. C. W. (1971). Size modes in biogenic carbonate sediment, southeastern Alaska. *J. Sediment. Petrol.* **41,** 1026–1037.

Illing, L. V. (1954). Bahaman calcareous sands. *Bull. Am. Assoc. Pet. Geol.* **38,** 1–59.

Ilyin, A. V. (1998) Phosphorites of the Russian Craton. *Earth Sci. Rev.* **45,** 89–101

Jackson, M. P. A., ed. (1991). "Salt Diapirs of the Great Kavir, Central Iran," Mem. No. 177. Geol. Soc. Am. Boulder, CO. 140pp.

Jackson, M. P. A., Roberts, D. G., and Snelson, S., eds. (1996). "Salt Tectonics: A Global Perspective," Mem. No. 65. Am. Assoc. Pet. Geol., Tulsa, OK. 454pp.

Jefferies, R. (1963). The stratigraphy of the *Actinocamax plenus* subzone (Turonian) in the Anglo-Paris Basin. *Proc. Geol. Assoc.* **74,** 1–34.

Kendall, A. C. (1984). Evaporites. *In* "Facies Models" (R. G. Walker, ed.), Repr. Ser. No. 1, pp. 259–298. Geoscience Canada. Toronto.

Kendall, C. G. St. C. (1969). An environmental re-interpretation of the Permian evaporite/carbonate shelf sediments of the Guadalupe Mountains. *Geol. Soc. Am. Bull.* **80,** 2503–2526.

Kent, B. H. (1986). Evolution of thick coal deposits in the Powder River basin of northeastern Wyoming. *Spec. Pap. — Geol. Soc. Am.* **210,** 105–122.

Khvorova, I. V. (1968). Geosynclinal siliceous rocks and some problems of their origin. *Int. Geol. Congr., Rep. Sess., 23rd,* Prague, Sect. 8, pp. 105–112.

Kimberley, M. M. (1979). Origin of oolitic iron formations. *J. Sediment. Petrol.* **49,** 111–132.

Kirkland, D. W., and Evans, R. (1973). "Marine Evaporites: Origins, Diagenesis and Geochemistry," Benchmark Papers in Geology. Dowden, Hutchinson & Ross, Stroudsburg, PA. 426pp.

Kyle, J. R., and Posey, H. H. (1991). Halokinesis, cap rock development and salt mineral resources. *In* "Evaporites, Petroleum and Mineral Resources" (J. L. Melvin, ed.), pp. 413–476. Elsevier, Amsterdam.

Langres, G. L., Robertson, J. O., and Chilingar, G. V. (1972). "Secondary Recovery and Carbonate Reservoirs." Elsevier, Amsterdam. 250pp.

Larsen, G., and Chilingar, G. V. (1967). "Diagenesis in Sediments." Elsevier, Amsterdam. 551pp.

Leighton, M. W., and Pendexter, G. (1962). Carbonate rock type. *Mem. — Am. Assoc. Pet. Geol.* **1,** 33–61.

Levorsen, A. I. (1934). Relation of oil and gas pools to unconformities in the mid-continent region. *In* "Problems of Petroleum Geology" (R. E. Wrather and L. H. Lahee, eds.), pp. 761–784. Am. Assoc. Pet. Geol., Tulsa, OK.

Lindstrom, M. (1979). Diagenesis of Lower Ordovician hardgrounds in Sweden. *Geol. Palaeontol.* **13,** 9–30.

Llewellyn, P. G., Backhouse, J., and Hoskin, I. R. (1969). Lower-Middle Tournaisian miospores from the Hathern Anhydrite Series, Carboniferous Limestone, Leicestershire. *Proc. Geol. Soc.* **1655,** 85–92.

Longman, M. W. (1980). Carbonate diagenetic textures from nearshore diagenetic environments. *AAPG Bull.* **64,** 461–487.

Loucks, R. G., and Sarg, J. F., eds. (1993). "Carbonate Sequence Stratigraphy," Mem. No. 57. Am. Assoc. Pet. Geol., Tulsa, OK. 545pp.

Lowenstam, H. A. (1963). Biologic problems relating to the composition and diagenesis of sediments. *In* "The Earth Sciences: Problems and Progress in Current Research" (T. W. Donelly, ed.), pp. 137–195. University of Chicago Press, Chicago.

Lucia, F. J. (1999). "Carbonate Reservoir Characterization." Springer-Verlag, Berlin. 226pp.

Lyons, P. C., and Rice, C. L., eds. (1986). "Paleoenvironmental and Tectonic Controls in Coal forming Basins of the United States," Spec. Pap. No. 210. Geol. Soc. Am., Boulder, CO. 191pp.

MacDonald, G. J. F. (1953). Anhydrite-gypsum equilibrium relations. *Am. J. Sci.* **251,** 89–898.

MacIntyre, I. G. (1985). Submarine cements — the peloidal question. *Spec. Publ. — Soc. Econ. Paleontol. Mineral.* **36,** 109–116.

McBride, E. F. (1979). Silica in sediments: Nodular and bedded cherts. *Spec. Publ. — Soc. Econ. Paleontol. Mineral.* **8,** 1–184.

McKee, E. D., and Gutschick, R. C. (1969). "History of Redwall Limestone of Northern Arizona," Mem. No. 114. Geol. Soc. Am., Boulder, CO. 726pp.

McKelvey, V. E. (1967). Phosphate deposits. *Geol. Surv. Bull. (U.S.)* **1252-D,** 1–21.

McKenzie, J. (1981). Holocene Dolomitization of Calcium Carboate Sediments from the Coastal Sabkhas of Abu Dhabi, U.A.E.: A Stable Isotope Study. *J. Geol.* **89,** 185–198.

Mel'nik, Y. P. (1982). "Precambrian Banded Iron Formations." Elsevier, Amsterdam. 310pp.

Melvin, J. L., ed. (1991). "Evaporites, Petroleum and Mineral Resources." Elsevier, Amsterdam. 556pp.

Merritt, R. D., and McGee, D. L. (1986). Depositional environment and resource potential of Cretaceous coal-bearing strata at Chignik and Herendeen Bay, Alaska peninsula. *Sediment. Geol.* **49,** 21–50.

Milliman, J. D. (1974). "Marine Carbonates." Springer-Verlag, Berlin. 375pp.

Milliman, J. D., Freile, D., Steiner, R. P., and Wilbur, R. J. (1993). Great Bahama Bank aragonite mud: mostly inorganically precipitated. *J. Sediment. Petrol.* **63,** 589–695.

Morrow, D. W. (1982a). Dolomite. Part I. *Geosci. Can.* **9,** 5–13.

Morrow, D. W. (1982b). Dolomite. Part II. *Geosci. Can.* **9,** 95–107.

Murray, R. C. (1960). Origin of porosity in carbonate rocks. *J. Sediment. Petrol.* **30,** 59–84.

Murray, R. C. (1964). Origin and diagenesis of gypsum and anhydrite. *J. Sediment. Petrol.* **34,** 512–523.

Notholt, A. J. G., Sheldon, R. P., and Davidson, D. F., eds. (1989). "Phosphate Deposits of the World," Vol. 2. Cambridge University Press, Cambridge, UK. 753pp.

Nriagu, A., and Moore, P. B. (1985). "Phosphate Minerals." Springer-Verlag, Berlin. 442pp.

Nunn, J. F. (1998). Evolution of the atmosphere. *Proc. Geol. Assoc.* **109,** 1–14.

Park, W., and Schot, D. H. (1968). Stylolitization in carbonate rocks. In "Carbonate Sedimentology in Central Europe" (G. M. Friedman and G. Muller, eds.), pp. 66–74. Springer-Verlag, Berlin.

Parker, R. J., and Siesser, W. G. (1972). Petrology and origin of some phosphorites from the Southern African continental margin. *J. Sediment. Petrol.* **42,** 431–440.

Parrish, J. T. (1995). Paleogeography of C_{org} — rich rocks and the preservation versus production controversy. *Stud. Geol. (Tulsa, Okla.)* **40,** 1–21.

Patijn, R. J. H. (1964). Die Entstehung von Erdgas — Erdol und Kohle. *Erdgas Petrochem.* **17,** 2–9.

Pedley, H. M., and Bennett, S. M. (1985). Phosphorites, hardgrounds and syndepositional solution subsidence: A palaeoenvironmental model for the Miocene of the Maltese Islands. *Sediment. Geol.* **45,** 1–34.

Peterson, M. N. A., and von der Borch, C. C. (1965). Chert: Modern inorganic deposition in a carbonate-precipitating locality. *Science* **149,** 1501–1503.

Purser, B. (1978). Early diagenesis and the preservation of porosity in Jurassic limestones. *JPT, J. Pet. Geol.* **1,** 83–94.

Purser, B. H., Tucker, M. E., and Zenger, D. H., eds. (1994). "Dolomites." Blackwell, Oxford. 464pp.

Rahman, R. A., and Flores, R. M. (1984). "Sedimentology of Coal and Coal-bearing Sequences," Spec. Publ. No. 7. Int. Assoc. Sedimentol. Blackwell, Oxford. 417pp.

Reeckmann, A., and Friedman, G. M. (1982). "Exploration for Carbonate Petroleum Reservoirs." Wiley, Chichester. 213pp.

Reijers, T. J. A., and Hsu, K. J. (1986). "Manual of Carbonate Sedimentology: A Lexicographical Approach." Academic Press, London. 302pp.

Robin, P. F. (1978). Pressure solution at grain-to-grain contacts. *Geochim. Cosmochim. Acta* **2,** 1383–1389.

Sawlowicz, Z. (1986). Comment on 'Oil shales, evaporites and ore deposits' by Hans P. Eugster. *Geochim. Cosmochim. Acta* **50,** 1829–1830.

Schneidermann, N., and Harris, P. M., eds. (1985). "Carbonate Cements," Spec. Publ. No. 36. Soc. Econ. Paleontol. Mineral., Tulsa, OK. 379pp.

Scholle, P. A. (1977). Chalk diagenesis and its relation to petroleum exploration: Oil from chalks, a modern miracle. *AAPG Bull.* **61,** 982–1009.

Scholle, P. A. (1978). Carbonate rock constituents, textures, cements and porosities. *Mem. — Am. Assoc. Pet. Geol.* **27,** 1–241.

Scholle, P. A., Arther, M. A., and Elkdale, A. A. (1983). Pelagic environment. *Mem. — Am. Assoc. Pet. Geol.* **33,** 619–692.

Scholle, P.A., Albrecht, T., and Tirsgaard, H. (1998). Form and diagenesis of bedding cycles in uppermost Cretaceous Chalks of the Dan Field, Danish North Sea. *Sedimentology* **45,** 223–245.

Schroder, J., and Purser, B. H. (1986). "Reef Diagenesis." Springer-Verlag, Berlin. 330pp.

Scoffin, T. P. (1987). "An Introduction to Carbonate Sediments and Rocks." Blackie, London. 274pp.

Scott, A. C. (1987). "Coal and Coal-bearing Strata: Recent Advances," Spec. Publ. No. 32. Geol. Soc. London, London. 340pp.

Shearman, D. J. (1966). Origin of marine evaporites by diagenesis. *Trans. — Inst. Min. Metall., Ser. B* **75,** 208–215.

Shearman, D. J., Khouri, J., and Taha, S. (1961). On the replacement of dolomite by calcite in some Mesozoic limestones from the French Jura. *Proc. Geol. Assoc.* **72,** 1–12.

Shlanger, S. O. (1964). Petrology of limestones of Guam. *Geol. Surv. Prof. Pap. (U.S.)* **403,** D1–D52.

Sieveking, G. de G., and Hart, M.B., eds. (1986). "The Scientific Study of Flint and Chert." Cambridge University Press, Cambridge, UK. 290pp.

Slansky, M. (1986). "Geology of Sedimentary Phosphates." North Oxford Academic, London. 211pp.

Sloss, L. L. (1969). Evaporite deposition from layered solutions. *Am. Assoc. Pet. Geol. Bull.* **53,** 776–789.

Smosna, R. (1987). Compositional maturity of limestones — a review. *Sediment. Geol.* **51,** 137–146.

Sonnenfeld, P. (1984). "Brines and Evaporites." Academic Press, London. 624pp.

Sonnenfeld, P. (1985). Evaporites as source rocks of oil and gas. *J. Pet. Geol.* **8,** 253–271.

Sorby, H. C. (1856). On the origin of the Cleveland Hill Ironstone. *Proc. Geol. Polytech. Soc. West Riding Yorks.* **3,** 257–461.

Spring, D., and Hansen, O. P. (1998). The influence of platform morphology and sea level on the development of a carbonate sequence: the Harash Formation, Eastern Sirt Basin, Libya. *In* Petroleum Geology of North Africa. (Macgregor, D. S., Moody, R. T. J., and D. D. Clark-Lowes, eds.) pp. 335–353. Geol. Soc. Lond. Sp. Pub. No. 132.

Steinitz, G. (1970). Chert "dike" structures seen in the Senonian chert beds, southern Negev, Israel. *J. Sediment. Petrol.* **28,** 205–208.

Stopes, M. C. (1919). On the four visible ingredients in banded bituminous coal. *Proc. R. Soc. London, Ser. B* **90,** 69–87.

Taylor, J. C. M., and Illing, L. V. (1969). Holocene intertidal calcium carbonate cementation, Qatar, Persian Gulf. *Sedimentology* **12,** 6–107.

Taylor, J. H. (1949). Petrology of Northampton sand ironstone formation. *Mem. Geol. Surv. G.B. Engl. Wales, Explan. Sheet,* pp. 1–111.

Tooms, J. S., Summerhayes, C. P., and McMaster, R. L. (1971). Marine geological studies on the north-west African margin: Rabat-Dakar. *Rep. — Inst. Geol. Sci. (U.K.)* **70/16,** 11–25.

Trask, C. B., and Palmer, J. E. (1986). Structure and depositional history of the Pennsylvanian system in Illinois. *Spec. Pap. — Geol. Soc. Am.* **210,** 63–78.

Trendall, A. F. (1968). Three great basins of PreCambrian banded iron formation deposition: A systematic comparison. *Bull. Geol. Soc. Am.* **79,** 1527–1544.

Tucker, M. E., and Wright, V. P. (1990). "Carbonate Sedimentology." Blackwell Science, Oxford. 496pp.

Turk, T. M., Doughri, A. K., and Banerjee, S. (1980). A review of the recent investigation on the Wadi as Shati iron ore deposits, northern Fazzan, Libya. *In* "The Geology of Libya" (M. J. Salem and M. T. Busrewil, eds.), pp. 1045–1050. Academic Press, London.

UNESCO (1973). Genesis of PreCambrian iron and manganese deposits. *Proc. Kiev. Symp., 1970,* pp. 1–382.

Usiglio, J. (1849). Analyse de l'eau de la Méditerranée sur les Côtes de France. *Ann. Chim. Phys.* **27,** 92–107.

Van Houten, F. B., and Arthur, M. A. (1989). Temporal patterns among Phanerozoic oolitic ironstones and oceanic anoxia. *In* "Phanerozoic Ironstones" (T. P. Young and W. E. G. Taylor, eds.), Spec. Publ. No. 46, Geol. Soc. London, London. pp. 33–49.

Van't Hoff, J. H., and Weigert, F. (1901). Untersuchungen uber die Bildungsverhalnisse der oceanischen Salzablagerungen, unsbesondere des Stass furter Salzlagers. *Sitzungsbor. Preuss. Akad. Wiss.* **23,** 114–1148.

von der Borch, C. C. (1976). Stratigraphy and formation of Holocene dolomitic carbonate deposits of the Coorong area, South Australia. *J. Sediment. Petrol.* **46,** 952–966.

Wanless, H. R., Baroffio, I. R., and Trescott, P. C. (1969). Conditions of deposition of Pennyslvanian coal beds. *Spec. Pap.— Geol. Soc. Am.* **114,** 105–142.

Wardlaw, N. C., and Schwerdtner, W. M. (1966). Halite-anhydrite seasonal layers in the Middle Devonian Prairie Evaporite Formation. Saskatchewan, Canada. *Geol. Soc. Am. Bull.* **77,** 331–342.

Watson, H. J. (1982). Casablanca Field, offshore Spain, a paleogeomorphic trap. *Mem.— Am. Assoc. Pet. Geol.* **32,** 237–250.

Wells, A. J. (1986). The dolomite enigma. *In* "Geology in the Real World" (R. W. Nesbitt and I. Nichol, eds.), pp. 465–473. Inst. Min. Metall., London.

Wells, A. J., and Illing, L. V. (1964). Present day precipitation of calcium carbonate in the Persian Gulf. *In* "Deltaic and Shallow Marine Deposits" (L. M. J. U. Van Straaten, ed.), pp. 429–435. Elsevier, Amsterdam.

Wenk, E. (1949). Die Assoziation von Radiolarienhornsteinen mit ophiolithischen Erstarrungsgesteinen als petrogenetisches. *Probl. Exp.* **6,** 226–232.

West, I. (1964). Evaporite diagenesis in the Lower Purbeck Beds of Dorset. *Proc. Yorks. Geol. Soc.* **34,** 315–330.

Weyl, P. K. (1958). The solution kinetics of calcite. *J. Geol.* **66,** 163–176.

Wood, G. V., and Wolfe, M. J. (1969). Sabkha cycles in Arab/Darb formation of the Trucial Coast of Arabia. *Sedimentology* **12,** 165–191.

Young, T. P., and Taylor, W. E. G., eds. (1989). "Phanerozoic Ironstones," Spec. Publ. No. 46. Geol. Soc. London, London. 251pp.

10 Sedimentary Basins

10.1 INTRODUCTION

10.1.1 Sediment Supply and Accommodation Space

This book has told the tale of how rock is broken down to sediment, how sediment is transported and deposited, and then turned back to rock again. This, the concluding chapter, describes the sedimentary basins within which the last acts of the great saga unfold. This chapter attempts to answer these questions: Why should sedimentation occur in one place at a particular time? What is the spatial organization of large volumes of sediment? What are the factors that control their facies? And how do petroleum and mineralizing fluids move within basins?

Chapter 6 introduced the concepts of environments of erosion, equilibrium, and deposition. Recall that erosion normally occurs in mountainous regions. Equilibrial surfaces characterize many deserts and continental shelves. Depositional environments occur in both open marine basins and closed continental ones.

For sediment to accumulate in a marine basin there must be a vertical interval, termed the **accommodation space,** between sea level and sea bed, the top of the accumulation (Jervey, 1988). When examined closely, however, the concept of accommodation space becomes complex. In marine environments, tide, wind, and wave-generated currents may be so strong that sediment can never accumulate to sea level, but is constantly redistributed across the shelf or into deeper water. In a closed continental basin the concept of accommodation space becomes still more complicated. In lake basins the accommodation space may still be defined as the interval between the lake bed and the surface. In continental environments, however, the accommodation space is defined as the interval between the top of the accumulation and the equilibrial surface. In eolian environments this may often be taken as the water table (Kocurek and Havholm, 1993). This is because sand can only accumulate and be preserved below the water table. Dry sand above this surface will be constantly reworked and transported. For enclosed fluvial basins the equilibrial surface is a plane level with the spill point of the basin (i.e., the lowest col through which sediment will be transported once accumulation reaches the height of the col).

The rate at which the accommodation space is infilled is a function of several variables in the basin and in the drainage catchment area. Variables within the basin include the environment and its energy level, that is, the ability of wind, wave, and tidal currents to transport, segregate, and deposit sediment. Other critical intra-basinal variables include changes in sea level, which affect the top of the accommodation space, and tectonism, principally rate of subsidence, which affects the floor of the accommodation space.

The rate of sediment supply into a basin is dependent on many variables. These include the type and intensity of weathering, erosion, and transportation within the drainage catchment area. Most of these variables are closely related to climate (Leeder *et al.*, 1998). Glacial, temperate, arid, and tropical climates all generate characteristic weathering profiles and have their own erosional and transportational characteristics, as described in Chapters 2 and 4.

Vegetation is another important controlling parameter, closely related to climate. Prior to the colonization of the land by plants in the Devonian Period, runoff was uninhibited by plants, their roots, and thick humus-rich soil. Except in arid climates, runoff has been much slower in post-Devonian times, due to the presence of vegetation. This is demonstrated by the fact that all Pre-Cambrian and pre-Devonian Phanerozoic alluvium shows the characteristics of deposition from ephemeral braided channel systems. Characteristic features of deposition from meandering channel systems occur only in post-Silurian alluvium (refer back to Section 6.3.2.2.3).

The chemical composition and physical parameters of the rock in the source area will also play an important part in controlling both the volume and mineralogy of the sediment supplied to a basin. For instance, a hinterland of igneous and metamorphic rocks will supply a diverse range of detrital minerals to a basin, together with sodium, potassium, calcium, magnesium, chloride, sulfate, phosphate, and carbonate ions, carried in solution in groundwater. By contrast, a hinterland of pure quartz sand can only contribute a detrital load of the same, or, to take another extreme situation, a limestone terrain will contribute little detritus to a basin, though the runoff will be rich in calcium, magnesium, and carbonate ions.

Figure 10.1 attemps to display the various factors just outlined that control the generation of sediment in a catchment area, and its dispersal within the accommodation space of a basin.

10.1.2 Basic Concepts and Terminology

The geographic distribution and organization of sediments is now described. This topic can be loosely referred to as basin analysis (Miall, 1984; Klein, 1987; Leeder, 1999). Sedimentary rocks cover a large part of the earth's surface, including some 75% of the land areas. Yet sedimentary rocks make up only 5% of the lithosphere (data from Pettijohn, 1957, p. 7). From this it follows that sedimentary rocks cover the earth only as a thin and superficial veneer. This cover is not evenly distributed. Thick sedimentary deposits were laid down in localized areas, loosely termed "sedimentary basins." Large areas of the continents lack a thick sedimentary cover; Pre-Cambrian igneous and metamorphic rocks crop out at or near the surface. These stable continental cores are termed "cratons." Sedimentary basins are of many types, ranging from small alluvial intermontane valleys to vast mountain ranges of contorted sediment, kilometers thick. Before proceeding to describe these various basins, some terms and concepts are first defined.

Basins are of three types: topographic, structural, and sedimentary. Topographic basins are low-lying areas of the earth's surface naturally surrounded by higher areas. Topographic basins are both subaerial and subaqueous. Subaerial topographic basins range from bolsons (intermontane plains) to transcontinental alluvial valleys such as the

SEDIMENT SUPPLY CONTROLLED BY:

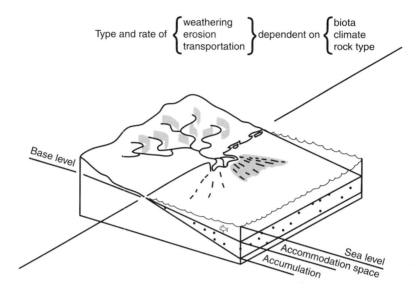

Type and rate of $\left\{\begin{array}{l}\text{weathering}\\ \text{erosion}\\ \text{transportation}\end{array}\right\}$ dependent on $\left\{\begin{array}{l}\text{biota}\\ \text{climate}\\ \text{rock type}\end{array}\right.$

ACCOMMODATION SPACE CONTROLLED BY:
- Eustatic sea level change
- Tectonic subsidence/uplift
  ability of currents to distribute sediment
- Rate of sediment supply — extrabasinal terrigenous input from the land,
                              intrabasinal formation of autochthonous sediment

Fig. 10.1. Diagram showing the interplay between the variables that control sediment supply to, and accommodation space within, a marine sedimentary basin. For complexities of the concept of accommodation space in marine and nonmarine basins, see text.

Amazon basin. Subaqueous topographic basins range from periglacial pingos to oceans. The existence of a topographic basin is necessary for the genesis sedimentary basin.

A sedimentary basin is an area of gently folded centripetally dipping strata. It is a matter of great importance to distinguish tectonic, or postdepositional, basins from synsedimentary basins. In tectonic basins facies trends and paleocurrents are unrelated to the basin architecture, indicating that subsidence took place after deposition of the deformed strata. In syndepositional basins, by contrast, the facies trends, paleocurrents, and depositional thinning of strata toward the basin margin all indicate contemporaneous movement (Fig. 10.2).

Sedimentary basins generally cover tens of thousands of square kilometers, but their sizes are not diagnostic. Distinction is made, however, between basins, embayments, and troughs (Fig. 10.3). Basins, in the restricted sense of the term, are saucer shaped and subcircular in plan. Embayments are basins that are not completely closed structurally, but which open out into a deeper area. Troughs are linear basins. These three basin types are essentially unmetamorphosed and undeformed by tectonism. These features distinguish them from the linear metamorphosed tectonized sedimentary troughs that were once termed "geosynclines," but are now more usually termed fore-arc basins. The thinning

Post - depositional basin   Syndepositional basin

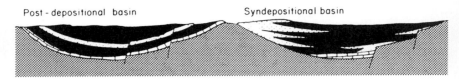

Fig. 10.2. Section illustrating the differences between postdepositional basins and syndepositional basins. The sedimentary facies of syndepositional basins reflect the position of the basement margin and the movement of faults. In postdepositional basins stratigraphy is discordant with structure.

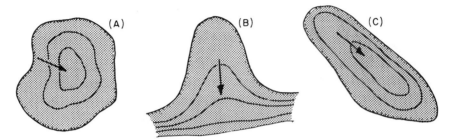

Fig. 10.3. Diagrams illustrating the nomenclature of basins. Lines indicate structure contours, arrows indicate depositional paleoslope. (A) A basin in the strictest sense of the term, subcircular in plan and structurally closed. (B) An embayment, opening out at one side to a structurally lower area. (C) A trough, structurally closed and elongated.

of the sedimentary cover toward the basin margin may be erosional or nondepositional, as shown by intraformational thinning. These margin types differentiate postdepositional from syndepositional tectonism. The axis of a basin is a line connecting the lowest structural points of the basin, as in a synclinal axis. Similarly the axes of troughs may plunge. The depocenter is the part of the basin with the thickest sedimentary fill.

It is very important to note that the depocenter and basin axis need not coincide with one another, nor indeed with the topographic axis (Fig. 10.4). This is particularly true of asymmetric basins with large amounts of terrigenous sedimentation on the limb of maximum uplift. In gently subsiding basins, with pelagic fine-grained and turbidite fill, depocenter, axis, and topographic nadir may coincide. These points should be considered when deciding whether a regional isopach map gives a valid picture of syndepositional basin architecture. It is a common feature of many basins that the depocenter moves across the basin in time (Fig. 10.5). This may reflect a migration of the

Topographic axis

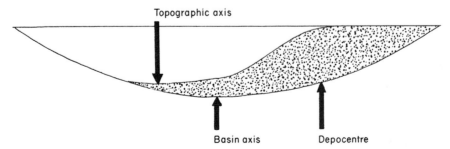

Basin axis   Depocentre

Fig. 10.4. Cross-section illustrating that topographic and structural axes need be neither coincident with each other, nor with the site of maximum sedimentation (the depocenter). Isopach maps are not necessarily, therefore, reliable measure of paleotopography and subsidence, especially in carbonate basins.

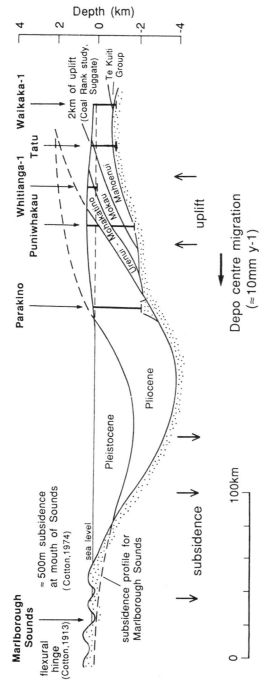

Fig. 10.5. Cross-section through the Wanganui basin, New Zealand, illustrating the westerly migration of the depocenter with time. [Reprinted from Stern, T. A., Quinlan, G. M., and Holt, W. E. 1993. Crustal dynamics associated with the formation of the Wanganui basin, New Zealand. *In* "South Pacific Sedimentary Basins" (P. F. Ballance, ed.), pp. 213–224, with permission from Elsevier Science.]

topographic axis of the basin, or merely a lateral progradation of the main locus of deposition across an essentially stable basin floor. The danger of using an isopach map as an measure of subsidence is particularly acute in carbonate basins. As discussed in Chapter 6, carbonate sedimentation rates are relatively slow inshore and in the open sea, and are at their fastest in between, forming an accretionary ramp. Thus carbonate depocenters are not coincident with basin axis.

Basins are separated one from another by raised linear areas where the sediment cover is thin or absent. These are variously termed arches, paleo-highs, schwelle, axes of uplift, or positive areas. Similarly, major basins are commonly divisible into sub-basins, troughs, and embayments by smaller positive features.

10.1.3 Basin-Forming Mechanisms

Accommodation space may be created by three tectonic processes (Stoneley, 1969). Subsidence may occur where subcrustal displacement of the mantle leads to down-dragging and compressional warping of the crust. This occurs principally at what are called zones of subduction, linear features that are the site of extensive sedimentation. Sedimentation may also occur on a large scale where changes in the mantle cause foundering and subsidence of the crust. This process is responsible for intracratonic basins. Conversely these changes can cause the crust to dome. Thick volcanic and sedimentary sequences may form in crestal rift basins. Finally, thick sedimentary sequences may form where the weight of the sediment itself causes isostatic depression of the crust. This process obviously requires an outside mechanism to create an initial crustal void, since it poses the old problem of which came first, the hen or the egg? The most likely place for such a process is the continental margin, where a whole ocean basin waits to be infilled. Sedimentation at the foot of the continental slope may cause isostatic depression of the crust (Drake *et al.*, 1968).

Considerable attention has been paid to the geomechanics of basin subsidence, a problem requiring geophysics and structural analysis for its elucidation (e.g., McKenzie, 1978; Neugebauer, 1987; Allen and Allen, 1990; Busby and Ingersoll, 1995; Kearey and Vine, 1996; Condie, 1997; Lillie, 1999).

Once upon a time geology students were taught that the continents were composed largely of silica and alumina (sial) and floated isostatically on denser oceanic crust composed largely of silica and magnesia (sima). Mountain chains occurred where continents bumped together pushing up folded belts of sediments deposited in the troughs between the continents. The compression of the deposits of the Tethys Ocean to form the Alpine mountain chain was the classic example, the motive power in this case being the convergence of the European and African shields (Pfiffner *et al.*, 1998). Mountain chains such as the Appalachians and the Andes were hard to fit into so simple a scheme, as half of the vice was absent. One explanation offered was that the continent had foundered on the oceanic side of such mountain chains, in apparent defiance of the principles of isostasy.

An alternative proposal was that continents could drift horizontally across the face of the earth. The close geographic and geologic fit of the circum-Atlantic continents was the keystone of this thesis (Wegener, 1924; du Toit, 1937). These ideas, rejuvenated

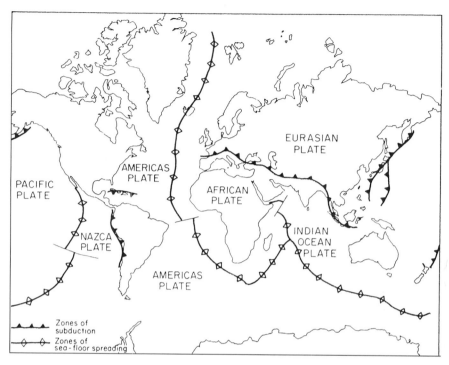

Fig. 10.6. World map showing the major zones of sea floor spreading and zones of subduction. New crust is generated at the former by upwelling along axial volcanic rifts, and drawn down into the mantle at the latter (simplified from various sources).

by advances in oceanography and geophysics in the 1960s, led to the formulation of a refined model variously referred to as "plate tectonics" or the "new global tectonics." The literature on this topic is vast. Source books include Wyllie (1971), Seyfert and Sirkin (1973), Tarling and Runcorn (1973), Fischer and Judson (1975), Davies and Runcorn (1980), Van Andel (1985), Cox and Hart (1986), Meissner (1986), and Klein (1987). The following brief account of these concepts is given to illustrate their relevance to basin analysis and sedimentology.

Plate tectonics is based on the evidence that the surface of the earth is made up of a mosaic of rigid plates (Fig. 10.6). These rigid plates, termed the lithosphere, consist of a continuous layer of basaltic rocks above which is a discontinuous layer of granitic and sedimentary continental crust. The upper mantle layer corresponds essentially to "sima" and forms the floor of the ocean basins. The granitoid continental crust corresponds to "sial."

The evidence suggests not that the crust moves over the upper mantle, but that the lithosphere (crust and upper mantle together) may ride over the deeper asthenosphere (lower mantle). Geophysical data suggest that the boundary between rigid lithosphere and plastic asthenosphere lies between 100 and 150 km beneath the earth's surface (Fig. 10.7). New lithospheric material is added to each plate at zones of sea floor spreading. These are the mid-oceanic ridges. These ridges are seismically and volcanically active, they have a high rate of heat flow, and are topographically and paleomagnetically

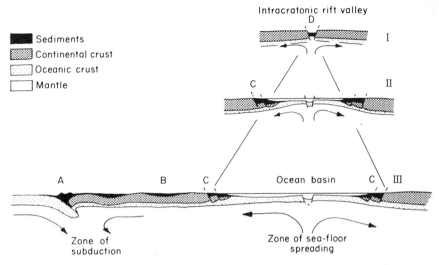

Fig. 10.7. Cross-sections illustrating the basic concepts of plate tectonics. The crust and upper part of the mantle (the lithospere) are rigid, and move over the plastic lower mantle (asthenospere). I: An axis of sea floor spreading develops, in this instance beneath the continental crust. Updoming occurs and a volcanic rift valley is formed (D). The East African rifts are a modern example. II: With continued upwelling the two new plates are torn apart. New oceanic crust is formed by extensive vulcanicity between two flanking half-graben basins (C). The Red Sea is now at this stage. III: New crust continues to be formed at the axial zone of sea floor spreading. The two continental shields, now far apart, have ocean margin basins on their opposing shores (C). At far left is a zone of subduction down which crust is drawn. This is the site of a fore-deep trough (A). Intracratonic basins occur on the continental crust (B). A cross-section from the Andes across the Atlantic Ocean to Africa is similar to this sketch.

symmetrical on either side of axial rift valleys. The ridges are composed of young basalts, which are overlain by progressively older pelagic sediment away from the ridge.

The new lithosphere generated at these zones of sea floor spreading is carried across the plate, and is ultimately drawn down into the earth along zones of subduction on the far side of the plate. Some zones of subduction occur at the junction of continental and oceanic crustal plates, as for example under Central America, and off the central and southern Andes. Island arcs are formed where two plates of oceanic crust meet in a zone of subduction. Continents move with the plates like lost luggage on an airport carousel.

The concept of plate tectonics is far more complex than the preceding few sentences suggest. Nevertheless, it is one of the most stimulating concepts impinging on all branches of geology. Plate tectonics has important implications for petroleum geology and for mineral exploration. Different types of mineral deposits are associated with different locations and phases of plate evolution (Sawkins, 1984). Figure 10.8 shows, for example, the distribution of Pre-Cambrian banded ironstone formations (Section 9.4.2). Today they are scattered worldwide. When the continental masses are restored to their Pre-Cambrian position, however, it is apparent that their original distribution was not random, but curvilinear upon the Pre-Cambrian crust (Goodwin, 1973).

Similarly, an understanding of plate tectonics helps petroleum exploration. The evolution of the different types of sedimentary basin can be modeled to establish the location of petroleum source beds, the heat flow to which they are subjected, and the migration pathways and loci of entrapment of petroleum fluids (McKenzie, 1981).

Basins formed as a result of crustal thinning and rifting are of particular interest to the petroleum industry because they are an important habitat for petroleum. Many theories have been advanced to explain their formation. Of the many models proposed, three are particularly significant. Salveson (1976, 1979) proposed a model of passive crustal separation in which the continental crust was deemed to deform by brittle failure, while the subcrustal lithosphere is thinned by ductile necking. This model was largely based on studies of the Red Sea and Gulf of Suez rift system.

McKenzie (1978) proposed a model that assumed that both the crust and subcrustal lithosphere deformed by brittle failure. This model was largely based on studies of the North Sea basin.

Wernicke (1981, 1985) proposed a model for crustal thinning by means of simple shear, in which a low-angle fault extends from the surface right through the lithosphere. This model was largely based on studies of the basin and range tectonic province of North America.

These three models are illustrated in Fig. 10.9, left. The McKenzie model has received particular interest in the oil industry because it offers a means of predicting the history of heat flow in a sedimentary basin. This is a prerequisite to the accurate modeling of petroleum generation and sediment diagenesis. Basin formation commences after a prerift idyll phase, during which the crust is in isostatic balance and a state of thermal equilibrium. As rifting develops, the crust thins, heat flux increases, and the temperature of the shallow rocks rises. After rifting has ceased, the crust cools, shrinks, and collapses. Sedimentation continues, but now infills a gently subsiding basin. Faults die out at the top of the syn-rift sediments. The crust returns to thermal equilibrium and a postrift idyll phase. The resultant basin is colloquially referred to as a "steer's head" basin, because it is reminiscent of a Texas Longhorn or Highland Cattle (Fig. 10.9, right).

McKenzie demonstrated mathematically that the heat flow within a basin was related to the amount of crustal stretching, termed the β value. The higher the rate of stretching, the higher the heat flux during the initial phase of rifting (Fig. 10.10, upper). The β value for a basin may be discovered by constructing a burial history curve for the basin and comparing it with known curves calculated for given β factors (Fig. 10.10, lower). The accurate prediction of the history of heat flow in a sedimentary basin is a prerequisite to the accurate modeling of petroleum generation (Dore *et al.,* 1991; Helbig, 1994).

10.2 SEDIMENTARY BASINS CLASSIFIED AND DESCRIBED

10.2.1 Classification of Sedimentary Basins

Attempts to classify the various types of sedimentary basins have been made by many geologists, notably Weeks (1958), Halbouty *et al.* (1970), Perrodon (1971), Klemme (1980), and Allen and Allen (1990). These classifications vary according to the defining parameters which have been chosen, and according to the purpose for which a scheme

BEDDED IRON ORE DEPOSITS

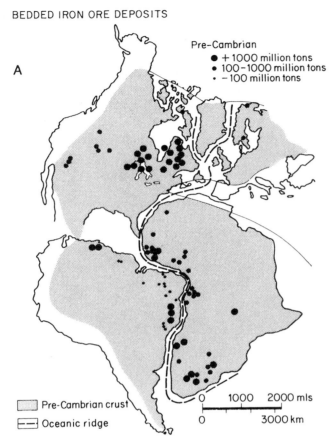

Fig. 10.8. (A, B) Atlantic and Gondwana continental reconstructions showing the distribution of Pre-Cambrian banded ironstone formations. (C) Global reconstruction of the Pre-Cambrian continent, Pangea, showing how the banded ironstone deposits have a curvilinear distribution. (From Goodwin, 1973. Courtesy of Academic Press.)

was drawn up. As with most geological phenomena, basins can be broadly grouped into several fairly well-established families, whose limits are ill-defined. The attempt at basin classification given in Table 10.1 owes much to those already cited. This scheme will be used as a framework for the following description and discussion of sedimentary basins. Figure 10.11 shows how the major types of sedimentary basin can be related to plate tectonic setting.

10.2.2 Crustal Sag Basins

A sedimentary basin, in the restricted sense of the term, is an essentially saucer-shaped area of sedimentary rocks. It is, therefore, subrounded in plan view. Strata dip and thicken centripetally toward the center of the basin. It is an interesting exercise, however, to draw a cross-section of a basin to scale, allowing for the curvature of the earth.

BEDDED IRON ORE DEPOSITS

Pre-Cambrian

● +1000 million tons
● 100-1000 million tons
· -100 million tons

▦ Pre-Cambrian crust
〓 Oceanic ridge

B

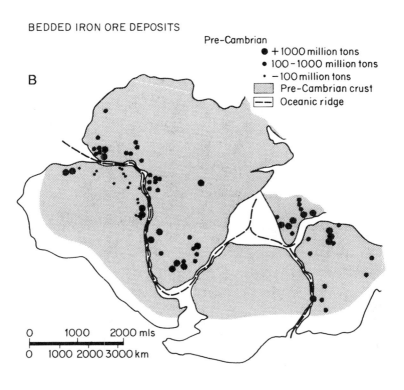

0   1000   2000 mls

0   1000 2000 3000 km

PANGAEA

C

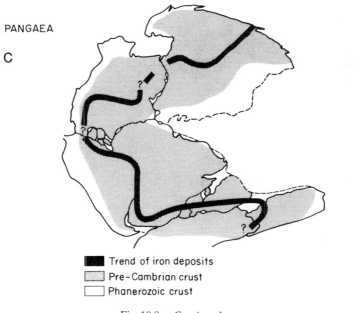

■ Trend of iron deposits
▦ Pre-Cambrian crust
▢ Phanerozoic crust

Fig. 10.8. — *Continued*

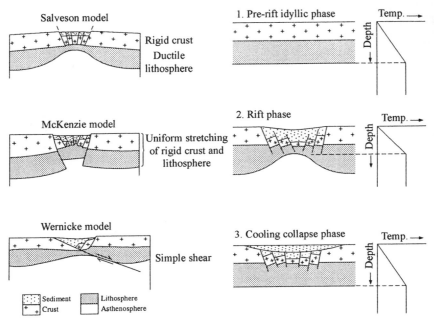

Fig. 10.9. (Left) Geophantasmograms illustrating three popular explanations of basin formation by litho-spheric stretching. For explanations and sources see text. (Right) Geophantasmograms illustrating the Mc-Kenzie model for the formation of basins by lithospheric stretching. The sequence begins with the crust at rest and in thermal equilibrium. Crustal thinning and uplift of the asthenosphere are associated with high heat flow and the formation of a rift basin. Subsequent cooling and shrinkage cause the crust to collapse, gently result-ing in the "steer's head" basin form. Thermal equilibrium is finally reestablished. Peace returns.

This reveals that basins are in fact veneers of sediment on the earth's surface, which are convex to the heavens (Dallmus, 1958).

Simple basins of this type are divisible into two groups. Intracratonic crustal sag ba-sins lie within the continental crust. Epicratonic basins lie on continental crust but are partially open to an ocean basin. These two types often occur adjacent to one another with little fundamental difference in genesis or fill. Descriptions now follow.

10.2.2.1 Intracratonic Crustal Sag Basins

Intracratonic basins are the classic type of sedimentary basin. Modern intracratonic sag basins include the Hudson Bay and the Baltic Sea, which lie on the Canadian and Scandinavian shields, respectively. Notable ancient examples include the Williston and Michigan basins of North America, and the Murzuk and Kufra basins of the Sahara. All of these basins are broadly comparable in shape, scale, and intracratonic setting. They show differences, however, in time of formation and type of fill. The North American examples are predominantly syndepositional carbonate basins. The Saharan examples are postdepositional terrigenous basins.

The Williston basin is a classic example of an intracratonic basin (Dallmus, 1958; Smith *et al.*, 1958; Darling and Wood, 1958). It contains some 3 km of rock of all periods from

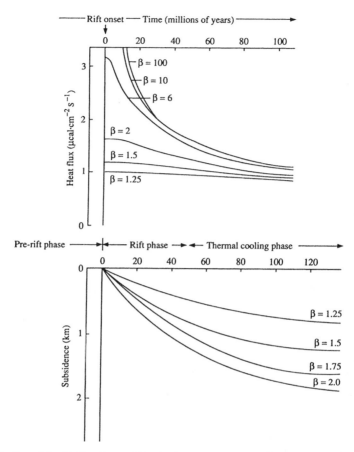

Fig. 10.10. Illustration of the McKenzie model for basin formation by lithospheric stretching. (Upper) Graph of heat flux against time for various β factors. (Lower) Graph of subsidence against time for various β factors. (Developed from McKenzie, 1978.)

Table 10.1
Classification of Sedimentary Basins

Generating process	Basin type	Plate tectonic setting
Crustal sag	Intracratonic basin	Intraplate collapse
Tension	⎰ Epicratonic downwarp ⎱ ⎰ ↑ ⎱ ⎰ Rift ⎱	Passive plate margin Sea floor spreading
Compression	⎰ Trench ⎱ ⎰ Fore-arc ⎱ ⎰ Back-arc ⎱	Subduction (active plate margin)
Wrenching	Strike-slip	Lateral plate movement

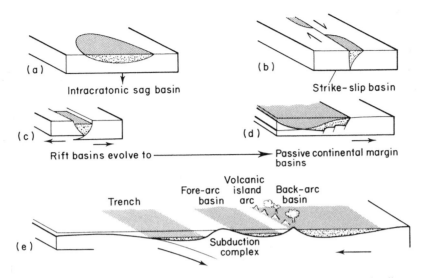

Fig. 10.11. Illustrations of the morphology and plate tectonic setting of the major types of sedimentary basin. (A) Intracratonic sags develop in the middle of continental crust. (B) Strike-slip basins develop where plates slide past one another. A sequence of basins is associated with crustal separation, beginning with rifts (C) and ending in passive ocean margin basins (D). (E) Elongated troughs associated with volcanic island arcs develop at compressive plate boundaries.

Cambrian to Tertiary, with notable gaps only in the Permian and Triassic (Fig. 10.12). Sedimentation spanned a range of environments including fluvial and marine sands, reefal carbonates, evaporites, and subwave-base pelagic muds. Deep-sea, turbidite, and deltaic deposits, igneous activity, and shallow syndepositional faulting are all absent.

The lengthy history, diverse facies and structural simplicity of the Williston basin are also found in the other basins cited as examples of intracratonic basins. It is important to note that the sedimentary facies, though diverse in lithology and environment, are seldom indicative of deep water or abrupt subsidence of the basin floor. Deposition took place close to sea level. Subsidence was thus a gradual, if erratic event, with sedimentation being sufficiently rapid to keep the basin nearly filled at any point in time.

The adjacent Michigan basin is similar to the Williston basin (Cohee and Landes, 1958). It is also infilled with some 4 km of sediment of Cambrian to Late Carboniferous (Pennsylvanian) age (Fig. 10.13). The concentric rim of Silurian reefs, with a corresponding basin fill of evaporites is particulary noteworthy, demonstrating the syndepositional subsidence of the basin (Sears and Lucia, 1979). Deep seismic surveys and boreholes reveal, however, that beneath the saucer-shaped Phanerozoic fill, there is a Pre-Cambrian rift infilled with deep-water sediments (Fowler and Kuenzi, 1978).

The Williston and Michigan basins lie well within the present limit of the continental margins, but it is obvious that they were frequently connected to the open sea. This is shown by their intermittent phases of marine carbonate and evaporite sedimentation. During these periods they might, therefore, be more truly termed embayments rather than basins. Intermittent uplift of the open rim of the embayment closes the basin off from oceanic influence. Evaporite or continental sedimentation follows.

An interesting example of this scenario is provided by the Murzuk and Kufra basins

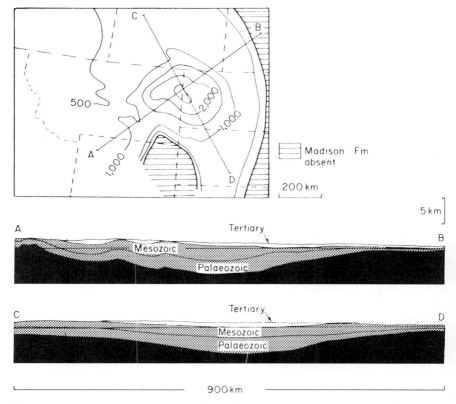

Fig. 10.12. Structure contour map on the top of the Madison Limestone (Mississippian) showing the subcircular shape of the Williston basin. Cross-sections show the many ages of the rocks which infill it. (From Dallmus, 1958. American Association of Petroleum Geologists Bulletin, AAPG © 1958, reprinted by permission of the American Association of Petroleum Geologists whose permission is required for further use.)

of southern Libya (MacGregor *et al.*, 1998). These are both classic intracratonic sag basins with inward-dipping strata at the present day. They share a common stratigraphy of Paleozoic and Mesozoic strata that can be correlated, not only within these two basins, but across most of the Sahara and Arabia (Selley, 1997a). The sequence consists of a mix of continental and shallow marine clastics and minor carbonates. Paleocurrent analysis of fluvial deposits within the Murzuk and Kufra basins reveals a northerly paleoslope, interrupted only by the intervening Tibesti-Sirte arch (Fig. 10.14). These data show that both basins were northerly plunging embayments open to the Tethyan Ocean. They did not become closed basins, separated from the ocean to the north, until the mid-Cretaceous (Selley, 1997b).

This brief review shows that some sag basins, such as the Michigan basin, with its circular rim of Silurian limestone reefs, are clearly syndepositional in origin. Others, however, such as the Murzuk and Kufra basins just described, are clearly postdepositional in origin.

The genesis of circular sag basins has long attracted attention. Several modes of origin have been postulated. One of the most popular proposes that thermal doming over

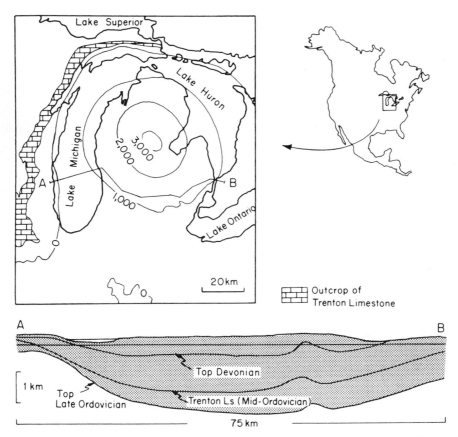

Fig. 10.13. Map and cross-section of the Michigan basin of the Great Lakes, North America. The subcircular basin shape is shown by a structure contour map of the top of the Trenton Limestone, Middle Ordovician. Deep seismic surveys and borehole data reveal the existence of a Pre-Cambrian rift beneath the saucer-shaped Phanerozoic fill. (From Cohee and Landes, 1958. American Association of Petroleum Geologists Bulletin, AAPG © 1958, reprinted by permission of the American Association of Petroleum Geologists whose permission is required for further use.)

a mantle "hot spot" leads to the erosion of uplifted crustal rocks, followed by cooling and crustal collapse, initially into a rift, followed by gentle sag subsidence (Allen and Allen, 1990). It has also been suggested that crustal sags may result from "cold spots" due to mantle cooling, resulting in downwelling and a dignified gentle sagging of the crust (Hartley and Allen, 1994). This model implies that sag basins may lack a precursor rifting phase, and an early high heat flux, important considerations when modeling basins for petroleum generation studies.

10.2.2.2 Epicratonic Basins and Continental Margin Downwarps

Sedimentary basins often occur near the boundary of continental and oceanic crust. Morphologically they range from embayments that plunge toward the ocean, to laterally extensive continental margin downwarps. The latter are typical of passive continental margins that develop at the end of the rift–drift sequence, in which context they will subsequently be described.

Epicratonic embayments, as the first type may be called, tend to be embayed and open toward the adjacent ocean basin. The axis of an epicratonic basin may plunge to the floor of the ocean or be interrupted by a sill-like feature at the rim of the continental margin. Epicratonic basins include those predominantly infilled with clastics, such as the Mississippi Gulf coast, and the Niger delta basin (Weber, 1971; Burke, 1972; Walcott, 1972; Wilhelm and Ewing, 1972; Evamy *et al.*, 1978; Reijers *et al.*, 1997), and those with predominant carbonate fill, such as the Sirte basin of North Africa (Salem *et al.*, 1996a,b,c; Selley, 1997c).

The Mississippi and Niger delta basins are very similar (Salvador, 1991; Reijers *et al.*, 1997). Both originated toward the end of the Mesozoic Era and continue to be sites of active clastic sedimentation until the present day. Both basins contain a basal layer of salt, which diapirically intrudes younger sedimentary rocks. The Louann salt of the Gulf basin is of Jurassic age. The basal salt of the Niger delta basin is believed to be of Albian-Aptian age (Mascle *et al.*, 1973). The Mississippi and Niger delta basins are infilled by prisms of terrigenous clastics which were deposited in a range of environments. On the landward sides of the embayments alluvial deposits predominate. These pass basinward into diverse shoreline facies, which include both barrier and deltaic deposits. These thin and grade seaward into marine slope muds, with some development of turbidite sand facies at the base of the slope (Fig. 10.15).

Geophysical data suggest a gradual seaward thinning of the continental crust beneath both the Mississippi Gulf coast and Niger delta basins (Walcott, 1972, and Burke, 1972, respectively).

The Sirte basin of North Africa shares many features with the Gulf coast and Niger basins. It, too, is essentially an embayment which opens out to an oceanic basin (the Mediterranean). The area now occupied by the Sirte basin was a major north–south uplift that controlled sedimentation throughout the Paleozoic era. This regional arch collapsed in the middle of the Cretaceous Period to form the Sirte embayment (Fig. 10.16).

Basal sands and thin evaporites are overlain by deep-water Upper Cretaceous and Paleocene shales. These are thickest in intra-basinal troughs, while reefal carbonates were deposited on adjacent horsts. Throughout the Eocene the Sirte embayment was infilled by nearly a kilometer of interbedded carbonates and evaporites. The final phase of basin infilling during the Oligocene and Miocene involved terrigenous and carbonate sedimentation in both marine and continental environments. The active history of the basin was concluded by an epsode of basaltic volcanism in the Pleistocene (Salem *et al.*, 1996a,b,c; Selley, 1997c).

These brief reviews of three epicratonic basins show how they differ from the previously described intracratonic examples. Epicratonic basins tend to be much less stable than intracratonic ones, due to their situation at continent margins. Initial basin subsidence can be rapid, resulting in an early phase of deep-water sedimentation. The floor of the Sirte basin was extensively faulted and there was some igneous activity. Like intracratonic basins, however, epicratonic embayments can be infilled by both carbonate

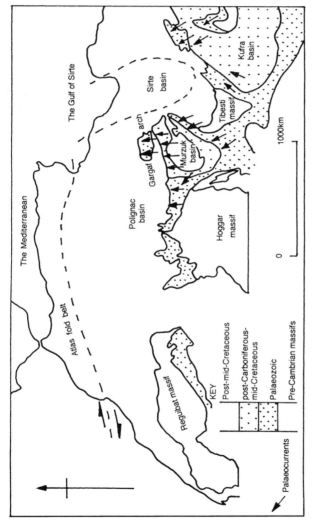

Fig. 10.14. (Upper) Map showing the paleocurrent directions for part of northwest Africa. These include over 3000 readings taken from slope-controlled fluvial deposits of the Hassaouna Formation (Cambro-Ordovician), the Tadrart Formation (Devonian), diverse Carboniferous fluvial sandstones, and the Continental Mesozoic sandstones. (Lower left) Outcrop map of the Paleozoic sediments of southern Libya and environs showing regional paleostrike lines, and depositional slope (black arrows), based on the data in the upper figure. Note the northerly paleoslope over the Gargaf arch, indicating that it was not a positive feature during the deposition of Paleozoic sediments. Note the northeasterly dip on the southwestern flank of the Kufra basin, indicating the existence of the Tibesti-Sirte arch. Paleocurrent data show that neither the Murzuk nor the Kufra basins were structurally closed during the Paleozoic Era. (Lower Right) Outcrop map of Continental Mesozoic sediments of southern Libya and environs showing regional paleostrike lines, and depositional slope (black arrows), based on the data in the upper figure. Note the northerly paleoslope on the north flank of the Murzuk basin, indicating that it was not a closed basin during the deposition of these sediments. Note the northeasterly dip on the southwestern flank of the Kufra basin, indicating the existence of the Tibesti-Sirte arch. Paleocurrent data show that neither the Murzuk, nor the Kufra basins were structurally closed during the deposition of the Continental Mesozoic sediments. [Reprinted from Selley, R. C., 1997b. The basins of Northwest Africa: Structural evolution. *In* "African Basins" (R. C. Selley, ed.), pp. 17–26. With permission from Elsevier Science.]

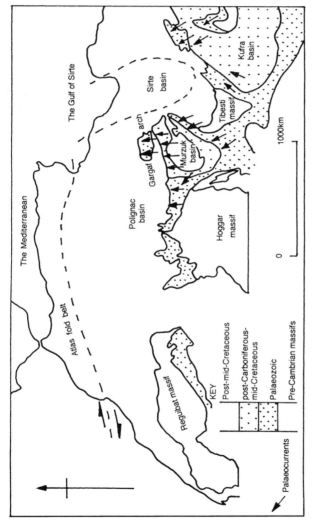

484

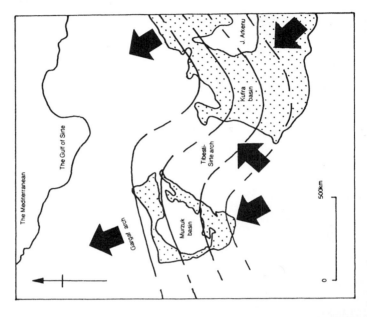

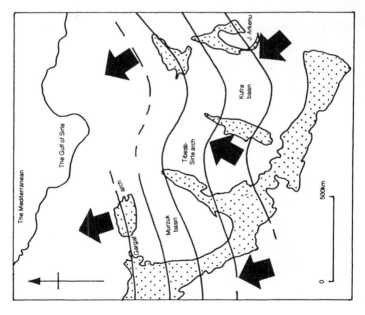

Fig. 10.14.— *Continued*

485

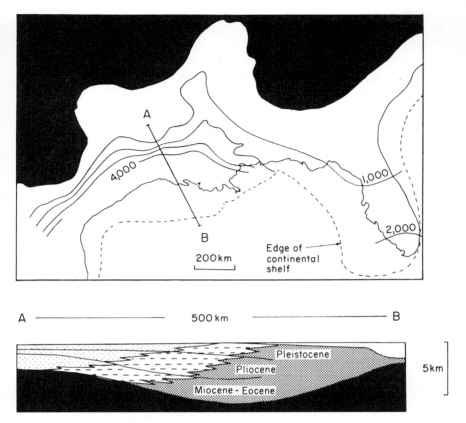

Fig. 10.15. (Upper) Isopach map of the Tertiary–Recent sedimentary fill of the Mississippi embayment. (From Murray, 1960. American Association of Petroleum Geologists Bulletin, AAPG © 1960, reprinted by permission of the American Association of Petroleum Geologists whose permission is required for further use.) (Lower) Cross-section of the embayment showing prograding sedimentary motif. Stippled, alluvial sands; dashed, deltaic sands, shales, and coals; blank, prodelta and shelf shales.

and terrigenous sediments. This differentiation is a function of the degree of uplift of the adjacent crust.

10.2.3 Basins Caused by Plate Convergence

Reference back to Fig. 10.11 and Table 10.1 shows that several types of basin are related to zones of crustal subduction at compressive plate boundaries. Before the advent of plate tectonics these troughs were termed **geosynclines.** Many varieties of geosynclinal basin were named and classified. Today the term geosyncline is obsolete. A new terminology has evolved using the terms **trench, fore-arc basin,** and **back-arc basin** (Fig. 10.17). The various types of geosynclinal trough are now seen, therefore, to reflect the nature of the plate boundaries. Subduction may occur where two continental masses converge (e.g., the modern Himalayas), where a continental and an oceanic plate converge (e.g., the Pacific coast of the Americas), or where two oceanic plates converge (e.g., the Japanese island arc) (Dickinson and Seely, 1979).

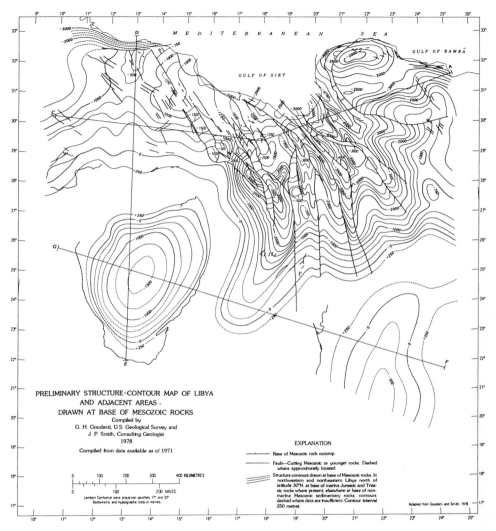

Fig. 10.16. Structure contour map on base of Mesozoic unconformity of the Sirte embayment. By contrast to the Mississippi embayment, the Sirte is infilled largely by carbonates, evaporites, and shales, with only minor amounts of sand in the initial and terminal phases. (From Goudarzi and Smith, 1978.)

The concept of the geosyncline was born from the work of Hall (1859) and Dana (1873a,b) in the Appalachians. Here a mountain range composed of a vast thickness of shallow-water sediment pointed to the continued subsidence of a linear trough over a long time span, followed by tectonism and uplift. The geosynclinal theory was introduced to Europe and applied to the Alps by Haug (1900). Subsequently the term geosyncline became so widely used and ill defined that it embraced all types of sedimentary basin. Significant papers on geosynclines and their classification were published by Schuchert (1923), Stille (1936), Kay (1944, 1947), and Glaessner and Teichert (1947). Aubouin's definitive monograph (1965) expounded the geosynclinal concept as it was

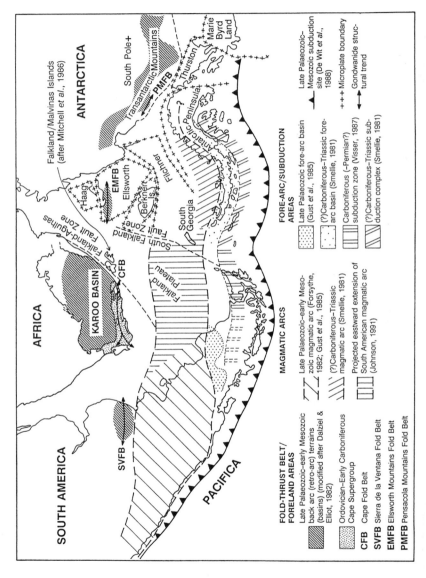

Fig. 10.17. (Left) Reconstruction of the plate tectonic setting of the basins associated with subduction of the Gondwana orogen. (Above) Cartoons to show the evolution of the Cape Fold belt of the Gondwana orogen in South Africa. [Reprinted from Johnson, M. R., van Vuuren, C. J., Visser, J. N. J., Cole, D. I., Wickens, H. de V., Christie, A. D. M., and Roberts, D. L. 1997. The Foreland Karoo basin, South Africa. *In* "African Basins" (R. C. Selley, ed.), pp. 269–317, with permission from Elsevier Science.]

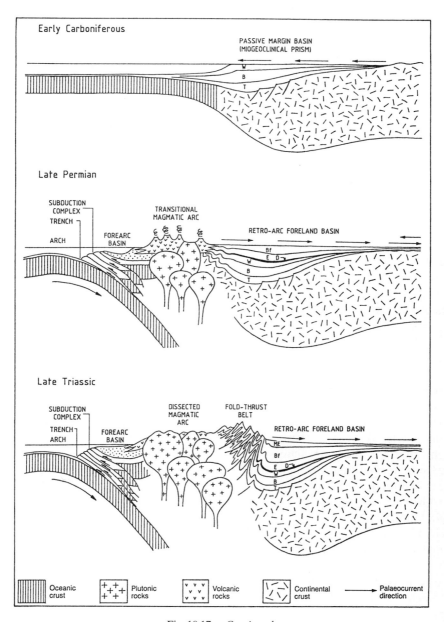

Fig. 10.17.— *Continued*

commonly understood by most geologists, and documented a model with several case histories. Aubouin's studies, largely based on the Hellenide fold belt of the eastern Mediterranean, divided troughs into several tectonomorphic zones. A stable foreland on the craton passes laterally into a shallow trough termed the miogeosynclinal furrow.

This is separated by a positive axis, the miogeanticline, from the eugeosynclinal furrow. This is the main active and unstable trough. The eugeosyncline is separated from the open oceanic basin by a second ridge, the eugeanticline. This second positive feature is an island arc of rising volcanics. The evolution of a geosyncline commences with the development of a tensional sag on the site of the future eugeosyncline. Pelagic limestones and radiolarian cherts develop in the basin synchronously with carbonates and sands on the adjacent shelf. The sag deepens to form the eugeosynclinal furrow. Subaqueous igneous activity causes the cherts to be interbedded with pillow lavas, ophiolites, and spilites (e.g., Hynes *et al.,* 1972). This is the Steinmann Trinity discussed in Section 8.5.2.3. Simultaneously, the miogeosynclinal furrow forms, separated from the eugeosyncline by the miogeanticline. Sedimentation in the miogeosyncline is still largely of shelf and shallow marine facies because subsidence is generally sufficiently gradual for deposition to keep pace. The sediments of the miogeosyncline differ from those of the foreland in thickness rather than facies. Miogeosynclines are in many ways similar to epicratonic basins. Continued tectonic activity leads to uplift of the eugeanticline. This becomes a major source of terrigenous sediment, immature in texture and mineralogy and often volcaniclastic. These deposits are laid down within the eugeosyncline largely as turbidites to form flysch facies.

The term **flysch** has had a lengthy and much abused history in the literature of geology. It has a high preservation potential. Detailed accounts of flysch facies have been given by Dzulinski and Walton (1965) and Lajoie (1970). Hsu (1970) discussed flysch semantics and Reading (1972) reviewed flysch facies within the context of plate tectonics. Flysch is generally regarded as the synorogenic deposit of geosynclinal troughs. Sedimentologically most fiysch show the features of turbidites. Flysch is petrographically diverse but in older Paleozoic and Proterozoic examples it is characteristically of greywacke type. Carbonate flysch is known, however, and, in the Hellenide geosyncline, troughs of detrital carbonate turbidites were supplied from the adjacent shelf. The main orogenic phase of the geosyncline is diverse and beyond the scope of this text. It may range from the high-angle thrust faulting and tight synclinal folding of the Rocky Mountains, to the recumbent folds and low-angle thrusts of the Alps (DeCelles and Mitra, 1999). Orogenesis also involves regional metamorphism and plutonic activity within the trough. During mountain building the tectonic front gradually migrates away from the trough axis. This forces the depositional axis of the flysch facies in the same direction, spilling sediment into the fore-deep. During this migration the deposits change from flysch to molasse facies.

The term **molasse,** like flysch, has had a chequered career since was first used in the French Alps by Bertrand in 1897. Molasse deposits are generally taken to be "late orogenic clastic wedges deposited in the linear fore-deep on the flank a craton" (Van Houten, 1973). Molasse facies are largely coarse terrigenous clastics with abundant conglomerates, few shales, and negligible limestones. These beds are generally laid down in lacustrine, fluvial, and fanglomerate environments. Grain size decreases and beds thin away from the mountain front.

The final phase of mountain building is regional epeirogenic uplift. This is typically accompanied by block faulting within the new mountain chain. Intramontane troughs are infilled with thick nonmarine clastic sequences, often accompanied by basalt effusion along the faults. The range and basin province of the Rocky Mountains are of this type. The Devonian rocks of the North Atlantic environs also provide good examples of postorogenic sedimentary basins. These continental red beds include both postorogenic molasse deposited in fore-deeps, such as the Pocono and Catskill facies of the Appalachians, and intramontane rift basins such as the Midland Valley of Scotland (Friend, 1969).

The geosynclinal model outlined here has been reviewed and modified in the light of plate tectonics. The concept has been criticized by Ahmad (1968) and Coney (1970), largely on the grounds that the time connotations of the geosynclinal cycle with beginning, a middle, and an end, are hard to reconcile with the continuous processes of subduction and sea floor spreading postulated by plate theory. More constructive attempts to integrate geosynclines and plate tectonics have been made by Mitchell and Reading (1969, 1978), Schwab (1971), Roberts (1972), Crostella (1977), and Dickinson and Seely (1979). Mitchell and Reading have shown that most geosynclines, now perceived as zones of subduction at compressive plate boundaries, fall into three main categories.

The various types of geosynclinal trough are now seen, therefore, to reflect the nature of the plate boundaries. Subduction may occur where two continental masses converge (e.g., the modern Himalayas), where a continental and an oceanic plate converges (e.g., the Pacific coast of the Americas), or where the two oceanic plates converge (e.g., the Japanese island arc) (Dickinson and Seely, 1979). This scheme successfully integrates the geosynclinal and plate theories and helps to explain the several different types of ancient troughs and their associated mountain chains.

In summary, this brief review of sedimentation associated with plate convergence describes the genesis of four different types of trough. Each trough has a characteristic morphology, plate tectonic setting, time of development, and type of fill. The eugeosynclinal (now fore-arc) basin is composed of preorogenic pelagic sediments overlain by synrogenic flysch facies. These are generally tectonically deformed and metamorphosed as the fore-arc basin migrates outward over the trench. The main flysch trough is often host to minor rifted troughs filled with postorogenic continental clastics.

The former miogeosyncline, with its fill of preorogenic shelf sediments and postorogenic molasse is now termed a "back-arc basin." Where this is developed on continental crust, and it is encroached by an active thrust system, it is termed a **foreland basin** (Beaumont, 1981; Dorobek and Ross, 1995; Mascle, 1998).

10.2.4 Basins Caused by Plate Divergence

There is an evolutionary series of basins related to crustal separation at boundaries of plate divergence. This may be termed the rift–drift suite of basins (refer back to Fig. 10.11).

Rift basins are long fault-bounded troughs which occur in various plate settings and all show a corresponding diversity of sediment fill. They are of considerable economic importance as sources of hydrocarbons, evaporites, and metals. Furthermore, rift basins contain many clues to the understanding of global tectonics. Important publications on this topic include works by Quennell (1985), Artyushkov (1987), Frostick *et al.* (1987), Olsen (1995), and Purser and Bosence (1998). Essentially there are four types of rift basin, each characterized by a particular type of tectonic setting and sedimentary fill.

The postorogenic intramontane rifts were described in the previous section. They are genetically distinct from all other rift basins save in their tensional origin. Typically they are infilled by continental clastics and volcanics. The three other types of rift basin all develop at zones of upwelling of the mantle which may ultimately become belts of sea floor spreading. More or less continuous lines of rift valleys are found along the mid-ocean ridge systems of the world and can be traced into the continental crust of Africa and Europe. The sedimentary basins of the mid-ocean ridge rifts are filled by sediment interbedded with the volcanic assemblage of new oceanic crust which is generated in these zones of intense igneous and seismic activity.

Rifts are often initiated by an updoming of the crust. The fracture system that results consists of a triradiate rift system whose node is referred to as a **triple-rift** (or colloquially triple-R) **junction.** Some rifts, such as the Red Sea, show straight subparallel faults. Recent detailed studies of the East African rifts show, however, that they are not composed of two parallel normal faults. They are, in fact, made up of an alternating series of listric faults dipping in opposing directions. The actual rift is formed by an antithetic fault that dips toward the opposing master fault. Thus though the rift is linear, it is asymmetric in cross-section, the asymmetry tilting from side to side along the length of the basin (Frostick *et al.*, 1987).

When a zone of sea floor spreading develops beneath a rift the crust separates and the floor of the rift subsides to be infilled by a regular sequence of sedimentary facies (Schneider, 1972). Initially the rift lie above sea level, so it is infilled with coarse fluvial and lacustrine sediments, often associated with volcanics. When the rift floor reaches sea level it provides a shallow restricted trough which (given the right climate) favors evaporite development, followed by restricted marine sedimentation as the rift subsides permanently below the sea. This favors oil source rock formation. With continued subsidence and separation the rift becomes infilled with open marine sediments, clastic or carbonate.

As the two sides of the new zone of sea floor spreading move apart, two of the rift arms develop into incipient oceans, each flanked with half-graben sedimentary basins. The third branch of the triple-R junction will not separate, though it may be infilled by sediments similar in facies and thickness to those in the other two rifts. This is referred to as an aborted rift or failed arm (Fig. 10.18). The North Sea rift is a fine example of a failed arm (Woodland, 1975; Illing and Hobson, 1981; Parker, 1993; Glennie, 1997).

Thus rift basins form a continuous spectrum related to the progressive breaking up and lateral drift of continental margins. The various basin types are now described in more detail.

10.2.4.1 Intracontinental Rift Basins

The first phase of rift development is seen within cratonic areas of continental crust. Localized updoming generates crestal tensional rifts. Triradial rift valley systems commonly diverge from the culmination of the dome. The rifts become infilled with continental fluviolacustrine deposits shed from the faulted basin margins. Igneous activity contributes lavas and volcaniclastic detritus to the basin fill. The best known of these intracontinental rift systems are those of Lake Baikal in Russia, of the Rhine Valley in Germany, and of East Africa.

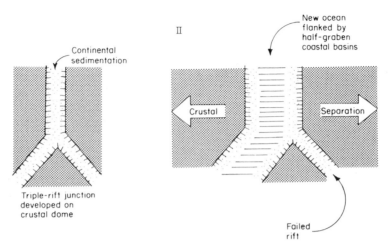

Fig. 10.18. Diagrams illustrating (left) the formation of a triple-rift junction in continental crust, (right) followed by crustal separation and the development of an aborted rift.

The Baikal rift developed in a basement of Paleozoic and Pre-Cambrian rocks. It is some 400 km in length and 50 km wide. Downwarping began in the Miocene; active rifting in the Pliocene. The Baikal rift basin now has a fill of some 5 km of nonmarine clastics (Salop, 1967).

The Rhine graben transects the Rhine shield with the Vosges mountains and the Black mountains on its west and east flank, respectively (Fig. 10.19). Rifting appears to have begun in the Middle Eocene. The Rhine graben contains over 3 km of diverse nonmarine facies and is still seismically active and presumably subsiding. Where best developed the basin is about 40 km wide and some 250 km long (Illies, 1970). Enthusiasts, however, can demonstrate that the Rhine graben is actually part of a line of rift basins traceable from the North Sea, via the Rhine, to the Rhone Valley, and thence across the Mediterranean to the Hon graben of Libya.

A second line of rifts subparallel to the first runs from the Gulf of Suez down the Red Sea and up to the great lakes of East Africa (Baker *et al.*, 1972; Darcott *et al.*, 1973; Frostick, 1997; Frostick *et al.*, 1987). These East African rift valleys are intracontinental basins analogous to the Rhine graben (Fig. 10.20). They are at present seismically and volcanically active. Let down into the African Pre-Cambrian metamorphic shield, they are infilled by Late Tertiary and Recent Volcanics and fluviolacustrine deposits (Fig. 10.21).

10.2.4.2 Intercratonic Rifts

Further tension and thinning of continental crust beneath a rift depresses its floor to sea level. Seawater floods the rift basin from time to time, but intermittent seismic uplift during net subsidence can trap saline water within the trough. Thus the continental clastic deposits may be overlain by evaporites in arid climates.

The Suez graben is an example of this next stage of rift basin evolution. Prerift basement rocks are overlain by some 4 km of Miocene sediments, which consist of a lower

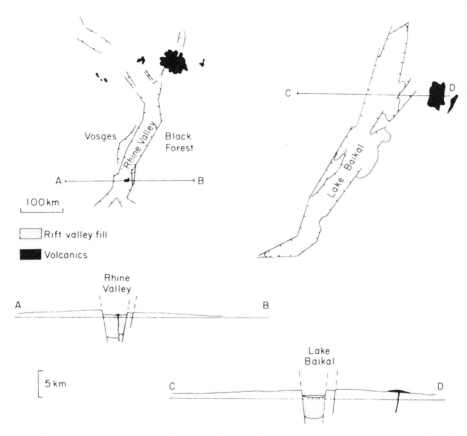

Fig. 10.19. Maps and cross-sections of (left) the Rhine Valley (Germany) and (right) Lake Baikal (Russia) rift basins drawn to the same scale. Both rifts are largely infilled with Tertiary to Recent continental clastic sediments and volcanics.

clastic part, the Gharandal Group, overlain by the Evaporite Group. The latter consists of a diverse assemblage of interbedded gypsum, anhydrite, marls, rocksalt, dolomites, and algal limestones (Heybroek, 1965).

The Suez gulf opens out southward into the Red Sea. This is a true intercratonic rift basin. It was formed by the crustal rifting of the Arabo-Nubian Pre-Cambrian craton (Fig. 10.22). The axis of the Red Sea consists of thin pelagic sediments, which overlie volcanics. Strange "hot spots" on the sea floor are infilled with brines and are areas where native metals are precipitated on the sea floor (Degens and Ross, 1969). Volcanic rocks lie at shallow depths and magnetic and gravity data suggest that oceanic crust is not far below (Lowell and Genik, 1972, 1975; Moretti and Chenet, 1987; Purser and Bosence, 1998).

The margins of the Red Sea are marked by two parallel sedimentary basins. They dip regionally away from the axis of the Red Sea and have complex horst and graben floors. Essentially these marginal basins of the intercratonic rift are stratigraphically contiguous with the Suez graben. Basement is overlain by red beds, which are in turn overlain

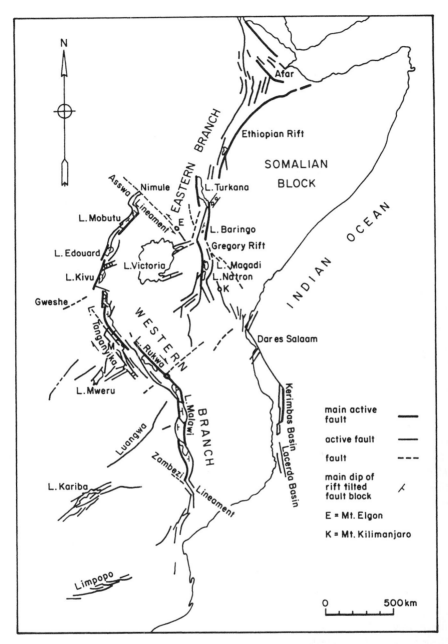

Fig. 10.20. Map of the East African Rift valley system. These rifts began to form in the Early Miocene and are infilled by thick sequences of volcanics and continental clastics. [Reprinted from Frostick, L. E. 1997. East African Rift basins. *In* "African Basins" (R. C. Selley, ed.), pp. 187–210, with permission from Elsevier Science.]

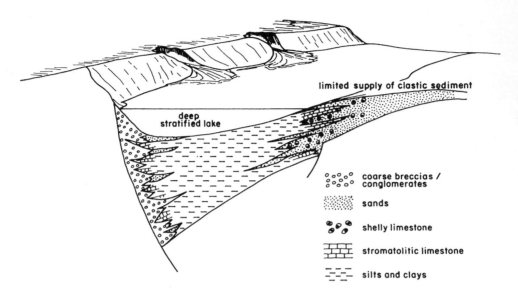

Fig. 10.21. Geophantasmogram of depositional environments in the Tanganyika rift valley. [Reprinted from Frostick, L. E. 1997. East African Rift basins. *In* "African Basins" (R. C. Selley, ed.), pp. 187–210, with permission from Elsevier Science.]

by thick evaporites in the grabens and reefal carbonates on the horsts (Lowell and Genik, 1972, 1975). These are Miocene in age and are unconformably overstepped by Pliocene and younger sediments, marine, nonmarine, terrigenous, and carbonate. The southern entrance of the Red Sea is continuous with the Carlsberg Ridge of the Indian Ocean. Thus the intracratonic and intercratonic rift basins are genetically related to the oceanic ridges and axial rifts that are the hallmarks of zones of sea floor spreading.

From the evidence of sea floor spreading seen in its early stages in the Red Sea and adjacent rifts, it follows that analogous half-graben basins with similar facies sequences should be present, for example, along the Atlantic coasts. This is indeed the case in the Southern Atlantic (Ala and Selley, 1997; Cameron *et al.,* 1999). The coasts of Brazil and of Africa, from Senegal to Gabon, possess a series of tensional horst and graben coastal basins. In each basin basal Cretaceous nonmarine clastics are overlain by evaporites, which diapirically intrude thick wedges of younger marine deposits (Fig. 10.23).

10.3 BASIN EVOLUTION, METALLOGENY, AND THE PETROLEUM SYSTEM

Having reviewed the various types of basins, it is appropriate to conclude with an analysis of how they are genetically related in time and space. It is now appreciated that many basins undergo a regular sequence of structural phases. These result in the deposition of a regular sequence of sedimentary facies. Recognition of this sequence of events and facies has important implications for understanding metallogeny and petroleum generation, migration, and entrapment. This pattern is most clearly seen in the rift–drift sequence of basins as just outlined. The evolution of a hypothetical asymmetric rift basin is now described to illustrate the relationship between structural process, sedimen-

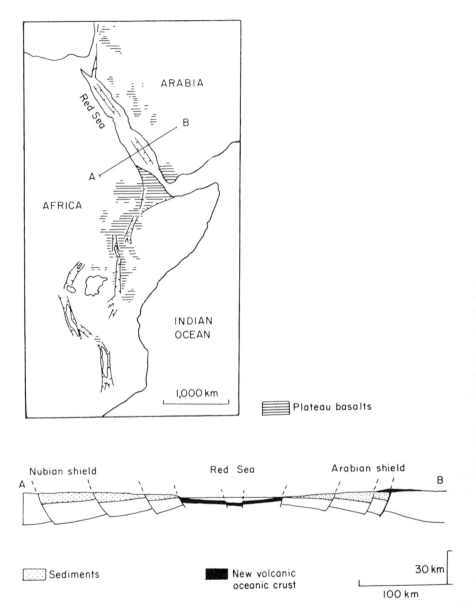

Fig. 10.22. (Upper) Map showing the extent of the East African rift valley basins and their continuation into the Red Sea. (Lower) Cross-section of the Red Sea showing the coastal sedimentary basins and the axial rift where volcanic activity forms new oceanic crust. (From Lowell and Genik, 1972. American Association of Petroleum Geologists Bulletin, AAPG © 1972, reprinted by permission of the American Association of Petroleum Geologists whose permission is required for further use.)

tary response, and the migration of basin fluids and their effects on petroleum migration and metallogeny.

An integral part of basin analysis includes consideration of the petroleum system. The **petroleum system** integrates the sedimentary and structural history of a basin with

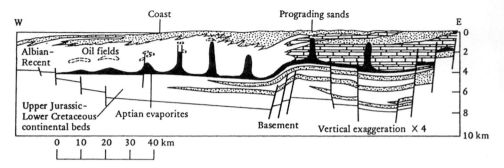

Fig. 10.23. Cross-section of the Gabon basin, west Africa. This is a typical example of an ocean margin basin, the end product of a rift–drift sequence. The characteristic basin fill shows a vertical sequence of syn-rift continental clastics and evaporites overlain by postrift open marine sediments. (Compiled from Belmonte *et al.*, 1965; Brink, 1974, and other sources.)

its petroleum characteristics, in terms of the richness, volume, and maturity of source rocks. The petroleum system has been variously defined as the relationship between "a pod of active source rock and the resulting oil and gas accumulations" (Magoon and Dow, 1994) or as "a dynamic petroleum generating and concentrating physico-chemical system functioning in a geologic space and time" (Demaison and Huizinga, 1991, 1994).

These last authors classify petroleum systems according to three parameters: the charge factor, the style of migration (vertical or lateral), and the entrapment style. The charge factor is calculated on the basis of the richness and volume of the source rocks in a basin. This is measured according to a parameter termed the source potential index:

$$SPI = \frac{h(S^1 + S^2)p}{1000},$$

where

SPI = the maximum quantity of hydrocarbons that can be generated within a column of source rock under 1 m² of surface area

h = thickness of source rock (m)

$S^1 + S^2$ = average genetic potential in kilograms hydrocarbons/metric ton of rock

p = source rock density in metric tons/cubic meter.

The volumes of petroleum generated can be modeled in one, two, and three dimensions. One-dimensional modeling deals with a single point, such as a well location, two-dimensional modeling deals with a vertical cross-section through a basin, and three-dimensional modeling considers a whole basin. Diverse complex computer programs are available for these exercises (Dore *et al.*, 1991; Helbig, 1994). Figure 10.24 illustrates some of the basic concepts.

Returning to the wider aspects of the petroleum system, the style of migration drainage is subdivided according to whether migration is principally lateral or vertical, though this can, of course, vary in time and space within the history of one basin. The style of entrapment is dependent on the length and continuity of permeable carrier beds, the

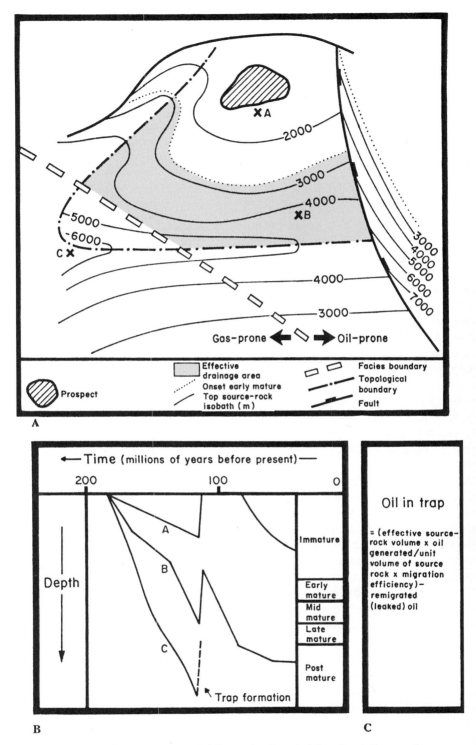

Fig. 10.24. Sketch map (A) and burial curve (B) illustrating the principles of modeling petroleum generation and migration. The volume of petroleum generated and whether it is oil and/or gas depend on the type of source rock and its level of maturity. The latter is largely governed by its cooking history. (From Cornford, R. 1990. Source rocks and hydrocarbons of the North Sea. *In* "Introduction to the Petroleum Geology of the North Sea" (W. K. Glennie, ed.), 3rd. ed., pp. 294–361. Courtesy of Blackwell Science.]

distribution and effectiveness of seals, and tectonic style. All of these together control the degree of resistance of the basin to petroleum migration. One of the major controls of petroleum generation is the heat flow to which the basin is subjected. As described earlier, McKenzie (1978) produced a mathematical analysis of the relationship between crustal extension (quantified as the β factor), crustal thinning, heat flow, and the rate of rift basin subsidence. The model begins with a dramatic phase of rapid crustal extension, subsidence, and high heat flow. This is followed by a subsequent phase of slower subsidence and a return to normal heat flow, as the thinned crust reestablishes its thermal equilibrium.

In the first phase faults extend up to the earth's surface causing the formation of rift valleys. Subsequent slower subsidence allows sediment to blanket the faults and gentler basin slopes to prevail. Neugebauer (1987) has reviewed the subsequent evolution of models of basin formation. As currently perceived the sequence sometimes, but not invariably, begins with a prerift phase of crustal doming (Fischer, 1975). The prebasinal phase of uplift is attributed to localized heating and expansion in the underlying asthenosphere (Fig. 10.25A). Crustal doming is followed by erosion of the crest and then by rifting. These domes cause regional negative gravity anomalies, and may be up to 1 km high and 2000 km in diameter (Kazmin, 1987). Neugebauer (1987) cites the Hoggar and Tibesti domes of the Sahara as modern examples. The Tertiary Rhine graben is an example of a rift developed on a regional crustal dome (Fig. 10.19). An ancient example is provided by the Sirte embayment of Libya. This resulted from the mid-Cretaceous collapse of a persistent Paleozoic crustal uplift, the Tibesti-Sirte arch (Section 10.2.2.1). Not all basin-forming mechanisms are preceded by a domal phase, however. Kazmin (1987) differentiates rifting associated with pulsating phases of high heat flow, and rifts related to plate separation. The latter are unassociated with doming, and their high heat flow is not the cause of rifting, but the result of the associated crustal thinning.

Whether preceded by crustal doming or not, the resultant rift basins are initially infilled by continental sediments and are associated with volcanic activity and high heat flow. The East African rifts illustrate this phase of development (Section 10.2.4). The early basin fill sediments are laid down in continental environments because the rift floor is still substantially above sea level. Fluvial sediments along the rift margins pass into finer grained sediments in the basin center. In humid climates this will be a permanent lake in which organic matter may be preserved to evolve into petroleum source beds. In arid climates the center of the rift may contain ephemeral lakes and sabkhas, such as the Quattara depression in the Western Desert of Egypt.

Meteoric fluids move downward and laterally through the sedimentary pile hydrodynamically in response to gravity. These fluids are normally oxidizing and acidic as discussed elsewhere (see Section 8.5.3.3). The volcanic activity that is associated with rifting has an important role to play in metallogeny. It facilitates the emplacement of numerous minerals within the sedimentary cover. Some of these may remain as primary ore deposits in the volcanogenic pile, but others may undergo remobilization. The flow of acid oxidizing groundwater leaches out soluble minerals, transports them in solution, and reprecipitates them elsewhere during the first or subsequent sedimentary cycles. Uranium carnotite ores are emplaced in this manner (see Section 6.3.2.2.4). In arid areas sodium-rich evaporites form in ephemeral desert lakes. Copper ore deposits are frequently emplaced in continental red beds during this early phase of basin development (Flint, 1986). Insoluble heavy minerals may weather out of the volcanogenic rocks and become segregated into placer ores. These may then undergo repeated phases of erosion and redeposition throughout subsequent sedimentary cycles within the basin (Fig. 10.25B).

McKenzie's model, supported by field evidence, shows that crustal stretching, basin subsidence, and heat flow gradually diminish with time. Faulting continues, but the slower rate of movement permits sediment to blanket them, and they are no longer manifest on the surface. As the rift loses its surface expression it evolves into a gently sloping elongate or subcircular basin.

With continuing subsidence the basin floor may descend to sea level. In arid climates this may permit vast coastal sabkhas to form, from which evaporites may be deposited with associated limestone reefs (Fig. 10.25C). The modern Gulf of California exemplifies this stage. The Miocene evaporites and reefs of the Red Sea are an ancient analog. In humid climates, however, evaporites will be absent, but the conditions are suitable for the development of marshes and swamps. The resultant peat deposits may be preserved and evolve into coals during burial.

With continued subsidence the sea will invade the trough permanently. Restricted access and circulation at the commencement of this phase may favor the development of thermally or salinity induced density layering in the seawater. This may result in anoxic bottom conditions that favor the preservation of organic matter. The earlier continental and evaporitic deposits thus become overlain by potential petroleum source beds (Fig. 10.25D).

As subsidence continues the basin becomes infilled with open marine sediments. If the hinterland is of low relief the basin may be infilled by carbonates. Where the hinterland is high, and subsidence active, however, deep-sea sands may be deposited at the foot of rapidly prograding deltas (Fig. 10.25E). Now all sorts of exciting things begin to happen. If the basin fill included a phase of evaporite deposition, then evaporite-derived brines may mobilize, concentrate, and precipitate Mississippi Valley-type lead–zinc ore in adjacent carbonate reefs (see Section 9.6.5). The overlying organic-rich shales may now reach thermal maturity, aided by the thermal blanket of rapidly deposited and often overpressured muds (Section 7.2). Petroleum is generated and migrates up to be trapped in reservoirs around the basin margin (McKenzie, 1981). Simultaneously, the evaporites may undergo halokinesis to form traps for the migrating petroleum (Fig. 10.25F).

Now the basin gradually enters a new phase. Subsidence declines and stops, heat flow wanes, and overpressured connate fluids bleed off at the surface. Pore pressures adjust to normal hydrostatic conditions. Deep within the basin, connate fluids may move in thermally driven convection cells (Neglia, 1979; Bonham, 1980). If the basin is in isostatic equilibrium, the land surface may be intensely weathered, leading to the formation of gossans over sulfide ore bodies and to diverse residual pedogenic and placer mineral deposits (Fig. 10.25G).

Finally, and this may happen after many millions of years of equilibrium, some basins undergo a subsequent phase of uplift. The causes of inversion are several, and include crustal compression and wrenching. Uplift initiates a new series of processes, the most obvious of which is the erosion of the upper part of the sedimentary fill. Hydrostatic

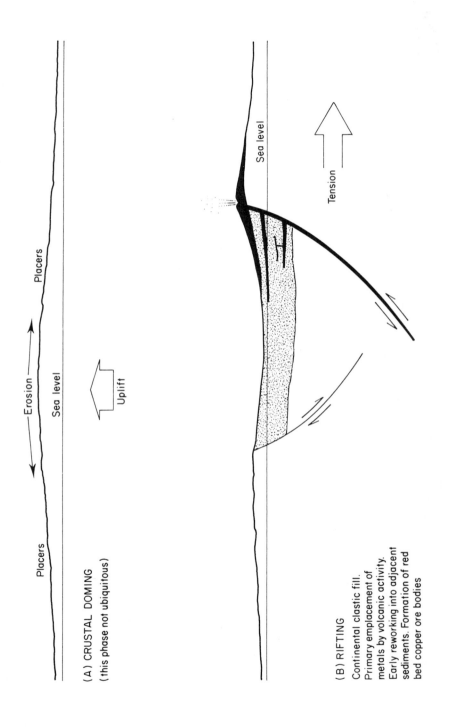

Placers   Placers

Erosion

Sea level

Uplift

(A) CRUSTAL DOMING
(this phase not ubiquitous)

Sea level

Tension

(B) RIFTING

Continental clastic fill.
Primary emplacement of
metals by volcanic activity.
Early reworking into adjacent
sediments. Formation of red
bed copper ore bodies

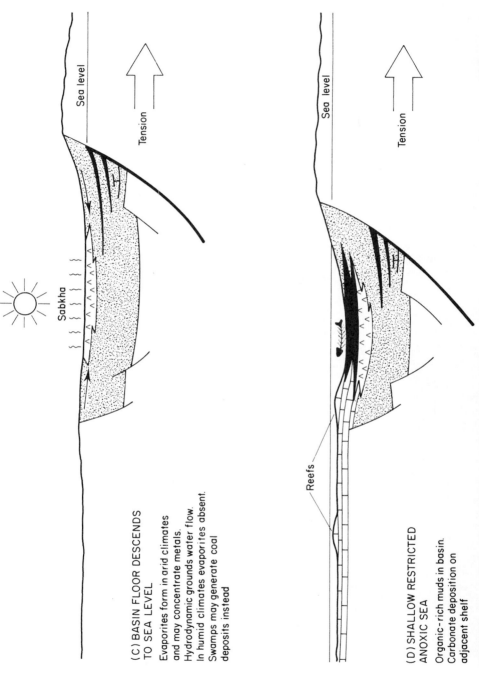

(C) BASIN FLOOR DESCENDS
TO SEA LEVEL

Evaporites form in arid climates
and may concentrate metals.
Hydrodynamic grounds water flow.
In humid climates evaporites absent.
Swamps may generate coal
deposits instead

(D) SHALLOW RESTRICTED
ANOXIC SEA

Organic-rich muds in basin.
Carbonate deposition on
adjacent shelf

Fig. 10.25. Cross-sections illustrating the evolution of an hypothetical asymmetric rift basin. This series of illustrations demonstrates how structure controls sedimentation, and how metals and petroleum are generated, mobilized, and emplaced by migrating pore fluids.

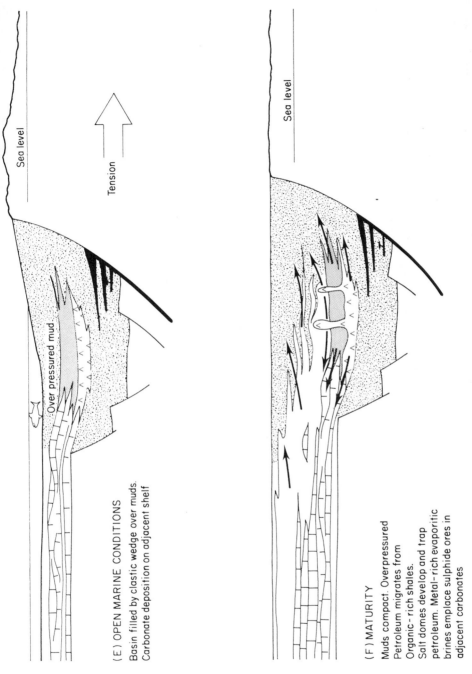

(E) OPEN MARINE CONDITIONS

Basin filled by clastic wedge over muds.
Carbonate deposition on adjacent shelf

Sea level

Tension

Over pressured mud

Sea level

(F) MATURITY

Muds compact. Overpressured
Petroleum migrates from
Organic-rich shales.
Salt domes develop and trap
petroleum. Metal-rich evaporitic
brines emplace sulphide ores in
adjacent carbonates

Fig. 10.25.—*Continued*

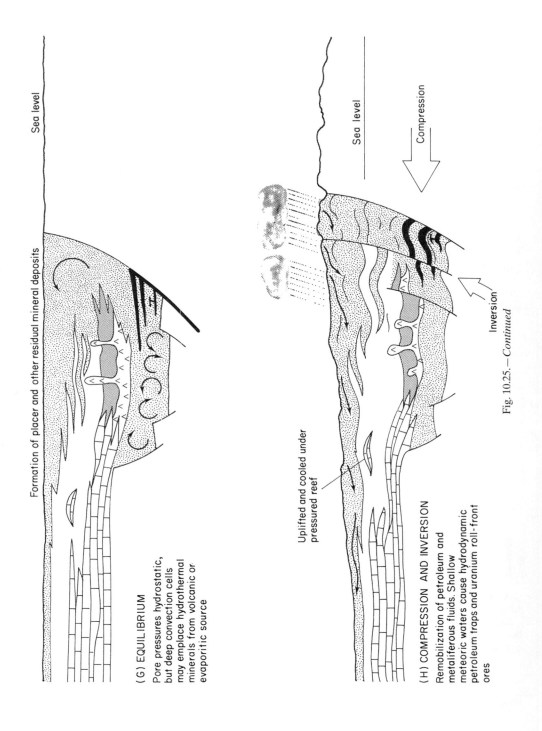

Sea level

Formation of placer and other residual mineral deposits

(G) EQUILIBRIUM

Pore pressures hydrostatic,
but deep convection cells
may emplace hydrothermal
minerals from volcanic or
evaporitic source

Sea level

Compression

Uplifted and cooled under
pressured reef

Inversion

(H) COMPRESSION AND INVERSION

Remobilization of petroleum and
metaliferous fluids. Shallow
meteoric waters cause hydrodynamic
petroleum traps and uranium roll-front
ores

Fig. 10.25. — *Continued*

505

pore pressure may now be modified by hydrodynamic flow in permeable formations, as meteoric waters displace connate fluids. In closed systems, however, cooling may cause subnormal pore pressures.

These changes in pore pressure and fluid chemistry affect both petroleum and mineralization. Remobilized petroleum may be trapped in hydrodynamic traps (see Section 7.2) and may undergo shallow bacterial degradation. Meteoric flow may cause further remobilization of minerals, particularly any uranium ores that have previously been emplaced within the system. Continuing erosion will rework placer ores at the surface and, as the rate of erosion declines, gossans and enriched zones will again develop on the weathered surfaces of previously emplaced sulfide ores (Fig. 10.25H).

For a hypothetical asymmetric rift basin of the type just described the relationship between structure, sedimentation, metallogeny, and petroleum migration is summarized in Fig. 10.26. This figure demonstrates the main theme of this book — the integrated nature of sedimentary processes and their control on the genesis of petroleum and other minerals.

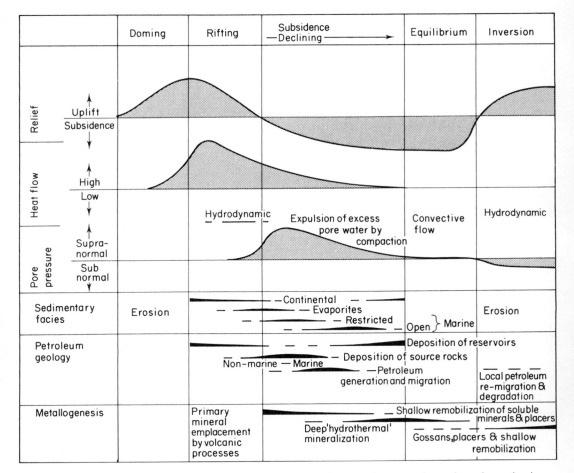

Fig. 10.26. Summary of the evolution of structure, sedimentation, metallogeny and petroleum formation in the hypothetical basin illustrated in Fig. 10.25.

SELECTED BIBLIOGRAPHY

Allen, P. A., and Allen, J. R. (1990). "Basin Analysis Principles and Applications." Blackwell, Oxford. 451pp.

Busby, C., and Ingersoll, R. (1995). "Tectonics of Sedimentary Basins." Blackwell, Oxford. 580pp.

Condie, K. C. (1997). "Plate Tectonics and Crustal Evolution," 4th ed. Butterworth-Heinemann, London.

Dore, A. G., Augustson, J. H., Hermanrud, C., Stewart, D. J., and Sylta, O., eds. (1991). "Basin Modelling: Advances and Applications." Elsevier, Amsterdam.

Helbig, K., ed. (1994). "Modelling the Earth for Oil Exploration." Elsevier, Amsterdam.

Kearey, P., and Vine, F. J. (1996). "Global Tectonics," 2nd ed. Blackwell Science, Oxford. 352pp.

Lillie, R. J. (1999). "Whole Earth Geophysics." Prentice Hall, London. 361pp.

REFERENCES

Ahmad, F. (1968). Orogeny, geosynclines and continental drift. *Tectonophysics* **15,** 177–189.

Ala, M. A., and Selley, R. C. (1997). The West African Coastal Basins. *In* "African Basins" (R. C. Selley, ed.), pp. 173–186. Elsevier, Amsterdam.

Allen, P. A., and Allen, J. R. (1990). "Basin Analysis Principles and Applications." Blackwell, Oxford. 451pp.

Artyushkov, E. V. (1987). Rifts and grabens. *Tectonophysics* **133,** 321–331.

Aubouin, J. (1965). "Geosynclines," Dev. Geotectonics, Vol. 1. Elsevier, New York. 335pp.

Baker, B. H., Mohr, P. A., and Williams, L. A. J. (1972). Geology of the eastern rift system of Africa. *Spec. Pap.— Geol. Soc. Am.* **136,** 145pp.

Beaumont, C. (1981). Foreland basins. *Geophys. J. R. Astr. Soc.,* **65,** 291–329.

Belmonte, Y., Hirtz, P., and Wenger, R. (1965). The salt basins of the Gabon and Congo (Brazzaville). *In* "Salt Basins Around Africa," pp. 55–78. Inst. Petrol., London.

Bertrand, M. (1897). Structure des Alpes Francaise et recurrence de certaines facies sedimentaires. *Int. Geol. Congr., Rep. Sess., 6th,* pp. 161–177.

Bonham, L. C. (1980). Migration of hydrocarbons in compacting basins. *AAPG Bull.* **64,** 549–567.

Brink, A. H. (1974). Petroleum geology of Gabon basin. *AAPG Bull.* **58,** 216–235.

Burke, K. (1972). Longshore drift, submarine canyons, and submarine fans in development of Niger Delta. *Bull. Am. Assoc. Pet. Geol.* **56,** 1975–1983.

Busby, C., and Ingersoll, R. (1995). "Tectonics of Sedimentary Basins." Blackwell, Oxford. 580pp.

Cameron, N., Bate, R., and Clure, V., eds. (1999). "The Oil and Gas Habitats of the South Atlantic," Spec. Publ. No. 153. Geol. Soc. London, London. 474pp.

Cohee, G. V., and Landes, K. K. (1958). Oil in the Michigan basin. *In* "The Habitat of Oil" (L. G. Weeks, ed.), pp. 473–493. Am. Assoc. Pet. Geol., Tulsa, OK.

Condie, K. C. (1997). "Plate Tectonics and Crustal Evolution," 4th ed. Butterworth-Heinemann, London.

Coney, P. J. (1970). The geotectonic cycle and the new global tectonics. *Geol. Soc. Am. Bull.* **81,** 739–747.

Cornford, R. (1990). Source rocks and hydrocarbons of the North Sea. *In* "Introduction to the Petroleum Geology of the North Sea" (W. K. Glennie, ed.), 3rd ed., pp. 294–361. Blackwell, Oxford.

Cox, A., and Hart, R. B. (1986). "Plate Tectonics: How it Works." Blackwell, Oxford.

Crostella, A. (1977). Geosynclines and plate tectonics in Banda Arcs, Eastern Indonesia. *AAPG Bull.* **61,** 2063–2081.

Dallmus, K. F. (1958). Mechanics of basin evolution and its relation to the habitat of oil. *In* "The Habitat of Oil" (L. G. Weeks, ed.), pp. 883–931. Am. Assoc. Pet. Geol., Tulsa, OK.

Dana, J. D. (1873a). On some results of the earth's contraction from cooling, including a discussion of the origin of mountains and the nature of the earth's interior. *Am. J. Sci.* **5,** 423–443.

Dana, J. D. (1873b). On some results of the earth's contraction from cooling, including a discussion of the origin of mountains and the nature of the earth's interior. *Am. J. Sci.* **6,** 6–14, 104–115, 161–172.

Darcott, B. W., Girdler, R. W., Fairhead, J. D., and Hall, S. A. (1973). The East African rift system. In "Implications of Continental Drift to the Earth Sciences" (D. H. Tarling and S. K. Runcorn, eds.), pp. 757–766. Academic Press, London and New York.

Darling, G. B., and Wood, P. W. J. (1958). Habitat of oil in the Canadian portion of the Williston basin. In "The Habitat of Oil" (L. G. Weeks, ed.), pp. 129–148. Am. Assoc. Pet. Geol., Tulsa, OK.

Davies, P. A., and Runcorn, S. A. (1980). "Mechanics of Continental Drift and Plate Tectonics." Academic Press, London and New York.

DeCelles, P. G., and Mitra, G. (1999). "Thrust Belts and Synorogenic Sediments." Blackwell, Oxford. 256pp.

Degens, E. T., and Ross, D. A., eds. (1969). "Hot Brines and Recent Heavy Metal Deposits in the Red Sea." Springer-Verlag, Berlin. 600pp.

Demaison, G., and Huizinga, H. J. (1991). Genetic classification of petroleum systems. AAPG Bull. 75, 1626–1643.

Demaison, G., and Huizinga, H. J. (1994). Genetic classification of petroleum systems using three factors: Charge, migration and entrapment. Mem. — Am. Assoc. Pet. Geol. 60, 73–92.

Dickinson, W. R., and Seely, D. R. (1979). Structure and stratigraphy of fore-arc regions. AAPG Bull. 63, 2–31.

Dore, A. G., Augustson, J. H., Hermanrud, C., Stewart, D. J., and Sylta, O., eds. (1991). "Basin Modelling: Advances and Applications." Elsevier, Amsterdam.

Dorobek, S. L., and Ross, G. M. (1995). "Stratigraphic Evolution of Foreland Basins," Spec. Publ. No. 52. Soc. Econ. Paleontol. Mineral., Tulsa, OK.

Drake, C. L., Ewing, J. I., and Stokand, H. (1968). The continental margin of the United States. Can. J. Earth Sci. 5, 99–110.

du Toit, A. L. (1937). "Our Wandering Continents." Oliver & Boyd, Edinburgh. 361pp.

Dzulinski, S., and Walton, E. K. (1965). "Sedimentary Features of Flysch and Greywacke." Elsevier, Amsterdam. 300pp.

Evamy, B. D., Haremboure, J., Kamerling, P., Knapp, W. A., Malloy, F. A., and Rowlands, P. H. (1978). Hydrocarbon habitat of Tertiary Niger delta. AAPG Bull. 62, 1–39.

Fischer, A. G. (1975). Origin and growth of basins. In "Petroleum and Global Tectonics" (A. G. Fischer and S. Judson, eds.), pp. 47–82. Princeton University Press, Princeton, NJ.

Fischer, A. G., and Judson, S., eds. (1975). "Petroleum and Global Tectonics." Princeton University Press, Princeton, NJ. 322pp.

Flint, S. (1986). Sedimentary and diagenetic controls on red bed ore genesis: The Middle Tertiary San Bartolo copper deposit, Antofagasto Province, Chile. Econ. Geol. 42, 41–48.

Fowler, J. H., and Kuenzi, W.D. (1978). Keweenawan turbidites in Michigan (deep borehole red beds): A foundered basin sequence developed during evolution of a protoceanic rift. J. Geophys. Res. 83, 5833–5843..

Friend, P. F. (1969). Tectonic features of Old Red Sedimentation in North Atlantic borders. Mem. — Am. Assoc. Pet. Geol. 12, 703–710.

Frostick, L. E. (1997). East African Rift basins. In "African Basins" (R. C. Selley, ed.), pp. 187–210. Elsevier, Amsterdam.

Frostick, L. E., Renaut, R. W., Reid, I., and Tiercelin, J. J., eds. (1987). "Sedimentation in the African Rifts," Spec. Publ. No. 25. Geol. Soc. London, London. 323pp.

Glaessner, M. F., and Teichert, C. (1947). Geosynclines: A fundamental concept in geology. Am. J. Sci. 245, 465–482, 571–591.

Glennie, E. K., ed. (1997). "Petroleum Geology of the North Sea: Basic Concepts and Recent Advances," 3rd ed. Blackwell, Oxford. 544pp.

Goodwin, A. M. (1973). Plate tectonics and the evolution of the continental crust. In "Implications of Continental Drift to the Earth Sciences" (D. H. Tarling and S. K. Runcorn, eds.), Vol. 2, pp. 1047–1070. Academic Press, London.

Goudarzi, G. H., and Smith, J. P. (1978). "Preliminary Structure Contour Map of the Libyan Arab Republic and Adjacent Areas, Misc. Geol. Invest., map I-350C. U.S. Geol. Surv., Washington, D.C.

Halbouty, M. T., Meyerhoff, A. A., King, R. E., Dott, R. H., Klemme, H. D., and Shabad, T.

(1970). World's giant oil and gas fields, geologic factors affecting their formation and basin classification. *Mem.—Am. Assoc. Pet. Geol.* **14,** 502–555.

Hall, A. J. (1859). "Natural History of New York," Vol. 3, pp. 1–96. Appleton-Century-Crofts, New York.

Hartley, R. W., and Allen, P. A. (1994). Interior cratonic basins of Africa: Relation to continental break-up and role of mantle convection. *Basin Res.* **6,** 95–113.

Haug, E. (1900). Les geosynclinaux et les aires continentales. *Geol. Soc. Fr. Bull.* **28,** 617–711.

Helbig, K., ed. (1994). "Modelling the Earth for Oil Exploration." Elsevier, Amsterdam.

Heybroek, F. (1965). The Red Sea Miocene Evaporite basin. *In* "The Salt Basins Around Africa," pp. 17–40. Inst. Petrol., London.

Hsu, K. J. (1970). The meaning of the word flysch, a short historical search. *Spec. Pap.—Geol. Assoc. Can.* **7,** 1–11.

Hynes, A. J., Nisbet, E. G., Smith, A. G., Welland, M. J. P., and Rex, D. P. (1972). Spreading and emplacement age of some ophiolites in the Othris region, eastern central Greece. *Z. Parasitenkd. Geol. Ges.* **123,** 455–468.

Illies, J. H. (1970). Graben tectonics as related to crust-mantle interaction. *In* "Graben Problems" (J. H. Illies and E. St. Mueller, eds.), pp. 4–27. Schweizerbart'sche Verlagsbuchhandlung, Stuttgart.

Illing, L. V., and Hobson, G. D., eds. (1981). "Petroleum Geology of the Continental Shelf of Northwest Europe." Heyden, London. 521pp.

Jervey, M. T. (1988). Quantitative geological modelling of siliciclastc rocks sequences and their seismic expression. *Spec. Publ.—Soc. Econ. Paleontol. Mineral.* **42,** 47–69.

Johnson, M. R., van Vuuren, C. J., Visser, J. N. J., Cole, D. I., Wickens, H. de V., Christie, A. D. M., and Roberts, D. L. (1997). The Foreland Karoo basin, South Africa. *In* "African Basins" (R. C. Selley, ed.), pp. 269–317. Elsevier, Amsterdam.

Kay, M. (1944). Geosynclines in continental development. *Science N.Y.* **99,** 461–462.

Kay, M. (1947). Geosynclinal nomenclature and the craton. *Bull. Am. Assoc. Pet. Geol.* **31,** 1289–1293.

Kazmin, V. (1987). Two types of rifting: Dependence on the condition of extension. *Tectonophysics* **143,** 85–92.

Kearey, P., and Vine, F. J. (1996). "Global Tectonics," 2nd ed. Blackwell Science, Oxford. 352pp.

Klein, G. de Vries (1987). Current aspects of basin analysis. *Sediment. Geol.* **50,** 95–118.

Klemme, H. D. (1980). Petroleum basins—Classification and characteristics. *J. Petrol. Geol.* **3,** 187–207

Kocurek, G., and Havholm, K. G. (1993). Eolian sequence stratigraphy—A conceptual framework. *Mem.—Am. Assoc. Pet. Geol.* **58,** 393–410.

Lajoie, J., ed. (1970). "Flysch Sedimentology in North America," Spec. Pap. No. 7. Geol. Soc. Can. Toronto. 272pp.

Leeder, M., (1999). "Sedimentology and Sedimentary Basins." Blackwell, Oxford. 512pp.

Leeder, M. Harris, T., and Kirby, M. J. (1998). Sediment supply and climate change: Implications for basin stratigraphy. *Basin Res.* **10,** 7–18.

Lillie, R. J. (1999). "Whole Earth Geophysics." Prentice Hall, London. 361pp.

Lowell, J. D., and Genik, G. J. (1972). Seafloor spreading and structural evolution of the southern Red Sea. *Am. Assoc. Pet. Geol. Bull.* **56,** 247–259.

Lowell, J. D., and Genik, G. J. (1975). Geothermal gradients, heat flow, and hydrocarbon recovery. *In* "Petroleum and Global Tectonics" (A. G. Fischer and S. Judson, eds.), pp. 129–156. Princeton University Press, Princeton, NJ.

Macgregor, D. S., Moody, R. T. J., and Clark-Lowes, D. D. (1998). "Petroleum Geology of North Africa," Spec. Publ. No. 132. Geol. Soc. London, London. 442pp.

Magoon, L. B., and Dow, W. G., eds. (1994). "The Petroleum System from Source to Trap," Am. Mem. No. 60. Assoc. Pet. Geol., Tulsa, OK.

Mascle, J. R., Bornhold, B. D., and Renard, V. (1973). Diapiric structures off Niger delta. *Am. Assoc. Pet. Geol. Bull.* **57,** 1672–1678.

McKenzie, D. (1978). Some remarks on the development of sedimentary basins. *Earth Planet. Sci. Lett.* **40,** 25–32.

McKenzie, D. (1981). The variation of temperature with time and hydrocarbon maturation in sedimentary basins formed by extension. *Earth Planet. Sci. Lett.* **55,** 87–98.

Meissner, R. (1986). "The Continental Crust." Academic Press, London. 448pp.

Miall, A. D. (1984). "Principles of Sedimentary Basin Analysis." Springer-Verlag, Berlin. 490pp.

Mitchell, A. H. G., and Reading, H. G. (1969). Continental margins, geosynclines, and ocean floor spreading. *J. Geol.* **77,** 629–646.

Mitchell, A. H. G., and Reading, H. G. (1978). Sedimentation and tectonics. *In* "Sedimentary Environments and Facies" (H. G. Reading, ed.), pp. 439 476. Blackwell Scientific, Oxford.

Moretti, I., and Chenet, P. Y. (1987). The evolution of the Suez Rift: A combination of stretching and secondary convection. *Tectonophysics* **133,** 229–234.

Murray, G. E. (1960). Geologic framework of Gulf Coastal Province of United States. *In* "Recent Sediments, Northwest Gulf of Mexico" (F. P. Shepard, F. B. Phleger, and T. H. Van Andel, eds.), pp. 5–33. Am. Assoc. Pet. Geol., Tulsa, OK.

Neglia, S. (1979). Migration of fluids in sedimentary basins. *AAPG Bull.* **63,** 573–597.

Neugebauer, H. J. (1987). Models of lithospheric thinning. *Annu. Rev. Earth Planet. Sci.* **15,** 421–444.

Olsen, K. H., ed. (1995). "Continental Rifts: Evolution, Structure and Tectonics." Elsevier, Amsterdam. 466pp.

Parker, J. R., ed. (1993). "Petroleum Geology of Northwest Europe," 2 vols. Geol. Soc. London, London. 1537pp.

Perrodon, A. (1971). Classification of sedimentary basins: An essay. *Sci. Terre* **16,** 193–227.

Pettijohn, F. F. (1957). "Sedimentary Rocks." Harper Bros., New York. 718pp.

Pfiffner, O. A., Lehner, P., Heitzmann, P., St. Mueller, and Steck, A., eds. (1998). "Deep Structures of the Swiss Alps." Birkhaeuser, Basel.

Purser, B. H., and Bosence, D. W. J., eds. (1998). "Sedimentation and Tectonics in Rift Basins. Red Sea: Gulf of Aden." Chapman & Hall, London. 663pp.

Quennell, A. M. (1985). "Continental Rifts," Benchmark Papers in Geology Series. Van Nostrand, New York. 349pp.

Reijers, T. J. A., Petters, S. W., and Nwajide, C. S. (1997). The Niger Delta basin. *In* "African Basins" (R.C. Selley, ed.), pp. 151–172. Elsevier, Amsterdam.

Roberts, R. J. (1972). Evolution of the Cordilleran fold belt. *Geol. Soc. Am. Bull.* **83,** 1989–2004.

Salem, M. J., Mouzoughi, A. J., and Hammuda, O. S., eds. (1996a). "The Geology of the Sirte Basin," Vol. 1. Elsevier, Amsterdam, 564 pp.

Salem, M. J., El-Hawat, A. S., and Sbeta, A. M., eds. (1996b). "The Geology of the Sirte Basin," Vol. 2. Elsevier, Amsterdam, 578 pp.

Salem, M. J., Busrewil, M. T., Misallati, A. A., and Sola, M., eds. (1996c). "The Geology of the Sirte Basin," Vol. 3. Elsevier, Amsterdam. 380 pp.

Salop, L. I. (1967). "Geology of the Baikal Region," Vol. 2. Izd. Nedra, Moscow (in Russian).

Salvador, A., ed. (1991). "The Gulf of Mexico Basin." The Geology of North America, Vol. J. Geol. Soc. Am., Boulder, CO. 452pp.

Salveson, J. O. (1976). "Variations in the Oil and Gas Geology of Rift Basins," 5th Pet. Explor Semin., pp.15–17. Egyptian General Petroleum Corporation, Cairo.

Salveson, J. O. (1979).Variations in the geology of rift basins — a tectonic model. *Rio Grande Rift: Tecton. Magmat., Sel. Pap. Int. Symp.,* Santa Fe, NM, *1978,* pp.11–28.

Sawkins, F. J. (1984). "Metal Deposits in Relation to Plate Tectonics." Springer-Verlag, Berlin. 325pp.

Schneider, E. D. (1972). Sedimentary evolution of rifted continental margins. *Mem. — Geol. Soc. Am.* **132,** 109–118.

Schuchert, C. (1923). Sites and nature of the North American geosynclines. *Geol. Soc. Am. Bull.* **34,** 151–230.

Schwab, F. L. (1971). Geosynclinal compositions and the new global tectonics. *J. Sediment. Petrol.* **41,** 928–938.

Sears, F. O., and Lucia, F. J. (1979). Reef growth model for Silurian pinnacle reefs, northen Michigan reef trend. *Geology* **3,** 299–302.

Selley, R. C. (1997a). The basins of Northwest Africa: Stratigraphy and sedimentation. *In* "African Basins" (R. C. Selley, ed.), pp. 3–16. Elsevier, Amsterdam.

Selley, R. C. (1997b). The basins of Northwest Africa: Structural evolution. *In* "African Basins" (R. C. Selley, ed.), pp. 17–26. Elsevier, Amsterdam.

Selley, R. C. (1997c). The Sirte basin. *In* "African Basins" (R. C. Selley, ed.), pp. 27–38. Elsevier, Amsterdam.

Seyfert, C. K., and Sirkin, L. A. (1973). "Earth History and Plate Tectonics." Harper & Row, New York. 544pp.

Smith, G. W., Summers, G. E., Wallington, D., and Lee, J. L. (1958). Mississippian oil reservoirs in Williston basin. *In* "The Habitat of Oil" (L. G. Weeks, ed.), pp. 149–177. Am. Assoc. Pet. Geol., Tulsa, OK.

Stern, T. A., Quinlan, G. M., and Holt, W. E. (1993). Crustal Dynamics Associated with the formation of the Wanganui basin, New Zealand. *In* "South Pacific Sedimentary Basins" (P. F. Ballance, ed.), pp. 213–224. Elsevier, Amsterdam.

Stille, H. (1936). "Wege unde Ergebnisse der geologisch-tectonischen Forschung," Wiss. Förh, Gesell. 25 Jahr Kaiser Wilhelm, Bd. 2, pp. 84–85.

Stoneley, R. (1969). Sedimentary thicknesses in orogenic belts. *In* "Time and Space in Orogeny," pp. 215–238. Geol. Soc. London, London.

Tarling, D. H., and Runcorn, S. K. (1973). "Implications of Continental Drift to the Earth Sciences," Vols. 1 and 2. Academic Press, London and New York. 1184pp.

Van Andel, Tj. H. (1985). "New Views on an Old Planet." Cambridge University Press, Cambridge, UK. 318pp.

Van Houten, F. B. (1973). Meaning of molasse. *Geol. Soc. Am. Bull.* **84,** 1973–1976.

Walcott, R. I. (1972). Gravity flexure and the growth of sedimentary basins at a continental edge. *Geol. Soc. Am. Bull.* **83,** 1845–1848.

Weber, K. J. (1971). Sedimentological aspects of oil fields in the Niger delta. *Geologie Mijnb.* **50,** 559–576.

Weeks, L. G. (1958). Factors of sedimentary basin development that control oil occurrence. *Bull. Am. Assoc. Pet. Geol.* **32,** 1093–1160.

Wegener, A. (1924). "The Origin of the Continents and Oceans." Methuen, London. 212pp.

Wernicke, B. (1981). Low-angle normal thrust faults in the Basin and Range Province; nappe tectonics in an extending orogen. *Nature (London)* **291,** 645–648.

Wernicke, B. (1985). Uniform-sense normal shear of the continental lithosphere. *Can. J. Earth Sci.* **22,** 108–125.

Wilhelm, O., and Ewing, M. (1972). Geology and history of the Gulf of Mexico. *Geol. Soc. Am. Bull.* **83,** 575–600.

Woodland, A. W., ed. (1975). "Petroleum and the Continental Shelf of North West Europe," Vol. 1. Applied Science, London. 501pp.

Wyllie, P. J. (1971). "The Dynamic Earth." Wiley-Interscience, New York. 416pp.

Index

Where there is more than one index entry for a term, the bold face number refers to the page where it is illustrated, and/or defined, or described fully. Where a term is defined in the text, it is set in bold face. Where a term is illustrated in a color plate, this is indicated in the index.

513

About the Author

Richard Selley has spent most of his career at the Royal School of Mines, Imperial College, London University, where he is a Professor of Applied Sedimentology. Between 1969 and 1975, however, he worked for oil companies in Libya, Greenland, and the North Sea.

He has published many papers on sedimentology and its applications to petroleum exploration and production. His five textbooks run to several editions and have been translated into many languages.

Richard Selley has had assignments in Australia, Bahrain, Belize, Canada, France, Germany, Greece, Greenland, Holland, Hungary, India, Indonesia, Ireland, Jamaica, Jordan, Malaysia, Morocco, Libya, Norway, the North Sea, Sao Tome and Principe, Saudi Arabia, Singapore, South Africa, Spain, the UAE, the United States, Vietnam, and the former Yugoslavia.

He is a Chartered Geologist, active in professional affairs, having served on the councils of the European Federation of Geologists, the Council of Science and Technology Institutes, the Petroleum Exploration Society of Great Britain, and the Geological Society of London (of which he has held the post of Foreign Secretary).

Richard Selley has been awarded the Murchison Fund of the Geological Society. He has been a Distinguished Lecturer of the Petroleum Exploration Society of Australia and has been awarded a Certificate of Merit and a Survivor Certificate by the American Association of Petroleum Geologists.